Confirmatory Factor Analysis for Applied Research

Methodology in the Social Sciences

David A. Kenny, Founding Editor
Todd D. Little, Series Editor
www.guilford.com/MSS

This series provides applied researchers and students with analysis and research design books that emphasize the use of methods to answer research questions. Rather than emphasizing statistical theory, each volume in the series illustrates when a technique should (and should not) be used and how the output from available software programs should (and should not) be interpreted. Common pitfalls as well as areas of further development are clearly articulated.

RECENT VOLUMES

APPLIED MISSING DATA ANALYSIS
Craig K. Enders

PRINCIPLES AND PRACTICE OF STRUCTURAL EQUATION MODELING, THIRD EDITION
Rex B. Kline

APPLIED META-ANALYSIS FOR SOCIAL SCIENCE RESEARCH
Noel A. Card

DATA ANALYSIS WITH Mplus
Christian Geiser

INTENSIVE LONGITUDINAL METHODS: AN INTRODUCTION
TO DIARY AND EXPERIENCE SAMPLING RESEARCH
Niall Bolger and Jean-Philippe Laurenceau

DOING STATISTICAL MEDIATION AND MODERATION
Paul E. Jose

LONGITUDINAL STRUCTURAL EQUATION MODELING
Todd D. Little

INTRODUCTION TO MEDIATION, MODERATION, AND CONDITIONAL
PROCESS ANALYSIS: A REGRESSION-BASED APPROACH
Andrew F. Hayes

BAYESIAN STATISTICS FOR THE SOCIAL SCIENCES
David Kaplan

CONFIRMATORY FACTOR ANALYSIS FOR APPLIED RESEARCH, SECOND EDITION
Timothy A. Brown

Confirmatory Factor Analysis for Applied Research

SECOND EDITION

Timothy A. Brown

Series Editor's Note by Todd D. Little

THE GUILFORD PRESS
New York   London

© 2015 The Guilford Press
A Division of Guilford Publications, Inc.
370 Seventh Avenue, Suite 1200, New York, NY 10001
www.guilford.com

Printed in the United States of America

This book is printed on acid-free paper.

Last digit is print number: 9 8 7 6 5

Library of Congress Cataloging-in-Publication Data
Brown, Timothy A., 1960–
    Confirmatory factor analysis for applied research / Timothy A. Brown. —
Second edition.
        pages cm. — (Methodology in the social sciences)
    Includes bibliographical references and index.
    ISBN 978-1-4625-1779-4 (hardcover) — ISBN 978-1-4625-1536-3 (paperback)
    1. Factor analysis.   I. Title.
    BF39.2.F32B76 2015
    150.1′5195354—dc23
                                                            2014040687

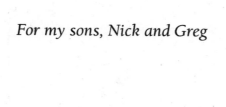

For my sons, Nick and Greg

Series Editor's Note

I have taught structural equation modeling (SEM) for over 25 years. For most of those years, I relied on a set of readings and other resources because I was dissatisfied with the lack of depth and breadth afforded to the measurement model in the works that were available at the time. The first edition of Timothy Brown's *Confirmatory Factor Analysis for Applied Research* (2006) was the first book that I required for my courses related to SEM. In 2006, there was no other book that covered the nuts and bolts of confirmatory factor analysis (CFA). Not only did Tim's book fill a crucial void; it did so with acumen and elegance. Both editions are full treatments that are very accessible, but the second edition accomplishes two critical goals that make it a must-have. First, it retains the accessible nature and applied focus that made the first edition a great read and thorough resource. Second, as a scholar of methodology and statistics, Tim has provided critical updates and integrated the most recent advances in the area. A lot of new ideas have emerged in the past decade, and Tim has gleaned the important advances from the literature and woven them into the fabric of his masterful tapestry.

The material covered in Tim's book represents about 90% of the effort and knowledge that are needed and utilized in conducting an SEM analysis. Somewhat ironically, it represents only about 10% of what is actually presented in the results section of a paper (i.e., all the foundational issues associated with properly specifying and estimating the measurement model). This material is the stuff behind the scenes that shines the spotlight on your analyses—like all the credits that scroll at the end of an award-winning movie. Tim's treatise is itself award worthy.

As an amazingly accomplished applied researcher, Tim understands the needs of researchers who are in the trenches doing the hard work of behavioral science research. Graduate students just learning advanced statistical techniques will relish the comprehensive coverage of all things CFA (e.g., Bayesian approaches, non-normal and categorical indicators, multitrait–multimethod models, factorial invariance—heck, just check out the subject index, and you'll be hard pressed to find an important topic that is missing!). As a gold standard for specifying and estimating the measurement model for a subsequent SEM analysis, this book will serve as a constant source for solving an applied problem.

When you read Tim's book, you'll notice he makes some recommendations that I don't make in my book, *Longitudinal Structural Equation Modeling*, and vice versa. I think it is important to understand that there is no one correct way to do SEM. It is always a matter of wise judgment and principled reasoning. Tim Brown is wise, and his reasoning is clear. Unlike too many pundits, Tim does not try to teach you *what* to think about CFA, but rather *how* to think about CFA. Learning to have civil dialogues with data, analysis models, and theories is the ultimate learning goal. Tim Brown's book is a Communication 101 requirement for applied researchers using CFA.

As the founding steward of this series, David A. Kenny, intimated in his introduction to the first edition, Tim does not lead with formulas, equations, and mathematics— but he does not shy away from explaining essential complexity. His ability to reach you as an applied researcher is a balance of the calculus of detail versus clarity. Tim Brown's balance of these two forces in his second edition is perfect.

TODD D. LITTLE
Texas Tech University
Founding Director of the Institute for Measurement, Methodology,
Analysis, and Policy (IMMAP)
On assignment in San José, Costa Rica

Preface

The first edition of this book was written for the simple reason that no other book of its kind had been published before. This statement continues to be true at this writing. Whereas the number of texts on structural equation modeling (SEM) continues to expand, this volume remains the first and only book devoted solely to the topic of confirmatory factor analysis (CFA). In the first edition, many important topics were brought into one volume, including the similarities and differences of exploratory factor analysis (EFA) and CFA; methods to move successfully from the EFA to the CFA framework; diagnosing and rectifying the various sources for the ill fit of measurement models; analysis of mean structures; modeling with multiple groups or multiple assessment points; CFA-based scale reliability evaluation; formative indicator models; and higher-order factor analysis. After covering the fundamentals of specifying, revising, and interpreting various types of CFA measurement models in the first several chapters, subsequent chapters dealt with issues germane to latent variable models of any type (e.g., modeling with non-normal or categorical indicators, dealing with missing data, and power analysis/sample size determination). This material was included not only because of its importance to the proper conduct of CFA, but because these topics have not received adequate coverage in other SEM sourcebooks. Thus this book should continue to serve as a useful guide to researchers working with a latent variable model of any type.

Although the aforementioned topics remain essential to applied CFA research, it was important to revise this book to incorporate the latest developments in latent variable modeling. Accordingly, existing chapters have been updated and expanded to cover several new methodological advances. These new topics include exploratory SEM (a more contemporary and advantageous method of moving from EFA to CFA, or as a better approach to measurement model evaluation when traditional CFA specification is not realistic); bifactor modeling (an alternative approach to evaluating the hierarchical/

general structure of a measurement scale that possesses several conceptual and statistical advantages over traditional higher-order CFA); measurement invariance evaluation of scales with categorical outcomes (with comparison of the theta and delta parameterizations in Mplus); and a new method for setting the measurement scales of latent variables (an approach that defines latent variables by a nonarbitrary and more interpretable metric than the more commonly used marker indicator approach). In addition to substantial revisions of existing chapters, this edition contains an entirely new chapter focused on the fundamentals, specification, and interpretation of Bayesian measurement models and multilevel measurement models. Among many other advantages (e.g., better performance in small samples or with non-normal or missing data), Bayesian estimation offers the applied researcher a more reasonable and flexible approach to model testing that is more closely aligned to substantive theory. For example, it allows for the specification of a measurement model avoiding the unnecessarily strict hypothesis that the factor cross-loadings or the indicator error covariances are exactly zero. Multilevel measurement models should be employed when data have been obtained by cluster or unequal probability sampling (e.g., data collected from students across multiple classrooms). In these situations, multilevel models not only can correctly adjust for the non-independence of observations, but can be used to learn more about the factor structure (and predictors of the factors) at the various levels of the data. Like the first edition of this book, this edition is not tied to a specific latent variable software package, and several popular programs are featured throughout (LISREL, Mplus, EQS, SAS/CALIS). All chapters have been revised to be up to date with the most recent releases of these major latent variable software programs. Moreover, several new syntax examples are provided for topics that were covered in the first edition (e.g., EFA, scale reliability evaluation, and multiple imputation in Mplus).

The target readers of this book are applied researchers and graduate students working within any domain of the social and behavioral sciences (e.g., psychology, education, political science, management/marketing, sociology, public health). In the classroom, this book can serve as a primary or supplemental text in courses on advanced statistics, factor analysis, SEM, or psychometrics/scale development. For applied researchers, this book can be used either as a resource for learning the procedures of CFA, or (for more experienced readers) as a reference guide for dealing with a more complex CFA model or data issue. The specific coverage of each chapter in the book is described in Chapter 1. Essentially, the first five chapters cover the fundamentals of CFA: what a researcher needs to know to conduct a CFA of any type. Thus, especially for readers new to CFA, it is recommended that the first five chapters be read in order, as this material serves as the foundation for the remainder of the book. The final six chapters address specific types of CFA and other issues, such as dealing with missing or categorical data and power analysis. As each of these latter chapters serves as a guide to a specific type of CFA model or data issue, it is less important that the reading order of these chapters be retained.

Quantitative methodological advances are often slow to be picked up by applied researchers, because such methods are usually disseminated in a manner that is inaccessible to many end users (e.g., formula-driven articles in mathematical/statistical journals). In reality, any multivariate statistic can be readily and properly employed by any researcher, provided that the test's assumptions, steps, common pitfalls, and so on, are laid out clearly. In keeping with that philosophy, this book was written as a user-friendly guide to conducting CFA with real data sets, with more of an emphasis on conceptual and practical aspects than on quantitative equations. Several strategies were used to help meet this goal: (1) Every key concept is exemplified with an applied data set, accompanied by syntax and output from the leading latent variable software packages; (2) tables are included to recap the procedures or steps of various methods (e.g., how to conduct an EFA, how to write up the results of a CFA study); (3) numerous figures are provided that graphically illustrate more complicated concepts or procedures (e.g., EFA factor rotation, forms of measurement invariance, types of nonpositive definite matrices, identification of formative indicator models, formation of posterior distributions in Bayesian CFA); (4) many chapters contain appendices that provide user-friendly illustrations of seemingly complex quantitative operations (e.g., data generation in Monte Carlo simulation research, calculation of matrix determinants and their role in model fit and improper solutions); and (5) a website (*www.guilford.com/brown3-materials*) offers data and computer files for most examples and other materials (e.g., updates, links to other CFA resources). I hope that through the use of these various devices, even the most complicated CFA model or data issue has been demystified and can now be readily tackled by any reader.

In closing, I would like to thank the people who were instrumental to the realization of this volume. First, I wish to extend my sincere gratitude to C. Deborah Laughton, my editor at The Guilford Press, who provided many excellent suggestions, positive feedback, and support throughout the process of preparing both editions of this book. Second, my thanks go out to Todd D. Little not only for his invaluable oversight of the Methodology in the Social Sciences series at Guilford, but for his programmatic work focused on developing and refining latent variable methodologies (some of the advances emanating from his research are covered in this book). Third, I am grateful to the following reviewers, whose uniformly constructive and invaluable feedback strengthened the first edition of this book considerably: Larry Price, Texas State University–San Marcos; Christopher Federico, University of Minnesota; and Ke-Hai Yuan, University of Notre Dame. In addition, I would like to thank the following reviewers who offered many excellent suggestions for revisions and additions to the second edition: Scott J. Peters, University of Wisconsin–Whitewater; Akihito Kamata, University of Oregon; Jim Anderson, Purdue University; Randall MacIntosh, California State University–Sacramento; G. Leonard Burns, Washington State University; Lesa Hoffman, University of Nebraska; and Noel Card, University of Arizona. I would also like to express my appreciation to Senior Production Editor Anna Nelson for her work in bringing a technically complex manuscript to press. And finally, special thanks to my wife, Bonnie, for her endless encouragement and support.

Contents

The companion website *www.guilford.com/brown3-materials*
provides downloadable data and program syntax files for most
of the book's examples and links to CFA-related resources.

1

Introduction

USES OF CONFIRMATORY FACTOR ANALYSIS

Confirmatory factor analysis (CFA) is a type of structural equation modeling (SEM) that deals specifically with measurement models—that is, the relationships between observed measures or *indicators* (e.g., test items, test scores, behavioral observation ratings) and latent variables or *factors*. A fundamental feature of CFA is its hypothesis-driven nature. Unlike its counterpart, exploratory factor analysis (EFA), CFA requires the researcher to prespecify all aspects of the model. Thus the researcher must have a firm a priori sense, based on past evidence and theory, of the number of factors that exist in the data, of which indicators are related to which factors, and so forth. In addition to its greater emphasis on theory and hypothesis testing, the CFA framework provides many other analytic possibilities that are not available in EFA (e.g., evaluation of method effects, examination of the stability or invariance of the factor model over time or informants). Moreover, for the reasons discussed below, CFA should be conducted prior to the specification of a structural equation model.

CFA has become one of the most commonly used statistical procedures in applied research. This is because CFA is well equipped to address the types of questions that researchers often ask. Some of the most common uses of CFA are discussed below.

Psychometric Evaluation of Test Instruments

CFA is almost always used during the process of scale development to examine the latent structure of a test instrument (e.g., a questionnaire). In this context, CFA is used to verify the number of underlying dimensions of the instrument (factors) and the pattern of item–factor relationships (*factor loadings*). CFA also assists in the determination of how a test should be scored. When the latent structure is multifactorial (i.e., two or

1

more factors), the pattern of factor loadings supported by CFA will designate how a test may be scored by using subscales; that is, the number of factors is indicative of the number of subscales, and the pattern of item–factor relationships (which items load on which factors) indicates how the subscales should be scored. Depending on other results and extensions of the analysis, CFA may support the use of total scores (composite of all items) in addition to subscale scores (composites of subsets of items). For example, the viability of a single total score might be indicated when the relationships among the latent dimensions (subscales) of a test can be accounted for by one higher-order factor, and when the test items are meaningfully related to the higher-order factor (see the discussion of higher-order CFA and bifactor modeling in Chapter 8). CFA is an important analytic tool for other aspects of psychometric evaluation. It can be used to estimate the *scale reliability* of test instruments in a manner that avoids the problems of traditional methods (e.g., Cronbach's alpha; see Chapter 8). Given recent advances in the analysis of categorical data (e.g., binary true–false test items), CFA now offers a comparable analytic framework to item response theory (IRT). In fact, in some ways, CFA provides more analytic flexibility than the traditional IRT model (see Chapter 9).

Construct Validation

Akin to a factor in CFA, a *construct* is a theoretical concept. In clinical psychology and psychiatry, for example, the mental disorders (e.g., major depression, schizophrenia) are constructs manifested by various clusters of symptoms that are reported by the patient or observed by others. In sociology, juvenile delinquency may be construed as a multi-dimensional construct defined by various forms of misconduct (e.g., property crimes, interpersonal violence, participation in drugs, academic misconduct). CFA is an indispensable analytic tool for construct validation in the social and behavioral sciences. The results of CFA can provide compelling evidence of the *convergent* and *discriminant validity* of theoretical constructs. Convergent validity is indicated by evidence that different indicators of theoretically similar or overlapping constructs are strongly interrelated; for example, symptoms purported to be manifestations of a single mental disorder load on the same factor. Discriminant validity is indicated by results showing that indicators of theoretically distinct constructs are not highly intercorrelated; for example, behaviors purported to be manifestations of different types of delinquency load on separate factors, and the factors are not so highly correlated as to indicate that a broader construct has been erroneously separated into two or more factors. One of the most elegant uses of CFA in construct validation is the analysis of multitrait–multimethod (MTMM) matrices (see Chapter 6). A fundamental strength of CFA approaches to construct validation is that the resulting estimates of convergent and discriminant validity are adjusted for measurement error and an error theory (see the "Method Effects" section, below). Thus CFA provides a stronger analytic framework than traditional methods that do not account for measurement error (e.g., ordinary least squares approaches such as correlation/multiple regression assume that variables in the analysis are free of measurement error).

Method Effects

Often some of the covariation of observed measures is due to sources other than the substantive latent variables. For instance, consider the situation where four measures of employee morale have been collected. Two indicators are the employees' self-reports (e.g., questionnaires); the other two are obtained from supervisors (e.g., behavioral observations). It may be presumed that the four measures are intercorrelated, because each is a manifest indicator of the underlying construct of morale. However, it is also likely that the employee self-report measures are more highly correlated with each other than with the supervisor measures, and vice versa. This additional covariation is not due to the underlying construct of morale, but reflects shared method variance. A *method effect* exists when additional covariation among indicators is introduced by the measurement approach. Method effects can also occur within a single assessment modality. For example, method effects are usually present in questionnaires that contain some combination of positively and negatively worded items (e.g., see Chapters 3 and 6). Unfortunately, traditional EFA is incapable of estimating method effects. In fact, the use of EFA when method effects exist in the data can produce misleading results—that is, can yield additional factors that are not substantively meaningful, but instead stem from artifacts of measurement. In CFA, however, method effects can be specified as part of the error theory of the measurement model. The advantages of estimating method effects within CFA include the ability to (1) specify measurement models that are more conceptually viable; (2) determine the amount of method variance in each indicator; and (3) obtain better estimates of the relationships of the indicators to the factors, and the relationships among latent variables (see Chapters 5 and 6).

Measurement Invariance Evaluation

Another key strength of CFA is its ability to determine how well measurement models generalize across groups of individuals or across time. *Measurement invariance* evaluation is an important aspect of test development. If a test is intended to be administered in a heterogeneous population, it should be established that its measurement properties are equivalent in subgroups of the population (e.g., gender, race). A test is said to be biased when some of its items do not measure the underlying construct comparably across groups. *Test bias* can be serious, such as in situations where a given score on a cognitive ability or job aptitude test does not represent the same true level of ability/aptitude in male and female respondents. Stated another way, the test is biased against women if, for a given level of true intelligence, men tend to score several IQ units higher on the test than women. These questions can be addressed in CFA by multiple-groups solutions and "multiple indicators, multiple causes" (MIMIC) models (see Chapter 7). For instance, in a multiple-groups CFA solution, the measurement model is estimated simultaneously in various subgroups (e.g., men and women). Other restrictions are placed on the multiple-groups solution to determine the equivalence of the measurement model across groups; for instance, if the factor loadings are equivalent, the magnitude of the relationships between the test items and the underlying construct (e.g., cognitive ability) are the same in men and women. Multiple-groups CFA solutions are also used to examine longitudinal measurement invariance. This is a very important aspect of latent variable analyses of repeated measures designs. In the absence of such evaluation, it cannot be determined whether temporal change in a construct is due to true change or to changes in the structure or measurement of the construct over time. Multiple-groups analysis can be applied to any type of CFA or SEM model. For example, these procedures can be incorporated into the analysis of MTMM data to examine the generalizability of construct validity across groups.

WHY A BOOK ON CFA?

It also seems appropriate to begin this volume by addressing this question: "Is there really a need for a book devoted solely to the topic of CFA?" On my bookshelf sit dozens of books on the subject of SEM. Why not go to one of these SEM books to learn about CFA? Given that CFA is a form of SEM, virtually all of these books provide some introduction to CFA. However, this coverage typically consists of a chapter at best. As this book will attest, CFA is a very broad and complex topic, and extant SEM books only scratch the surface. This is unfortunate, because in applied SEM research, most of the work deals with measurement models (CFA). Indeed, many applied research questions are addressed by using CFA as the primary analytic procedure (e.g., psychometric evaluation of test instruments, construct validation). Another large proportion of SEM studies focus on structural regression models—that is, the manner in which latent variables are interrelated. Although CFA is not the ultimate analysis in such studies, a viable measurement model (CFA) must be established prior to evaluating the structural (e.g., regressive) relationships among the latent variables of interest. When poor model fit is encountered in such studies, it is more likely that it will stem from misspecifications in the measurement portion of the model (i.e., the manner in which observed variables are related to factors) than from the structural component that specifies the interrelationships of the factors. This is because there are usually more things that can go wrong in the measurement model than in the structural model (e.g., problems in the selection of observed measures, misspecified factor loadings, additional sources of covariation among observed measures that cannot be accounted for by the latent variables). Existing SEM resources do not provide sufficient details on the sources of ill fit in CFA measurement models or on how such models can be diagnosed and respecified. Moreover, advanced applications of CFA are rarely discussed in general SEM books (e.g., CFA with categorical indicators, scale reliability evaluation, MIMIC models, formative indicators, multilevel measurement models, Bayesian CFA).

Given the popularity of CFA, this book was written to provide an in-depth treatment of the concepts, procedures, pitfalls, and extensions of this methodology. Although the overriding objective of the book is to provide critical information on applied CFA that has not received adequate coverage in the past, it is important to note that the topics pertain to SEM in general (e.g., sample size/power analysis, missing data, non-normal or categorical data, formative indicators, multilevel modeling, Bayesian analysis). Thus it is hoped that this book will also provide a useful resource to researchers using any form of SEM.

COVERAGE OF THE BOOK

The first five chapters of this book present the fundamental concepts and procedures of CFA. Chapter 2 introduces the reader to the concepts and terminology of the common factor model. The common factor model is introduced in the context of EFA. This book is not intended to be a comprehensive treatment of the principles and practice of EFA. However, an overview of the concepts and operations of EFA is provided in Chapter 2 for several reasons: (1) Most of the concepts and terminology of EFA generalize to CFA; (2) this overview fosters the discussion of the similarities and differences of EFA and CFA in later chapters (e.g., Chapter 3); and (3) in programmatic research, an EFA study is typically conducted prior to a CFA study to develop and refine measurement models that are reasonable for CFA (thus the applied CFA researcher must also be knowledgeable about EFA). An introduction to CFA is provided in Chapter 3. After providing a detailed comparison of EFA and CFA, this chapter presents the various parameters, unique terminology, and fundamental equations of CFA models. Many other important concepts are introduced in this chapter that are essential to the practice of CFA and that must be understood in order to proceed to subsequent chapters; these include model identification, model estimation (e.g., maximum likelihood or ML), and goodness of model fit. Chapter 4 illustrates and extends these concepts, using a complete example of a CFA measurement model. In this chapter, the reader will learn how to program and interpret basic CFA models, using several of the most popular latent variable software packages (LISREL, Mplus, EQS, SAS/CALIS). The procedures for evaluating the acceptability of the CFA model are discussed. In context of this presentation, the reader is introduced to other important concepts, such as model misspecification and Heywood cases. Chapter 4 concludes with a section on the material that should be included in the report of a CFA study. Chapter 5 covers the important topics of model respecification and model comparison. It deals with the problem of poor-fitting CFA models and the various ways a CFA model may be misspecified. This chapter also presents EFA within the CFA framework and exploratory SEM—methods of developing more viable CFA measurement models on the basis of EFA findings. The concepts of nested models, equivalent models, and method effects are also discussed.

The second portion of the book focuses on more advanced or specialized topics and issues in CFA. Chapter 6 discusses how CFA can be conducted to analyze MTMM data in the validation of social or behavioral constructs. Although the concepts of method effects, convergent validity, and discriminant validity are introduced in earlier chapters (e.g., Chapter 5), these issues are discussed extensively in context of MTMM models in Chapter 6. Chapter 7 discusses CFA models that contain various combinations of equality constraints (e.g., estimation of a CFA model with the constraint of holding two or more parameters to equal the same value), multiple groups (e.g., simultaneous CFA in separate groups of males and females), and mean structures (CFAs that entail the estimation of the intercepts of indicators and factors). These models are discussed and illustrated in context of the analysis of measurement invariance; that is, is the measurement model equivalent in different groups or within the same group across time? Two different approaches to evaluating CFA models in multiple groups are presented in detail: multiple-groups solutions and MIMIC models. This chapter also illustrates another method of scaling of latent variables, called *effects coding*, that can foster the interpretation of the unstandardized solution in single-group and multiple-groups analyses.

Chapter 8 presents four other types of CFA models: higher-order CFA, bifactor analysis, CFA approaches to scale reliability estimation, and CFA with formative indicators. Higher-order factor analysis is conducted in situations where the researcher can posit a more parsimonious conceptual account for the interrelationships of the factors in the initial CFA model. Bifactor modeling is another form of hierarchical analysis, but unlike higher-order factor analysis, the overarching dimension (or dimensions) exerts direct effects on the indicators. The section on scale reliability evaluation shows that the unstandardized parameter estimates of a CFA solution can be used to obtain point estimates and confidence intervals of the reliability of test instruments (i.e., reliability estimate = the proportion of the total observed variance in a test score that reflects true score variance). This approach has important advantages over traditional estimates of internal consistency (Cronbach's alpha). Models with formative indicators contain observed measures that "cause" the latent construct. In the typical CFA, indicators are defined as linear functions of the latent variable, plus error; that is, indicators are considered to be the effects of the underlying construct. In some situations, however, it may be more plausible to view the indicators as causing a latent variable; for example, socioeconomic status is a concept determined by one's income, education level, and job status (not the other way around). Although formative indicators pose special modeling challenges, Chapter 8 shows how such models can be handled in CFA.

The next two chapters consider issues that must often be dealt with in applied CFA research, but are rarely discussed in extant SEM sourcebooks. Chapter 9 addresses data set complications such as how to accommodate missing data, and how to conduct CFA when the distributions of continuous indicators are non-normal. Various methods of handling each issue are discussed and illustrated (e.g., missing data: multiple imputation, direct ML; non-normal data: alternative statistical estimators, bootstrapping, item parceling). Chapter 9 also includes a detailed treatment of CFA with categorical outcomes (e.g., tests with binary items such as true–false scales). In addition to illustrating the estimation and interpretation of such models, this section of the chapter demonstrates the parallels and extensions of CFA to traditional IRT analysis. The section also contains a detailed discussion and illustration of measurement invariance evaluation with categorical outcomes. Chapter 10 deals with the often overlooked topic of determining the sample size necessary to achieve sufficient statistical power and precision of the parameter estimates in a CFA study. Two different approaches to this issue are presented (Satorra–Saris method, Monte Carlo method).

The final chapter of this book (Chapter 11) presents and illustrates two relatively new modeling possibilities involving CFA: Bayesian analysis and multilevel factor models. Among the many advantages of Bayesian CFA is the ability to specify approximate zeroes for parameters that are fixed to zero in traditional CFA (e.g., cross-loadings). This allows for the testing of more reasonable measurement models that may be more closely aligned with substantive theory. Multilevel factor models should be conducted in data sets that violate the assumption of independence of observations (e.g., clustered data, such as in a study of families where more than one member per family has contributed data). In addition to properly adjusting for the dependency in the data, multilevel models allow the researcher to study the structure and nature of relationships at each level of the data.

OTHER CONSIDERATIONS

This book was written with applied researchers and graduate students in mind. It is intended to be a user-friendly guide to conducting CFA with real data sets. To achieve this goal, conceptual and practical aspects of CFA are emphasized, and quantitative aspects are kept to a minimum (or separated from the main text; e.g., Chapter 3). Formulas are not avoided altogether, but are provided in instances where they are expected to foster the reader's conceptual understanding of the principles and operations of CFA. Although this book does not require a high level of statistical acumen, a basic understanding of correlation/multiple regression will facilitate the reader's passage through the occasional more technically oriented section.

It is important that a book of this nature not be tied to a specific latent variable software program. For this reason, most of the examples provided in this book are accompanied with input syntax from each of the most widely used software programs (LISREL, Mplus, EQS, SAS/CALIS). Although Amos Basic was included in the first edition of this book, Amos is not included in the current edition because the vast majority of Amos users rely on the graphical interface for model specification (Amos Basic syntax can still be found on the book's companion website, *www.guilford.com/brown3-materials*). Several comments about the examples are in order. First, readers will note that many of the syntax examples are first discussed in context of the LISREL program. This is not intended to imply a preference for LISREL over other software programs. Rather, this is more reflective of the historical fact that the etiological roots of SEM are strongly tied to LISREL. For instance, the widely accepted symbols of the parameters and computational formulas of a CFA model stem from LISREL notation (e.g., λ = factor loading). The illustrations of LISREL matrix programming allow interested readers to understand computational aspects of CFA more quickly (e.g., by using the provided formula, how the model-implied covariance of two indicators can be calculated on the basis of the

CFA model's parameter estimates). Knowledge of this notation is also useful to readers who are interested in developing a deeper quantitative understanding of CFA and SEM in more technical sources (e.g., Bollen, 1989). On the other hand, the output from the Mplus program is relied on heavily in various examples in the book. Again, a preference for Mplus should not be inferred. This reflects the fact that the results of CFA are provided more succinctly by Mplus than by other programs (concise output = concise tables in this book), as well as the fact that some analytic features are only available in Mplus.

Another potential pitfall of including computer syntax examples is the high likelihood that soon after a book is published, another version of the software will be released. When this book was written, the following versions of the software programs were current: LISREL 9.1, Mplus 7.11, EQS 6.2, and SAS/CALIS 9.4. For all examples, the associated computer syntax was revised to be up to date with these versions of these software programs. New releases typically introduce new features to the software, but do not alter the overall programming framework. In terms of this book, the most probable consequence of new software releases is that some claims about the (in)capabilities of the programs will become outdated. However, the syntax examples should be upwardly compatible (i.e., fully functional) with any subsequent software releases (e.g., although LISREL 7 does not contain many of the features of LISREL 9.1, syntax written in this version is fully operational in subsequent LISREL releases).

Especially in earlier chapters of this book, the computer syntax examples contain few, if any, programming shortcuts. Again, this is done to foster the reader's understanding of CFA model specification. This is another reason why LISREL is often used in the programming examples: in LISREL matrix-based programming, the user must specify every aspect of the CFA model. Thus CFA model specification is more clearly conveyed in LISREL than in some programs where these specifications occur "behind the scenes" (e.g., Mplus contains a series of defaults that automatically specify marker indicators, free and fixed factor loadings, factor variances and covariances, etc., in a standard CFA model). On a related note, many latent variable software programs (e.g., Amos, LISREL, EQS) now contain graphical interfaces that allow the user to specify the CFA model by constructing a path diagram with a set of drawing tools. Indeed, graphical interfaces are an increasingly popular method of model programming, particularly with researchers new to CFA and SEM. The primary reason why graphical input is not discussed in this book is that it does not lend itself well to the written page. Yet there are other reasons why syntax programming can be more advantageous. For instance, it is often quicker to generate a CFA solution from a syntax program than from constructing a path diagram in a drawing editor. In addition, many of the advanced features of model specification are more easily invoked through syntax. Users who understand the logic of syntax programming (either matrix- or equation-based syntax operations) are able to move from one latent variable software package to another much more quickly and easily than users who are adept only in the use of a graphical interface of a given software program.

In attempt to make the illustrations more provocative to the applied researcher, most of the examples in this book are loosely based on findings or test instruments in the extant literature. The examples are drawn from a variety of domains within the social and behavioral sciences—clinical psychology, personality psychology, social psychology, industrial/organizational psychology, and sociology. In some instances, the examples use actual research data, but in many cases the data have been artificially generated strictly for the purposes of illustrating a given concept. Regardless of the origin of the data, the examples should not be used to draw substantive conclusions about the research domain or test instrument in question.

Many of the examples in this book use a variance–covariance matrix as input data (specifically, the correlations and standard deviations of the indicators are inputted, from which the program generates the sample variance–covariance matrix). This was done to allow interested readers to replicate examples directly from the information provided in the figures and tables of the book. Although matrix input is used as a convenience feature in this book, it is not necessarily the best method of reading data into an analysis. All leading latent variable programs are capable of reading raw data as text files, and many can read data saved in other software formats (e.g., SPSS .sav files, Microsoft Excel files). There are several advantages of using raw data as input. First, it is more convenient, because the user does not need to compute the input matrices prior to conducting the latent variable analysis. Second, the input data are more precise when the software program computes the input matrices from raw data (user-generated matrices usually contain rounding error). Third, there are some situations where raw data must be analyzed—for instance, models that have missing, non-normal, or categorical data. Some sections of this book (e.g., Chapter 9) illustrate how raw data are read into the analysis. The interested reader can download the files used in these and other examples from the book's companion website (*www.guilford.com/brown3-materials*).

SUMMARY

This chapter has provided a general overview of the nature and purposes of CFA, including some of the fundamental differences between EFA and CFA. The ideas introduced in this chapter provide the background for a more detailed discussion of the nature of the common factor model and EFA, the subject of Chapter 2. This book is intended to be a user-friendly guide to conducting CFA in real data sets, aimed at students and applied researchers who do not have an extensive background in quantitative methods. Accordingly, practical and conceptual aspects of CFA are emphasized over mathematics and formulas. In addition, most of the chapters are centered on data-based examples drawn from various realms of the social and behavioral sciences. The overriding rationale of these examples is discussed (e.g., use of software programs, method of data input) to set the stage for their use in subsequent chapters.

2

The Common Factor Model and Exploratory Factor Analysis

This chapter introduces the reader to the concepts, terminology, and basic equations of the common factor model. Both exploratory factor analysis (EFA) and confirmatory factor analysis (CFA) are based on the common factor model. In this chapter, the common factor model is discussed primarily in the context of EFA. Nonetheless, most of the concepts and terminology (e.g., common and unique variances, factor loadings, communalities) of EFA are also used in CFA. This chapter discusses some of the fundamental similarities and differences of EFA and CFA. In applied research, EFA and CFA are often conducted in conjunction with one another. For instance, CFA is frequently used in the later stages of scale development, after the factor structure of a testing instrument has been explored and refined by EFA. Thus, because the applied CFA researcher must have a working knowledge of EFA, the methods of conducting an EFA are reviewed in this chapter. This overview is also provided to allow more detailed comparisons of EFA and CFA in later chapters.

OVERVIEW OF THE COMMON FACTOR MODEL

Since its inception over a century ago (Spearman, 1904, 1927), factor analysis has become one of the most widely used multivariate statistical procedures in applied research endeavors across a multitude of domains (e.g., psychology, education, sociology, management, political science, public health). The fundamental intent of factor analysis is to determine the number and nature of latent variables or *factors* that account for the variation and covariation among a set of observed measures, commonly referred to as *indicators*. Specifically, a factor is an unobservable variable that influences more than one observed measure and that accounts for the correlations among these observed measures. In other words, the observed measures are intercorrelated because they share a common cause (i.e., they are influenced by the same underlying construct); if the latent construct is partialed out, the intercorrelations among the observed measures will be zero.[1] Thus factor analysis attempts a more parsimonious understanding of the covariation among a set of indicators because the number of factors is less than the measured variables.

In applied research, factor analysis is most commonly used in psychometric evaluations of multiple-item testing instruments (e.g., questionnaires; cf. Floyd & Widaman, 1995). For example, a researcher may have generated 20 questionnaire items that he or she believes are indicators of the unidimensional construct of self-esteem. In the early stages of scale development, the researcher may use factor analysis to examine the plausibility of this assumption (i.e., the ability of a single factor to account for the intercorrelations among the 20 indicators) and to determine if all 20 items are reasonable indicators of the underlying construct of self-esteem (i.e., how strongly is each item related to the factor?). In addition to psychometric evaluation, other common uses for factor analysis include construct validation (e.g., obtaining evidence of convergent and discriminant validity by demonstrating that indicators of selected constructs load onto separate factors in the expected manner; e.g., Brown, Chorpita, & Barlow, 1998) and data reduction (e.g., reducing a larger set of intercorrelated indicators to a smaller set of composite variables, and using these composites—i.e., factor scores—as the units of analysis in subsequent statistical tests; e.g., Cox, Walker, Enns, & Karpinski, 2002).

These concepts emanate from the *common factor model* (Thurstone, 1947), which postulates that each indicator in a set of observed measures is a linear function of one or more common factors and one unique factor. Thus factor analysis partitions the variance of each indicator (derived from the sample correlation/covariance matrix which is used as input for the analysis) into two parts: (1) *common variance*, or the variance accounted for by the factor, which is estimated on the basis of variance shared with other indicators in the analysis; and (2) *unique variance*, which is a combination of reliable variance that is specific to the indicator (i.e., systematic factors that influence only one indicator) and random error variance (i.e., measurement error or unreliability in the indicator). There are two main types of analyses based on the common factor model: *exploratory factor analysis* (EFA) and *confirmatory factor analysis* (CFA; Jöreskog, 1969, 1971a). Both EFA and CFA aim to reproduce the observed relationships among a group of indicators with a smaller set of latent variables, but they differ fundamentally by the number and nature of a priori specifications and restrictions made on the factor model. EFA is a data-driven approach, such that no specifications are made in regard to the number of factors (initially) or the pattern of relationships between the common factors and the indicators (i.e., the *factor loadings*). Rather, a researcher employs EFA as an exploratory or descriptive technique to determine the appropriate number of common factors, and to uncover which measured variables are reasonable indicators of the various latent dimensions (e.g., by the size and differential magnitude of factor loadings). In CFA, the researcher specifies the number of factors and the pattern of indicator–factor loadings in advance, as well as other parameters such as those bearing on the independence or covariance of the factors and indicator unique variances.[2] The prespecified factor solution is evaluated in terms of how well it reproduces the sample correlation (covariance) matrix of the measured variables. Thus, unlike EFA, CFA requires a strong empirical or conceptual foundation to guide the specification and evaluation of the factor model. Accordingly, EFA is typically used earlier in the process of scale development and construct validation, whereas CFA is used in later phases after the underlying structure has been established on prior empirical (EFA) and theoretical grounds. Other important differences between EFA and CFA are discussed in Chapter 3.

A brief, applied example is used to illustrate some of the key concepts of the common factor model. In this basic example, four behavioral observation ratings (O1–O4) have been collected on 300 individuals admitted to an inpatient psychiatric facility. The four ratings are hopelessness (O1), feelings of worthlessness/guilt (O2), psychomotor retardation (O3), and sleep disturbance (O4). As shown in Table 2.1, these four clinical ratings (indicators) are moderately intercorrelated. It is conjectured that each of these ratings is a manifest indicator of the latent construct of Depression; that is, each of the observed symptoms (e.g., hopelessness, worthlessness) has the shared influence of Depression, the single latent variable (factor) that accounts for the intercorrelations among these observed measures. The only reason the indicators are correlated is that they share the common cause of Depression; if this latent variable is partialed out, no relationship among these indicators will be seen.

With the sample correlations presented in Table 2.1 as input, a factor analysis is conducted by using the EFA routines provided in SPSS (FACTOR), SAS (PROC FACTOR), and Mplus (see Table 2.2). For reasons noted later in this chapter, only a one-factor solution can be pursued. Because EFA typically uses correlations as the units of analysis, it can be run in SPSS and SAS by embedding the sample correlation matrix in the body of the syntax (as shown in Table 2.2), although both programs can generate this matrix by reading raw input data files. In Mplus, the input data must be read in from an external file (in text format). The procedures of EFA are discussed later in this chapter (e.g., methods of factor extraction and selection), but for purposes of this illustration, consider the selected results of the analysis presented in Table 2.2. In the selected output from SPSS, of particular interest is the output under the heading "Factor Matrix," which provides the factor loadings for the four clinical ratings. In EFA, the factor loadings are completely standardized estimates of the regression slopes for predicting the indicators from the factor, and thus are interpreted along the lines of standardized regression (β)

TABLE 2.1. Intercorrelations among Four Behavioral Observation Ratings of Depression

	O1	O2	O3	O4
O1	1.00			
O2	0.70	1.00		
O3	0.65	0.66	1.00	
O4	0.62	0.63	0.60	1.00

Note. (N = 300) O1, hopelessness; O2, feelings of worthlessness/guilt; O3, psychomotor retardation; O4, sleep disturbance.

TABLE 2.2. SPSS, SAS, and Mplus Syntax and Selected Output for a Basic One-Factor Model

SPSS syntax

```
MATRIX DATA VARIABLES=ROWTYPE_ O1 O2 O3 O4.
BEGIN DATA.
N 300 300 300 300
COR 1.0
COR .70 1.0
COR .65 .66 1.0
COR .62 .63 .60 1.0
END DATA.
FACTOR
  /MATRIX=IN(COR=*)
  /MISSING LISTWISE
  /ANALYSIS O1 O2 O3 O4
  /PRINT INITIAL EXTRACTION
  /CRITERIA FACTORS(1) ITERATE(25)
  /EXTRACTION ML
  /ROTATION NOROTATE.
```

SAS syntax

```
data sm (type=CORR);
    input _type_ $ _name_ $ O1 O2 O3 O4;
cards;
n    .  300    .      .      .
corr  O1 1.00   .      .      .
corr  O2 0.70  1.00    .      .
corr  O3 0.65  0.66   1.00    .
corr  O4 0.62  0.63   0.60   1.00
;
proc factor data=sm method=ml nfactors=1 scree;
run;
```

Mplus syntax

```
TITLE: MPLUS EXAMPLE OF EFA
DATA:     FILE IS tab2.2.dat;
          NOBSERVATIONS = 300;
          TYPE=CORRELATION;
VARIABLE: NAMES ARE O1 O2 O3 O4;
ANALYSIS: TYPE = EFA 1 1;
          ESTIMATOR=ML;
OUTPUT:   FSDETERMINACY;
```

Selected output (SPSS)

```
Initial Statistics:
```

Variable	Communality	*	Factor	Eigenvalue	Pct of Var	Cum Pct
		*				
O1	.57811	*	1	2.93118	73.3	73.3
O2	.59175	*	2	.41039	10.3	83.5
O3	.53077	*	3	.35924	9.0	92.5
O4	.48795	*	4	.29919	7.5	100.0

(continued)

TABLE 2.2. *(continued)*

Test of fit of the   1-factor model:

Chi-square statistic: .2031, D.F.:   2, Significance: .9035

Factor Matrix:

              Factor 1

O1   .82822
O2   .84090
O3   .78766
O4   .75228

Final Statistics:

Variable	Communality	*	Factor	SS Loadings	Pct of Var	Cum Pct
		*				
O1	.68595	*	1	2.57939	64.5	64.5
O2	.70712	*				
O3	.62040	*				
O4	.56592	*				

Selected output (Mplus)

RESULTS FOR EXPLORATORY FACTOR ANALYSIS

          EIGENVALUES FOR SAMPLE CORRELATION MATRIX
              1   2   3   4
           --------   --------   --------   --------
      1   2.931   0.410   0.359   0.299

MODEL FIT INFORMATION

Number of Free Parameters   8

Chi-Square Test of Model Fit
          Value   0.206
          Degrees of Freedom   2
          P-Value   0.9023

RMSEA (Root Mean Square Error Of Approximation)
          Estimate   0.000
          90 Percent C.I.   0.000 0.048
          Probability RMSEA <= .05   0.952

CFI/TLI
          CFI   1.000
          TLI   1.009

(continued)

TABLE 2.2. *(continued)*

SRMR (Standardized Root Mean Square Residual)
            Value   0.003

            GEOMIN ROTATED LOADINGS (* significant at 5% level)
                1

	1
O1	0.828*
O2	0.841*
O3	0.788*
O4	0.752*

            ESTIMATED RESIDUAL VARIANCES

	O1	O2	O3	O4
1	0.314	0.293	0.380	0.434

            S.E. GEOMIN ROTATED LOADINGS
                1

	1
O1	0.024
O2	0.023
O3	0.027
O4	0.030

            Est./S.E. GEOMIN ROTATED LOADINGS
                1

	1
O1	34.141
O2	35.963
O3	28.926
O4	25.112

            FACTOR DETERMINACIES
                1

	1
1	0.940

or correlation (*r*) coefficients as in multiple regression/correlational analysis (cf. Cohen, Cohen, West, & Aiken, 2003).[3] For instance, the factor loading estimate for O1 (hopelessness) is .828, which is interpreted as indicating that a standardized score increase in the factor (Depression) is associated with an .828 standardized score increase in tearfulness. Squaring the factor loadings provides the estimate of the amount of variance in the indicator accounted for by the latent variable (e.g., $.828^2$ = 68.5% variance explained). In factor analysis, the amount of variance in the indicator explained by the common factors is often referred to as the *communality* (shown at the bottom of the SPSS output in Table 2.2). Thus, for the O1 (hopelessness) indicator, the factor model estimates that 68.5% of its total variance is *common variance* (variance explained by the latent variable of Depression), whereas the remaining 31.5% (i.e., 1 − .685 = .315) is *unique variance*. As stated earlier, unique variance is some combination of specific factor and measurement error variance. It is important to note that EFA and CFA do not provide separate estimates of specific variance and error variance.

In addition, Table 2.2 provides selected output from the Mplus program. As would be expected, many of the results are identical to those generated by SPSS (e.g., eigenvalues, factor loadings). However, Mplus also provides other useful output, including an expanded set of goodness-of-fit statistics; standard errors and significance tests for the factor loadings (as well as for the residual variances, not shown in Table 2.2); and an estimate of factor determinacy (if requested by the user on the OUTPUT line; see Table 2.2). Each of these additional aspects of the Mplus output is discussed later in this book.

Path diagrams of the one-factor measurement model are provided in Figure 2.1. The first diagram presents the solution, using common symbols for the various elements of

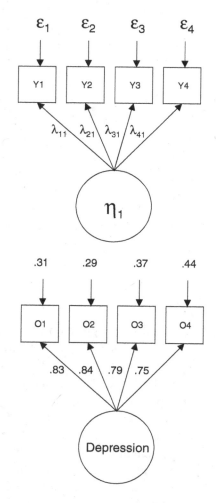

FIGURE 2.1. Path diagram of the one-factor model.

factor models (and LISREL latent Y variable notation); the second diagram replaces these elements with the sample estimates obtained from the EFA presented in Table 2.1. Following the conventions of factor analysis and structural equation modeling (SEM), the latent variable (factor) of Depression is depicted by a circle or an oval, whereas the four clinical ratings (indicators) are represented by squares or rectangles. The unidirectional arrows ($\rightarrow$) represent the *factor loadings* (λ, or *lambda*), which are the regression slopes (direct effects) for predicting the indicators from the factor (η, or *eta*). These arrows are also used to relate the unique variances (ε, or *epsilon*) to the indicators.[4]

A fundamental equation of the common factor model is

$$y_j = \lambda_j1\eta_1 + \lambda_j2\eta_2 + \ldots + \lambda_{jm}\eta_m + \varepsilon_j \tag{2.1}$$

where y_j represents the jth of p indicators (in the case $p = 4$; O1, O2, O3, O4) obtained from a sample of n independent participants (in this case, $n = 300$); λ_{jm} represents the factor loading relating variable j to the mth factor η (in the case $m = 1$; the single factor of Depression); and ε_j represents the variance that is unique to indicator y_j and is independent of all ηs and all other εs. As will be seen in subsequent chapters, similar notation is used to represent some of the equations of CFA. In this simple factor solution entailing a single factor (η_1) and four indicators, the regression functions depicted in Figure 2.1 can be summarized by four separate equations:

$$O1 = \lambda_{11}\eta_1 + \varepsilon_1 \tag{2.2}$$
$$O2 = \lambda_{21}\eta_1 + \varepsilon_2$$
$$O3 = \lambda_{31}\eta_1 + \varepsilon_3$$
$$O4 = \lambda_{41}\eta_1 + \varepsilon_4$$

This set of equations can be summarized in a single equation that expresses the relationships among observed variables (y), factors (η), and unique variances (ε):

$$y = \Lambda_y\eta + \varepsilon \tag{2.3}$$

or in expanded matrix form:

$$\Sigma = \Lambda_y\Psi\Lambda'_y + \Theta\varepsilon \tag{2.4}$$

where Σ is the $p \times p$ symmetric correlation matrix of p indicators; Λ_y is the $p \times m$ matrix of factor loadings λ (in this case, a 4×1 vector); Ψ is the $m \times m$ symmetric correlation matrix of the factor correlations (1×1); and $\Theta\varepsilon$ is the $p \times p$ diagonal matrix of unique variances ε ($p = 4$). In accord with matrix algebra, matrices are represented in factor analysis and SEM by uppercase Greek letters (e.g., Λ, Ψ, and Θ), and specific elements of these matrices are denoted by lowercase Greek letters (e.g., λ, ψ, and ε). With minor variations, these fundamental equations can be used to calculate various aspects of the sample data from the factor analysis parameter estimates, such as the variances, covariances, and means of the input indicators (the latter can be conducted in context of CFA with mean and covariance structures; see Chapter 7). For example, the following equation reproduces the variance in the O1 indicator:

$$VAR(O1) = \sigma_{11} = \lambda_{11}{}^2\psi_{11} + \varepsilon_1 \qquad (2.5)$$
$$= .828^2(1) + .315$$
$$= 1.00$$

where ψ_{11} is the variance of the factor η_1, and ε_1 is the unique variance of O1. Note that both ψ_{11} and σ_{11} equal 1.00 because the EFA model is completely standardized; that is, when variables are standardized, their variances equal 1.00. Similarly, the model estimate of the covariance (correlation) of O1 and O2 can be obtained from the following equation:

$$COV(O1, O2) = \sigma_{21} = \lambda_{11}\psi_{11}\lambda_{21} \qquad (2.6)$$
$$= (.828)(1)(.841)$$
$$= .696$$

Because the solution is completely standardized, this covariance is interpreted as the factor model estimate of the sample correlation of O1 and O2. In other words, the model-implied correlation of the indicators is the product of their completely standardized factor loadings. Note that the sample correlation of O1 and O2 is .70, which is very close to the factor-model-implied correlation of .696. As discussed in further detail in Chapter 3, the acceptability of factor analysis models is determined in large part by how well the parameter estimates of the factor solution (e.g., the factor loadings) are able to reproduce the observed relationships among the input variables. The current illustration should exemplify the point made earlier that common variance (i.e., variance explained by the factors as reflected by factor loadings and communalities) is estimated on the basis of the shared variance among the indicators used in the analysis. EFA generates a matrix of factor loadings (Λ) that best explain the correlations among the input indicators.

PROCEDURES OF EFA

Although a full description of EFA is beyond the scope of this book, an overview of its concepts and procedures is helpful to make later comparisons to CFA. The reader is referred to papers by Fabrigar, Wegener, MacCallum, and Strahan (1999); Floyd and Widaman (1995); and Preacher and MacCallum (2003) for detailed guidelines on conducting EFA in applied data sets.

As stated earlier, the overriding objective of EFA is to evaluate the dimensionality of a set of multiple indicators (e.g., items from a questionnaire) by uncovering the smallest number of interpretable factors needed to explain the correlations among them. Whereas the researcher must ultimately specify the number of factors, EFA is an "exploratory" analysis because no a priori restrictions are placed on the pattern of relationships between the observed measures and the latent variables. This is a key difference between EFA and CFA. In CFA, a researcher must specify in advance several key aspects of the factor model (e.g., number of factors, patterns of indicator–factor loadings).

After determining that EFA is the most appropriate analytic technique for the empirical question at hand, the researcher must decide which indicators to include in the analysis, and determine if the size and the nature of the sample are suitable for the analysis (for more details on these issues, see Chapters 9 and 10). Other procedural aspects of EFA include (1) selection of a specific method to estimate the factor model; (2) selection of the appropriate number of factors; (3) in the case of models that have more than one factor, selection of a technique to rotate the initial factor matrix to foster the interpretability of the solution; and (4) if desired, selection of a method to compute factor scores.

Factor Extraction

There are many methods that can be used to estimate the common factor model, such as maximum likelihood (ML), principal factors (PL), weighted least squares, unweighted least squares, generalized least squares, imaging analysis, minimum residual analysis, and alpha factoring, to name just some. For EFA with continuous indicators (i.e., observed measures that approximate an interval-level measurement scale), the most frequently used factor extraction methods are ML and PF. ML is also the most commonly used estimation method in CFA, and its fundamental properties are discussed in Chapter 3. A key advantage of the ML estimation method is that it allows for a statistical evaluation of how well the factor solution is able to reproduce the relationships among the indicators in the input data. That is, how closely do the correlations among the indicators predicted by the factor analysis parameters approximate the relationships seen in the input correlation matrix (see Eq. 2.6)? This feature is very helpful for determining the appropriate number of factors. However, as discussed in Chapter 9, ML estimation requires the assumption of multivariate normal distribution of the variables. If the input data depart substantially from a multivariate normal distribution, important aspects of the results of ML-estimated EFA model can be distorted and not trustworthy (e.g., goodness of model fit, significance tests of model parameters).

Another potential disadvantage of ML estimation is its occasional tendency to produce *improper solutions*. An improper solution exists when a factor model does not converge on a final set of parameter estimates, or produces an *out-of-range* estimate such as an indicator with a communality above 1.0. On the other hand, PF has the strong advantages of being free of distributional assumptions and of being less prone to improper solutions than ML (Fabrigar et al., 1999). Unlike ML, PF does not provide goodness-of-fit indices useful in determining the suitability of the factor model and the number of latent variables. Thus PF might be preferred in instances where marked non-normality is evident in the observed measures, or perhaps when ML estimation produces an improper solution. However, as discussed later in this book, the presence of improper solutions may be a sign of more serious problems, such as a poorly specified factor model or a poorly behaved input data matrix. If distributional assumptions hold, ML may be favored because of its ability to produce a wide range of fit indices that guide other important aspects of the factor analytic procedure. As noted in Chapter 3, ML is a full information estimator that provides standard errors that can be used for statistical significance testing and confidence intervals of key parameters such as factor loadings and factor correlations. Strategies for dealing with non-normal, continuous outcomes and categorical indicators are discussed in Chapter 9.

Although related to EFA, principal components analysis (PCA) is frequently mistaken to be an estimation method of common factor analysis. Unlike the estimators discussed in the preceding paragraphs (ML, PF), PCA relies on a different set of quantitative methods that are not based on the common factor model. PCA does not differentiate common and unique variance. Instead, PCA aims to account for the *variance* in the observed measures rather than explain the *correlations* among them. Thus PCA is more appropriately used as a data reduction technique to reduce a larger set of measures to a smaller, more manageable number of composite variables for use in subsequent analyses. However, some methodologists have argued that PCA is a reasonable or perhaps superior alternative to EFA, in view of the fact that PCA possesses several desirable statistical properties. For instance, it is computationally simpler; it is not susceptible to improper solutions; it often produces results similar to EFA; and PCA is able to calculate a participant's score on a principal component, whereas the indeterminate nature of EFA complicates such computations. Although debate on this issue continues, Fabrigar et al. (1999) provide several reasons in opposition to the argument for the place of PCA in factor analysis. These authors underscore the situations where EFA and PCA produce dissimilar results (e.g., when communalities are low or when there are only a few indicators of a given factor; cf. Widaman, 1993). Regardless, if the overriding rationale and empirical objectives of an analysis are in accord with the common factor model, then it is conceptually and mathematically inconsistent to conduct PCA; that is, EFA is more appropriate if the stated objective is to reproduce the intercorrelations of a set of indicators with a smaller number of latent dimensions, recognizing the existence of measurement error in the observed measures. Floyd and Widaman (1995) make the related point that estimates based on EFA are more likely to generalize to CFA than are those obtained from PCA (because, unlike PCA, EFA and CFA are based on the common factor model). This is a noteworthy consideration in light of the fact that EFA is often used as a precursor to CFA in scale development and construct validation. A detailed demonstration of the computational differences between PCA and EFA can be found in multivariate and factor analytic textbooks (e.g., Tabachnick & Fidell, 2013).

Factor Selection

Next, the factor analysis is run with the selected estimation method (e.g., ML, PF). The results of the initial analysis are used to determine the appropriate number of factors to be extracted in subsequent analyses. This is often considered to be the most crucial decision in EFA because *underfactoring* (selecting too few factors) and *overfactoring* (selecting too many factors) can severely compromise the validity of the factor model and its resulting estimates (e.g., can introduce considerable error in the factor loading estimates), although some research suggests that the consequences of overfactoring are less severe than those of underfactoring (cf. Fabrigar et al., 1999). Despite the fact that EFA is an exploratory or descriptive technique by nature, the decision about the appropriate number of factors should be guided by substantive considerations, in addition to the statistical guidelines discussed below. For instance, the validity of a given factor should be evaluated in part by its interpretability (e.g., does a factor revealed by the EFA have substantive importance?). A firm theoretical background and previous experience with the variables will strongly foster the interpretability of factors and the evaluation of the overall factor model. Moreover, factors in the solution should be well defined (i.e., composed of several indicators that strongly relate to it). Factors that are represented by two or three indicators may be underdetermined (have poor determinacy; see below) and highly unstable across replications. The solution should also be evaluated with regard to whether "trivial" factors exist in the data—for instance, factors based on differential relationships among indicators that stem from extraneous or methodological artifacts (e.g., method effects arising from subsets of very similarly worded or reverse-worded items; see Chapter 5).

It is also important to note that the number of factors (m) that can be extracted by EFA is limited by the number of observed measures (p) that are submitted to the analysis. The upper limit on the number of factors varies across estimation techniques. For instance, in EFA using PF, the maximum number of factors that can be extracted is $p - 1$.[5] In ML EFA, the number of elements in the input correlation or covariance matrix (a) must be equal to or greater than the number of parameters that are estimated in the factor solution (b) (i.e., $a \geq b$). As the number of factors (m) increases, so does the number of estimated parameters (b) in the solution. The fact that the maximum number of factors is mathematically limited by the input data can be problematic for ML analyses that use a small set of indicators; that is, the data may not support extraction of the number of factors that are posited to exist on conceptual grounds. For example, because only four observed measures ($p = 4$) are involved, it is possible to extract only one factor ($m = 1$) in the EFA presented in Table 2.2. Although a two-factor solution may be conceptually viable (e.g., Cognitive Depression: O1, O2; Somatic Depression: O3, O4), the number of parameters associated with a two-factor model (b) would exceed the number of pieces of information in the input correlation matrix (a). a and b can be readily calculated by the following equations:

$$a = [p * (p + 1)] / 2 \qquad (2.7)$$

$$b = (p * m) + [(m * (m + 1)) / 2] + p - m^2 \qquad (2.8)$$

where p is the number of observed variables (indicators), and m is the number of factors.

Solving for a indicates that input matrix contains 10 pieces of information (see Table 2.1), corresponding to the 6 correlations in the off-diagonal and the 4 standardized variances on the diagonal; that is, $a = (4 * 5) / 2 = 10$. Solving for b (when $m = 1$) indicates that there are 8 parameters estimated in a one-factor solution; that is, $b = (4 * 1) + [(1 * 2) / 2] + 4 - 1 = 4 + 1 + 4 - 1 = 8$. Because the number of elements of the input matrix ($a = 10$) is greater than the number of parameter estimates ($b = 8$), a single factor can be extracted from the data. (As seen in Table 2.2, the degrees of freedom associated with the χ^2 fit statistic is 2, corresponding to the difference $a - b$, $10 - 8 = 2$; see Chapter 3.) However, two factors cannot be extracted because the number of parameters to be estimated in this model exceeds the number of elements of the input matrix by one; that is, $b = (4 * 2) + [(2 * 3) / 2] + 4 - 4 = 8 + 3 + 4 - 4 = 11$.

Each aspect of the equation used to solve for b corresponds to specific parameters and mathematical restrictions in the EFA model (cf. Eq. 2.4). The first aspect, $(p * m)$, indicates the number of factor loadings (Λ_y). The second aspect, $([m * (m + 1)] / 2)$, indicates the number of factor variances and covariances (Ψ). The third aspect, p, corresponds to the number of residual variances (θ_ε). The final aspect, m^2, reflects the number of restrictions that are required to identify the EFA model (e.g., mathematically convenient restrictions, which include fixing factor variances to unity). For example, as depicted in Figure 2.1, in the one-factor model there are 4 factor loadings ($p * m$), 1 factor variance ($[m * (m + 1)] / 2$), and 4 indicator residuals (p); however, for identification purposes, the factor variance is fixed to 1.0 ($m^2 = 1^2 = 1$), and thus the model contains 8 estimated parameters. A two-factor solution would entail 8 factor loadings ($4 * 2$), 2 factor variances and 1 factor covariance [($2 * 3) / 2$], and 4 residual variances (total number of parameters = 15). After subtracting the identifying restrictions ($m^2 = 2^2 = 4$; $15 - 4 = 11$), the number of parameters to be estimated in the two-factor model ($b = 11$) still exceeds the pieces in the input matrix ($a = 10$). Thus two factors cannot be extracted from the data by ML when $p = 4$.

Especially when an estimation procedure other than ML is used (e.g., PF), factor selection is often guided by the *eigenvalues* generated from either the *unreduced correlation matrix* ($\mathbf{R}$; i.e., the input correlation matrix with unities—1.0s—in the diagonal) or the *reduced correlation matrix* ($\mathbf{R}_r$; i.e., the correlation matrix with communality estimates in the diagonal). For example, the selected SPSS output in Table 2.2 provides eigenvalues from the unreduced correlation matrix under the heading "Initial Statistics."[6] Most multivariate procedures such as EFA rely on eigenvalues and their corresponding *eigenvectors* because they summarize variance in a given correlation or variance–covariance matrix. The calculation of eigenvalues and eigenvectors is beyond the scope of this chapter (for an informative illustration, see Tabachnick & Fidell, 2013), but for practical purposes, it is useful to view eigenvalues as representing the variance in the indicators explained by the successive factors. This is illustrated in the final two sections of Table 2.2—specifically, the eigenvalue corresponding to the single factor that was extracted to account for the interrelationships of the four ratings of clinical depression. In the SPSS printout, this eigenvalue is listed under the heading "SS Loadings" and equals 2.579. Calculating the sum of squares of the four factor loadings (i.e., $.82822^2 + \ldots + .75228^2 = 2.579$) provides the eigenvalue for this factor. Dividing this eigenvalue by the total variance of the input matrix (because indicators are standardized, total variance is equal to the number of input measures, p) yields the proportion of variance in the indicators that is accounted for by the factor model (i.e., 2.579 / 4 = .645), as also denoted under the heading "Pct of Var" (64.5%) in the "Final Statistics" section of the SPSS printout in Table 2.2.

The previous paragraph has discussed eigenvalues (e.g., 2.579) that are derived from the reduced correlation matrix (R_r) produced by the EFA solution. The SPSS printout (Table 2.2) also presents eigenvalues for R, listed under the "Initial Statistics" heading (i.e., 2.93, .410, .359, .299). In line with the notion that eigenvalues communicate variance, note that the sum of the eigenvalues for R is 4 (i.e., total variance = number of input indicators, p). As was the case for eigenvalues associated with R_r, dividing the eigenvalue by 4 yields an estimate of explained variance (e.g., 2.93 / 4 = .733; see Table 2.2). Thus eigenvalues guide the factor selection process by conveying whether a given factor explains a considerable portion of the total variance of the observed measures.

Three commonly used factor selection procedures are based on eigenvalues. They are (1) the Kaiser–Guttman rule, (2) the scree test, and (3) parallel analysis. The *Kaiser– Guttman rule* (also referred to as the *Kaiser criterion* or the *eigenvalues > 1.0 rule*) is very straightforward: (1) Obtain the eigenvalues derived from the input correlation matrix, R (as noted by Fabrigar et al., 1999, researchers occasionally make the mistake of using eigenvalues of the reduced correlation matrix, R_r); (2) determine how many eigenvalues are greater than 1.0; and (3) use that number to determine the number of nontrivial latent dimensions that exist in the input data. As seen in the "Initial Statistics" section of the selected SPSS output provided in Table 2.2, a single eigenvalue from the input correlation matrix (R) is above 1.0 (i.e., 2.93); thus the Kaiser–Guttman rule suggests a unidimensional latent structure.

The logic of the Kaiser–Guttman rule is that when an eigenvalue is less than 1.0, the variance explained by a factor is less than the variance of a single indicator. Recall that eigenvalues represent variance, and that EFA standardizes both the latent and observed variables (e.g., the variance that each standardized input variable contributes to the factor extraction is 1.0). Thus, because a goal of EFA is to reduce a set of input indicators (the number of factors should be smaller than the number of input indicators), if an eigenvalue is less than 1.0, then the corresponding factor accounts for less variance than the indicator (whose variance equals 1.0). The Kaiser–Guttman rule has wide appeal (and in fact is the default in popular statistical software such as SPSS) because of its simplicity and objectivity. Nevertheless, many methodologists have criticized this procedure because it can result in either overfactoring or underfactoring, and because of its somewhat arbitrary nature. For example, sampling error in the input correlation matrix may result in eigenvalues of .99 and 1.01, but nonetheless the Kaiser–Guttman rule indicates that the latter is an important factor, whereas the former is not.

Another popular approach, called the *scree test* (Cattell, 1966), also uses the eigenvalues that can be taken from either the input or reduced correlation matrix (although Fabrigar et al., 1999, note reasons why scree tests based on R_r might be preferred). To provide a more realistic illustration of this procedure, a larger data set is used ($p = 20$). As shown in Figure 2.2, the scree test employs a graph in which the eigenvalues form the vertical axis and the factors form the horizontal axis. The graph is inspected to determine the last substantial decline in the magnitude of the eigenvalues—or the point where lines drawn through the plotted eigenvalues change slope. A limitation of this approach is that the results of the scree test may be ambiguous (e.g., there is no clear shift in the slope) and open to subjective interpretation. This is evident in Figure 2.2, where the results can be interpreted as indicating either a four- or five-factor solution. However, as noted by Gorsuch (1983), the scree test performs reasonably well under conditions when the sample size is large and well-defined factors are present in the data (i.e., factors defined by multiple items with high communalities).

Another eigenvalue-based procedure for guiding factor selection is *parallel analysis* (Horn, 1965; Humphreys & Montanelli, 1975). The approach is based on a scree plot of the eigenvalues obtained from the sample data against eigenvalues that are estimated from a data set of random numbers (i.e., the means of eigenvalues produced by multiple sets of completely random data).[7] Both the observed sample and random data eigenvalues are plotted, and the appropriate number of factors is indicated by the point where the two lines cross. Thus factor selection is guided by the number of real eigenvalues greater than the eigenvalues generated from the random data; that is, if the "real" factor explains less variance than the corresponding factor obtained from random numbers, it should not be included in the factor analysis. The term *parallel analysis* refers to the fact that the random data set(s) should parallel aspects of the actual research data (e.g., sam-

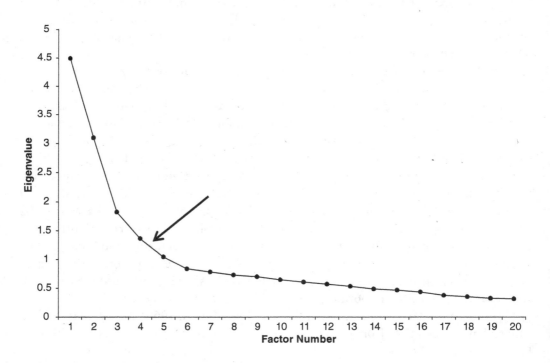

FIGURE 2.2. Scree test of eigenvalues from the unreduced correlation matrix. Arrow indicates region of curve where slope changes.

ple size, number of indicators). The rationale of parallel analysis is that the factor should account for more variance than is expected by chance (as opposed to more variance than is associated with a given indicator, according to the logic of the Kaiser–Guttman rule). Using the 20-item data set, parallel analysis suggests four factors (see Figure 2.3). After the eigenvalue for the fourth factor, the eigenvalues from the randomly generated data (averages of 50 replications) exceed the eigenvalues of the research data. Although parallel analysis frequently performs well, like the scree test it is sometimes associated with somewhat arbitrary outcomes (e.g., chance variation in the input correlation matrix may result in eigenvalues falling just above or below the parallel analysis criterion). A practical drawback of the procedure is that it is not available in major statistical software packages such as SAS and SPSS, although parallel analysis is an option in the Mplus and Stata software programs, and in various shareware programs found on the Internet (e.g., O'Connor, 2001). In addition, Hayton, Allen, and Scarpello (2004) have provided syntax for conducting parallel analysis in SPSS, although the user must save and summarize the eigenvalues generated from random data outside of SPSS.

As noted above, when a factor estimation procedure other than ML is employed, eigenvalue-based procedures such as the Kaiser–Guttman rule, the scree test, and par-

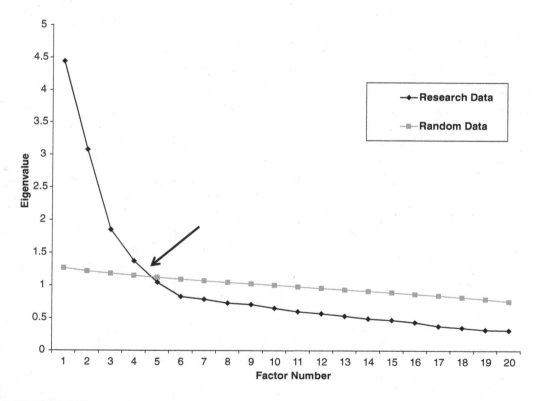

FIGURE 2.3. Parallel analysis using eigenvalues from research and random data (average of 50 replications). Arrow indicates that eigenvalues from random data exceed the eigenvalues from research data after the fourth factor.

allel analysis can be used to assist in factor selection. Although these methods can also assist in determining the appropriate number of factors in ML factor analysis, ML has the advantage of being a full information estimator that allows for goodness-of-fit evaluation and statistical inference such as significance testing and confidence interval estimation. ML is covered extensively in later chapters, so only a brief overview relevant to EFA is provided here. It is helpful to consider ML EFA as a special case of SEM. For example, like CFA and SEM, ML EFA provides goodness-of-fit information that can be used to determine the appropriate number of factors. Various goodness-of-fit statistics (such as χ^2 and the root mean square error of approximation, or RMSEA; Steiger & Lind, 1980) provide different pieces of information about how well the parameter estimates of the factor model are able to reproduce the sample correlations. As seen earlier in this chapter, the factor loadings of O1 and O2 yield a predicted correlation of .696 (i.e., Eq. 2.6), which is very similar to the correlation of these indicators in the sample data (i.e., .70; see correlation between O1 and O2 in Table 2.1). If the remaining observed relationships in the input matrix are reproduced as well by the factor loading estimates in this solution, descriptive fit statistics such as χ^2 and RMSEA will indicate that the one-factor model provides a good fit to the data. As shown in Table 2.2, the SPSS and Mplus output provides a χ^2 test of the fit of the one-factor solution (as seen in Table 2.2, Mplus also provides several other goodness-of-fit statistics that are discussed in Chapter 3). Because the χ^2 is statistically nonsignificant, $\chi^2(2) = .20$, $p = .90$, it can be concluded that the one-factor model provides a reasonable fit to the data. The nonsignificant χ^2 test suggests that the correlation matrix predicted by the factor model parameter estimates does not differ from the sample correlation matrix. However, it will be seen in Chapter 3 that χ^2 has serious limitations, and thus it should not be used as the sole index of overall model fit.

The goal of goodness-of-fit approaches is to identify the solution that reproduces the observed correlations considerably better than more parsimonious models (i.e., models involving fewer factors), but that is able to reproduce these observed relationships equally or nearly as well as more complex solutions (i.e., models with more factors). Accordingly, a researcher conducting ML EFA is apt to estimate the factor model several times (specifying different numbers of factors) to compare the fit of the solutions. As in other approaches (e.g., eigenvalue-based methods), factor selection should not be determined by goodness of fit alone, but should be strongly assisted by substantive considerations (e.g., prior theory and research evidence) and other aspects of the resulting solution. Although a factor solution may provide a reasonable fit to the data, it may be unacceptable for other reasons—such as the presence of factors that have no strong conceptual basis or utility (e.g., factors arising from methodological artifacts—see Chapter 5); poorly defined factors (e.g., factors in which only one or two indicators have strong primary loadings); indicators that do not have salient loadings on any factor; or indicators that have high loadings on multiple factors. Again, EFA is a largely exploratory procedure, but substantive and practical considerations should strongly guide the factor analytic process. Because of this and other issues (e.g., the role of sampling error), the results of an initial EFA should be interpreted cautiously and should be cross-validated (additional EFAs or CFAs should be conducted with independent data sets).

Factor Rotation

Once the appropriate number of factors has been determined, the extracted factors are rotated, to foster their interpretability. In instances when two or more factors are involved (rotation does not apply to one-factor solutions), rotation is relevant because of the indeterminate nature of the common factor model. That is, for any given multiple-factor model, there exist an infinite number of equally good-fitting solutions (each represented by a different factor loading matrix). The term *simple structure* was coined by Thurstone (1947) to refer to the most readily interpretable solutions, in which (1) each factor is defined by a subset of indicators that load highly on the factor; and (2) each indicator (ideally) has a high loading on one factor (often referred to as a *primary loading*) and has a trivial or close to zero loading on the remaining factors (referred to as a *cross-loading* or *secondary loading*). In applied research, factor loadings greater than or equal to .30 or .40 are often interpreted as *salient* (i.e., the indicator is meaningfully related to a primary or secondary factor), although explicit or widely accepted guidelines do not exist, and the criteria for salient and nonsalient loadings often depend on the empirical context. Thus, for models that contain two or more factors (where an infinite number of equally fitting solutions is possible), rotation is conducted to produce a solution with the best simple structure. It is important to emphasize that rotation does not alter the fit of the solution; for example, in ML EFA, model χ^2 is the same before and after factor rotation. Rather, factor rotation is a mathematical transformation (i.e., rotation in multidimensional space) that is undertaken to foster interpretability by maximizing larger factor loadings closer to one and minimizing smaller factor loadings closer to zero. For a mathematical demonstration of this procedure, the reader is referred to Comrey and Lee (1992).

There are two major types of rotation: *orthogonal* and *oblique*. In orthogonal rotation, the factors are constrained to be uncorrelated (i.e., factors are oriented at 90° angles in multidimensional space); in oblique rotation, the factors are allowed to intercorrelate (i.e., to permit factor axis orientations of less than 90°). The correlation between two factors is equal to the cosine of the angle between the rotational axes. Because cos(90) = 0, the factors are uncorrelated in orthogonal rotation. In oblique rotations, the angle of the axis is allowed to be greater or less than 90°, and thus the cosine of the angle may yield a factor correlation between zero and one.

In applied social sciences research, orthogonal rotation is used most often, perhaps because it has historically been the default in major statistical programs such as SPSS (*varimax rotation*), and researchers have traditionally perceived that orthogonally rotated solutions are more easily interpreted because the factor loadings represent correlations between the indicators and the factors (e.g., squaring the factor loadings provides the proportion of variance in the indicator that the factor solution explains). In oblique solutions, factor loadings usually do not reflect simple correlations between the indicators and the factors unless the factors themselves have no overlap. Because oblique rotations allow the factors to intercorrelate, the correlations between indicators and factors may be inflated by the covariation of the factors; that is, an indicator may correlate with one factor in part through its correlation with another factor. However, orthogonal rotation may produce misleading solutions in situations where the factors are expected to be intercorrelated (e.g., a questionnaire whose latent structure entails several interrelated dimensions of a broader construct). In other words, although substantial correlations may exist among factors, orthogonal rotation constrains the solution to yield uncorrelated latent variables.

Thus, in most cases, oblique rotation is preferred because it provides a more realistic representation of how factors are interrelated. If the factors are in fact uncorrelated, oblique rotation will produce a solution that is virtually the same as one produced by orthogonal rotation. On the other hand, if the factors are interrelated, oblique rotation will yield a more accurate representation of the magnitude of these relationships. In addition, estimation of factor correlations provides important information, such as the existence of redundant factors or a potential higher-order structure. Factor intercorrelations above .80 or .85 may imply poor discriminant validity, and suggest that a more parsimonious solution could be obtained (see Chapter 5). If all factors in the solution are moderately intercorrelated at roughly the same magnitude, a single higher-order factor may account for these relationships (see Chapter 8). Moreover, when EFA is used as a precursor to CFA (see Chapter 5), oblique solutions are more likely to generalize to CFA than orthogonal solutions (i.e., constraining factors to be uncorrelated in CFA will typically result in poor model fit).

Several forms of oblique rotation have been developed (e.g., promax, geomin, quartamin, orthooblique). When oblique rotation is requested, most software programs (such as SPSS) output both a *pattern matrix* and a *structure matrix*. The loadings in the pattern matrix convey the *unique* relationship between a factor and an indicator. They are interpreted in the same fashion as partial regression coefficients in standard multiple regression; that is, the coefficient represents the relationship between the predictor (factor) and outcome (indicator), while controlling for the influence of all other predictors (other factors). Thus indicator variance that is explained by more than one factor is omitted from the loadings in the pattern matrix. The structure matrix is calculated by multiplying the pattern matrix by the factor correlation matrix (oblique rotation produces a factor correlation matrix, but orthogonal rotation does not). Hence loadings in the structure matrix reflect both the unique relationship between the indicator and factor (as in the pattern matrix) *and* the relationship between the indicator and the shared variance among the factors. In other words, the loadings in the structure matrix reflect a zero-order relationship between the indicator and a given factor, without holding the other factors in the solution constant. Unless the correlations among factors are minimal, loadings in the structure matrix will typically be larger than those in the pattern matrix, because they are inflated by the overlap in the factors (akin to zero-order correlations vs. partial regression coefficients in standard multiple regression). Although there is some debate about whether the pattern matrix or structure matrix should be used, the pattern matrix is by far more often interpreted and reported in applied research. In fact, some popular latent variable software programs only provide the pattern matrix. As noted above, the mathematical operations for generating the structure matrix are quite straightforward when the pattern matrix and factor intercorrelation are available. Thus, with the aid of software (e.g., SAS PROC IML), either matrix can be readily computed on the basis of the other (i.e., structure matrix = pattern matrix multiplied by the factor correlation matrix; pattern matrix = structure matrix multiplied by the inverse of factor correlation matrix).

The factor intercorrelations produced by an oblique rotation of a PCA solution are often lower than those of obliquely rotated solutions based on the common factor model (e.g., PF, ML; Fabrigar et al., 1999; Widaman, 1993). This is because in common factor analysis, random error is removed from the factors. In PCA, random error is included in the components (i.e., PCA does not differentiate common and unique variance). Thus another potential adverse consequence of PCA is a mistaken conclusion that components share modest variance when in fact the intercorrelations have been attenuated by random error (or a conclusion that components are distinct when in fact the error-disattenuated correlations would be above .80). Because factor correlations arising from common factor analysis are more likely to be closer to population values, this is another reason why methodologists usually advocate EFA over PCA.

Factor rotation is illustrated in Figure 2.4, using a real data set of eight indicators collected from a sample of 500 participants. A scree test and parallel analysis suggest a two-factor solution. Results indicate that the first four indicators (Y1–Y4) load on Factor 1 and the remaining four indicators (Y5–Y8) load on Factor 2. Figure 2.4 displays a geometric representation of unrotated, orthogonally rotated (varimax), and obliquely rotated (promax) factor matrices. ML estimation produces the unrotated factor loadings presented in Figure 2.4A. Figure 2.4B shows the results of the varimax rotation. The factor axes remain at 90° angles, but are rotated in the most optimal fashion to maximize high factor loadings and minimize low loadings. Rotation produces a transformation matrix. With matrix algebra, the unrotated factor loading matrix is multiplied by the transformation matrix to produce the rotated factor loading matrix. In this data set, the varimax transformation matrix is as follows:

	Factor 1	Factor 2
Factor 1	.93347	.35867
Factor 2	−.35867	.93347

These values convey how much the axes are rotated to foster simple structure. Specifically, the values on the diagonal (.93347) are cosines, and the values on the off-diagonal (.35867, −.35867) are sines and −sines. As shown in Figure 2.4B, the axes are rotated 21° to better transect the clusters of indicators. Within rounding error, the cos(21) equals .933 and the sin(19) equals .359, the same as the transformation coefficients shown above. Because orthogonal rotation is used, the axes of Factor 1 and Factor 2 remain at right angles, and thus the factors are constrained to be uncorrelated; that is, cos(90) = 0.

To witness the effects of rotation on maximizing and minimizing factor loadings, consider the fifth indicator, Y5. Before rotation, the loadings of Y5 on Factor 1 and Factor 2 are very similar (.386 and .329, respectively; Figure 2.4A). A 21° rotation of the

A. Unrotated Factor Matrix

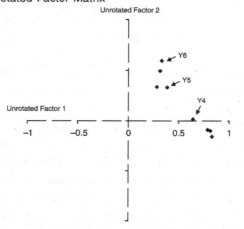

	Factor	
	1	2
Y1	.834	−.160
Y2	.813	−.099
Y3	.788	−.088
Y4	.642	.015
Y5	.386	.329
Y6	.333	.593
Y7	.313	.497
Y8	.284	.336

B. Orthogonally Rotated Factor Matrix (Varimax)

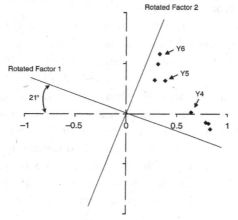

	Factor	
	1	2
Y1	.836	.150
Y2	.794	.199
Y3	.767	.201
Y4	.594	.244
Y5	.242	.445
Y6	.098	.673
Y7	.114	.576
Y8	.145	.416

C. Obliquely Rotated Factor Matrix (Promax)

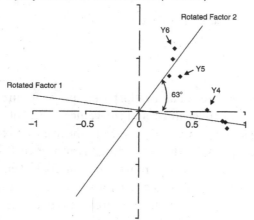

	Factor	
	1	2
Y1	.875	−.062
Y2	.817	.003
Y3	.788	.012
Y4	.588	.106
Y5	.154	.418
Y6	−.059	.704
Y7	−.018	.595
Y8	.055	.413

FIGURE 2.4. Geometric representations of unrotated, orthogonally rotated, and obliquely rotated factor matrices.

factor axes raises Y5's position on the Factor 2 axis (.445), and decreases this indicator's position on the Factor 1 axis (.242) (Figure 2.4B). Although this transformation fosters the interpretability of the solution, it does not alter the communality of Y5 or any other indicator. In a solution entailing more than one latent variable, communalities in an orthogonal EFA are calculated by taking the sum of squared loadings for a given indicator across all factors.[8] Before and after rotation, the proportion of variance explained in Y5 is .257; unrotated solution: $.386^2 + .329^2 = .257$, rotated solution: $.242^2 + .445^2 = .257$. Thus rotation does not alter the fit of the factor solution.

Figure 2.4B also suggests that oblique rotation may be more appropriate. A factor solution is best defined when the indicators are clustered around the upper end of their respective factor axes. The higher up the axis, the higher the factor loading; if an indicator is in close proximity to one factor axis, it does not load highly on another factor. As shown in Figure 2.4B, orthogonal rotation moves the Factor 2 axis closer to the Y5–Y8 indicators. As a result, the rotation has an overall effect of increasing the primary loadings of Y5–Y8 on Factor 2, and decreasing their cross-loadings on Factor 1 (compared to the unrotated solution, Figure 2.4A). However, orthogonal rotation moves the Factor 1 axis away from the Y1–Y4 indicators, which has the general effect of increasing the magnitude of the cross-loadings of Y1–Y4 on Factor 2 (e.g., Y4: .015 vs. .244 for the unrotated and rotated solutions, respectively). Indeed, in instances where all the indicators fall in between the factor axes after orthogonal rotation (as seen in Figure 2.4B), the restriction of maintaining a 90° orientation of the factor axes may not be tenable. Oblique rotations such as promax begin with an orthogonal rotation, but then "break" the 90° angle to allow the factor axes to pass through the clusters of indicators better. The angle of the factor axes reflects the factor correlation. If factors are uncorrelated, the angle of factor axes will remain close to 90°. If the factors are correlated, the angle of factor axes will deviate from 90°.

Figure 2.4C provides a geometric depiction of oblique rotation (promax) of the two-factor solution. Note that the axes of Factor 1 and Factor 2 are both turned inward somewhat to better transect the two clusters of indicators. Compared to orthogonal rotation, the oblique rotation increases the values of most primary loadings further. A more notable impact of oblique rotation is its success at moving the cross-loadings closer to zero (but, as before, the overall fit of the solution is the same). To accomplish this, the angle of the factor axes is shifted from 90° to 63° (see Figure 2.4C). The results of the analysis indicate that the correlation of Factor 1 and Factor 2 is .45. This corresponds to the cosine of the factor angle; that is, cos(63) = .45.

Factor Scores

After an appropriate factor solution has been established, a researcher may wish to calculate factor scores by using the factor loadings and factor correlations. Factor scores are used for various purposes (e.g., to serve as proxies for latent variables, to determine a participant's relative standing on the latent dimension). Conceptually, a factor score is the score that would have been observed for a person if it had been possible to measure the factor directly. In applied research, factor scores are often computed by creating *coarse factor scores*, which are simple unweighted composites of the raw scores of indicators (e.g., averaging or summing) found to have salient loadings on the factor. However, there are many reasons why coarse factor scores may poorly represent factors (e.g., they may be highly intercorrelated even when the factors are truly orthogonal; Grice, 2001). Alternatively, factor scores can be estimated by multivariate methods that use various aspects of the reduced or unreduced correlation matrix and factor analysis coefficients. The resulting values are called *refined factor scores*. A frequently used method of estimating refined factor scores is Thurstone's (1935) least squares regression approach, although several other strategies have been developed (e.g., Bartlett, 1937; Harman, 1976; McDonald, 1981). Most statistical software packages provide options to compute refined factor scores by one or more of these methods. In the majority of instances, refined factor scores have less bias than coarse factor scores and thus are favored over coarse factor scores as proxies for the factors (Grice, 2001). However, a complicating issue in factor score estimation is the indeterminate nature of the common factor model. With respect to factor scores, this indeterminacy means that an infinite number of sets of factor scores can be computed from any given factor analysis that will be equally consistent with the same factor loadings (Grice, 2001). The degree of indeterminacy depends on several aspects, such as the ratio of items to factors and the size of the item communalities (e.g., factors defined by several items with strong communalities have better determinacy). If a high degree of indeterminacy is present, the sets of factor scores can vary so widely that an individual ranked high on the dimension in one set may receive a low ranking on the basis of another set. In such scenarios, the researcher has no way of discerning which set of scores or rankings is most accurate.

Thus, although typically neglected in applied factor analytic research, the degree of factor score indeterminacy should be examined as part of EFA, especially in instances when factor scores are to be computed for use in subsequent statistical analyses. Grice (2001) has specified three criteria for evaluating the quality of factor scores: (1) *validity coefficients*, or correlations between the factor score estimates and their respective factor scores; (2) *univocality*, or the extent to which the factor scores are excessively or insufficiently correlated with other factors in the same analysis; and (3) *correlational accuracy*, or how closely the correlations among factor scores correspond to the correlations among the factors. For instance, Gorsuch (1983) has recommended that validity coefficients should be at least .80, although higher values (e.g., >.90) may be required in some situations (e.g., when factor scores are used as dependent variables). Unfortunately, procedures for evaluating factor scores are not standard options in most software packages. As shown in Table 2.2, an exception is the Mplus program, where validity coefficients can be requested as part of EFA and CFA by using the FSDETERMINACY option of the OUTPUT command (the validity coefficient is .94, which indicates an acceptable level of factor score determinacy; see also Chapter 5). In addition, Grice (2001) has developed SAS PROC IML computer code for assessing the degree of factor score indeterminacy (validity coefficients, univocality, correlational accuracy) in the context of EFA (this syntax can be downloaded from *http://psychology.okstate.edu/faculty/jgrice/factorscores*).

SUMMARY

Procedural recommendations for conducting applied EFA are summarized in Table 2.3. In addition to providing a practical overview of EFA (e.g., procedural considerations for factor estimation, selection, rotation, and interpretation), the goal of this chapter has been to introduce key concepts that are carried forward in the subsequent chapters on CFA (e.g., observed vs. latent variables, factor loadings, factor correlations, common and unique variance, basic equations and notation). Some fundamental differences of EFA and CFA have been described. Unlike EFA, in CFA the number of factors and the pattern of indicator–factor loadings are specified in advance on the basis of strong empirical knowledge or theory. The acceptability of the CFA model is evaluated in part by descriptive fit statistics that convey the ability of the solution to reproduce the observed relationships among the input indicators (although similar testing can be applied in EFA when the ML estimator is used). As will be seen in Chapter 3, EFA and CFA differ in several other important manners.

NOTES

1. Exceptions to this rule are discussed in Chapters 3 and 4 (e.g., when indicator measurement errors are correlated).
2. As discussed in Chapter 5, hybrid models that combine the features of EFA and CFA have been developed recently (e.g., exploratory structural equation modeling).
3. For instance, in the current example, which entails a single factor, the factor loadings can be interpreted as zero-order correlation coefficients between the factor and the observed measures (i.e., factor loading = the standardized regression slope = zero-order correlation). In solutions involving multiple, correlated factors (oblique rotation), factor loadings from the factor pattern matrix are interpreted as partial regression coefficients.
4. The reader will encounter many variations in this notational system across factor analysis and SEM texts. For instance, because indicator unique variances (ε) are not observed, it is common to see these parameters depicted as circles in path diagrams. In Chapter 3, this notation is expanded by differentiating latent X (exogenous) and latent Y (endogenous) solutions.
5. In PCA, the limit on the number of components is equal to p.
6. Because eigenvalues are drawn from the unreduced correlation matrix (**R**), PCA is always conducted initially, regardless of the type of factor analysis requested (e.g., PF).
7. Some researchers (e.g., Glorfeld, 1995) have recommended that the 95th percentile of eigenvalues from random data be used in place of average eigenvalues, in part to adjust for parallel analysis's slight tendency to overfactor (regardless of the method used, research has shown that parallel analysis is accurate in the vast majority of cases; e.g., Humphreys & Montanelli, 1975; Zwick & Velicer, 1986).
8. Although communalities can also be hand-calculated from the estimates of an obliquely rotated EFA solution, this computation is less straightforward, because the factors are permitted to be intercorrelated and thus the factor loadings are partial regression coefficients. Later chapters (e.g., Chapter 3) discuss the tracing rules necessary to compute these estimates.

TABLE 2.3. Fundamental Steps and Procedural Recommendations for EFA

Factor extraction

- Use an estimator based on the common factor model, such as:
  - *Principal factors*: No distributional assumptions; less prone to improper solutions than maximum likelihood
  - *Maximum likelihood*: Assumes multivariate normality, but provides goodness-of-fit evaluation and, in some cases, significance tests and confidence intervals of parameter estimates

Factor selection

- Determine the appropriate number of factors by:
  - *Scree plot* of eigenvalues from the reduced correlation matrix,
  - *Parallel analysis*, and/or
  - *Goodness of model fit* (e.g., χ^2, RMSEA; see Chapter 3)

Factor rotation

- In multifactorial models, rotate the solution to obtain simple structure by:
  - Using an *oblique rotation* method (e.g., promax, geomin)

Interpret the factors and evaluate the quality of the solution

- Consider the meaningfulness and interpretability of the factors:
  - Factors should have substantive meaning and conceptual/empirical relevance
  - Rule out nonsubstantive explanations such as method effects (e.g., factors composed of reverse- and non-reverse-worded items; see Chapters 3 and 5)

- Eliminate poorly defined factors, such as:
  - Factors on which only two or three items have salient loadings
  - Factors defined by items that have small loadings (i.e., *low communalities*)
  - Factors with low *factor determinacy* (poor correspondence between the factors and their factor scores; see Grice, 2001)

- Eliminate poorly behaved items (indicators), such as:
  - Items with high loadings on more than one factor (i.e., *cross-loadings*)
  - Items with small loadings on all factors (i.e., *low communalities*)

Rerun and (ideally) replicate the factor analysis

- If items or factors are dropped in preceding step, rerun the EFA in the same sample

- Replicate the final EFA solution in an independent sample

- Consider further replications/extensions of the factor solution by:
  - Developing tentative CFA models (e.g., exploratory SEM; see Chapter 5)
  - Larger-scale CFA investigations
  - Measurement invariance evaluation in population subgroups (e.g., equivalence of solution between sexes; see Chapter 7)

Note. EFA, exploratory factor analysis; RMSEA, root mean square error of approximation; CFA, confirmatory factor analysis; SEM, structural equation modeling.

3

Introduction to CFA

The purpose of this chapter is to introduce the reader to the purposes, parameters, and fundamental equations of CFA. Now that the common factor model and EFA have been described in Chapter 2, CFA and EFA are compared more thoroughly. On the basis of these comparisons, the advantages and purposes of CFA will become apparent. The notation and computation of the parameters of the CFA model are presented. This chapter also deals with the important concepts of model identification, maximum likelihood (ML) estimation, and goodness-of-fit evaluation. The concepts introduced in this chapter should be studied carefully, as they are germane to all examples of CFA presented in subsequent chapters of this book.

SIMILARITIES AND DIFFERENCES OF EFA AND CFA

Common Factor Model

Like that of EFA, the purpose of CFA is to identify factors that account for the variation and covariation among a set of indicators. Both EFA and CFA are based on the common factor model, and thus many of the concepts and terms discussed in Chapter 2 apply to CFA (such as factor loadings, unique variances, communalities, and residuals). However, while EFA is generally a descriptive or exploratory procedure, in CFA the researcher must prespecify all aspects of the factor model: the number of factors, the pattern of indicator–factor loadings, and so forth. As noted in Chapter 2, CFA requires a strong empirical or conceptual foundation to guide the specification and evaluation of the factor model. Accordingly, CFA is typically used in later phases of scale development or construct validation—after the underlying structure has been tentatively established by prior empirical analyses using EFA, as well as on theoretical grounds.

EFA and CFA often rely on the same estimation methods (e.g., *maximum likelihood*, or ML). When a full information estimator such as ML is used, the factor models arising from EFA and CFA can be evaluated in terms of how well the solution reproduces the observed variances and covariances among the input indicators (i.e., goodness-of-fit evaluation). In addition, the quality of EFA and CFA models is determined in part by the size of resulting parameter estimates (e.g., magnitude of factor loadings and factor intercorrelations) and how well each factor is represented by observed measures (e.g., number of indicators per factor, size of indicator communalities, factor determinacy).

Standardized and Unstandardized Solutions

The tradition in EFA is to *completely standardize* all variables in the analysis. Specifically, a correlation matrix is used as input in EFA, and both the factors and indicators are completely standardized: Factor variances equal 1.0; factor loadings are interpreted as correlations or standardized regression coefficients.[1] Although CFA also produces a completely standardized solution, much of the analysis does not standardize the latent or observed variables. Instead of using a correlation matrix (i.e., a completely standardized variance–covariance matrix), CFA typically analyzes a variance–covariance matrix (needed to produce an unstandardized CFA solution) or raw data that are used by the software program to produce an input variance–covariance matrix. Thus the CFA input matrix is composed of indicator variances on the diagonal (a variance equals the indicator's standard deviation squared; i.e., $VAR = SD^2$), and indicator covariances in the off-diagonal (a covariance can be calculated by multiplying the correlation of two indicators by their standard deviations; i.e., $COV_{xy} = r_{xy}SD_xSD_y$). In addition to a completely standardized solution, the results of CFA include an *unstandardized solution* (parameter estimates expressed in the original metrics of the indicators), and possibly a *partially standardized solution* (relationships involving unstandardized indicators and standardized latent variables, or vice versa). Unstandardized, partially standardized, and completely standardized solutions are discussed in more detail in Chapter 4. Of particular note here is the fact that many key aspects of CFA are based on unstandardized estimates, such as the standard errors and significance testing of model parameters.[2] As will be seen in Chapter 7, the various forms of measurement invariance evaluation (e.g., equivalence of parameter estimates within and across groups) are also based on the unstandardized solution.

The unstandardized means of the indicators can also be included in the CFA. Thus, in contrast to EFA, which focuses on completely standardized values, CFA may entail the analysis of both unstandardized variance–covariance structures and mean structures (as the result of standardization in EFA, indicator means are presumed to be zero). As discussed in Chapter 7, when indicator means are included as input in CFA, the analysis can estimate the means of the factors and the intercepts of the indicators. Akin to multiple regression, an indicator intercept is interpreted as the predicted value of the indicator when the factor—or predictor—is zero. The analysis of mean structures is particularly relevant to multiple-groups CFA models, where a researcher may be interested in comparing groups on the latent means (the SEM parallel to analysis of variance, or ANOVA) or determining the equivalence of a testing instrument's measurement properties across groups (e.g., inequality of item intercepts is indicative of test bias/differential item functioning; see Chapter 7).

The outcome of EFA is reported as a completely standardized solution. In applied CFA research, completely standardized solutions are most commonly reported. However, SEM methodologists often express a strong preference for reporting unstandardized solutions because the analysis itself is based on unstandardized variables, and completely standardized values are potentially misleading. For instance, the true nature of the variance and relationships among indicators and factors can be masked when these variables have been standardized; when the original metric of variables is expressed in meaningful units, unstandardized estimates more clearly convey the importance or substantive significance of the effects (cf. Willett, Singer, & Martin, 1998). Chapter 4 demonstrates that through basic equations, one can readily calculate a (completely) standardized solution from the unstandardized solution, and vice versa.

Indicator Cross-Loadings/Model Parsimony

In addition, EFA and CFA differ markedly in the manner by which indicator cross-loadings are handled in solutions entailing multiple factors (in unidimensional models, the issues of cross-loadings and factor rotation are irrelevant). As noted in Chapter 2, all indicators in EFA freely load on all factors, and the solution is rotated to maximize the magnitude of primary loadings and minimize the magnitude of cross-loadings. Factor rotation does not apply to CFA. This is because the identification restrictions associated with CFA are achieved in part by fixing most or all indicator cross-loadings to zero (see Figure 3.1). In other words, rotation is not necessary in CFA because simple structure is obtained by specifying indicators to load on only one factor. CFA models are typically more parsimonious than EFA solutions because, while primary loadings and factor correlations are freely estimated, no other relationships are specified between the indicators and factors (e.g., no loadings that relate indicators Y1–Y4 to Factor 2, and indicators Y5–Y8 to Factor 1; see Figure 3.1A). Thus, with few exceptions (cf. EFA in the CFA framework; Chapter 5), CFA attempts to reproduce the observed relationships among input indicators with fewer parameter estimates than EFA.

Table 3.1 presents the factor loading matrices of three analyses of the same data set (N = 1,050 adolescents): (1) a CFA (Model A), (2) an EFA with oblique rotation (Model B), and (3) an EFA with orthogonal rotation (Model C). Eight antisocial behaviors are used as indicators in the analyses. Each analysis entails two factors: Property Crimes (e.g., shoplifting, vandalism) and Violent Crimes (e.g., fighting, aggravated assault). The path diagrams of Models A and B in Figure 3.1 correspond to Models A and B in Table 3.1. The Model B path diagram can be edited to conform to an orthogonal EFA by removing the double-headed curved arrow reflecting the factor correlation. As can be seen in Table 3.1, each indicator in EFA loads on all factors. Rotation (either orthogonal or oblique) is used to foster the interpretability of the factor loadings (i.e., to maximize large loadings, to minimize small loadings). For the reasons stated in Chapter 2, rotation does not affect the fit of the EFA solution; that is, the indicator communalities are

Model A: Confirmatory Factor Model (all measurement error is random)

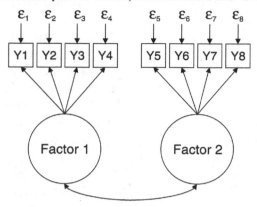

Model B: Exploratory Factor Model (oblique rotation)

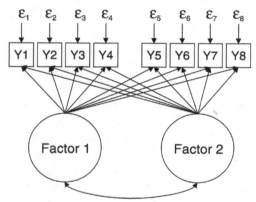

Model C: Confirmatory Factor Model (with a correlated measurement error)

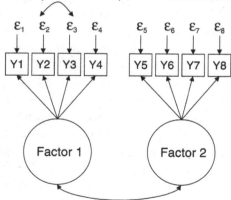

FIGURE 3.1. Path diagrams of confirmatory and exploratory factor models.

TABLE 3.1. Factor Loading Matrices from EFA and CFA of Adolescent Antisocial Behaviors

A. CFA (factor correlation = .6224)

	Property Crimes	Violent Crimes	Communality
	Factor		
Y1	.7996	.0000	.64
Y2	.6451	.0000	.42
Y3	.5699	.0000	.32
Y4	.4753	.0000	.23
Y5	.0000	.7315	.53
Y6	.0000	.5891	.35
Y7	.0000	.7446	.55
Y8	.0000	.5803	.34

B. EFA (oblique rotation, factor correlation = .5722)

	Property Crimes	Violent Crimes	Communality
	Factor		
Y1	.9187	−.0958	.75
Y2	.5422	.1045	.37
Y3	.5300	.0372	.30
Y4	.4494	.0103	.21
Y5	.0434	.7043	.53
Y6	−.1178	.6999	.41
Y7	.1727	.6106	.52
Y8	.0264	.5756	.35

C. EFA (orthogonal rotation, factor correlation = 0)

	Property Crimes	Violent Crimes	Communality
	Factor		
Y1	.8493	.1765	.75
Y2	.5509	.2574	.37
Y3	.5185	.1898	.30
Y4	.4331	.1408	.21
Y5	.2587	.6826	.53
Y6	.1032	.6314	.41
Y7	.3535	.6312	.52
Y8	.2028	.5552	.35

Note. N = 1,050. Y1 = shoplifting, Y2 = vandalism, Y3 = theft, Y4 = broke into building/vehicle, Y5 = fighting, Y6 = aggravated assault, Y7 = hit family/teachers, Y8 = threatened others.

identical in orthogonal and oblique EFA (see Table 3.1 and Appendix 3.1). The CFA model is more parsimonious than the EFA models because all indicator cross-loadings are prespecified to equal zero; that is, Y1–Y4 on Violent Crimes = 0, Y5–Y8 on Property Crimes = 0 (see Model A in Table 3.1 and Figure 3.1). Thus there are only 8 factor loading estimates in the CFA, compared to 16 factor loading estimates in the EFAs. Accordingly, rotation in CFA is not required.

Another consequence of fixing cross-loadings to zero in CFA is that factor correlation estimates in CFA are usually of higher magnitude than analogous EFA solutions. This can be seen in the current data set, where the factor correlations between Property Crimes and Violent Crimes are .57 and .62 for the oblique EFA and CFA, respectively (Table 3.1). The reason for this outcome, along with an explanation of the communality and model-implied estimates in EFA and CFA, is provided in Appendix 3.1.

Unique Variances

Unlike EFA, CFA offers the researcher the ability to specify the nature of relationships among the measurement errors (unique variances) of the indicators. Although both EFA and CFA differentiate common and unique variance, within EFA the relationships among unique variances are not specified. Because CFA typically entails a more parsimonious solution (i.e., CFA usually attempts to reproduce the observed relationships among indicators with fewer parameter estimates than EFA), it is possible to estimate such relationships when this specification is substantively justified and other identification requirements are met (see Chapter 5). Consequently, because of EFA's identification restrictions, factor models must be specified under the assumption that measurement error is random. In contrast, correlated measurement error can be modeled in a CFA solution. The CFA model presented in Model A of Figure 3.1 (i.e., Figure 3.1A) depicts a two-factor measurement model where all measurement error is presumed to be random. The underlying assumption of this specification is that the observed relationship between any two indicators loading on the same factor (e.g., Y7 and Y8) is due entirely to the shared influence of the latent dimension; that is, if Factor 2 is partialed out, the intercorrelations between these indicators will be zero. The model presented in Figure 3.1C depicts the same CFA measurement model, with the exception that a correlated error has been specified between Y2 and Y3. This specification assumes that whereas indicators Y2 and Y3 are related in part because of the shared influence of the latent dimension (Factor 1), some of their covariation is due to sources other than the common factor. In measurement models, the specification of correlated errors may be justified on the basis of source or method effects that reflect additional indicator covariation resulting from common assessment methods (e.g., observer ratings, questionnaires); reversed or similarly worded test items; or differential susceptibility to other influences, such as response set, demand characteristics, acquiescence, reading difficulty, or social desirability (cf. Brown, 2003; Marsh, 1996). The specification in Figure 3.1C depicts a *correlated uniqueness* approach to modeling error covariances (i.e., zero-order relationships are freely estimated between pairs of indicators). As noted in the discussion of multitrait–multimethod (MTMM) CFA solutions (Chapter 6), the relationships among indicator errors can also be modeled by using a *correlated methods* approach that entails the specification of methods factors in addition to the latent variables of substantive interest.

The inability to specify correlated errors (i.e., the nature of the relationships among unique variances) is a very significant limitation of EFA. For instance, in applied factor analytic research on questionnaires composed of a combination of positively and negatively worded items, a common consequence of this EFA limitation is the tendency to extract and interpret methods factors that have little substantive basis (cf. Brown, 2003; Marsh, 1996). For example, a very extensive psychometric literature exists on the Rosenberg (1965) Self-Esteem Scale (SES), a questionnaire that consists of four positively worded items (e.g., "I feel good about myself") and three negatively worded items (e.g., "At times I think I am no good at all"). Early EFA research routinely produced two SES factors composed of negatively and positively worded items that were interpreted as substantively meaningful (e.g., Positive Self-Evaluation vs. Negative Self-Evaluation). However, as compellingly argued by Marsh (1996), a strong conceptual basis did not exist in support for distinct dimensions of positive and negative self-esteem. Instead, Marsh (1996) noted that these two-factor solutions were artifacts of response styles associated with the wording of the items (e.g., response biases such as acquiescence). Using CFA, Marsh (1996) evaluated various SES measurement models corresponding to previously reported solutions (e.g., one-factor model without error covariances, two-factor models) and correlated uniqueness (residual) models. Results indicated the superiority of a unidimensional solution (Global Self-Esteem) with method effects (correlated residuals) associated with the negatively worded items. Although having a compelling substantive basis (i.e., the existence of a single dimension of self-esteem, but need for an error theory to account for the additional covariation among similarly worded items), this model could not be estimated in EFA because EFA does not allow for the specification of correlated indicator errors.

Model Comparison

The preceding sections of this chapter have documented some of the key ways that CFA offers greater modeling flexibility than EFA (i.e., prespecification of the number of factors, patterns of item–factor relationships, presence or absence of error covariances). In addition to these aspects, the CFA framework allows a researcher to impose other restrictions on the factor solution, such as constraining all the factor loadings or all the unique variances to be equal (e.g., as in the evaluation of the conditions of tau equivalence or parallel tests; see Chapter 7). The viability of these constraints can be evaluated by statistically comparing whether the fit of the more restricted solution is worse than a comparable solution without these constraints. Direct statistical comparison of alternative solutions is possible when the models are nested. As discussed further in Chapter 5, a *nested model* contains a subset of the free parameters of another model (which is often referred to as the *parent model*). For example, consider the following two models: (1) *Model P*, a one-factor model composed of six indicators allowed to load freely onto the factor; and (2) *Model N*, a one-factor model identical to Model P, except that the factor loadings are constrained to load equally onto the factor. Although the models are structurally the same (i.e., they consist of one factor and the same six indicators), they differ in their number of freely estimated versus constrained parameters. When parameters are *freely estimated*, the researcher allows the analysis to find the values for the parameters in the CFA solution (e.g., factor loadings, factor correlations, unique variances) that optimally reproduce the variances and covariances of the input matrix. In the case of *fixed parameters*, the researcher assigns specific values (e.g., fixes cross-loadings to zero to indicate no relationship between an indicator and a factor; cf. lack of an arrow between Y1 and Factor 2 in Figure 3.1A). When parameters are *constrained*, the researcher does not specify the parameters' exact values, but places other restrictions on the magnitude these values can take on. For instance, in the case of Model N, the researcher instructs the analysis to optimally reproduce the input matrix under the condition that all factor loadings are the same. Thus Model N is nested under Model P (the parent model) because it contains a subset of Model P's free parameters. Accordingly, the fit of Model N can be statistically compared to the fit of Model P (through methods such as the χ^2 difference test; see Chapter 4) to directly evaluate the viability of the condition of equal factor loadings (i.e., tau equivalence = do the six indicators relate equally to the latent factor?). Because EFA entails only freely estimated parameters (fixed parameters cannot be specified), comparative model evaluation of this nature is not possible.[3]

These procedures (e.g., χ^2 difference testing) can be used to statistically compare other forms of nested models in CFA. For instance, CFA can be used to statistically determine whether the various measurement parameters of a factor model (e.g., factor loadings) are the same in two or more groups (e.g., males and females; see Chapter 7).

PURPOSES AND ADVANTAGES OF CFA

From the preceding sections, the objectives and advantages of CFA in relation to EFA may now be apparent. Although both EFA and CFA are based on the common factor model and often use the same estimation method (e.g., ML), the specification of CFA is strongly driven by theory or prior research evidence. Thus, whereas in EFA the researcher can only prespecify the number of factors, the CFA researcher usually tests a much more parsimonious solution by indicating the number of factors, the pattern of factor loadings (and cross-loadings, which are usually fixed to zero), and an appropriate error theory (e.g., random or correlated indicator error). In contrast to EFA, CFA allows for the specification of relationships among the indicator uniquenesses (error variances), which may have substantive importance (e.g., correlated errors due to method effects). Thus every aspect of the CFA model is specified in advance. The acceptability of the specified model is evaluated by goodness of fit and by the interpretability and strength of the resulting parameter estimates (overall goodness of fit also applies to EFA when the ML estimator is used). As noted previously, CFA is more appropriate than EFA in the later stages of construct validation and test construction, when prior evidence and theory support more "risky" a priori predictions regarding latent structure. For example, the modeling flexibility and capabilities of CFA (e.g., specification of an error theory) afford sophisticated analyses of construct validity, such as in the MTMM approach (see Chapter 6), where the convergent and discriminant validity of dimensions are evaluated in context of (partialing out the influence of) varying assessment methods.

In addition, CFA offers a very strong analytic framework for evaluating the equivalence of measurement models across distinct groups (e.g., demographic groups such as sexes, races, or cultures). This is accomplished by either multiple-groups solutions (i.e., simultaneous CFAs in two or more groups) or "multiple indicators, multiple causes" (MIMIC) models (i.e., the factors and indicators are regressed onto observed covariates representing group membership; see Chapter 7). Although some methods of examining the concordance of factor structures within EFA are available (e.g., Ahmavaara, 1954), the CFA framework is superior in terms of its modeling flexibility (e.g., ability to specify partial invariance models; cf. Byrne, Shavelson, & Muthén, 1989) and its ability to examine every potential source of invariance in the factor solution, including latent means and indicator intercepts. These capabilities permit a variety of important analytic opportunities in applied research, such as the evaluation of whether a scale's measurement properties are invariant across population subgroups (e.g., are the number of factors, factor loadings, item intercepts, etc., that define the latent structure of a questionnaire equivalent in males and females?). Measurement invariance is an important aspect of scale development, as this endeavor determines whether a testing instrument is appropriate for use in various groups (Chapter 7). Indeed, multiple-groups CFA can be used to evaluate the generalizability of a variety of important constructs (e.g., are the diagnostic criteria sets used to define mental disorders equivalent across demographic subgroups such as race and gender?). Moreover, this approach can be used to examine group differences in the means of the latent dimensions. Although CFA is analogous to ANOVA, it is superior to ANOVA because group comparisons are made in the context of measurement invariance (unlike ANOVA, which simply assumes that a given observed score reflects the same level of the latent construct in all groups).

Similarly, another advantage of CFA and SEM is the ability to estimate the relationships among variables adjusting for measurement error. A key limitation of ordinary least squares (OLS) approaches such as correlational and multiple regression analysis is the assumption that variables have been measured without error (i.e., they are perfectly reliable, meaning that all of an observed measure's variance is true score variance). However, this assumption rarely holds in the social and behavioral sciences, which rely heavily on variables that have been assessed by questionnaires, independent observer ratings, and so forth. Consequently, estimates derived from OLS methods (e.g., correlations, regression coefficients) are usually attenuated to an unknown degree by measurement error in the variables used in the analysis.[4] On the other hand, CFA and SEM allow for such relationships to be estimated after adjustments for measurement error and an error theory (extent of random and correlated measurement error). For example, in the CFA model presented in Figure 3.1A, the relationship between the two constructs is reflected by their factor intercorrelation (r between Factor 1 and Factor 2), as opposed to the observed relationships among the indicators that load on these factors. Indeed, this factor correlation is a better estimate of the population value of this relationship than any two indicator pairings (e.g., r between Y1 and Y4) because it has been adjusted for measurement error; that is, shared variance among the factor's indicators is operationalized as true-score variance, which is passed on to the latent variable.

In Chapter 2, the problems with the computation and use of factor scores in EFA have been briefly noted. On occasion, a researcher will wish to relate the factors revealed by EFA to other variables. Typically (but see Gorsuch, 1997), this requires the researcher to compute factor scores to serve as proxies for the factors in subsequent analyses. However, this practice is limited by the issue of factor score indeterminacy: for any given EFA, an infinite number of sets of factor scores can be computed that are equally consistent with the factor loadings (see Chapter 2). In CFA and SEM, indeterminacy of factor scores is not a problem because this analytic framework eliminates the need to compute factor scores; that is, the latent variables themselves are used in the analysis. Unlike EFA, CFA and SEM offer the researcher considerable modeling flexibility, so that additional variables can be readily brought into the analysis to serve as correlates, predictors, or outcomes of the latent variables (e.g., see the discussion of MIMIC models in Chapter 7).

Frequently, CFA is used as a precursor to SEM, which specifies structural relationships (e.g., regressions) among the latent variables. A structural equation model can be broken down into two major components: (1) the *measurement model*, which specifies the number of factors, how the various indicators are related to the factors, and the relationships among indicator errors (i.e., a CFA model); and (2) the *structural model*, which specifies how the various factors are related to one another (e.g., direct or indirect effects, no relationship, spurious relationship). Consider the two basic path diagrams in Figure 3.2. Whereas both diagrams depict models entailing the same set of indicators and the same factors, the first diagram (A) represents a measurement model (a CFA model entailing three intercorrelated factors), and the second diagram (B) reflects a structural model to indicate that the relationship between Factor X and Factor Y is fully mediated by Factor Z (as with factor loadings, direct effects among latent variables are depicted by unidirectional arrows in Figure 3.2B). Thus, whereas the relationships among the latent variables are allowed to intercorrelate freely in the CFA model (analogous to an oblique EFA solution), the exact nature of the relationships is specified in the structural model; that is, Factor X has a direct effect on Factor Z, Factor Z has a direct effect on Factor Y, and Factor X has an indirect effect on Factor Y. Note that in the measurement (CFA) model, there are three parameters relating the factors to one another: factor correlations between X and Y, X and Z, and Y and Z (depicted by double-headed, curved arrows in Figure 3.2A). In the structural model, there are only two structural parameters, X → Y and Y → Z. As discussed later in this chapter, the structural portion of this solution is *overidentified*, meaning that there exist fewer structural parameters (i.e., two: X → Y and Y → Z) in the model than the number of possible relationships among the factors (i.e., three: correlations between X and Y, X and Z, and Y and Z). Thus the structural model is more parsimonious than the measurement model because it attempts to reproduce the relationships among the latent variables with one less freely estimated parameter. Because of the overidentified nature of the structural portion of this model, its goodness of fit may be poorer than that of the measurement model. As illustrated by a tracing rule presented later in this chapter (e.g., Eq. 3.16), the structural portion of this model will result in poor fit if the product of the Factor X → Factor Z path and Factor Z → Factor Y path does not closely approximate the correlation between Factors X and Y estimated in the measurement model. Indeed, the indirect effects structural model in Figure 3.2B will be poor-fitting because the product of the X → Z and Z → Y direct effects [(.40)(.50) = .20] does not approximate the correlation between Factors X and Y (.60; see Figure 3.2A).

The purpose of this discussion is to illustrate that goodness of model fit is determined by how adequately both the measurement and structural portions of a model are specified. A key aspect of CFA evaluation is the ability of the parameters from the measurement model (e.g., factor loadings and factor correlations) to reproduce the observed relationships among the indicators. If the CFA model is misspecified (e.g., failure to specify the correct number of factors, pattern of factor loadings), a poor-fitting solution will result. However, poor fit may also arise from a misspecified structural model which, like the model depicted in Figure 3.2B, often possesses fewer freely estimated parameters than its corresponding measurement model. Because there are various potential sources of poor fit in CFA models involving multiple indicators, the researcher should establish a viable measurement model prior to pursuing a structural solution. If test-

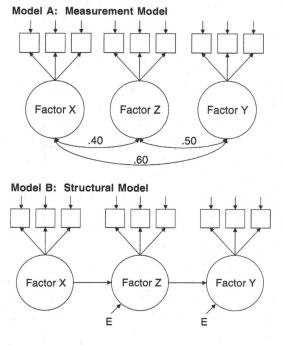

FIGURE 3.2. Path diagrams of measurement and structural models.

ing is initiated with a structural model, it is usually difficult to determine the extent to which poor fit is attributable to the measurement or structural aspects of the solution. For instance, consider the scenario where the measurement model in Figure 3.2A is well specified (i.e., good-fitting, strong, and interpretable factor loadings and factor correlations), but the researcher begins by testing the structural model shown in Figure 3.2B. Although poor fit would be due to the misspecified structural model (i.e., inability to reproduce the Factor X and Factor Y relationship), the researcher may falsely suspect the measurement aspect of the model. In most cases, poor fit cannot arise from the structural portion of a CFA measurement model because the factors are usually specified as freely intercorrelated. Thus CFA solutions are a useful prelude to SEM solutions, which aim to reproduce the relationships among latent variables with a more parsimonious set of structural parameter estimates.

PARAMETERS OF A CFA MODEL

All CFA models contain factor loadings, unique variances, and factor variances. *Factor loadings* are the regression slopes for predicting the indicators from the latent variable. *Unique variance* is variance in the indicator that is not accounted for by the latent variables. Unique variance is typically presumed to be measurement error and is thus often referred to as such (other synonymous terms include *error variance* and *indicator unreliability*). In an unstandardized solution, a *factor variance* expresses the sample variability or dispersion of the factor—that is, the extent to which sample participants' relative standing on the latent dimension is similar or different. If substantively justified, a CFA may include *error covariances* (referred to as *correlated uniquenesses*, *correlated residuals*, or *correlated errors*), which suggest that two indicators covary for reasons other than the shared influence of the latent variable (e.g., see Figure 3.1C). In CFA solutions, error covariances are often specified on the basis of method effects (e.g., the indicators are measured by a common method), although other sources of these relationships are possible. When the CFA solution consists of two or more factors, a factor covariance (a *factor correlation* being the completely standardized counterpart) is usually specified to estimate the relationship between the latent dimensions (although it is possible to fix factor covariances to zero, akin to an orthogonal EFA solution).

CFA is often confined to the analysis of variance–covariance structures. In this instance, the aforementioned parameters (factor loadings, error variances and covariances, factor variances and covariances) are estimated to reproduce the input variance–covariance matrix. The analysis of covariance structures is based on the implicit assumption that indicators are measured as deviations from their means (i.e., all indicator means equal zero). However, the CFA model can be expanded to include the analysis of mean structures, in which case the CFA parameters also strive to reproduce the observed sample means of the indicators (which are included along with the sample variances and covariances as input data). Accordingly, such CFA models also include parameter estimates of the indicator intercepts (predicted value of the indicator when the factor is zero) and the latent variable means, which are often used in multiple-groups

CFA to test whether distinct groups differ in their relative standing on latent dimensions (see Chapter 7).

Latent variables in CFA may be either exogenous or endogenous. An *exogenous variable* is a variable that is not caused by other variables in the solution (such as Factor X in Figure 3.2B). Conversely, an *endogenous variable* is caused by one or more variables in the model (i.e., other variables in the solution exert direct effects on the variable, as in Factor Y in Figure 3.2B). Thus exogenous variables can be viewed as synonymous to X, independent, or predictor (causal) variables. Similarly, endogenous variables are equivalent to Y, dependent, or criterion (outcome) variables. However, in the case of structural models, an endogenous variable may be the cause of another endogenous variable, as is the case of Factor Z in Figure 3.2B.

CFAs are typically considered to be exogenous (latent X) variable solutions (e.g., Figure 3.2A). However, when a CFA includes covariates (i.e., predictors of the factors or indicators as in MIMIC models; see Chapter 7) or higher-order factors (see Chapter 8), the latent factors are endogenous (latent Y). Even when the CFA is a pure measurement model, some researchers (e.g., methodologists using LISREL software) choose to specify the analysis as a latent Y solution. There are various reasons for this including the ability to accomplish useful specification tricks in LISREL (e.g., LISREL specifications for scale reliability evaluation; see Chapter 8), greater simplicity, and the fact that many statistical papers use latent Y specifications to present information. The issue of "latent X" versus "latent Y" CFA specification is not relevant to many latent variable software programs (e.g., Mplus, EQS, Amos), which on the surface do not rely on matrix operations. In LISREL, specifying a pure CFA measurement model as a latent X or latent Y solution has no impact on the fit and parameter estimates of the solution.

Figures 3.3 and 3.4 present the LISREL notation for the parameters and matrices of a CFA solution for latent X and latent Y specifications, respectively. As noted in the preceding paragraph, it is not necessary to understand this notation in order to specify CFA models in most software packages. However, knowledge of this notational system is useful because most sourcebooks and quantitative papers rely on it to describe the parameters and equations of CFA and SEM. Consistent with material presented in context of EFA in Chapter 2, lowercase Greek symbols correspond to specific parameters (i.e., elements of a matrix such as λ), whereas uppercase Greek letters reflect an entire matrix (e.g., the full matrix of factor loadings, Λ). As in EFA, factor loadings are symbolized by lambdas (λ) with x and y subscripts in the case of exogenous and endogenous latent variables, respectively. The unidirectional arrows ($\rightarrow$) from the factors (e.g., ξ_1, η_1) to the indicators (e.g., X1, Y1) depict direct effects (regressions) of the latent dimensions onto the observed measures; the specific regression coefficients are the lambdas (λ). Thetas (Θ) represent matrices of indicator error variances and covariances—theta-delta (Θ_δ) in the case of indicators of latent X variables, theta-epsilon (Θ_ε) for indicators of latent Y variables. For notational ease, the symbols δ and ε are often used in place of θ_δ and θ_ε, respectively, in reference to elements of Θ_δ and Θ_ε (as is done throughout this book). Although unidirectional arrows connect the thetas to the observed measures (e.g., X1–X6), these arrows do not depict regressive paths; that is, Θ_δ and Θ_ε are symmetric variance–covariance matrices consisting of error variances on the diagonal, and

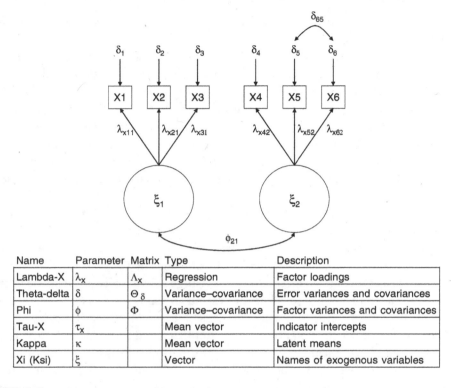

Name	Parameter	Matrix	Type	Description
Lambda-X	λ_x	Λ_x	Regression	Factor loadings
Theta-delta	δ	Θ_δ	Variance–covariance	Error variances and covariances
Phi	ϕ	Φ	Variance–covariance	Factor variances and covariances
Tau-X	τ_x		Mean vector	Indicator intercepts
Kappa	κ		Mean vector	Latent means
Xi (Ksi)	ξ		Vector	Names of exogenous variables

FIGURE 3.3. Latent X notation for a two-factor CFA model with one error covariance. Factor variances, factor means, and indicator intercepts are not depicted in the path diagram.

error covariances, if any, in the off-diagonal. Less commonly, some notational systems do not use directional arrows in the depiction of error variances in order to avoid this potential source of confusion (one notational variation is to symbolize error variances with ovals because, like latent variables, measurement errors are not observed).

Factor variances and covariances are notated by phi (ϕ) and psi (ψ) in latent X and latent Y models, respectively. Curved, bidirectional arrows are used to symbolize covariances (correlations); in Figures 3.3 and 3.4, curved arrows indicate the covariance between the factors (ϕ_{21}, ψ_{21}) and the error covariance of the X5 and X6 indicators (δ_{65}, ε_{65}). When relationships are specified as covariances, the researcher is asserting that the variables are related (e.g., ξ_1 and ξ_2). However, this specification makes no claims about the nature of the relationship, due to either the lack of knowledge regarding the directionality of the association (e.g., $\xi_1 \rightarrow \xi_2$) or the unavailability to the analysis of variables purported to account for this overlap (e.g., ξ_1 and ξ_2 are related because they share a common cause that is not represented by observed measures or latent variables in the analysis). Nonetheless, as discussed in Chapter 8, higher-order factor analysis is a useful approach for explaining the covariances among factors when a strong theory exists in regard to the patterning of the factor interrelationships.

The parameters in Figures 3.3 and 3.4 also possess numerical subscripts to indicate the specific elements of the relevant matrices. For example, λ_{x11} (Figure 3.3) indicates

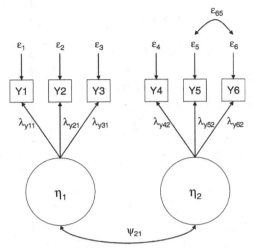

Name	Parameter	Matrix	Type	Description
Lambda-Y	λ_y	Λ_y	Regression	Factor loadings
Theta-epsilon	ε	Θ_ε	Variance–covariance	Error variances and covariances
Psi	ψ	Ψ	Variance–covariance	Factor variances and covariances
Tau-Y	τ_y		Mean vector	Indicator intercepts
Alpha	α		Mean vector	Latent means
Eta	η		Vector	Names of endogenous variables

FIGURE 3.4. Latent Y notation for a two-factor CFA model with one error covariance. Factor variances, factor means, and indicator intercepts are not depicted in the path diagram.

that the X1 measure loads on the first exogenous factor (ξ_1), and λ_{x21} indicates that X2 also loads on ξ_1. This numeric notation assumes that the indicators are ordered X1, X2, X3, X4, X5, and X6 in the input variance–covariance matrix. If the input matrix is arranged in this fashion, the lambda X matrix (Λ_x) in Figure 3.3 will be as follows:

$$
\begin{array}{ccc}
 & \xi_1 & \xi_2 \\
X1 & \lambda_{x11} & 0 \\
X2 & \lambda_{x21} & 0 \\
X3 & \lambda_{x31} & 0 \\
X4 & 0 & \lambda_{x42} \\
X5 & 0 & \lambda_{x52} \\
X6 & 0 & \lambda_{x62}
\end{array}
\tag{3.1}
$$

where the first numerical subscript refers to the row of Λ_x (i.e., the positional order of the X indicator), and the second numerical subscript refers to the column of Λ_x (i.e., the positional order of the exogenous factors, ξ). For example, λ_{x52} conveys that the fifth indicator in the input matrix (X5) loads on the second latent X variable (ξ_2). Thus Λ_x and Λ_y are full matrices whose dimensions are defined by p rows (number of indicators) and m columns (number of factors). The zero elements of Λ_x (e.g., λ_{x12}, λ_{x41}) indicate the absence of cross-loadings (e.g., the relationship between X1 and ξ_2 is fixed to zero). This is also depicted in Figures 3.3 and 3.4 by the absence of directional arrows between certain indicators and factors (e.g., no arrow connecting ξ_2 to X1 in Figure 3.3).

A similar system is used for variances and covariances among factors (ϕ in Figure 3.3, ψ in Figure 3.4) and indicator errors (δ and ε in Figures 3.3 and 3.4, respectively). However, because these aspects of the CFA solution reflect variances and covariances, they are represented by $m \times m$ symmetric matrices with variances on the diagonal and covariances in the off-diagonal. For example, the phi matrix (Φ) in Figure 3.3 will look as follows:

$$
\begin{array}{c c c}
 & \xi_1 & \xi_2 \\
\xi_1 & \phi_{11} & \\
\xi_2 & \phi_{21} & \phi_{22}
\end{array}
\tag{3.2}
$$

where ϕ_{11} and ϕ_{22} are the factor variances, and ϕ_{21} is the factor covariance. Similarly, the theta-delta matrix (Θ_δ) in Figure 3.3 is the following $p \times p$ symmetric matrix:

$$
\begin{array}{c c c c c c c}
 & X1 & X2 & X3 & X4 & X5 & X6 \\
X1 & \delta_{11} & & & & & \\
X2 & 0 & \delta_{22} & & & & \\
X3 & 0 & 0 & \delta_{33} & & & \\
X4 & 0 & 0 & 0 & \delta_{44} & & \\
X5 & 0 & 0 & 0 & 0 & \delta_{55} & \\
X6 & 0 & 0 & 0 & 0 & \delta_{65} & \delta_{66}
\end{array}
\tag{3.3}
$$

where δ_{11} through δ_{66} are the indicator errors, and δ_{65} is the covariance of the measurement errors of indicators X5 and X6. For notational ease, the diagonal elements are indexed by single digits in Figures 3.3 and 3.4 (e.g., δ_6 is the same as δ_{66}). The zero elements of Θ_δ (e.g., δ_{21}) indicate the absence of error covariances (i.e., these relationships are fixed to zero).

In CFA with mean structures (see Chapter 7), indicator intercepts are symbolized by tau (τ), and latent exogenous and endogenous means are symbolized by kappa (κ) and alpha (α), respectively. Because the focus has been on CFA, only parameters germane to measurement models have been discussed thus far. LISREL notation also applies to structural component of models that entail directional relationships among exogenous and endogenous variables. For instance, gamma (γ, matrix: Γ) denotes regressions between latent X and latent Y variables, and beta (β, matrix: B) symbolizes directional effects among endogenous variables. Most of the CFA illustrations provided in this book do not require gamma or beta parameters. Exceptions include CFA with covariates (e.g., MIMIC models, Chapter 7), where the measurement model is regressed on observed background measures (e.g., a dummy code indicating male versus female); higher-order CFA (Chapter 8); and models with formative indicators (Chapter 8).

FUNDAMENTAL EQUATIONS OF A CFA MODEL

CFA aims to reproduce the sample variance–covariance matrix by the parameter estimates of the measurement solution (e.g., factor loadings, factor covariances, etc.). To illustrate, Figure 3.3 has been revised: Parameter estimates have been inserted for all factor loadings, factor correlation, and indicator errors (see now Figure 3.5). For ease of illustration, completely standardized values are presented, although the same concepts and formulas apply to unstandardized solutions. The first three measures (X1, X2, X3) are indicators of one latent construct (ξ_1), whereas the next three measures (X4, X5, X6) are indicators of another latent construct (ξ_2). It can be said, for example, that indicators X4, X5, and X6 are *congeneric* (cf. Jöreskog, 1971a) because they share a common factor (ξ_2).[5] An indicator is not considered congeneric if it loads on more than one factor.

In the case of congeneric factor loadings, the variance of an indicator is reproduced by multiplying its squared factor loading by the variance of the factor, and then summing this product with the indicator's error variance. The predicted covariance of two indicators that load on the same factor is computed as the product of their factor loadings times the variance of the factor. The model-implied covariance of two indicators

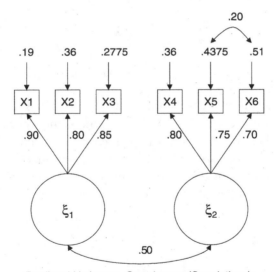

Predicted Variances–Covariances (Correlations):

	X1	X2	X3	X4	X5	X6
X1	1.00000					
X2	0.72000	1.00000				
X3	0.76500	0.68000	1.00000			
X4	0.36000	0.32000	0.34000	1.00000		
X5	0.33750	0.30000	0.31875	0.60000	1.00000	
X6	0.31500	0.28000	0.29750	0.56000	0.72500	1.00000

FIGURE 3.5. Reproduction of the input matrix from the parameter estimates of a two-factor measurement model (completely standardized solution).

that load on separate factors is estimated as the product of their factor loadings times the factor covariance. For example, based on the parameter estimates in the solution presented in Figure 3.5, the variance of X2 can be reproduced by the following equation (using latent X notation):

$$VAR(X2) = \sigma_{22} = \lambda_{x21}^2 \phi_{11} + \delta_2 \tag{3.4}$$
$$= .80^2(1) + .36$$
$$= 1.00$$

In the case of completely standardized solutions (such as the current illustration), one can reproduce the variance of an indicator by simply squaring its factor loading ($.80^2$) and adding its error ($.36$), because the factor variance will always equal 1.00 (however, the factor variance must be included in this calculation when one is dealing with unstandardized solutions). Note that the variance of ξ_2 also equals 1.00 because of the completely standardized model (e.g., variance of X6 = $\lambda_{x62}^2 \phi_{22} + \delta_6 = .70^2 + .51 = 1.00$).

The squared factor loading represents the proportion of variance in the indicator that is explained by the factor (often referred to as a *communality*; see Chapter 2). For example, the communality of X2 is

$$\eta^2_{22} = \lambda_{x21}^2 \tag{3.5}$$
$$= .80^2$$
$$= .64$$

indicating that ξ_1 accounts for 64% of the variance in X2. Similarly, in the completely standardized solution presented in Figure 3.5, the errors represent the proportion of variance in the indicators that is not explained by the factor; for example, $\delta_2 = .36$, indicating that 36% of the variance in X2 is unique variance (e.g., measurement error). These errors (residual variances) can be readily calculated as 1 minus the squared factor loading. Using the X2 indicator, the computation is:

$$\delta_2 = 1 - \lambda_{x21}^2 \tag{3.6}$$
$$= 1 - .80^2$$
$$= .36$$

The predicted covariance (correlation) between X2 and X3 is estimated as follows:

$$COV(X2, X3) = \sigma_{3,2} = \lambda_{x21} \phi_{11} \lambda_{x31} \tag{3.7}$$
$$= (.80)(1)(.85)$$
$$= .68$$

As before, in the case of completely standardized solutions the factor variance will always equal 1.00, so the predicted correlation between two congeneric indicators can be calculated by the product of their factor loadings; for instance, model-implied correlation of X4, X5 = .80(.75) = .60.

The predicted covariance (correlation) between X3 and X4 (indicators that load on separate factors) is estimated as follows:

$$COV(X3, X4) = \sigma_{4,3} = \lambda_{x31}\phi_{21}\lambda_{x42} \tag{3.8}$$
$$= (.85)(.50)(.80)$$
$$= .34$$

Note that the factor correlation (ϕ_{21}) rather than the factor variance is used in this calculation.

Figure 3.5 presents the 6 variances and 15 covariances (completely standardized) that are estimated by the two-factor measurement model. This model also contains a correlation between the errors of the X5 and X6 indicators (δ_{65} = .20). In this instance, the covariation between the indicators is not accounted for fully by the factor (ξ_2); that is, X5 and X6 share additional variance due to influences other than the latent construct (e.g., method effects). Thus the equation to calculate the predicted correlation of X5 and X6 includes the correlated error:

$$COV(X5, X6) = \sigma_{6,5} = (\lambda_{x52}\phi_{22}\lambda_{x62}) + \delta_{65} \tag{3.9}$$
$$= (.75)(1)(.70) + .20$$
$$= .725$$

CFA MODEL IDENTIFICATION

In order to estimate the parameters in CFA, the measurement model must be *identified*. A model is identified if, on the basis of known information (i.e., the variances and covariances in the sample input matrix), it is possible to obtain a unique set of parameter estimates for each parameter in the model whose values are unknown (e.g., factor loadings, factor correlations). Model identification pertains in part to the difference between the number of freely estimated model parameters and the number of pieces of information in the input variance–covariance matrix. Before this issue is addressed, an aspect of identification specific to the analysis of latent variables is discussed—scaling the latent variable.

Scaling the Latent Variable

In order for a researcher to conduct a CFA, every latent variable must have its scale identified. By nature, latent variables are unobserved and thus have no defined metrics (units of measurement). Thus these units of measurement must be set by the researcher. In CFA, this is most often accomplished in one of two ways.

In the first and by far the more popular method, the researcher fixes the metric of the latent variable to be the same as one of its indicators. The indicator selected to pass its metric on to the factor is often referred to as a *marker* or *reference indicator*. The guidelines for selecting and specifying marker indicators are discussed in Chapter 4. When a marker indicator is specified, a portion of its sample variance is passed on to the latent variable. Using Figure 3.5, suppose X1 is selected as the marker indicator for ξ_1 and has a sample variance (σ_{11}) of 16. Because X1 has a completely standardized factor loading on ξ_1 of .90, 81% of its variance is explained by ξ_1; $.90^2 = .81$ (cf. Eq. 3.5). Accordingly, 81% of the sample variance in X1 is passed on to ξ_1 to represent the factor variance of ξ_1:

$$\phi_{11} = \lambda_{x11}{}^2 \sigma_{11} \tag{3.10}$$
$$= (.81)16$$
$$= 12.96$$

As will be shown in Chapter 4, these estimates are part of the unstandardized CFA solution.

In the second method, the variance of the latent variable is fixed to a specific value, usually 1.00. Consequently, a standardized and a completely standardized solution are produced. Although the latent variables have been standardized (i.e., their variances are fixed to 1.00), the fit of this model is identical to that of the unstandardized model (i.e., models estimated using marker indicators). While it is useful in some circumstances (e.g., as a parallel to the traditional EFA model), this method is used less often than the marker indicator approach. The former strategy produces an unstandardized solution (in addition to a completely standardized solution), which is useful for several purposes, such as tests of measurement invariance across groups (Chapter 7) and evaluations of scale reliability (Chapter 8). However, in many instances this method of scale setting can be considered superior to the marker indicator approach, especially when the indicators have been assessed on an arbitrary metric, and when the completely standardized solution is of more interest to the researcher (coupled with the fact that some programs, like Mplus, now provide standard errors and significance tests for standardized parameter estimates).

More recently, Little, Slegers, and Card (2006) have introduced a third method of scaling latent variables that is akin to effects coding in ANOVA. In this approach, a priori constraints are placed on the solution, such that the set of factor loadings for a given construct average to 1.00 and the corresponding indicator intercepts sum to zero. Consequently, the variance of the latent variables reflects the average of the indicators' variances explained by the construct, and the mean of the latent variable is the optimally weighted average of the means for the indicators of that construct. Thus, unlike the marker indicator approach—where the variances and means of the latent variables will vary, depending on which indicator is selected as the marker indicator—the method developed by Little et al. (2006) has been termed *nonarbitrary* because the latent variable will have the same unstandardized metric as the average of all its manifest indicators. This approach is demonstrated in Chapter 7.

Statistical Identification

Besides scaling the latent variable, the parameters of a CFA model can be estimated only if the number of freely estimated parameters does not exceed the number of pieces of information in the input variance–covariance matrix. The concept has been introduced in Chapter 2 in context of the identification of EFA models estimated by ML. A model is *underidentified* when the number of unknown (freely estimated) parameters exceeds the number of known information (i.e., elements of the input variance–covariance matrix). An underidentified model cannot be solved because there are an infinite number of parameter estimates that result in perfect model fit. To illustrate, consider this basic equation:

$$x + y = 7 \qquad\qquad (3.11)$$

In this instance, there are 2 unknowns (x and y) and 1 known ($x + y = 7$). This equation is underidentified because the number of unknown parameters (x and y) exceeds the known information; consequently, x and y can take on an infinite number of pairs of values to solve for $x + y = 7$ ($x = 1$, $y = 6$; $x = 2$, $y = 5$; etc.). In context of CFA, consider the model depicted in Figure 3.6A. In CFA, the knowns are usually the sample variances and covariances of the input indicators. When the CFA involves the analysis of mean structures, the sample means of the indicators are also included in the count of known information, and the indicator intercepts and latent means are included in the count of parameter estimates (see Chapter 7). For the model depicted in Figure 3.6A, the input matrix is composed of 3 knowns (pieces of information): the variances of X1 and X2, and the covariance of X1 and X2. The unknowns of the CFA solution are the freely estimated model parameters. In the Figure 3.6A model, there are 4 freely estimated parameters: 2 factor loadings (λ_{x11}, λ_{x21}) and 2 indicator errors (δ_1, δ_2). In this example, the metric of ξ_1 is set by fixing its variance to 1.0. Thus, because the factor variance (ϕ_{11}) is fixed, it is not included in the count of unknowns. Alternatively, we may opt to define the metric of ξ_1 by choosing either X1 or X2 to serve as a marker indicator. In this case, the factor variance (ϕ_{11}) contributes to the count of freely estimated parameters, but the factor loading of the marker indicator is not included in this tally because it is fixed to pass its metric on to ξ_1 (see Chapter 4 for more details). Regardless of which method is used to define the units of measurement of ξ_1, the count of freely estimated parameters in Figure 3.6A equals 4.

Thus the CFA model in Figure 3.6A is underidentified because the number of unknowns (4 freely estimated parameters) exceeds the number of knowns (3 elements of the input matrix = 2 variances, 1 covariance). This model aims to reproduce the sample covariance of X1 and X2. Suppose, for example, this sample covariance corresponds to a correlation of .64. In this case, λ_{x11}, λ_{x21}, δ_1, and δ_2 can take on an infinite number of sets of values to reproduce an X1–X2 correlation of .64. Recall that in a completely standardized solution, the predicted correlation between two indicators that load on the same factor is the product of their factor loadings. Thus there are endless pairs of values that can be estimated for λ_{x11} and λ_{x21} that will produce a perfectly fitting model (e.g., $\lambda_{x11} = .80$, $\lambda_{x21} = .80$; $\lambda_{x11} = .90$, $\lambda_{x21} = .711$; $\lambda_{x11} = .75$, $\lambda_{x21} = .853$).

Model A: Underidentified ($df = -1$)

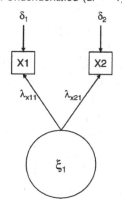

Input Matrix (3 elements)

	X1	X2
X1	σ_{11}	
X2	σ_{21}	σ_{22}

Freely Estimated Model Parameters = 4
(e.g., 2 factor loadings, 2 error variances)

Model B: Just-Identified ($df = 0$)

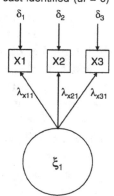

Input Matrix (6 elements)

	X1	X2	X3
X1	σ_{11}		
X2	σ_{21}	σ_{22}	
X3	σ_{31}	σ_{32}	σ_{33}

Freely Estimated Model Parameters = 6
(e.g., 3 factor loadings, 3 error variances)

Model C: Underidentified ($df = -1$)

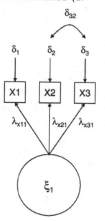

Input Matrix (6 elements)

	X1	X2	X3
X1	σ_{11}		
X2	σ_{21}	σ_{22}	
X3	σ_{31}	σ_{32}	σ_{33}

Freely Estimated Model Parameters = 7
(e.g., 3 factor loadings, 3 error variances, 1
error covariance)

FIGURE 3.6. Examples of underidentified and just-identified CFA models.

Incidentally, it is possible to identify the Figure 3.6A model if additional constraints are imposed on the solution. For instance, the researcher can add the restriction of constraining the factor loadings to equality. In this case, the number of knowns (3) will equal the number of unknowns (3), and the model will be *just-identified*. As discussed in further detail in the next paragraph, in just-identified models there exists one unique set of parameter estimates that perfectly fit the data. In the current illustration, the only factor loading parameter estimate that will reproduce the observed X1–X2 correlation (.64) is .80; λ_{x11} = .80 and λ_{x21} = .80, solved by imposing the equality constraint. Although these constraints may assist in model identification by reducing the number of freely estimated parameters (such as in the current example), such restrictions are often unreasonable on the basis of evidence or theory.

Unlike underidentified models, just-identified models can be "solved." In fact, because the number of knowns equals the number of unknowns, in just-identified models there exists a single set of parameter estimates that *perfectly* reproduce the input matrix. Before further applying this concept to CFA, consider this example from simultaneous equations algebra:

$$x + y = 7 \tag{3.12}$$
$$3x - y = 1$$

In this example, the number of unknowns (x, y) equals the number of knowns $(x + y = 7, 3x - y = 1)$. Through basic algebraic manipulation, it can be readily determined that $x = 2$ and $y = 5$; that is, there is only one possible pair of values for x and y.

Now consider the CFA model in Figure 3.6B. In this example, the input matrix consists of 6 knowns (3 variances, 3 covariances), and the model consists of 6 freely estimated parameters: 3 factor loadings and 3 indicator errors (again assume that the variance of ξ_1 is fixed to 1.0). This CFA model is just-identified and will produce a unique set of parameter estimates $(\lambda_{x11}, \lambda_{x21}, \lambda_{x31}, \delta_1, \delta_2, \delta_3)$ that perfectly reproduce the correlations among X1, X2, and X3 (see Appendix 3.2). Thus, although just-identified CFA models can be conducted on a sample input matrix, goodness-of-model-fit evaluation does not apply because, by nature, such solutions always have perfect fit. This is also why goodness of fit does not apply to traditional statistical analyses such as multiple regression; that is, these models are inherently just-identified. For instance, in OLS multiple regression, all observed variables are connected to one another either by direct effects, $X_1, X_2 \rightarrow Y$, or freely estimated correlations, $X_1 \leftrightarrow X_2$.

It is important to note that while a CFA model of a construct consisting of 3 observed measures may meet the conditions of identification (as in Figure 3.6B), this is only true if the errors of the indicators are not correlated with each other. For instance, the model depicted in Figure 3.6C is identical to that in Figure 3.6B, with the exception of a correlated residual between indicators X2 and X3. This additional parameter (δ_{32}) now brings the count of freely estimated parameters to 7, which exceeds the number of elements of the input variance–covariance matrix (6). Thus the Figure 3.6C model is underidentified and cannot be fit to the sample data.

A model is *overidentified* when the number of knowns (i.e., number of variances and covariances in the input matrix) exceeds the number of freely estimated model parameters. For example, the one-factor model depicted in Figure 3.7 (Model A) is structurally overidentified because there are 10 elements of the input matrix (4 variances for X1–X4, 6 covariances), but only 8 freely estimated parameters (4 factor loadings, 4 error variances; the variance of ξ_1 is fixed to 1.0). The difference in the number of knowns (a) and the number of unknowns (b; i.e., freely estimated parameters) constitutes the model's *degrees of freedom* (df). Overidentified solutions have positive df; just-identified models have 0 df (because the number of knowns equals the number of unknowns); and underidentified models have negative df (they cannot be solved or fit to the data). Thus the Figure 3.7A model is overidentified with $df = 2$:

$$df = a - b \qquad (3.13)$$
$$= 10 - 8$$
$$= 2$$

The second model in Figure 3.7 (Model B) is also overidentified with $df = 1$ (assuming that ξ_1 and ξ_2 have a nonzero correlation; see the discussion of empirical underidentification below). As in the Figure 3.7A solution, there are 10 elements of the input matrix. However, the Figure 3.7B model consists of 9 freely estimated parameters (4 factor loadings, 4 error variances, 1 factor covariance), thus resulting in 1 df.

As a final example of an overidentified solution, consider the measurement model presented in Figure 3.5. In this case, there are 21 pieces of information in the input matrix (6 variances, 15 covariances). The reader may note that it becomes cumbersome to count the elements of the input matrix as the number of variables increases. Fortunately, the following formula, initially presented in Chapter 2 (Eq. 2.8), readily provides this count:

$$a = p(p + 1) / 2 \qquad (3.14)$$

where a is the number of elements of the input matrix, and p is the number of indicators included in the input matrix. Thus, in the Figure 3.5 model involving 6 indicators (X1–X6),

$$a = 6(6 + 1) / 2 = 21$$

reflecting the 6 variances (p) and the 15 covariances [$p(p - 1) / 2$].

The specification of the Figure 3.5 model entails 14 freely estimated parameters (in this example, assume that the metric of ξ_1 and ξ_2 is set by selecting X1 and X4 as marker indicators): 4 factor loadings, 6 error variances, 1 error covariance, 2 factor variances, 1 factor covariance. The loadings of X1 and X4 are not included in this count because they are fixed in order to pass their metrics onto ξ_1 and ξ_2. Therefore, this model is overidentified with $df = 7$ (21 knowns minus 14 unknowns).

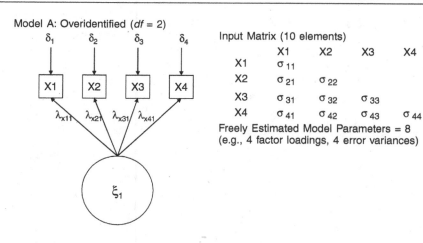

Model A: Overidentified ($df = 2$)

Input Matrix (10 elements)

	X1	X2	X3	X4
X1	σ_{11}			
X2	σ_{21}	σ_{22}		
X3	σ_{31}	σ_{32}	σ_{33}	
X4	σ_{41}	σ_{42}	σ_{43}	σ_{44}

Freely Estimated Model Parameters = 8
(e.g., 4 factor loadings, 4 error variances)

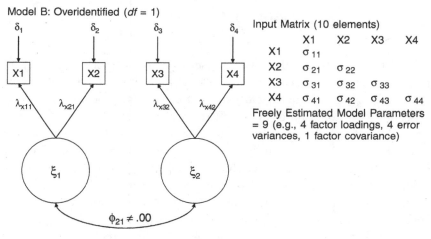

Model B: Overidentified ($df = 1$)

Input Matrix (10 elements)

	X1	X2	X3	X4
X1	σ_{11}			
X2	σ_{21}	σ_{22}		
X3	σ_{31}	σ_{32}	σ_{33}	
X4	σ_{41}	σ_{42}	σ_{43}	σ_{44}

Freely Estimated Model Parameters = 9 (e.g., 4 factor loadings, 4 error variances, 1 factor covariance)

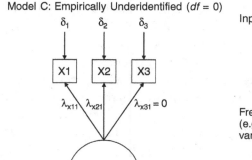

Model C: Empirically Underidentified ($df = 0$)

Input Matrix (6 elements)

	X1	X2	X3
X1	σ_{11}		
X2	σ_{21}	σ_{22}	
X3	σ_{31}^{*}	σ_{32}^{*}	σ_{33}

$^{*}\sigma_{31} = 0,\ ^{*}\sigma_{32} = 0$

Freely Estimated Model Parameters = 6 (e.g., 3 factor loadings, 3 error variances)

(continued)

FIGURE 3.7. Examples of overidentified and empirically underidentified CFA models. In all examples, the metric of the factor is defined by fixing its variance to 1.0.

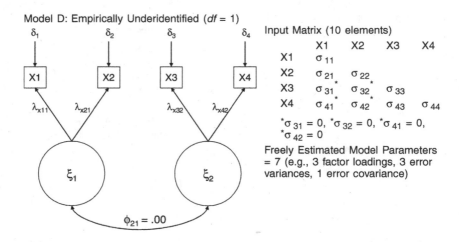

FIGURE 3.7. (*continued*)

Degrees of freedom are used in many descriptive indices of *goodness of model fit* (such as model χ^2; see the "Descriptive Goodness-of-Fit Indices" section of this chapter). Indeed, an important aspect of overidentified solutions is that goodness-of-fit evaluation applies—specifically, evaluation of how well the model is able to reproduce the input variances and covariances (i.e., the input matrix) with fewer unknowns (i.e., freely estimated model parameters). Thus, as in just-identified models, the available known information indicates that there is one best value for each freely estimated parameter in the overidentified solution. Unlike just-identified models, overidentified models rarely fit the data perfectly (i.e., a perfectly fitting model is one whose parameter estimates recreate the input variance–covariance matrix exactly). The principles of goodness of fit are discussed in fuller detail in the last major section of this chapter.

Specification of a model to have at least 0 *df* is a necessary but not sufficient condition for identification. Figure 3.7 also illustrates examples of *empirically underidentified* solutions (Kenny, 1979). In an empirically underidentified solution, the model is statistically just-identified or overidentified, but aspects of the input matrix or model specification prevent the analysis from obtaining a unique and valid set of parameter estimates. The most obvious example of empirical underidentification is the case where all covariances in the input matrix equal 0. However, empirical underidentification can result from other patterns of (non)relationships in the input data. For example, Figure 3.7C depicts a model that is identical to the just-identified model presented in Figure 3.6B, yet its input matrix reveals an absence of a relationship between X3 and X1 and X2. This aspect of the input matrix renders the Figure 3.7C model functionally equivalent to the underidentified solution presented in Figure 3.6A. Similarly, Figure 3.7D represents a model that is identical to the overidentified model in Figure 3.7B. However, because the factor correlation is 0 (due to a lack of relationship of X1 and X2 with X3 and X4), this solution is analogous to simultaneously attempting to estimate two Figure 3.6A solutions (underidentified). For these and other reasons (e.g., increased power and precision of parameter estimates; cf. Marsh, Hau, Balla, & Grayson, 1998), methodologists recommend that latent variables be defined by a minimum of three indicators to avoid this possible source of underidentification. For example, even if $\phi_{21} = 0$, a solution for the Figure 3.7D model can be obtained if ξ_1 and ξ_2 are measured by three indicators each.

Empirical underidentification will also occur if the marker indicator selected to define the metric of the latent variable is not correlated with the other indicators loading on that factor. For instance, suppose that in the Figure 3.7A model, X1 is used as the marker indicator for ξ_1 but is not correlated with X2, X3, and X4 (i.e., σ_{21}, σ_{31}, and σ_{41} are not significantly different from 0). Consequently, λ_{x11} will equal 0, and thus no variance will be passed on to ξ_1 ($\phi_{11} = \lambda_{x11}^2\sigma_{11} = 0\sigma_{11} = 0$; cf. Eq. 3.10), producing an empirically underidentified solution.

If an attempt is made to fit an empirically underidentified model, the computer software will fail to yield a solution or will provide an improper solution (usually accompanied by error messages or warnings). Among the possible consequences are so-called *Heywood cases*, which refer to parameter estimates that have out-of-range values (e.g., negative indicator error variances). Improper solutions may arise from a variety of causes, such as poorly specified models (e.g., in Figure 3.7C, X3 is not a reasonable indicator of ξ_1) and problems with the input matrix (e.g., pairwise deletion, collinearities, small sample size; cf. Wothke, 1993). These problems are discussed in detail in Chapter 5.

Guidelines for Model Identification

On the basis of the preceding discussion, some basic guidelines for model identification can be summarized:

1. Regardless of the complexity of the model (e.g., one factor vs. multiple factors, size of indicator set), latent variables must be scaled, most often by either specifying marker indicators or fixing the variance of the factor (usually to a value of 1.00).

2. Regardless of the model's complexity, the number of pieces of information in the input matrix (e.g., indicator variances and covariances) must equal or exceed the number of freely estimated model parameters (e.g., factor loadings, factor variances–covariances, indicator error variances–covariances).

3. In the case of one-factor models, a minimum of three indicators is required. When three indicators are used (and no correlated errors are specified; e.g., Figure 3.6B), the one-factor solution is just-identified and goodness-of-fit evaluation does not apply, although this model can still be evaluated in terms of the interpretability and strength of its parameter estimates (e.g., magnitude of factor loadings). When four or more indicators are used (and no correlated errors are specified; e.g., Figure 3.7A), the model is overidentified (i.e., there are more elements of the input matrix than freely estimated model parameters) and goodness of fit can be used in evaluating the acceptability of the solution.

4. In the case of models that entail two or more factors and two indicators per latent construct, the solution will be overidentified—provided that every latent variable

is correlated with at least one other latent variable, and the errors between indicators are uncorrelated (e.g., Figure 3.7B). However, because such solutions are susceptible to empirical underidentification (e.g., Figure 3.7D), a minimum of three indicators per latent variable is recommended.

These guidelines assume models where each indicator loads on only one factor and are free of correlated measurement error (i.e., Θ is a diagonal matrix, and thus all off-diagonal elements equal 0). As discussed in Chapter 5, the issue of identification is more complicated in models that involve double-loading indicators (i.e., *noncongeneric* indicators) or correlated indicator errors (e.g., Figure 3.6C).

ESTIMATION OF CFA MODEL PARAMETERS

The objective of CFA is to obtain estimates for each parameter of the measurement model (i.e., factor loadings, factor variances and covariances, indicator error variances and possibly error covariances) to produce a predicted variance–covariance matrix (symbolized as Σ) that resembles the sample variance–covariance matrix (symbolized as S) as closely as possible. For instance, in overidentified models (such as Figure 3.7A), perfect fit is rarely achieved (i.e., $\Sigma \neq S$). Thus, in the case of a CFA model such as Figure 3.7A, the goal of the analysis is to find a set of factor loadings ($\lambda_{x11}, \lambda_{x21}, \lambda_{x31}, \lambda_{x41}$) that yield a predicted covariance matrix (Σ) that best reproduces the input matrix (S)—for example, to find parameter estimates for λ_{x11} and λ_{x21} such that the predicted correlation between X1 and X2 ($\lambda_{x11}\phi_{11}\lambda_{x21}$) closely approximates the sample correlation of these indicators (σ_{21}) (although in the actual estimation process, this occurs simultaneously for all parameters and implied covariances). This process entails a *fitting function*, a mathematical operation to minimize the difference between Σ and S. By far, the fitting function most widely used in applied CFA research (and SEM in general) is ML. The fitting function that is minimized in ML is

$$F_{ML} = \ln|S| - \ln|\Sigma| + \text{trace}[(S)(\Sigma^{-1})] - p \qquad (3.15)$$

where $|S|$ is the determinant of the input variance–covariance matrix; $|\Sigma|$ is the determinant of the predicted variance–covariance matrix; p is the order of the input matrix (i.e., the number of input indicators); and ln is the natural logarithm. Although a full explication of this function is beyond the scope of this chapter (cf. Bollen, 1989; Eliason, 1993), a few observations are made in effort to foster its conceptual understanding (see also Appendix 3.3).[6] The determinant and trace summarize important information about matrices such as S and Σ. The *determinant* is a single number (i.e., a scalar) that reflects a generalized measure of variance for the entire set of variables contained in the matrix. The *trace* of a matrix is the sum of values on the diagonal (e.g., in a variance–covariance matrix, the trace is the sum of variances). The objective of ML is to minimize the differences between these matrix summaries (i.e., the determinant and trace) for S and Σ. The operations of this formula are most clearly illustrated in the context of a perfectly fitting model, as quoted from Jaccard and Wan (1996):

> In this case, the determinant of S will equal the determinant of Σ, and the difference between the logs of these determinants will equal 0. Similarly, $(S)(\Sigma^{-1})$ will equal an identity matrix with all ones in the diagonal. When the diagonal elements are summed (via the trace function), the result will be the value of p. Subtracting p from this value yields 0. Thus, when there is perfect model fit, F_{ML} equals 0. (pp. 85–86)

The calculation and use of F_{ML} are discussed in greater detail later in this chapter and in Appendix 3.3.

The underlying principle of ML estimation in CFA is to find the model parameter estimates that maximize the probability of observing the available data if the data are collected from the same population again. In other words, ML aims to find the parameter values that make the observed data most likely (or, conversely, that maximize the likelihood of the parameters, given the data). Finding the parameter estimates for an overidentified CFA model is an *iterative* procedure. That is, the computer program (such as LISREL, Mplus, or EQS) begins with an initial set of parameter estimates (referred to as *starting values* or *initial estimates*, which can be automatically generated by the software or specified by the user), and repeatedly refines these estimates in an effort to reduce the value of F_{ML} (i.e., to minimize the difference between Σ and S). Each refinement of the parameter estimates to minimize F_{ML} is an *iteration*. The program conducts internal checks to evaluate its progress in obtaining parameter estimates that best reproduce S (i.e., that result in the lowest F_{ML} value). *Convergence* of the model is reached when the program arrives at a set of parameter estimates that can not be improved upon to further reduce the difference between Σ and S. In fact, for a given model and S, the researcher may encounter minor differences across programs in F_{ML}, goodness-of-fit indices, and so forth, as the result of variations across software packages in minimization procedures and stopping criteria.

Occasionally, a latent variable solution will fail to converge. Convergence is often related to the quality and complexity of the specified model (e.g., the number of restrictions imposed on the solution) and the adequacy of the starting values. In the case of complex (yet properly specified) models, convergence may not be reached because the program has stopped at the maximum number of iterations (set by either the program's default or a number specified by the user). This problem may be rectified by simply increasing the maximum number of iterations or possibly using the preliminary parameter estimates as starting values. However, a program may also cease before the maximum number of iterations has been reached because its internal checks indicate that progress is not being made in obtaining a solution that minimizes F_{ML}. Although this can be an indication of more serious problems, such as a grossly misspecified model, this outcome may stem from more innocuous issues such as the scaling of the indicators and the adequacy of the starting values.[7]

Starting values affect the minimization process in a number of ways. If these initial values are similar to the final model parameter estimates, fewer iterations are required to reach convergence. On the other hand, if the starting values are quite different from the final parameter estimates, there is a greater likelihood that nonconvergence will occur or that the solution will contain Heywood cases (e.g., communalities > 1.00, negative error variances). Fortunately, the CFA researcher usually does not need to be concerned about starting values because most latent variable software programs have incorporated sophisticated methods for automatically generating these initial estimates (e.g., Version 8 of LISREL uses the instrumental variables and two-stage least squares methods to compute starting values). However, in more complex models (such as those involving nonlinear constraints; cf. the CFA approach to scale reliability evaluation described in Chapter 8), it is sometimes necessary for the user to provide these values. The strategy for selecting starting values varies somewhat, depending on the type of model (e.g., multiple-groups solution, a solution with nonlinear constraints), and thus only a few broad recommendations are provided here.

It is useful to ensure that starting values are consistent with the metric of the observed data. For example, starting values for error variances should not be greater than the variance of the indicator; initial estimates of factor variances should not be greater than the observed variance of the marker indicator. Often the most important parameters to provide starting values are the error variances. Initial values can also be generated by noniterative statistical procedures conducted on the data set (e.g., OLS estimates) or basic algebraic calculations (cf. Appendix 3.2). If such estimates are not available, these initial estimates may be gleaned from prior research involving similar models or indicator sets. In addition, some SEM sourcebooks contain guidelines for calculating starting values when the researcher can make a reasonable guess about whether the parameter estimates in question constitute a weak, moderate, or strong effect (e.g., Bollen, 1989). In the case of complex models, it can be fruitful to begin by fitting only portions of the model initially, and then to use the resulting parameter estimates as starting values in the larger solution.

One reason why ML is widely used in CFA model estimation is that it possesses desirable statistical properties such as the ability to provide standard errors (SEs) for each of the model's parameter estimates. These standard errors are used for conducting statistical significance tests of the parameter estimates (i.e., z = parameter estimate divided by its standard error) and for determining the precision of these estimates (e.g., 95% confidence interval [CI] = parameter estimate $\pm SE*1.96$). Moreover, F_{ML} is used in the calculation of many goodness-of-fit indices (see below).

However, it is important to note that ML is only one of many methods that can be used to estimate CFA and other types of structural equation models. Indeed, ML has several requirements that render it an unsuitable estimator in some circumstances. Compared to some estimators, ML is more prone to Heywood cases. In addition, ML is more likely to produce markedly distorted solutions if minor misspecifications have been made to the model. Some key assumptions of ML are that (1) the sample size is large (asymptotic); (2) the indicators have been measured on continuous scales (i.e., approximate interval-level data); and (3) the distribution of the indicators is multivariate normal. The latter two assumptions apply to indicators of latent variables, not to other observed measures that may exist in the analysis, such as nominal variables that serve as covariates (e.g., see the discussion of MIMIC models in Chapter 7). Although the actual parameter estimates (e.g., factor loadings) may not be affected, non-normality in ML analysis can result in biased standard errors (and hence faulty significance tests) and a poorly behaved χ^2 test of overall model fit. If non-normality is extreme (e.g., marked floor effects, as would occur if the majority of the sample responded to items using the lowest response choice—such as 0 on a 0–12 scale), then ML will produce incorrect parameter estimates (i.e., the assumption of a linear model is invalid). Thus, in the case of non-normal, continuous indicators, it is better to use a different estimator, such as ML with robust standard errors and χ^2 (MLM; Bentler, 1995) or the MLR estimator in the Mplus program. MLM and MLR provide the same parameter estimates as ML, but both the model χ^2 and standard errors of the parameter estimates are corrected for non-normality in large samples. For example, MLM produces the Satorra–Bentler scaled (mean adjusted) χ^2, where the typical normal-theory χ^2 is divided by a scaling correction to better approximate χ^2 under non-normality. If one or more of the factor indicators is categorical (or non-normality is extreme), normal theory ML should not be used. In this instance, estimators such as weighted least squares (WLS; also known as asymptotic distribution free, or ADF), robust weighted least squares (e.g., WLSMV), and unweighted least squares (ULS) are more appropriate. WLS can also be used for non-normal, continuous data, although MLM or MLR is often preferred, given the ability of both to outperform WLS in small and medium-sized samples (Curran, West, & Finch, 1996; Hu, Bentler, & Kano, 1992). Other limitations of WLS are discussed in Chapter 9, in context of a detailed presentation of the most widely used estimators of non-normal and categorical data.

Because of its widespread popularity, ML is used in most of the examples in this book. However, the reader should be aware that the vast majority of the principles and procedures discussed in the remaining chapters will apply, regardless of the type of fitting function that is employed. Exceptions are noted when important procedural differences exist across estimators (e.g., χ^2 difference testing with the Satorra–Bentler scaled χ^2; see Chapter 9).

Illustration

To illustrate the concepts of parameter estimation and F_{ML} minimization, consider this simple example. As shown in Figure 3.8, a basic path model is tested with single indicators of behavioral inhibition (x), school refusal (y), and social anxiety (z) in a group of school-age children ($N = 200$). Of particular interest in this example is whether the relationship between behavioral inhibition (x) and school refusal (y) is fully mediated by social anxiety (z). Although the model is somewhat unrealistic (e.g., it assumes no measurement error in x, y, and z, and does not conform to the typical strategy for evaluating mediated effects; cf. MacKinnon, 2008), its simplified nature will foster the illustration of the concepts and calculations introduced in the preceding and subsequent sections. Note that two of the effects in this model are *tautological*, meaning that the paths between x and z, and z and y, must equal their observed relationships; in other words, given the way that the model is specified (e.g., x and z, and z and y, are linked by direct effects), full reproduction of their observed covariances (correlations) is guaranteed (for algebraic proof of this fact, see Jaccard & Wan, 1996). However, the model also possesses one *nontautological* (i.e., overidentified) relationship involving x and y. In other words, although the model will generate a unique set of parameter estimates (i.e., paths between x and z and between z and y, variance of x, residual variances of z and y), complete reproduction of the observed relationship between x and y is not ensured. Specifically, when a simple tracing rule is used (Bollen, 1989; Loehlin, 2004; Wright, 1934), the predicted correlation (and covariance) between x and y will be the product of the paths between x and z and between z and y; that is,

$$r_{xy} = (p_{xz})(p_{zy}) \qquad (3.16)$$

(although ML uses a different approach to estimate the model and Σ, the results are the same). The model-implied relationship between x and y will not necessarily equal the observed relationship between these variables. Thus the proximity of Σ to S depends entirely on the ability of the path model to reproduce the observed zero-order relationship between x and y. The model is thus overidentified with 1 df corresponding to the nontautological relationship between x and y. As discussed earlier, another way to determine whether the model has 1 df is to take the difference between the number of elements of the input matrix ($a = 6 = 3$ variances, 3 covariances) and the number of freely estimated parameters ($b = 5 = 2$ regressive paths, the variance of x, and the 2 residual variances of y and z). Because the model is overidentified, goodness-of-fit evaluation will apply. A good-fitting model will be obtained if the model's parameter estimates reasonably approximate the observed relationship between x and y. If this model-implied relationship differs considerably from the observed x–y relationship, a poor-fitting model will result.

As shown in Figure 3.8, the model-implied correlation between x and y is .30; $r_{yx} = (p_{zx})(p_{yx}) = .6(.5) = .30$ (the predicted covariance of x and y is predicted to be 1.2). This predicted relationship differs from the observed correlation between x and y (.70), thus suggesting a poor-fitting model. Table 3.2 presents the use of SAS PROC IML to calculate the residual matrix (sample matrix minus the predicted matrix) and F_{ML} (see Appendix 3.3 for hand calculation and a conceptual overview of F_{ML}). Because the relationship between x and y is the only nontautological effect in this model, this is the only element of the residual matrix that can take on a value other than zero. As seen in Figure 3.8, the residual correlation and covariance for x and y are .40 and 1.6, respectively. Table 3.2 illustrates the calculation of F_{ML} on the basis of variance-covariance matrices, although the same F_{ML} value will be obtained if correlation matrices are used (yet, as noted earlier, variance–covariance matrices are often preferred in order to obtain unstandardized solutions and valid standard errors, and to permit other options such as multiple-groups evaluation). As shown in Table 3.2, the fitted model results in an F_{ML} value of 0.4054651, reflecting the discrepancy between S and Σ. If perfect fit were attained, all elements of

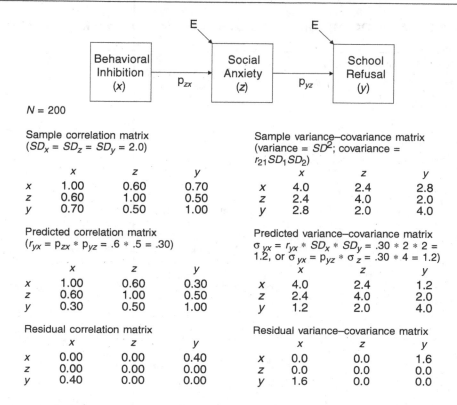

FIGURE 3.8. Illustration of model estimation of a simple path model.

the residual matrix and F_{ML} would be 0. Although this value in and of itself is not readily interpretable (unless F_{ML} = 0), it is used in the calculation of various goodness-of-fit indices.

DESCRIPTIVE GOODNESS-OF-FIT INDICES

The classic goodness-of-fit index is χ^2. Under typical ML model estimation, χ^2 is calculated as

$$\chi^2 = F_{ML}(N-1) \tag{3.17}$$

although latent variable software programs (e.g., Mplus, LISREL starting with Version 9.1) increasingly calculate χ^2 by multiplying F_{ML} by N instead of $N - 1$.[8] Using N, the Figure 3.8 model χ^2 is 81.093 (0.4054651 * 200). Because this model is associated with 1 *df*, the critical χ^2 value (α = .05) is 3.84 (i.e., $\chi^2 = z^2 = 1.96^2 = 3.8416$). The model χ^2 of 81.093 exceeds the critical value of 3.84, and thus the null hypothesis that $S = \Sigma$ is rejected. Thus a statistically significant χ^2 (latent variable software programs provide

TABLE 3.2. SAS PROC IML Syntax and Output for Computing a Residual Matrix and F_{ML}

SAS PROC IML syntax

```
PROC IML;
OPTIONS NOCENTER;
*SAMPLE COVARIANCE MATRIX;
S =    {4.0 2.4 2.8,
        2.4 4.0 2.0,
        2.8 2.0 4.0};
*PREDICTED COVARIANCE MATRIX;
SIGM = {4.0 2.4 1.2,
        2.4 4.0 2.0,
        1.2 2.0 4.0};
RES = S - SIGM;           *RESIDUAL COVARIANCE MATRIX;
SDET = DET(S);            *DETERMINANT OF THE SAMPLE COVARIANCE MATRIX;
SIGMDET = DET(SIGM);      *DETERMINANT OF THE PREDICTED COVARIANCE MATRIX;
LOGS = LOG(SDET);         *NATURAL LOG OF SAMPLE MATRIX DETERMINANT;
LOGSIGM = LOG(SIGMDET);   *NATURAL LOG OF PREDICTED MATRIX DETERMINANT;
SIGMINV = INV(SIGM);      *INVERSE OF PREDICTED MATRIX;
SDIV = S*SIGMINV;         *MULTIPLICATION OF SAMPLE MATRIX AND PREDICTED INVERSE;
STRACE = TRACE(SDIV);     *TRACE OF THE RESULTING SDIV MATRIX;
SORDER = NROW(S);         *ORDER OF SAMPLE MATRIX = NUMBER OF INDICATORS;
*CALCULATION OF FML;
FML = ABS((LOGS - LOGSIGM) + STRACE - SORDER);
PRINT S;
PRINT SIGM;
PRINT RES;
PRINT SDET;
PRINT SIGMDET;
PRINT LOGS LOGSIGM;
PRINT SDIV;
PRINT STRACE;
PRINT SORDER;
PRINT FML;
```

Annotated SAS PROC IML output

```
The SAS System

              S    (the sample variance-covariance matrix)

       4          2.4        2.8
     2.4           4           2
     2.8           2           4

              SIGM (the predicted variance-covariance matrix)

       4          2.4        1.2
     2.4           4           2
     1.2           2           4

    RES      (the residual variance-covariance matrix)

       0           0         1.6
       0           0           0
     1.6           0           0
```

(continued)

TABLE 3.2. *(continued)*

SDET   (the determinant of the sample variance-covariance matrix)

   20.48

SIGMDET   (the determinant of the predicted variance-covariance matrix)

   30.72

   LOGS   LOGSIGM (the natural logs of the sample and predicted matrices)

3.0194488 3.4249139

   SDIV   (matrix produced by multiplying the sample matrix and the
            inverse of the predicted matrix)

        1 -0.266667 0.5333333
-3.47E-17   1   0
    0.625   -0.375   1

STRACE   (trace of SDIV; sum of diagonal elements = 3)

        3

SORDER   (order of the sample matrix = number of input indicators)

        3

   FML   (F value reflecting minimization of the maximum likelihood criterion)

0.4054651

the exact probability value of the model χ^2) supports the alternative hypothesis that $S \neq \Sigma$, meaning that the model estimates do not sufficiently reproduce the sample variances and covariances (i.e., the model does not fit the data well).

Although χ^2 is steeped in the traditions of ML and SEM (e.g., it was the first fit index to be developed), it is rarely used in applied research as a sole index of model fit. Indeed, important criticisms of χ^2 include the following: (1) In many instances (e.g., small N, non-normal data), its underlying distribution is not χ^2-distributed (compromising the statistical significance tests of the model χ^2); (2) it is inflated by sample size (e.g., if N were to equal 100 in the Figure 3.8 model, $\chi^2 = 40.55$), and thus large-N solutions are routinely rejected on the basis of χ^2 even when differences between S and Σ are negligible; and (3) it is based on the very stringent hypothesis that $S = \Sigma$. As discussed below, many alternative fit indices are based on less stringent standards, such as "reasonable" fit and fit relative to an independence model.[9] Nevertheless, χ^2 is used for other purposes such as nested model comparisons (see Chapters 4, 5, and 7) and the calculation of other fit indices (e.g., the Tucker–Lewis index; see below). While χ^2 is routinely reported in CFA research, other fit indices are usually relied on more heavily in the evaluation of model fit.

Although many fit indices are available, only a handful are described and recommended here. These fit indices have been selected on the basis of their popularity in the applied literature and, more importantly, their favorable performance in Monte Carlo research. Other widely used indices, such as the goodness-of-fit index and adjusted goodness-of-fit index, are not included due to evidence of their poor behavior in simulation studies (e.g., Hu & Bentler, 1998; Marsh, Balla, & McDonald, 1988). However, it is repeatedly noted that this topic is surrounded by considerable controversy (e.g., what indices should be used under what contexts? what cutoff values should be used to indicate acceptable fit?). Interested readers are referred to Hu and Bentler (1995, 1998, 1999) for more information.

Fit indices can be broadly characterized as falling into three categories: absolute fit, fit adjusting for model parsimony, and comparative or incremental fit. This typology is not perfect, as some fit indices (such as the Tucker–Lewis index) have features of more than one category. Most latent variable software packages (e.g., LISREL, Mplus, EQS) provide each of the fit indices described below. Because each type of index provides different information about model fit, researchers are advised to consider and report at least one index from each category when evaluating the fit of their models.

Absolute Fit

Absolute fit indices assess model fit at an absolute level; in various ways, they evaluate the reasonability of the hypothesis that $S = \Sigma$ without taking into account other aspects (such as fit in relation to more restricted solutions). Thus χ^2 is an example of an absolute fit index. Another index that falls in this category is the standardized root mean square residual (SRMR). Conceptually, the SRMR can be viewed as the average discrepancy between the *correlations* observed in the input matrix and the correlations predicted by the model (though in actuality the SRMR is a positive square root average; see Eq. 3.18 below). Accordingly, it is derived from a residual correlation matrix (e.g., see Figure 3.8). A similarly named index, the root mean square residual (RMR), reflects the average discrepancy between observed and predicted *covariances*. However, the RMR can be difficult to interpret because its value is affected by the metric of the input variables; thus the SRMR is generally preferred. In most instances (e.g., models involving a single input matrix), the SRMR can be calculated by (1) summing the squared elements of the residual correlation matrix and dividing this sum by the number of elements in this matrix (on and below the diagonal); that is, $a = p(p + 1) / 2$ (Eq. 3.14), and (2) taking the square root of this result. For example, the SRMR of the Figure 3.8 solution would be computed as follows:

$$\text{SRMR} = \text{SQRT}[(0^2 + 0^2 + 0^2 + .4^2 + 0^2 + 0^2)/ 6] = .163 \qquad (3.18)$$

The SRMR can take a range of values between 0.0 and 1.0, with 0.0 indicating a perfect fit (i.e., the smaller the SRMR, the better the model fit).

Parsimony Correction

Although sometimes grouped under the category of absolute fit indices (e.g., Hu & Bentler, 1999), *parsimony correction indices* differ from χ^2, SRMR, and so forth by incorporating a penalty function for poor model parsimony (i.e., number of freely estimated parameters as expressed by model *df*). For example, consider a scenario where two different models, Model A and Model B, fit a sample matrix (*S*) equally well at the absolute level; yet the specification of Model B entails more freely estimated parameters than Model A (i.e., Model A has more *df*s than Model B). Indices from the parsimony class would thus favor Model A over Model B because the Model A solution fit the sample data with fewer freely estimated parameters.

A widely used and recommended index from this category is the root mean square error of approximation (RMSEA; Steiger & Lind, 1980). The RMSEA is a population-based index that relies on the *noncentral* χ^2 distribution, which is the distribution of the fitting function (e.g., F_{ML}) when the fit of the model is not perfect. The noncentral χ^2 distribution includes a *noncentrality parameter* (NCP), which expresses the degree of model misspecification. The NCP is estimated as $\chi^2 - df$ (if the result is a negative number, NCP = 0). When the fit of a model is perfect, NCP = 0 and a central χ^2 distribution holds. When the fit of the model is not perfect, the NCP is greater than 0 and shifts the expected value of the distribution to the right of that of the corresponding central χ^2 (cf. Figure 1 in MacCallum, Browne, & Sugawara, 1996). The RMSEA is an "error of approximation" index because it assesses the extent to which a model fits *reasonably* well in the population (as opposed to testing whether the model holds exactly in the population; cf. χ^2). To foster the conceptual basis of the calculation of RMSEA, the NCP is rescaled to the quantity *d*: $d = \chi^2 - df / (N)$. The RMSEA is then computed:

$$\text{RMSEA} = \text{SQRT}(d / df) \qquad (3.19)$$

where *df* is the model *df* (although slight variations exist in some programs; e.g., some programs use $N - 1$ instead of *N*). As can be seen in Eq. 3.19, the RMSEA compensates for the effect of model complexity by conveying discrepancy in fit (*d*) per each *df* in the model. Thus it is sensitive to the number of model parameters; being a population-based index, the RMSEA is relatively insensitive to sample size. The RMSEA from the Figure 3.8 solution would be

$$\text{RMSEA} = \text{SQRT}(.40 / 1) = 0.63$$

where $d = (81.093 - 1) / 200 = 0.40$.

Although its upper range is unbounded, it is rare to see the RMSEA exceed 1.00. As with the SRMR, RMSEA values of 0 indicate perfect fit (and values very close to 0 suggest good model fit).

The noncentral χ^2 distribution can be used to obtain confidence intervals for RMSEA (a 90% interval is typically used). The confidence interval indicates the precision of the RMSEA point estimate. Methodologists recommend including this confidence interval when reporting the RMSEA (e.g., MacCallum et al., 1996). However, researchers should be aware that the width of this interval is affected by sample size and the number of freely estimated parameters in the model (e.g., unless N is very large, complex models are usually associated with wide RMSEA confidence intervals).[10]

Moreover, to address the overly stringent nature of χ^2 (i.e., it tests for "perfect" fit), Browne and Cudeck (1993) have developed a statistical test of closeness of model fit using the RMSEA. Specifically, "close" fit (CFit) is operationalized as RMSEA values less than or equal to .05. This one-sided test appears in the output of most software packages as the probability value that RMSEA $\leq$ .05. Nonsignificant probability values (i.e., $p > .05$) may be viewed in accord with acceptable model fit (i.e., a "close-fitting" model), although some methodologists have argued for stricter guidelines (e.g., $p > .50$; Jöreskog & Sörbom, 1996a). As in any type of significance test, the power of the CFit test is adversely affected by small sample size, but also by model saturation (i.e., less power in models with fewer dfs).

Comparative Fit

Comparative fit indices (also referred to as *incremental fit indices*; e.g., Hu & Bentler, 1998) evaluate the fit of a user-specified solution in relation to a more restricted, nested baseline model. Typically, this baseline model is a "null" or "independence" model in which the covariances among all input indicators are fixed to zero, although no such constraints are placed on the indicator variances (however, other types of null models can and sometimes should be considered; cf. O'Boyle & Williams, 2011). As one might expect, given the relatively liberal criterion of evaluating model fit against a solution positing no relationships among the variables, comparative fit indices often look more favorable (i.e., more suggestive of acceptable model fit) than indices from the preceding categories. Nevertheless, some indices from this category have been found to be among the best behaved of the host of indices that have been introduced in the literature.

One of these indices, the comparative fit index (CFI; Bentler, 1990), is computed as follows:

$$\text{CFI} = 1 - \max[(\chi^2_{\text{T}} - df_{\text{T}}), 0] / \max[(\chi^2_{\text{T}} - df_{\text{T}}), (\chi^2_{\text{B}} - df_{\text{B}}), 0] \qquad (3.20)$$

where χ^2_{T} is the χ^2 value of the target model (i.e., the model under evaluation); df_{T} is the df of the target model; χ^2_{B} is the χ^2 value of the baseline model (i.e., the "null" model); and df_{B} is the df of the baseline model. Also, max indicates to use the largest value—for example, for the numerator, use $(\chi^2_T - df_T)$ or 0, whichever is larger. The χ^2_B and df_B of the null model are included as default output in most software programs. If the user wishes to obtain these values in programs that do provide this information, χ^2_B and df_B can be calculated by fixing all relationships to 0 (but freely estimating the indicator variances). The CFI has a range of possible values between zero and one, with values closer to one implying good model fit. Like the RMSEA, the CFI is based on the NCP (i.e., $\lambda = \chi^2_T - df_T$, included in standard output of some programs such as LISREL), meaning that it uses information from expected values of χ^2_T or χ^2_B (or both, in the case of the CFI) under the noncentral χ^2 distribution associated with $S \neq \Sigma$ (e.g., central χ^2 is a special case of the noncentral χ^2 distribution when $\lambda = 0$). Using the results of the Figure 3.8 model, the CFI would be

$$CFI = 1 - [(81.093 - 1) / (227.887 - 3)] = .644$$

Another popular and generally well-behaved index falling under this category is the Tucker–Lewis index (TLI; Tucker & Lewis, 1973), referred to as the *non-normed fit index* in some programs). In addition, the TLI has features that compensate for the effect of model complexity; that is, as does the RMSEA, the TLI includes a penalty function for adding freely estimated parameters that do not markedly improve the fit of the model. The TLI is calculated by the following formula:

$$TLI = [(\chi^2_B / df_B) - (\chi^2_T / df_T)] / [(\chi^2_B / df_B) - 1] \qquad (3.21)$$

where, as with the CFI, χ^2_T is the χ^2 value of the target model (i.e., the model under evaluation); df_T is the df of the target model; χ^2_B is the χ^2 value of the baseline model (i.e., the "null" model); and df_B is the df of the baseline model. Unlike the CFI, the TLI is non-normed, which means that its values can fall outside the range of zero to one. However, it is interpreted in a fashion similar to the CFI, in that values approaching one are interpreted in accord with good model fit. The TLI for the Figure 3.8 solution is

$$TLI = [(227.877 / 3) - (81.093 / 1)] / [(227.877 / 3) - 1] = -0.068$$

The goodness-of-fit indices from each category point to the poor fit of the Figure 3.8 solution: $\chi^2(1) = 81.093$, $p < .001$, SRMR = .163, RMSEA = 0.633, CFI = .644, TLI = −0.068. Although straightforward in the Figure 3.8 example, the issues and guidelines for using these descriptive indices of overall model fit are considered more fully in the next section of this chapter.

Guidelines for Interpreting Goodness-of-Fit Indices

As noted earlier, issues surrounding goodness-of-fit indices are hotly debated (e.g., which indices should be used? what cutoff criteria should be applied to indicate good and poor model fit?). If the reader were to peruse published books and journal articles on this topic, he or she would note few areas of consensus in regard to recommended fit index cutoffs. Thus it would be inappropriate to recommend cutoffs unequivocally in this book. Indeed, such endeavor is complicated by the fact that fit indices are often differentially affected by various aspects of the analytic situation, such as sample size, model complexity, estimation method (e.g., ML, WLS), amount and type of misspecification, normality of data, and type of data (e.g., TLI and RMSEA tend to reject models falsely when N is small—Hu & Bentler, 1999; SRMR does not appear to perform well in CFA models based on categorical indicators—Yu, 2002). The reader is referred to Hu and Bentler (1998, 1999) for evidence and a detailed discussion of how such aspects may affect the performance of these fit indices (although it should be again noted that the RMSEA, SRMR, CFI, and TLI have been selected in this book in part on the basis of their overall satisfactory performance in the Hu and Bentler simulations). It is also important to emphasize that goodness-of-fit indices constitute only one aspect of model evaluation. As discussed in Chapter 4, although model evaluation usually begins with the examination of these fit indices, it is equally important to examine a solution in terms of potential areas of localized strain (e.g., are there specific relationships the model does not adequately reproduce?) and the interpretability and strength of the resulting parameter estimates (e.g., absence of Heywood cases; statistical significance, direction, and size of parameter estimates all in accord with prediction). With these caveats fully in mind, a few prominent guidelines for fit indices are reviewed.

In one of the more comprehensive evaluations of cutoff criteria, the findings of simulation studies conducted by Hu and Bentler (1999) suggest the following guidelines. Support for contentions of reasonably good fit between the target model and the observed data (assuming ML estimation) is obtained in instances where (1) SRMR values are close to .08 or below; (2) RMSEA values are close to .06 or below; and (3) CFI and TLI values are close to .95 or greater. Hu and Bentler's (1999) use of the phrase "close to" is not accidental, because the recommended cutoff values were found to fluctuate as a function of modeling conditions (e.g., type of misspecified model) and whether or not an index was used in combination with other fit indices (e.g., the acceptability of Type I and Type II error rates often improved when a combination of indices was employed). Other methodologists have handled these complexities by providing descriptive anchors for various ranges of fit index values rather than specifying explicit cutoffs. For instance, Browne and Cudeck (1993) suggest, as a rule of thumb, that RMSEA values less than 0.08 suggest adequate model fit (i.e., a "reasonable error of approximation," p. 144); that RMSEA values less than 0.05 suggest good model fit; and that models with RMSEA ≥ 0.1 should be rejected. MacCallum et al. (1996) further elaborated on these guidelines by asserting that RMSEAs in the range of 0.08–0.10 suggest "mediocre" fit. Additional support for the fit of the solution would be evidenced by a 90% confidence interval of the RMSEA whose upper limit is below these cutoff values (e.g., 0.08). As noted earlier, a nonsignificant CFit (RMSEA < 0.05) could also be interpreted in accord with acceptable model fit. Similarly, methodologists have noted that while CFI and TLI values below .90

should lead the researcher to strongly suspect (reject) the solution, CFI and TLI values in the range of .90 and .95 may be indicative of acceptable model fit (e.g., Bentler, 1990).

In keeping with the notion that this is a contentious area of methodological research, some researchers have asserted that the Hu and Bentler (1999) guidelines are far too conservative for many types of models, including CFA models consisting of many indicators and several factors where the majority of cross-loadings and error covariances are fixed to zero (cf. Marsh, Hau, & Wen, 2004). Moreover, because the performance of fit statistics and their associated cutoffs have been shown to vary as a function of various aspects of the model (e.g., degree of misspecification, size of factor loadings, number of factors; e.g., Beauducel & Wittman, 2005), the fit statistic thresholds suggested by simulation studies may have limited generalizability to many types of measurement models in applied research.

Nonetheless, when fit indices fall into these "marginal" ranges, it is especially important to consider the consistency of model fit as expressed by the various types of fit indices in tandem with the particular aspects of the analytic situation (e.g., when N is somewhat small, an RMSEA = 0.08 may be of less concern if all other indices are strongly in a range suggesting "good" model fit). Again, this underscores the importance of considering fit indices from multiple fit categories (absolute fit, parsimony correction, comparative fit) in tandem with examining other relevant aspects of the solution (e.g., localized areas of ill fit; interpretability and size of parameter estimates). These aspects of model evaluation are discussed in Chapter 4.

SUMMARY

This chapter has provided a detailed comparison of EFA and CFA. Although EFA and CFA are based on the common factor model, CFA has a fundamental advantage over EFA in that it allows the researcher to control every aspect of the model specification (e.g., generate an unstandardized solution; specify correlated errors; and place various constraints on the solution, such as fixing cross-loadings to zero or holding model parameters to equality). As will be seen in later chapters, these features allow CFA to address important research questions for which EFA is not well suited (or incapable), such as the comparison of factor models across groups (e.g., measurement invariance of tests; Chapter 7), the analysis of MTMM data in construct validation (Chapter 6), the analysis of mean structures (Chapter 7), scale reliability evaluation (Chapter 8), and the inclusion of covariates in the factor model (e.g., MIMIC; Chapter 7).

In addition, this chapter has introduced the fundamental concepts of CFA, such as model notation, basic equations, model identification, model estimation (e.g., ML), and goodness-of-fit evaluation. The Appendices in this chapter are provided to foster the reader's understanding of each of these core principles through data-based illustrations. In Chapter 4, these concepts are exemplified and extended in context of a fully worked-through CFA of an applied data set.

NOTES

1. In some software programs, it is possible to conduct EFA by using a (unstandardized) variance–covariance matrix. In fact, use of a covariance matrix in ML EFA would be preferred when the data are not scale-free (i.e., EFAs based on a correlation and covariance matrix produce markedly different model χ^2 values).

2. Although this is not true for all software programs, some programs (e.g., Mplus) provide standard errors and significance tests for partially standardized and completely standardized parameter estimates.

3. However, beginning in Version 7.1, the Mplus program performs χ^2 difference testing to compare the fit of EFA solutions that differ in the number of factors as a guide to factor model selection.

4. When measurement error is systematic (e.g., method effects), it can also lead to positively biased parameter estimates.

5. Historically, the term *congeneric* has been used more strictly to refer to one-factor measurement models (i.e., a set of indicators that load on one factor). In this book, the term is used more generally in reference to sets of indicators that load on just one factor (e.g., multifactorial measurement models that contain no double-loading indicators).

6. An excellent, user-friendly introduction to ML can be found in Chapter 3 of Enders (2010).

7. Some programs (e.g., Mplus) prefer that the indicators submitted to the latent variable analysis be kept on a similar scale. For instance, problems with convergence may occur if the variances of the indicators are markedly heterogeneous. This can be resolved by rescaling the indicator(s) through a linear transformation. Say, for example, that X1, X2, and X3 are measured on 0–8 scales, but X4 has scores ranging from 1 to 70 (Figure 3.7A). To avoid potential problems with convergence, X4 can be rescaled simply by dividing its observed scores by 10.

8. The Mplus developers have provided justification for using N instead of $N - 1$ in a webnote: (*www.statmodel.com/download/N%20vs%20N-1.pdf*).

9. An initial attempt to address the shortcomings of χ^2 dates back to the seminal SEM study by Wheaton, Muthén, Alwin, and Summers (1977). In this study, the authors introduced the χ^2/df ratio in attempt to use χ^2 in a manner that would foster more realistic model evaluation. Although the χ^2/df ratio has become very popular in applied research, its use is strongly discouraged (in fact, even Wheaton, 1987, subsequently recommended that the χ^2/df ratio not be used). For instance, researchers frequently use this index under the false understanding that it adjusts for χ^2's excessive power to reject $S = \Sigma$ when in fact the χ^2/df ratio is equally sensitive to sample size (i.e., $N - 1$ is used in the calculation of both χ^2 and χ^2/df). Other reasons contraindicating the use of the χ^2/df ratio include the fact that it has not undergone the scrutiny of simulation research (cf. Hu & Bentler, 1998, 1999), and the lack of consistent standards of what χ^2/df value represents good or bad model fit.

10. The assumptions of the noncentral distribution and RMSEA do not hold if the model is grossly misspecified. For other cautions on the use of RMSEA (e.g., impact of non-normality and sample size), the reader is referred to Yuan (2005).

Appendix 3.1

Communalities, Model-Implied Correlations, and Factor Correlations in EFA and CFA

I. COMMUNALITY ESTIMATES

As noted in Chapter 2, the calculation of communalities in orthogonal EFA is straightforward. In these models, the communality is the sum of squared factor loadings for a given indicator across all factors. For instance, in the two-factor model of adolescent antisocial behavior presented in Table 3.1, the communality of the Y1 indicator is equal to its squared loading on Property Crimes plus its squared loading on Violent Crimes; that is, $.8493^2 + .1765^2 = .75$.

Communalities can also be easily computed in CFA. In orthogonal EFA, this computation is straightforward because the factors are not correlated (i.e., the factor loadings reflect the zero-order correlations between the indicator and the factors). In CFA, this computation is clear-cut because indicators are typically specified to load on one factor only (i.e., all cross-loadings are fixed to zero). Thus the communality is simply the factor loading squared. For instance, in the CFA, the communality of the Y1 indicator is $.7996^2 = .64$ (cf. Eq. 3.5).

In oblique EFA (or in a CFA where an indicator is specified to load on two or more factors), the calculation of a communality requires an understanding of the path diagram tracing rules and equations (e.g., in this chapter, see the "Fundamental Equations of a CFA Model" section). Because factors are intercorrelated in oblique EFA, the factor loadings are partial regression coefficients (pattern matrix). Thus computation of a communality in oblique EFA is a function of both the factor loadings and the factor correlations. Below is a path diagram of the oblique EFA solution from Table 3.1 for the Y1 and Y5 indicators only.

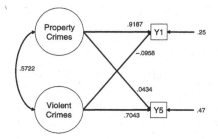

In this example, the communality of Y1 is

$$\eta^2_{y1} = \lambda_{11}^2 + \lambda_{12}^2 + 2\lambda_{11}\phi_{21}\lambda_{12} \tag{3.22}$$
$$= .9187^2 + -.0958^2 + 2(.9187)(.5722)(-.0958)$$
$$= .75$$

The communality of Y5 is

$$\eta^2_{y5} = \lambda_{52}^2 + \lambda_{51}^2 + 2\lambda_{52}\phi_{21}\lambda_{51} \tag{3.23}$$
$$= .7043^2 + .0434^2 + 2(.7043)(.5722)(.0434)$$
$$= .53$$

Thus, for example, the communality of Y1 is the sum of the unique direct effects of Property Crimes and Violent Crimes on Y1 (λ_{11}^2 and λ_{12}^2, respectively), and the variance in Y1 that Property Crimes and Violent Crimes jointly explain $[2(\lambda_{11})(\phi_{21})(\lambda_{12})]$. In other words, Property Crimes and Violent Crimes are correlated ($\phi_{21} = .5722$), and some portion of the variance in Y1 that is explained by Property Crimes is also explained by Violent Crimes.

II. MODEL-IMPLIED CORRELATIONS

The quality of the factor solutions is determined in part by how well the parameter estimates of the solution are able to reproduce the observed correlations of the indicators. The correlations predicted by the parameters of the factor solution are often referred to as *model-implied estimates*. The model-implied correlations from oblique EFA, orthogonal EFA, and CFA are illustrated below, again by using the Y1 and Y5 indicators as an example.

In the oblique EFA, the model-implied correlation between Y1 and Y5 is

$$Corr(Y5, Y1) = \lambda_{11}\lambda_{51} + \lambda_{52}\lambda_{12} + \lambda_{11}\phi_{21}\lambda_{52} + \lambda_{51}\phi_{21}\lambda_{12} \tag{3.24}$$
$$= (.9187)(.0434) + (.7043)(-.0958) + (.9187)(.5722)(.7043)$$
$$+ (.0434)(.5722)(-.0958)$$
$$= .34$$

The calculation of the model-implied correlation of Y1 and Y5 is a bit easier in orthogonal EFA because the factors are not correlated. Thus the predicted correlation is calculated solely from the summed cross-products of the factor loadings:

$$Corr(Y5, Y1) = \lambda_{11}\lambda_{51} + \lambda_{52}\lambda_{12} \tag{3.25}$$
$$= (.8493)(.2587) + (.6826)(.1765)$$
$$= .34$$

Again note that the model-implied correlation of Y1 and Y5 is the same in orthogonal and oblique EFA; the choice of rotation method does not affect the fit of the factor solution.

In CFA models where each indicator loads on only one factor (cf. Table 3.1, Figure 3.1A), computation of the model-implied correlation is very straightforward (i.e., although factors are intercorrelated, each indicator loads on a single factor). The correlation of Y1 and Y5 that is predicted from the CFA solution is (cf. Eq. 3.8)

$$\text{Corr(Y5, Y1)} = \lambda_{11}\phi_{21}\lambda_{52} \qquad (3.26)$$
$$= (.7996)(.6224)(.7315)$$
$$= .36$$

III. FACTOR CORRELATIONS

In this chapter, it has been stated that CFA factor correlations are often higher than factor correlations emanating from oblique EFA of the same data set. This occurs in the analyses presented in Table 3.1, where the factor correlations estimated by oblique EFA and CFA are .57 and .62, respectively. In an oblique EFA, the model-implied correlation of indicators with primary loadings on separate factors can be estimated in part by the indicator cross-loadings (cf. Eq. 3.24). In contrast, the model-implied correlation of indicators loading on separate factors in CFA is estimated solely by the primary loadings and the factor correlation (cf. Eq. 3.26). For example, compared to oblique EFA, in the CFA more burden is on the factor correlation to reproduce the correlation between Y1 and Y5 because there are no cross-loadings to assist in this model-implied estimate (cf. Figure 3.1A). Therefore, in the iterative process of establishing CFA parameter estimates that best reproduce the sample correlation matrix, the magnitude of the factor correlation estimate may be increased somewhat (relative to oblique EFA) to better account for the relationships of indicators that load on separate factors (assuming that these indicators are correlated to some degree).

Appendix 3.2

Obtaining a Solution for a Just-Identified Factor Model

Factor analysis typically entails matrix algebra and other mathematical operations that are very cumbersome to demonstrate and conduct by hand calculation. However, the solution for a basic, just-identified factor model such as the one depicted in Figure 3.6B can be calculated on the basis of principles discussed in this chapter and the help of the algebra of simultaneous equations. Consider the following input matrix and just-identified, one-factor model:

Input matrix:

	X1	X2	X3
X1	1.000		
X2	0.595	1.000	
X3	0.448	0.544	1.000

For ease of notation, let $a = \lambda_{x11}$, $b = \lambda_{x21}$, and $c = \lambda_{x31}$.

As discussed in this chapter, the number of knowns (6 elements of the input matrix) equals the number of unknowns (6 parameters = 3 factor loadings, 3 errors; the factor variance is fixed to 1.0). Thus the model is just-identified, and its parameter estimates will perfectly reproduce the input matrix. It has also been discussed that for two indicators loading on the same factor, multiplying their factor loadings provides the model estimate of their zero-order correlation. Because the current model is just-identified, the products of the loadings will perfectly reproduce the zero-order relationships among X1, X2, and X3. The systems of equations are as follows:

Equation 1. $ab = .595$

Equation 2. $ac = .448$

Equation 3. $bc = .544$

This problem is also just-identified, as there are 3 unknowns (a, b, c) and 3 knowns (the 3 equations). Systems of equations can be solved in various ways (e.g., substitution, elimination, matrices). In the example, substitution is used by first solving for a:

Step 1: Rearrange Equations 1 and 2, so that b and c are the outputs.

Equation 1.	$ab = .595$	$b = .595/a$
Equation 2.	$ac = .448$	$c = .448/a$
Equation 3.	$bc = .544$	

Step 2: Substitute Equations 1 and 2 in Equation 3.

Equation 3. $b \quad c \quad = .544$

$(.595/a)(.448/a) = .544$

Step 3: Solve Equation 3.

Equation 3.
$$(.595/a)(.448/a) = .544$$
$$.26656/a^2 = .544$$
$$.26656/.544 = a^2$$
$$.49 = a^2$$
$$.70 = a$$

Step 4: Now that we know that $a = .70$, it is straightforward to solve for b and c, using the original equations.

Equation 1.	$.70b = .595$	$.595/.70 = b$	$b = .85$
Equation 2.	$.70c = .448$	$.448/.70 = c$	$c = .64$

Thus the factor loadings are .70, .85, and .64, for λ_{x11}, λ_{x21}, and λ_{x31}, respectively. As an accuracy check, note that multiplying these loadings together reproduces the input correlations perfectly; that is, $.70(.85) = .595$, $.70(.64) = .448$, $.85(.64) = .544$.

Step 5: Because this is a completely standardized solution, the errors for X1, X2, and X3 can be obtained by squaring the loadings and subtracting the result from 1.0.

$$\delta_1 = 1 - .70^2 = .51$$
$$\delta_2 = 1 - .85^2 = .2775$$
$$\delta_3 = 1 - .64^2 = .5904$$

Now every unknown parameter in the factor model has been solved, and the solution is as follows:

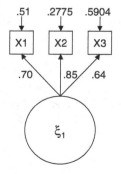

Hand Calculation of F_{ML} for the Figure 3.8 Path Model

Calculation of F_{ML} is best left to the computer because the computations needed to produce this value are quite tedious and complex in most data sets. However, given the simplicity of the Figure 3.8 model, hand calculation of F_{ML} is presented in effort to foster the reader's conceptual understanding of the ML fit function:

$$F_{ML} = \ln|S| - \ln|\Sigma| + \text{trace}[(S)(\Sigma^{-1})] - p$$

The presentation focuses primarily on the first half of this equation, $\ln|S| - \ln|\Sigma|$, given its relationship to other widely known statistics (e.g., F ratio, likelihood ratio), and the fact that in this example the remaining part of the equation will equal zero. The remaining portion of the F_{ML} formula, $\text{trace}[(S)(\Sigma^{-1})] - p$, pertains to the distances of the indicator variances in S and Σ. In many instances (as in the Figure 3.8 solution), this difference will be zero because the CFA model perfectly reproduces the observed variance of the input indicators; cf. $\sigma = \lambda_x^2 \phi_1 + \delta$, and thus $\text{trace}[(S)(\Sigma^{-1})] - p = 3 - 3 = 0$ (see Table 3.2). However, there are some instances where differences in the diagonal elements (indicator variances) of S and Σ do not equal zero and thus contribute to F_{ML} (e.g., in the CFA evaluation of tau equivalence; see Chapter 7). But, in the Figure 3.8 solution, the value of F_{ML} is determined solely by the difference in the natural logs of the determinants of S and Σ.

DETERMINANTS ($|S|$, $|\Sigma|$)

As noted in this chapter, a *determinant* is a matrix algebra term for a single number (i.e., a scalar) that reflects a generalized measure of variance for the entire set of variables contained in a matrix. Stated another way, it is an index of the amount of nonredundant variance that reflects the extent to which variables in the matrix are free to vary. This concept is illustrated in the following zero-order (2×2) and multivariate (3×3) examples, drawing from the sample variance–covariance matrix in Figure 3.8.

Consider the scenario where x and z share none of their variance (i.e., r_{xz} = .00). If x and z have SDs = 2.0, the 2×2 variance–covariance matrix will be as follows:

	x	z
x	4.0	0.0
z	0.0	4.0

The variance and (lack of) covariance of x and z are represented in the following Venn diagram:

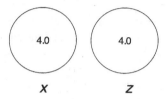

Using a basic principle of matrix algebra, the determinant of a 2×2 matrix,

$$A = \begin{bmatrix} a & b \\ c & d \end{bmatrix}$$

is $|A| = ad - bc$

In the present example, A is a variance–covariance matrix, and thus the equation $ad - bc$ represents subtracting the covariance (bc) from the variance (ad). Solving this equation, 4(4) – 0(0), results in a determinant of 16. This value indicates that all the variance in the matrix of x and z (4 * 4 = 16) is nonredundant; that is, x and z are entirely free to vary. Because the Venn diagram indicates no overlapping variance between x and z, the determinant can be calculated simply by multiplying the variance of x and z.

Consider the situation where x and z share all of their variance (i.e., r_{xz} = 1.0). The variance–covariance matrix and Venn diagram will look as follows:

	x	z
x	4.0	4.0
z	4.0	4.0

·XZ

In this case, the determinant equals 0, $ad - bc$ = 4(4) – 4(4) = 0, which reflects that all the variance in x and z (ad) is redundant (bc); in other words, there is no generalized variability in this matrix (no freedom to vary). Alternatively, with the Venn diagram, the determinant could be solved by multiplying the variance in x (4) by the unique variance in z (0): 4 * 0 = 0.

When a matrix has a determinant of zero, it is said to be *singular* (conversely, matrices with nonzero determinants are *nonsingular*), meaning that one or more rows or columns are linearly dependent on other rows or columns. (For instance, this will occur if a variance–covariance matrix is constructed by using the subscales and total score of a questionnaire if the total score is formed by summing the subscales.) Because determinants are used in calculating the *inverse* of a matrix (e.g., Σ^{-1}; see F_{ML} equation), a singular matrix is problematic because it has no inverse; the matrix algebra equivalent to division is multiplying a matrix by the inverse of another matrix—for example, $(S)(\Sigma^{-1})$ (see F_{ML} equation). Thus many statistics for the matrix cannot be computed. Users of latent variable software programs may occasionally encounter error messages stating that a matrix is singular or nonpositive definite. These errors often stem from linear dependency or some other problems in the input matrix (e.g., multicollinearity/one or more eigenvalues = 0; use of pairwise deletion as a missing data strategy; an N that exceeds the number of input indicators), although the matrix generated by the specified model must also be positive definite (Wothke, 1993).

Now consider the zero-order relationship of x and z that is observed in the Figure 3.8 model (r_{xz} = .60). The variance–covariance matrix and Venn diagram are as follows:

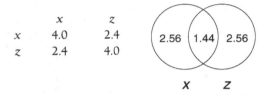

	x	z
x	4.0	2.4
z	2.4	4.0

The determinant of this matrix is 10.24; $ad - bc = 4(4) - 2.4(2.4) = 16 - 5.76 = 10.24$. Some of the total variance in x and z ($ad = 16$) is redundant ($bc = 5.76$); subtracting bc from ad provides the measure of generalized variability in this matrix. Also note that dividing bc by ad yields the proportion of shared variance in x and z: 5.76 / 16 = .36 = r^2. Alternatively, with the Venn diagram, the determinant can be computed by multiplying the variance of x (4) by the unique variance in z (2.56): 4 * 2.56 = 10.24. The unique variance in z is calculated by multiplying its variance (4) by its proportion of unique variance ($1 - r^2 = .64$): 4 * .64 = 2.56.

Computing the determinant for matrices of the order 3×3 and higher is more complicated. In order to begin with a computationally simpler example, consider the scenario where the correlation between x and z is .60, and z and y are correlated .50, but x and y share none of their variance (i.e., r_{xy} = 0.0). This variance–covariance matrix and Venn diagram will follow:

	x	z	y
x	4.0	2.4	0.0
z	2.4	4.0	2.0
y	0.0	2.0	4.0

Using another method from matrix algebra, we find that the determinant of a 3×3 matrix,

$$A = \begin{bmatrix} a & b & c \\ d & e & f \\ g & h & i \end{bmatrix}$$

is $|A| = (aei - bdi) + (bfg - ceg) + (cdh - afh)$.

Although there are different approaches to solving for the determinant of a 3×3 matrix (e.g., the "butterfly method," use of *minors* and their matrices of *cofactors*), the above-described method is computationally similar to the strategy used to solve for 2×2 matrices (although alternative approaches, such as the method of minors and cofactors, generalize better to matrices of the order of 4×4 and higher; cf. Bollen, 1989). Here we begin with the first element of the matrix (*a*) and multiply it by all other elements on its diagonal (*aei*). Then the element in the column adjacent to *a* is selected (*b*) and is multiplied by the remaining elements in the diagonal going in the opposite direction of *aei* (*bdi*; compare to the method used for 2×2 matrices). Note that the remaining element in *bdi* (i.e., *i*) is obtained by bringing the column [*c f i*] in front of the column [*a d g*]. *bdi* is then subtracted from *aei*, and the process is repeated two more times for the remaining elements in the first row of **A** (*bfg* – *ceg*, *cdh* – *afh*). The results are then summed.

Applying this method to the variance–covariance matrix above yields a determinant of 24.96:

aei – *bdi*:	(4.0 * 4.0 * 4.0) – (2.4 * 2.4 * 4.0) =	40.96
bfg – *ceg*:	(2.4 * 2.0 * 0.0) – (0.0 * 4.0 * 0.0) =	0.00
cdh – *afh*:	(0.0 * 2.4 * 2.0) – (4.0 * 2.0 * 2.0) =	−16.00
	Total:	24.96

Thus, of the total variance in the matrix of *x*, *z*, and *y* (4 * 4 * 4 = 64), a considerable portion is overlapping (39.04) and not included in the determinant. Note that dividing 39.04 by 64 yields .61, which will reflect R^2 if *z* is regressed onto *x* and *y* (it works out cleanly in this example because *x* and *y* do not correlate; $r_{xz}^2 = .36$, $r_{zy}^2 = .25$, .36 + .25 = .61). Indeed, using the Venn diagram method, we could readily solve for the determinant by multiplying the variance of *x* (4) by the variance of *y* (4) by the unique variance in *z* (1.56): 4 * 4 * 1.56 = 24.96; the unique variance in *z* is computed by subtracting its variance (4) with the variance explained by *x* (4 * .36 = 1.44) and *y* (4 * .25 = 1.0); 4 – 1.44 – 1 = 1.56 (again simplified by having no overlap between *x* and *y*).

Ideally, this explanation provides the reader with a better conceptual understanding of a determinant. Let's now turn to the sample (*S*) and predicted (Σ) matrices from Figure 3.8:

	S				Σ		
	x	*z*	*y*		*x*	*z*	*y*
x	4.0	2.4	2.8	*x*	4.0	2.4	1.2
z	2.4	4.0	2.0	*z*	2.4	4.0	2.0
y	2.8	2.0	4.0	*y*	1.2	2.0	4.0

Using the matrix algebra method presented earlier, we find that the determinant of *S* is

aei – *bdi*:	(4.0 * 4.0 * 4.0) – (2.4 * 2.4 * 4.0) =	40.96
bfg – *ceg*:	(2.4 * 2.0 * 2.8) – (2.8 * 4.0 * 2.8) =	−17.92
cdh – *afh*:	(2.8 * 2.4 * 2.0) – (4.0 * 2.0 * 2.0) =	−2.56
	Total:	20.48

Note that this is the same value calculated for S by SAS PROC IML in Table 3.2. Using the same method, we find that the determinant of Σ is 30.72. (The determinants of S and Σ can also be computed by Venn diagrams, although this will require more time-consuming steps of solving for the unique and overlapping areas in x, y, and z—regressing each variable onto the remaining two variables.) Because Σ has more generalized variance (30.72) than S (20.48), a poor-fitting model may result. The additional nonredundant variance in Σ is due to the fact that the Figure 3.8 model parameters predict less covariance between x and y ($\sigma_{xy} = 1.2$) than is observed in the sample data ($\sigma_{xy} = 2.8$).

DIFFERENCE OF NATURAL LOGS ($\ln|S| - \ln|\Sigma|$)

Returning to the equation for F_{ML}, note that the natural logs (ln) of the determinants are used rather than the raw values of the determinants themselves. The reason for this is as much computational as theoretical. Specifically, logs have several properties that make working with and interpreting them easier than other methods. For instance, consider the situation where the probability of a treatment success is 70% ($p = .7$), and the probability of no treatment success is 30% ($q = .3$). Converting the probability of success to an odds ratio, OR $= p / (1 - p)$, yields a value of 2.33 (.7/.3); that is, a treatment success is 2.3 times more likely than nonsuccess. However, the OR for nonsuccess is .429 (.3 / .7). Although it would be helpful to be able to interpret this outcome in a symmetrical fashion (i.e., the odds of success are opposite the odds of nonsuccess), ORs do not have this symmetry (i.e., 2.33, .43). However, taking the natural log of these ORs provides this symmetry, $\ln(2.33) = .846$, $\ln(.429) = -.846$. This is one reason why log odds are commonly used in logistic regression. Logs have other useful properties, such as

If:	$a = bc$	e.g., $20 = 5(4)$
Then:	$\ln(a) = \ln(b) + \ln(c)$	$2.996 = 1.6094 + 1.3863$

and thus

If:	$a/b = c$	e.g., $20/5 = 4$
Then:	$\ln(c) = \ln(a) - \ln(b)$	$1.3863 = 2.996 = 1.6094$
		(within rounding error)

The latter equation is reflected in the first part of the formula for F_{ML}. Specifically, taking the difference in $\ln|S|$ and $\ln|\Sigma|$ is equivalent to dividing the determinant of one of these matrices by the determinant of the other; in fact, in a few textbooks, this portion of the equation is written as $\ln(S\Sigma^{-1})$, which produces the same result as $\ln|S| - \ln|\Sigma|$. Thus the parallel of $\ln|S| - \ln|\Sigma|$ with the F ratio found in ANOVA (calculated by dividing one variance into another) should be apparent. This is also why the likelihood ratio (LR) is so called, when in fact it is calculated by taking the difference between two log-likelihoods: LR $= 2(LL_A - LL_O)$.

Using the results from Figure 3.8, we find that the natural logs of the determinants of S and Σ are as follows:

$$\ln(S) = \ln(20.48) = 3.01945$$
$$\ln(\Sigma) = \ln(30.72) = 3.42491$$

The difference between the natural logs of these determinants is .40546, which equals the value of F_{ML} obtained by SAS PROC IML (Table 3.2). The last part of the equation does not alter this value because the trace of $[(S)(\Sigma^{-1})]$ and order of the matrix (p) both equal 3 (actually, the result of $\ln|S| - \ln|\Sigma|$ is a negative, -0.40546, but the absolute value is obviously the same). Equivalently, we could solve for F_{ML} by dividing the raw score values of the determinants of S and Σ, and then derive the natural log of the result; for example, $30.72/20.48 = 1.5$, $\ln(1.5) = 0.40546$; $20.48/30.72 = 0.66667$, $\ln(.66667) = -0.40546$ (again illustrating the advantage of natural logs shown in the discussion of log odds).

MODEL χ^2

Finally, note from Chapter 3 that the model χ^2 is calculated by $F_{ML}(N - 1)$. This equation parallels a χ^2 formula commonly found in introductory statistics books for testing whether one sample variance ($\sigma_2{}^2$) differs from another variance or a population variance ($\sigma_1{}^2$):

$$\chi^2 = \frac{(N-1)\sigma_2{}^2}{\sigma_1{}^2}$$

This equation can be readily reexpressed as $(\sigma_2{}^2/\sigma_1{}^2)(N-1)$, of which the first part is equivalent to $\ln(\sigma_2{}^2) - \ln(\sigma_1{}^2)$, and of which, in its entirety, is equivalent to $\chi^2 = (\ln|S| - \ln|\Sigma|)(N - 1)$ or $F_{ML}(N - 1)$ (again, this holds in situations where the latter half of the F_{ML} equation equals 0).

4

Specification and Interpretation of CFA Models

This chapter uses a complete, applied example to demonstrate and extend the concepts of CFA presented in the previous chapters. A two-factor measurement model of personality (Neuroticism, Extraversion) is evaluated with four popular latent variable software programs (LISREL, Mplus, EQS, SAS/CALIS). The example is used to illustrate how the acceptability of the CFA solution should be evaluated (e.g., overall goodness of fit, areas of strain in the solution, interpretability/strength of parameter estimates). Moreover, the results of this analysis are used to demonstrate how to interpret and compute the unstandardized and standardized parameter estimates of the CFA solution. The aims and procedures of CFA models that include single indicators are described. The chapter concludes with a discussion and illustration of the material that should be included in reporting the results of a CFA study.

AN APPLIED EXAMPLE OF A CFA MEASUREMENT MODEL

The concepts introduced in Chapters 2 and 3 are now illustrated and extended in context of a full example of a CFA measurement model. The hypothesized model and sample data (correlations, standard deviations) are presented in Figure 4.1 (drawn loosely from the vast literature on the five-factor model of personality; e.g., Wiggins, 1996). In this example, a researcher has collected eight measures (subscales of a widely used personality measure) from a sample of 250 psychotherapy outpatients: anxiety (N1), hostility (N2), depression (N3), self-consciousness (N4), warmth (E1), gregariousness (E2), assertiveness (E3), and positive emotions (E4). A two-factor model is posited in which the observed measures of anxiety, hostility, depression, and self-consciousness are conjectured to load on a latent dimension of Neuroticism, and the observed measures of warmth, gregariousness, assertiveness, and positive emotions are predicted to load onto a distinct factor of Extraversion.

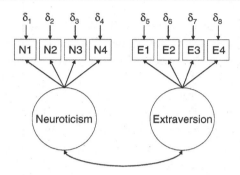

Correlations/Standard Deviations (SD):

	N1	N2	N3	N4	E1	E2	E3	E4
N1	1.000							
N2	0.767	1.000						
N3	0.731	0.709	1.000					
N4	0.778	0.738	0.762	1.000				
E1	-0.351	-0.302	-0.356	-0.318	1.000			
E2	-0.316	-0.280	-0.300	-0.267	0.675	1.000		
E3	-0.296	-0.289	-0.297	-0.296	0.634	0.651	1.000	
E4	-0.282	-0.254	-0.292	-0.245	0.534	0.593	0.566	1.000
SD:	5.700	5.600	6.400	5.700	6.000	6.200	5.700	5.600

FIGURE 4.1. Path diagram and input data for two-factor CFA model of Neuroticism and Extraversion. N1, anxiety; N2, hostility; N3, depression; N4, self-consciousness; E1, warmth; E2, gregariousness; E3, assertiveness; E4, positive emotions. All indicators measured on scales ranging from 0 to 32 (higher scores reflect higher levels of the assessed dimension). $N = 250$.

Although the Figure 4.1 model is basic, numerous predictions underlie this model specification. For instance, all measurement error is presumed to be unsystematic (i.e., there are no correlated measurement errors for any pairs of indicators). This implies that for indicators loading on the same factor, the observed covariance among these measures can be explained entirely by the underlying construct; that is, there is no reason for these observed relationships other than the factor. For example, anxiety (N1) and hostility (N2) are correlated ($r = .767$, see Figure 4.1) because they are manifest symptoms influenced (caused) by the same latent construct (Neuroticism); in other words, if Neuroticism is somehow partialed out, the relationship between anxiety and hostility will be reduced to zero ($r = .00$). In addition, this measurement model asserts that Neuroticism and Extraversion are correlated (as depicted by the bidirectional, curved arrow in Figure 4.1), although the nature of this relationship is unanalyzed. In CFA, latent variables are almost always permitted to be intercorrelated, and there is no claim about the directionality of such relationships (e.g., that Neuroticism has a direct effect on Extraversion or vice versa). Indeed, as noted in Chapter 3, if the researcher specifies directional relationships among factors, he or she has moved out of the CFA measurement model framework and into the realm of structural models, where any sources of ill fit may stem from both the measurement and structural portions of the solution (see Figure 3.2 in Chapter 3).[1] Allowing Neuroticism and Extraversion to be correlated also implies that there may be some relationship between indicators that load on separate factors. However, because indicators do not load on more than one factor, it is predicted that any such relationships can be accounted for by the correlation between the factors. For example, the observed correlation between anxiety (N1) and warmth (E1) is –.351 (see Figure 4.1). As illustrated by a formula introduced in Chapter 3 (Eq. 3.8), this model specification implies that the observed correlation of –.351 should be reproduced in the completely standardized solution by multiplying the factor loadings of N1 and E1 and the correlation between Neuroticism and Extraversion (i.e., $\lambda_{x11}\phi_{21}\lambda_{52}$). In other words, no cross-loadings (e.g., Neuroticism → E1) or error covariances are needed to reproduce these relationships; any covariation between N1–N4 and E1–E4 can be explained by overlap in Neuroticism and Extraversion in tandem with the relationships of the indicators with their factors.

With a simple formula provided in Chapter 3 (Eq. 3.14), it can be readily determined that the input matrix (S) contains 36 pieces of information: 8 variances (p) and 28 covariances [$p(p + 1) / 2$]; that is, 8(9) / 2 = 36. The measurement model presented in Figure 4.1 contains 17 freely estimated parameters: 6 factor loadings (N1 and E1 serve as marker indicators, and thus their factor loadings are fixed), 8 error variances, 2 factor variances, and 1 factor covariance. The model is overidentified with 19 df (df = 36 – 17), and goodness-of-fit evaluation will apply.

MODEL SPECIFICATION

Substantive Justification

A few important steps precede the actual CFA. As noted in preceding chapters, CFA requires specification of a measurement model that is well grounded by prior empirical evidence and theory. This is because CFA entails more constraints than other approaches, such as EFA (e.g., prespecification of indicator–factor relationships, fixing cross-loadings and error covariances to zero); the researcher is thus required to have a firm substantive and empirical basis to guide the model specification. The CFA specification depicted in Figure 4.1 is likely to be based on a strong conceptual framework (e.g., the Big Five model of personality) and prior research of a more exploratory nature (e.g., EFAs that indicate the number of factors and pattern of indicator–factor loadings, and that may have led to the refinement of an assessment instrument such as removal of poorly behaved indicators).

Defining the Metric of Latent Variables

Model specification also entails defining the metric of the latent variables. It has been shown in Chapter 3 that this is usually accomplished by setting one observed measure on each factor as a marker indicator or by fixing the variance of the factors to a specific value (most commonly to 1.0). In applied research, the marker indicator approach is most frequently used. When this method is used, the researcher must decide which observed measures will serve as marker indicators. In practice, marker indicators are often selected with little consideration or are determined by software defaults (e.g., unless the default is overridden by the user, Mplus automatically selects the first indicator to be the reference indicator; see Figure 4.2). Although this selection may be relatively trivial in some instances (e.g., CFA of a questionnaire with tau-equivalent items), there are many circumstances where marker indicators should be chosen carefully. For example, consider the scenario where a researcher wishes to define a latent construct of Depression with the following three indicators: (1) a single 0–8 clinical rating of depression severity (with little available evidence supporting its interrater reliability); (2) a widely used and psychometrically well-established and normed questionnaire (range of possible scores = 0–64); and (3) a newly developed but psychometrically promising self-report screening measure composed of dimensional ratings of the nine constituent symptoms of *DSM-5* (American Psychiatric Association, 2013) major depression (range of possible scores = 0–36). In this instance, a strong case can be made for using the second measure as the marker indicator: It is the psychometrically strongest measure; it is the most widely known in the applied literature; its metric is most meaningful (e.g., normative data exist); and it yields the most units of discrimination (e.g., 0–64 vs. 0–36

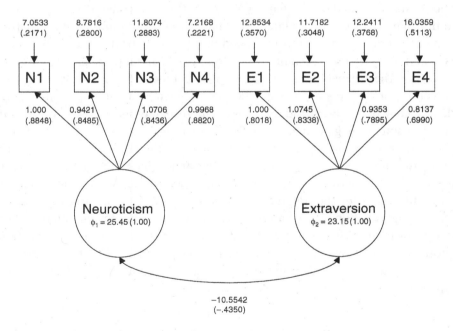

FIGURE 4.2. Unstandardized and completely standardized parameter estimates from the two-factor CFA model of Neuroticism and Extraversion. N1, anxiety; N2, hostility; N3, depression; N4, self-consciousness; E1, warmth; E2, gregariousness; E3, assertiveness; E4, positive emotions. Completely standardized parameter estimates are presented in parentheses. All freely estimated unstandardized parameter estimates are statistically significant ($p < .001$). Estimates obtained from LISREL output in Table 4.4.

or 0–8). As a result of choosing the second indicator, more variance is apt to be passed on to the latent variable of Depression, and the unstandardized solution should be more interpretable. The latter aspect is particularly germane to MIMIC models (Chapter 7) and structural models that may entail regressive effects involving the latent dimension of Depression. For example, a 1-unit increase in the predictor is associated with x-unit change in Depression, and this unstandardized path x is more easily interpreted if the metric of the marker indicator for Depression is widely known and understood.

DATA SCREENING AND SELECTION OF THE FITTING FUNCTION

The vast majority of CFA and SEM studies in the applied research literature are conducted by using maximum likelihood (ML) estimation. As discussed in Chapter 3, ML rests on several key assumptions, including sufficient sample size and the use of indicators that approximate interval-level scales and multivariate normality. In some instances, ML estimation is relatively robust to mild violations in these assumptions (see Chapter 9). Nevertheless, it is important to evaluate the sample size and sample data in regard to their suitability for CFA and ML estimation because marked departures from these assumptions would point to the need for alternative analytic approaches or estimators (e.g., robust ML, weighted least squares [WLS]). Sample size considerations are fully discussed in Chapter 10. The major latent variable software programs contain routines for screening the sample data for univariate and multivariate normality (i.e., skewness and kurtosis) and outliers (i.e., aberrant cases). Broader statistical packages such as SAS and SPSS also have these capabilities (cf. Tabachnick & Fidell, 2013). The specifics of these data-screening procedures are not detailed here, given that this topic falls somewhat outside the scope of this book and that thorough presentations of this issue are available in other SEM sources (e.g., Kline, 2011). Data screening is conducted on the raw sample data. If the data are deemed suitable for ML analysis, the researcher has the option of using the raw data, a correlation matrix, or a variance–covariance matrix as input data for the CFA. If a correlation matrix is used, the indicator standard deviations must also be provided in order for the program to convert the matrix into variances and covariances. If a fitting function other than ML is employed, a simple correlation or covariance matrix cannot be used (e.g., WLS requires that either the raw data or correlation and asymptotic covariance matrices be inputted; see Chapter 9). At this stage of the process, the researcher must also decide how missing data will be handled—a topic discussed in Chapter 9.

RUNNING CFA IN DIFFERENT SOFTWARE PROGRAMS

After model specification and data-screening issues have been settled, the CFA model can be fit to the data. Table 4.1 provides syntax programs for the Figure 4.1 model in LISREL, Mplus, EQS, and SAS/CALIS. The reader is encouraged to focus initially on

TABLE 4.1. Computer Syntax (LISREL, Mplus, EQS, SAS/CALIS) for a Two-Factor CFA Model of Neuroticism and Extraversion

LISREL 9.1

```
TITLE TWO FACTOR MODEL OF NEUROTICISM AND EXTRAVERSION
DA NI=8 NO=250 MA=CM
LA
N1 N2 N3 N4 E1 E2 E3 E4
KM
1.000
 0.767  1.000
 0.731  0.709  1.000
 0.778  0.738  0.762  1.000
-0.351 -0.302 -0.356 -0.318  1.000
-0.316 -0.280 -0.300 -0.267  0.675  1.000
-0.296 -0.289 -0.297 -0.296  0.634  0.651  1.000
-0.282 -0.254 -0.292 -0.245  0.534  0.593  0.566  1.000
SD
5.7 5.6 6.4 5.7 6.0 6.2 5.7 5.6
MO NX=8 NK=2 PH=SY,FR LX=FU,FR TD=SY,FR
LK
NEUROT EXTRAV
PA LX
0 0
1 0
1 0
1 0
0 0
0 1
0 1
0 1
VA 1.0 LX(1,1) LX(5,2)
PA TD
1
0 1
0 0 1
0 0 0 1
0 0 0 0 1
0 0 0 0 0 1
0 0 0 0 0 0 1
0 0 0 0 0 0 0 1
PA PH
1
1 1
OU ME=ML RS MI SC ND=4
```

Mplus 7.11

```
TITLE:      TWO FACTOR MODEL OF NEUROTICISM AND EXTRAVERSION
DATA:       FILE IS NEUROT.DAT;
            TYPE IS STDEVIATIONS CORRELATION;
            NOBSERVATIONS ARE 250;
VARIABLE:   NAMES ARE N1-N4 E1-E4;
ANALYSIS:   ESTIMATOR=ML;
```

(continued)

TABLE 4.1. *(continued)*

```
MODEL:        NEUROT BY N1-N4;
              EXTRAV BY E1-E4;
OUTPUT:       SAMPSTAT MODINDICES(3.84) STANDARDIZED RESIDUAL;
```

EQS 6.2

```
/TITLE
 two-factor model of neuroticism and extraversion
/SPECIFICATIONS
 CASES=250; VARIABLES=8; METHODS=ML; MATRIX=COR; ANALYSIS=COV;
/LABELS
 v1=anxiety; v2=hostil; v3=depress; v4=selfcon; v5=warmth; v6=gregar;
 v7=assert; v8=posemot; f1 = neurot; f2 = extrav;
/EQUATIONS
 V1 =     F1+E1;
 V2 =    *F1+E2;
 V3 =    *F1+E3;
 V4 =    *F1+E4;
 V5 =     F2+E5;
 V6 =    *F2+E6;
 V7 =    *F2+E7;
 V8 =    *F2+E8;
/VARIANCES
 F1 TO F2 = *;
 E1 TO E8 = *;
/COVARIANCES
 F1 TO F2 = *;
/MATRIX
1.000
 0.767  1.000
 0.731  0.709  1.000
 0.778  0.738  0.762  1.000
-0.351 -0.302 -0.356 -0.318  1.000
-0.316 -0.280 -0.300 -0.267  0.675  1.000
-0.296 -0.289 -0.297 -0.296  0.634  0.651  1.000
-0.282 -0.254 -0.292 -0.245  0.534  0.593  0.566  1.000
/STANDARD DEVIATIONS
5.7 5.6 6.4 5.7 6.0 6.2 5.7 5.6
/PRINT
 fit=all;
/LMTEST
/WTEST
/END
```

SAS 9.4 PROC CALIS

```
Title "CFA of Two-Factor Model of Neuroticism and Extraversion";
Data NEO (type=CORR);
 input _TYPE_ $ _NAME_ $ V1-V8;
 label V1 = 'anxiety'
       V2 = 'hostil'
       V3 = 'depress'
       V4 = 'selfcons'
```

(continued)

TABLE 4.1. *(continued)*

```
          V5 = 'warmth'
          V6 = 'gregar'
          V7 = 'assert'
          V8 = 'posemots';
cards;
mean  .    0       0       0       0       0       0       0       0
 std  .   5.7     5.6     6.4     5.7     6.0     6.2     5.7     5.6
   N  .   250     250     250     250     250     250     250     250
corr V1  1.000     .       .       .       .       .       .       .
corr V2  0.767   1.000     .       .       .       .       .       .
corr V3  0.731   0.709   1.000     .       .       .       .       .
corr V4  0.778   0.738   0.762   1.000     .       .       .       .
corr V5 -0.351  -0.302  -0.356  -0.318   1.000     .       .       .
corr V6 -0.316  -0.280  -0.300  -0.267   0.675   1.000     .       .
corr V7 -0.296  -0.289  -0.297  -0.296   0.634   0.651   1.000     .
corr V8 -0.282  -0.254  -0.292  -0.245   0.534   0.593   0.566   1.000
;
run;


proc calis data=NEO cov method=ml pall pcoves;
var = V1-V8;
lineqs
 V1 = 1.0  f1 + e1,
 V2 = lam2 f1 + e2,
 V3 = lam3 f1 + e3,
 V4 = lam4 f1 + e4,
 V5 = 1.0  f2 + e5,
 V6 = lam6 f2 + e6,
 V7 = lam7 f2 + e7,
 V8 = lam8 f2 + e8;
std
  f1-f2 = ph1-ph2,
  e1-e8 = td1-td8;
cov
  f1-f2 = ph3;
run;
```

the longhand, matrix-based LISREL programming, given its direct correspondence to the matrices, symbols, and equations introduced in Chapter 3—for example, lambda-X (LX; Λ_x, λ_x), theta-delta (TD; Θ_δ, δ), and phi (PH; Φ, ϕ). Although the particular aspects of each programming language are not detailed, a few comparisons are merited. In each analysis, anxiety and warmth are used as marker indicators for Neuroticism and Extraversion, respectively. However, in addition to many other syntax differences, the method of setting the marker indicator varies across software programs: (1) LISREL uses the Value (VA) command to fix the unstandardized loading to 1.0; (2) by default, Mplus selects the first indicator listed after the BY keyword as the marker variable (this default can be overridden by additional programming); (3) in EQS, an asterisk (*) is used to specify that a parameter should be freely estimated, and if the asterisk is omitted, the EQS default is to fix the unstandardized parameter to 1.0 (as in Mplus, this and any other system default can be overridden by the user); and (4) in SAS/CALIS, a 1 is placed before the factor name in the equation involving the marker indicator.

Mplus requires the fewest programming lines because the sample data cannot be embedded in the syntax file, and because the program includes a variety of defaults that correspond to frequently employed CFA specifications. For instance, Mplus automatically sets the first observed variable to be the marker indicator and freely estimates the factor loadings for the remaining indicators in the list. By default, all error variances are freely estimated, and all error covariances and indicator cross-loadings are fixed to 0; the factor variances and covariances are also freely estimated by default. These and other convenience features in Mplus are no doubt quite appealing to the experienced latent variable researcher. However, novice users are advised to become fully aware of these system defaults to ensure that their models are specified as intended. Note that starting values have not been provided in any of the syntax files. Thus, by default, each program is instructed to automatically generate initial estimates to begin the iterations to minimize F_{ML}.

MODEL EVALUATION

One of the most important aspects of model evaluation occurs prior to the actual statistical analysis—that is, providing a compelling rationale that the model is meaningful and useful on the basis of prior research evidence and theory. After substantive justification of the model is established, the acceptability of the fitted CFA solution should be evaluated on the basis of three major aspects: (1) overall goodness of fit; (2) the presence or absence of localized areas of strain in the solution (i.e., specific points of ill fit); and (3) the interpretability, size, and statistical significance of the model's parameter estimates. A common error in applied CFA research is to evaluate models exclusively on the basis of overall goodness of fit. However, descriptive fit indices are best viewed as providing information on the extent of a model's *lack of fit*. In other words, although these indices may provide conclusive evidence of a misspecified model, they cannot be used in isolation from other information to support the conclusion of a good-fitting model. Goodness-of-fit indices provide a global descriptive summary of the ability of the model to reproduce the input covariance matrix, but the other two aspects of fit evaluation (localized strain, parameter estimates) provide more specific information about the acceptability and utility of the solution.

Overall Goodness of Fit

Overall goodness of fit in model evaluation has been discussed in Chapter 3, where several goodness-of-fit indices have been recommended (e.g., SRMR, RMSEA, CFI). As noted in Chapter 3, at least one index from each fit class (absolute, parsimony, comparative) should be considered because each provides different information about the fit of the CFA solution. After ensuring that the model has been specified as intended (e.g., verifying model *df* and freely estimated, fixed, and constrained parameters), goodness-of-fit indices are then examined to begin evaluating the acceptability of the model. If these indices are consistent with good model fit, this provides initial (tentative) support for the notion that the model has been properly specified. The remaining aspects of model evaluation can be evaluated in this context; for example, is goodness of fit also supported by the lack of localized areas of strain in the solution? On the other hand, if indices point to poor fit, subsequent aspects of fit evaluation will be focused on diagnosing the sources of model misspecification (e.g., inspection of modification indices and standardized residuals; see next section). In addition, it will be erroneous to interpret the model's parameter estimates (e.g., size and significance of factor loadings and factor correlations) if the solution fit the data poorly because misspecified models produce biased parameter estimates. Occasionally, fit indices will provide inconsistent information about the fit of the model. In these instances, greater caution is needed in determining the acceptability of solution and in detecting the potential sources of misspecification. For example, the SRMR and CFI may suggest fit is acceptable at absolute level and in relation to a null solution, but an RMSEA > .08 may indicate a lack of parsimony (i.e., the researcher will need to look for the inclusion of freely estimated parameters that are unnecessary).

In the Figure 4.1 example, all of the overall goodness-of-fit indices suggest that the two-factor model does fit the data well: $\chi^2(19) = 13.285$, $p = .82$, SRMR = .019, RMSEA = 0.00 (90% CI = 0.00–.034, CFit = .99), TLI = 1.005, CFI = 1.00. Although providing initial support for the acceptability of the two-factor model, this tentative judgment must be verified by considering other aspects of the results.

Localized Areas of Strain

A limitation of goodness-of-fit statistics (e.g., SRMR, RMSEA, CFI) is that they provide a *global*, descriptive indication of the ability of the model to reproduce the observed relationships among the indicators in the input matrix. However, in some instances, overall goodness-of-fit indices suggest acceptable fit despite the fact that some relationships among indicators in the sample data have not been reproduced adequately (or, alternatively, some model-implied relationships may markedly exceed the associations seen in the data). This outcome is more apt to occur in complex models (e.g., models that entail an input matrix consisting of a large set of indicators) where the sample matrix is reproduced reasonably well on the whole, and the presence of a few poorly reproduced relationships has less impact on the global summary of model fit (as reflected by overall goodness-of-fit statistics). On the other hand, overall goodness-of-fit indices may indicate a model poorly reproduced the sample matrix. However, these indices do not provide information on the reasons why the model fits the data poorly (various forms of model misspecification are considered in Chapter 5). Two statistics that are frequently used to identify focal areas of misfit in a CFA solution are *residuals* and *modification indices*.

Residuals

As discussed in Chapter 3 (e.g., see Figure 3.8), there are three matrices associated with the typical CFA model: the sample variance–covariance matrix (S); the model-implied (predicted) variance–covariance matrix (Σ); and the residual variance–covariance matrix, which reflects the difference between the sample and model-implied matrices (i.e., residual matrix = $S - \Sigma$). Table 4.2 presents these matrices for the two-factor model depicted in Figure 4.1. Although a correlation matrix and indicator *SD*s are used as input (see Table 4.1), the latent variable software program uses this information to create the sample variance–covariance matrix; for example, the covariance of N1 and N2 = .767(5.7)(5.6) = 24.48. As discussed in Chapter 3, the predicted (model-implied) variance–covariance matrix is generated on the basis of the parameter estimates obtained in the minimization of F_{ML} (or some other fitting function such as WLS). For example, as will be shown later in this chapter, the completely standardized factor loadings of N1 and N2 are estimated to be .8848 and .8485, respectively. The model-implied correlation between these two indicators is .751; that is, $\sigma_{21} = \lambda_{x11}\phi_{11}\lambda_{x21} = .8848(1)(.8485)$. Thus the predicted covariance of N1 and N2 is 23.97; that is, .751(5.7)(5.6) (cf. Table 4.2). The fitted residual for the N1 and N2 is .518, the difference between their sample and model-implied covariance; using the unrounded values from Table 4.2: 24.4826 – 23.965 = .5176.

Thus, while goodness-of-fit statistics such as the SRMR provide a global summary of the difference between the sample and model-implied matrices (see Chapter 3), the residual matrix provides specific information about how well each variance and each covariance are reproduced by the model's parameter estimates. Indeed, there exists one residual for each indicator and each pair of indicators. However, fitted residuals can be difficult to interpret because they are affected by the raw metric and dispersion of the observed measures. It is particularly difficult to determine whether a fitted (unstandardized) residual is large or small when the units of measurement of the indicators are markedly disparate. This problem is addressed by *standardized residuals*, which are computed by dividing the fitted residuals by their estimated standard errors.[2] Accordingly, standardized residuals are analogous to standard scores in a sampling distribution and can thus be interpreted along the lines of z scores. Stated another way, these values can be conceptually considered as the number of standard deviations by which the fitted residuals differ from the zero-value residuals associated with a perfectly fitting model.

The standardized residuals of the two-factor CFA model are presented in Table 4.2. As this table shows, standardized residuals can have either positive or negative values. A *positive* standardized residual suggests that the model's parameters *underestimate* the zero-order relationship between two indicators to some degree. For example, the standardized residual for the N1–N2 relationship is 1.57; the sign of this residual is consistent with the fact that the sample covariance of these indicators (24.48) is larger than the model-implied covariance (23.96). Large, positive standardized residuals may indicate that additional parameters are needed in the model to better account for the covariance between the indicators. Conversely, a *negative* standardized residual suggests that the model's parameters *overestimate* the relationship between two indicators to some extent. As seen in Table 4.2, the standardized residual for the N1–N3 relationship is –1.46, due to the fact that the sample covariance (26.67) is smaller than the covariance predicted by the two-factor model (27.23).

Of course, CFA models specified in applied research will rarely produce standardized residuals that uniformly approximate zero. Thus the question arises as to how large standardized residuals should be to be considered salient. Because standardized residuals can be roughly interpreted as *z* scores, the *z*-score values that correspond to conventional statistical significance levels are often employed as practical cutoffs. For instance, researchers may scan for standardized residuals that are equal to or greater than the absolute value of 1.96 because this value corresponds to a statistically significant *z* score at $p < .05$ (although in practice, this critical value is often rounded up to 2.00). However, the size of standardized residuals is influenced by sample size. In general, larger *N*s are associated with larger standardized residuals because the size of the standard errors of the fitted residuals is often inversely related to sample size. For this reason, some methodologists recommend the use of larger cutoff values (e.g., 2.58, which corresponds to the .01 alpha level; Byrne, 2014). In any case, the researcher should be mindful of the potential impact of sample size when interpreting the salience of the standardized residual. Other information provided by the software output, such as expected parameter change, can often assist in these considerations (see next section). While a cutoff of 2.00 or 2.58 provides a general guideline for this aspect of model fit diagnostics, the researcher should especially look for standardized residuals with outlying values. This process is fostered in the LISREL software by the output of stemleaf and Q plots, which can also be useful for detecting non-normality or nonlinear associations in the indicators, in addition to specification errors.

As seen in Table 4.2, the standardized residuals range from –1.65 to 1.87. Using the guidelines just presented, this outcome would be consistent with the conclusion for the absence of localized areas of ill fit in the solution. No relationships among the indicators are substantially under- or overestimated by the model's parameter estimates.

Modification Indices

Another aspect of model evaluation that focuses on specific relationships in the solution is the *modification index* (Sörbom, 1989; referred to as the univariate *Lagrange multiplier* in the EQS program). Modification indices can be computed for each fixed parameter (e.g., parameters that are fixed to zero, such as indicator cross-loadings and error covariances) and constrained parameter in the model (e.g., parameters that are constrained to equal magnitudes, as in the test of tau equivalence; see Chapter 7). The modification index reflects an *approximation* of how much the overall model χ^2 will decrease if the fixed or constrained parameter is freely estimated. Indeed, if the parameter is freely estimated in a subsequent analysis, the actual decrease in model χ^2 may be somewhat smaller or larger than the value of the modification index. In other words, the modification index is roughly equivalent to the difference in the overall χ^2 between two models

TABLE 4.2. Sample, Model-Implied, Fitted Residual, and Standardized Residual Matrices for the Two-Factor CFA Model of Neuroticism and Extraversion

Sample variance–covariance matrix (S)

	N1	N2	N3	N4	E1	E2	E3	E4
N1	32.4900							
N2	24.4826	31.3600						
N3	26.6669	25.4106	40.9600					
N4	25.2772	23.5570	27.7978	32.4900				
E1	-12.0042	-10.1472	-13.6704	-10.8756	36.0000			
E2	-11.1674	-9.7216	-11.9040	-9.4358	25.1100	38.4400		
E3	-9.6170	-9.2249	-10.8346	-9.6170	21.6828	23.0063	32.4900	
E4	-9.0014	-7.9654	-10.4653	-7.8204	17.9424	20.5890	18.0667	31.3600

Model-implied variance–covariance matrix (Σ; also referred to as the fitted or predicted matrix)

	N1	N2	N3	N4	E1	E2	E3	E4
N1	32.4900							
N2	23.9650	31.3600						
N3	27.2314	25.6558	40.9600					
N4	25.3548	23.8878	27.1437	32.4900				
E1	-10.5542	-9.9436	-11.2988	-10.5202	36.0000			
E2	-11.3400	-10.6839	-12.1401	-11.3035	24.8700	38.4400		
E3	-9.8715	-9.3003	-10.5680	-9.8397	21.6493	23.2613	32.4900	
E4	-8.5876	-8.0907	-9.1934	-8.5599	18.8335	20.2358	17.6153	31.3600

Fitted residual matrix (also referred to as the unstandardized residual matrix)

	N1	N2	N3	N4	E1	E2	E3	E4
N1	0.0000							
N2	0.5177	0.0000						
N3	-0.5645	-0.2452	0.0000					
N4	-0.0776	-0.3309	0.6541	0.0000				
E1	-1.4500	-0.2036	-2.3716	-0.3554	0.0000			
E2	0.1726	0.9623	0.2361	1.8678	0.2400	0.0000		
E3	0.2545	0.0755	-0.2666	0.2227	0.0335	-0.2550	0.0000	
E4	-0.4139	0.1253	-1.2718	0.7395	-0.8911	0.3531	0.4515	0.0000

Standardized residual matrix

	N1	N2	N3	N4	E1	E2	E3	E4
N1	0.0000							
N2	1.5737	0.0000						
N3	-1.4640	-0.5189	0.0000					
N4	-0.2804	-0.9866	1.6636	0.0000				
E1	-1.2152	-0.1621	-1.6374	-0.2961	0.0000			
E2	0.1509	0.7872	0.1673	1.6208	0.5890	0.0000		
E3	0.2187	0.0619	-0.1898	0.1904	0.0710	-0.6210	0.0000	
E4	-0.3123	0.0922	-0.8152	0.5560	-1.3707	0.6172	0.6928	0.0000

Note. Matrices obtained from LISREL 9.1 printout.

where in one model, the parameter is fixed or constrained, and in the other model, the parameter is freely estimated. Thus modification indices are analogous to the χ^2 difference (with a single df) of *nested models* (nested models have been introduced in Chapter 3, and are discussed more fully in subsequent chapters—e.g., Chapters 5 and 7).

To illustrate this concept, modification indices from the Figure 4.1 solution are presented in Table 4.3. Consider the modification index for the E1 indicator (1.32) listed under the heading "Modification indices for Lambda-X." This value indicates that if the E1 indicator is freely estimated to cross-load on the Neuroticism factor, the overall model χ^2 is estimated to drop by 1.32 units (e.g., from 13.285 to 11.96). If this cross-loading is in fact specified in a subsequent analysis, the model df will decrease from 19 to 18 because an additional parameter has been freely estimated (Neuroticism $\rightarrow$ E1). Therefore, this modification index estimates a χ^2 difference (with 1 df) of 1.32 if this cross-loading is freely estimated.

In general, a good-fitting model should also produce modification indices that are small in magnitude. Because the modification index can be conceptualized as a χ^2 statistic with 1 df, indices of 3.84 or greater (which reflects the critical value of χ^2 at $p < .05$, 1 df) suggest that the overall fit of the model can be significantly improved ($p < .05$) if the fixed or constrained parameter is freely estimated (although in practice, this critical value is often rounded up to 4.00; cf. Jaccard & Wan, 1996). Thus, for example, the modification index corresponding to the E1 cross-loading (1.32) argues against freely estimating this parameter because this respecification will not result in a significant improvement in model fit.

Like overall model χ^2 and standardized residuals, modification indices are sensitive to sample size. For instance, when N is very large, a large modification index may suggest the need to add a given parameter—despite the fact that the magnitude of the parameter in question, if freely estimated, is rather trivial (e.g., a large modification index associated with a completely standardized cross-loading of .10). To address this problem, latent variable software programs provide *expected parameter change* (EPC) values for each modification index. The EPC values provided by these programs may be unstandardized, standardized, or completely standardized. Some programs, such as Mplus and LISREL, provide all three forms of EPC values. As the name implies, EPC values provide an approximation of how much a parameter is expected to change in a positive or negative direction if it is freely estimated in a subsequent analysis. Unstandardized EPC values are proportional to the scale of the observed measures in question. Thus fit diagnostics typically focus on completely standardized EPC values. For example, as shown in Table 4.3, the completely standardized cross-loading of E1 onto Neuroticism is predicted to be –.06, which further supports the contention that warmth (E1) has no meaningful direct relationship with the dimension of Neuroticism. Especially when sample size is large, the size and direction of EPC values should be considered in tandem with modification indices to assist in the determination of whether the respecification is conceptually and statistically viable (cf. Kaplan, 1989, 1990).

Indeed, a key principle of model respecification is that modification indices and standardized residuals should only prompt the researcher to relax a fixed or constrained parameter when there exists a compelling substantive basis for doing so (i.e., freeing the parameter is supported by empirical, conceptual, or practical considerations) and when the EPC value of the parameter can be clearly interpreted. Especially when sample size is large, modification indices and standardized residuals will often indicate the presence of parameters that, if freed, will improve the fit of the model. However, such parameters should not be freed with the sole intent of improving model fit. Rather, all model respecifications must be justified on the basis of prior research or theory.

Several studies have highlighted the problems that occur when models are respecified solely on the basis of modification indices or standardized residuals. For example, MacCallum (1986) examined how often and under what conditions that *specification searches* (i.e., post hoc revisions to a fitted model as determined by modification indices) are likely to lead to the discovery of the correct population model. This question was examined by generating simulated data for which the "true" model was known, fitting various misspecified models to the data, and then determining whether the specification search (e.g., freeing parameters with the highest modification indices one at a time in sequential order) led to finding the correct model. In many instances, the model revisions implied by the specification searches were incorrect; for example, they led to model specifications that did not correspond closely to the true model. This was particularly apparent when the initial model was markedly misspecified and when sample size was modest (e.g., $N = 100$). In a follow-up to this study, Silvia and MacCallum (1988) found that restricting modifications to those that could be justified on the basis of prior theory greatly improved the success of the specification search.

For these reasons, Jöreskog (1993) has recommended that model modification begin by freely estimating the fixed or constrained parameter with the largest modification index (and EPC) *if this parameter can be interpreted substantively*. If there does not exist a substantive basis for relaxing the parameter with the largest modification index, the researcher should consider the parameter associated with the second largest modification index, and so on. However, it is important to note that a poorly reproduced relationship may be reflected by high modification indices in more than one matrix (e.g., suggestive of a cross-loading in lambda-X; suggestive of an error covariance in theta-delta). Thus several high modification indices may be remedied by freeing a single parameter. Because the model-implied covariance matrix (Σ) is dependent on all model parameters, there are countless ways a single model revision can affect Σ. Therefore, revisions of a model should always focus exclusively on parameters justified by prior evidence or theory. This point is further underscored by the fact that some modification indices output by latent variable software programs appear nonsensical (e.g., a modification index suggesting a direct effect of a disturbance on a latent variable, or a method effect involving the measurement error of a single indicator).

Other potential adverse consequences of atheoretical specification searches are overfitting (i.e., adding unnecessary parameters to the model) and capitalization on chance associations in the sample data (i.e., accounting for weak effects in the data that stem from sampling error and are not apt to be replicated in independent data sets). MacCallum, Roznowski, and Necowitz (1992) have argued against modifying a good-

TABLE 4.3. Modification Indices and (Completely Standardized) Expected Parameter Change Values for the Two-Factor CFA Model of Neuroticism and Extraversion

Modification indices for LAMBDA-X

	NEUROT	EXTRAV
N1	--	0.2719
N2	--	0.1422
N3	--	0.8988
N4	--	1.0607
E1	1.3206	--
E2	1.1146	--
E3	0.0143	--
E4	0.0125	--

Completely standardized expected change for LAMBDA-X

	NEUROT	EXTRAV
N1	--	-0.0226
N2	--	0.0175
N3	--	-0.0443
N4	--	0.0448
E1	-0.0613	--
E2	0.0551	--
E3	0.0064	--
E4	-0.0065	--

No nonzero modification indices for PHI

Modification indices for THETA-DELTA

	N1	N2	N3	N4	E1	E2	E3	E4
N1	- -							
N2	3.2020	- -						
N3	2.7271	0.3096	- -					
N4	0.1201	1.2434	3.4815	- -				
E1	0.2541	0.6555	1.0965	0.0000	- -			
E2	0.4975	0.0043	0.1069	1.2708	0.4235	- -		
E3	0.9158	0.2207	0.2526	0.8349	0.0057	0.4580	- -	
E4	0.0162	0.0021	0.5849	0.4783	1.9711	0.4092	0.5005	- -

Completely standardized expected change for THETA-DELTA

	N1	N2	N3	N4	E1	E2	E3	E4
N1	- -							
N2	0.0482	- -						
N3	-0.0443	-0.0147	- -					
N4	-0.0097	-0.0300	0.0499	- -				
E1	-0.0119	0.0206	-0.0269	-0.0001	- -			
E2	-0.0160	-0.0016	0.0081	0.0257	0.0287	- -		
E3	0.0229	-0.0121	0.0131	-0.0220	0.0031	-0.0293	- -	
E4	-0.0034	0.0013	-0.0220	0.0184	-0.0545	0.0252	0.0274	- -

Maximum modification index is 3.48 for element (4, 3) of THETA-DELTA

Note. Matrices obtained from LISREL 9.1 printout. For space reasons, only completely standardized expected parameter change values are presented.

fitting model to achieve even better fit because this practice is likely to be "fitting small idiosyncratic characteristics of the sample" (p. 501). Thus sampling error, in tandem with the sensitivity of modification indices and standardized residuals to sample size, heightens the risk for model respecifications of a trivial nature. Researchers are perhaps most prone to model overfitting in instances where overall goodness-of-fit indices are on the border of conventional guidelines for acceptable fit. For example, as the result of examining modification indices, a researcher may determine that freeing some correlated errors in a CFA solution will improve a "borderline" value of a given fit index. However, in addition to lacking a substantive basis, this practice may introduce other problems such as biasing other parameters in the model (e.g., the magnitude of the factor loadings) and their standard errors.

Even when revisions to the initial solution can be defended with compelling empirically or theoretically based arguments, it is important to note that by pursuing respecifications of the initial model, one has moved out of a CFA framework. Indeed, as the name implies, a "specification search" (MacCallum, 1986) entails an exploratory endeavor to determine the nature of the misfit of the initial, hypothesized model. Accordingly, respecified models should be interpreted with caution. Especially in instances where substantial changes have been made to the initial model, modified solutions should be replicated in independent samples. Or if the original sample is sufficiently large, the sample can be split into derivation and holdback samples to cross-validate the revised model.

Unnecessary Parameters

Thus far, the discussion of model modification has focused primarily on the issue of adding new freely estimated parameters to the initial model. Although this is the most common form of model respecification, models can be revised by eliminating statistically nonsignificant parameters. It has been noted earlier that the presence of unnecessary parameters may be reflected by large, negative standardized residuals, which indicate that the model is overestimating the observed relationship between a pair of indicators. A similar statistic is the *univariate Wald test*. Converse to the modification index (i.e., the univariate Lagrange multiplier), the univariate Wald test provides an estimate of how much the overall model χ^2 will increase if a freely estimated parameter is fixed to zero.[3] A nonsignificant Wald test value (e.g., <3.84 in the case of a single parameter) indicates that removing the freely estimated parameter (e.g., fixing it to zero) will not result in a significant decrease in model fit. The necessity of existing parameter estimates can be evaluated by examining their statistical significance. As discussed in the next section, the statistical significance of a freely estimated parameter is indicated by a test statistic (which can be interpreted as a z statistic) that is calculated by dividing the parameter estimate by its standard error. At the .05 alpha level (two-tailed), parameter estimates associated with z values of ±1.96 or greater are statistically significant. Thus parameter estimates with z values less than 1.96 are statistically nonsignificant and may be considered unnecessary to the solution. As noted in Chapter 3, the z and χ^2 distributions are closely related; for example, the critical value of χ^2 with 1 df, $\alpha = .05$ is 3.842, which equals the squared critical value ($\alpha = .05$) of z (i.e., $1.96^2 = 3.842$). Accordingly, although the univariate Wald test is not provided by every program, it can be calculated from the output of all software packages by squaring the z value associated with the unstandardized parameter estimate (i.e., the squared z value provides an estimate of how much model χ^2 will increase if the freed parameter is removed). However, CFA and SEM users should be aware that the results of Wald tests and parameter significance tests are sensitive to how the metric of the latent variable is identified (e.g., selection of marker indicator, fixing variance of the latent variable to 1.0). This issue is discussed in Appendix 4.1.

In structural models, Wald tests suggest that some directional paths among latent variables may not be necessary. In CFA measurement models, nonsignificant parameters or Wald tests may point to a variety of model modifications, such as the removal of indicators (e.g., nonsignificant factor loadings) or specific parameters that have been freely estimated in the initial solution (e.g., correlated errors, cross-loadings, factor correlations). The various forms of CFA model misspecification are discussed in Chapter 5.

Interpretability, Size, and Statistical Significance of the Parameter Estimates

Returning to the Figure 4.1 example, recall that each of the overall goodness-of-fit indices suggests that the two-factor model does fit the data well. Goodness of model fit is further verified by the absence of large modification indices and standardized residuals (see Tables 4.2 and 4.3), thereby indicating no focal areas of ill fit in the solution. Thus model evaluation can proceed to inspecting the direction, magnitude, and significance of the parameter estimates—namely, the factor loadings, factor variances and covariance, and indicator errors. These results are presented in Table 4.4 (selected LISREL output) and Table 4.5 (selected Mplus output). The discussion of results to follow is taken primarily from the LISREL output in Table 4.4, although some comparisons with the Mplus output are also noted.

An initial step in this process is to determine whether the parameter estimates make statistical and substantive sense. From a statistical perspective, the parameter estimates should *not take on out-of-range values* such as completely standardized factor correlations that exceed 1.0, negative factor variances, or negative indicator error variances.[4] In Chapter 3 it has been noted that such out-of-range values, which are often referred to as *Heywood cases* (or *offending estimates*), may be indicative of model specification error or problems with the sample or model-implied matrices (e.g., a nonpositive definite matrix, small N). Thus the model and sample data must be viewed with caution to rule out the existence of more serious causes of these outcomes. A classic reference for dealing with improper solutions and nonpositive definite matrices is Wothke (1993). This issue is also discussed at length in Chapter 5. As seen in Table 4.4, all parameter estimates from the Figure 4.1 solution are statistically viable (e.g., no negative variances). Readers will note that the estimate of the factor covariance/correlation is negative (e.g., completely standardized estimate = −.435; see the Phi matrix in the "Completely Standardized Solu-

TABLE 4.4. Selected Results from the Two-Factor CFA Model of Neuroticism and Extraversion (LISREL 9.1)

```
LISREL Estimates (Maximum Likelihood
        LAMBDA-X

              NEUROT      EXTRAV
            --------    --------
    N1       1.0000       - -

    N2       0.9421       - -
            (0.0523)
            18.0173

    N3       1.0706       - -
            (0.0601)
            17.8267

    N4       0.9968       - -
            (0.0515)
            19.3511

    E1        - -        1.0000

    E2        - -        1.0745
                        (0.0786)
                        13.6635

    E3        - -        0.9353
                        (0.0722)
                        12.9530

    E4        - -        0.8137
                        (0.0722)
                        11.2697

        PHI

              NEUROT      EXTRAV
            --------    --------
NEUROT      25.4367
            (2.9058)
             8.7536

EXTRAV     -10.5542     23.1466
            (1.9236)    (3.1940)
            -5.4866      7.2469

        THETA-DELTA
```

	N1	N2	N3	N4	E1	E2
	--------	--------	--------	--------	--------	--------
	7.0533	8.7816	11.8074	7.2168	12.8534	11.7182
	(0.9087)	(1.0010)	(1.3307)	(0.9180)	(1.5837)	(1.6054)
	7.7619	8.7727	8.8729	7.8613	8.1162	7.2994

(continued)

TABLE 4.4. *(continued)*

	E3	E4
	12.2411	16.0359
	(1.4612)	(1.6693)
	8.3776	9.6063

Squared Multiple Correlations for X - Variables

	N1	N2	N3	N4	E1	E2
	0.7829	0.7200	0.7117	0.7779	0.6430	0.6952

	E3	E4
	0.6232	0.4887

Standardized Solution

LAMBDA-X

	NEUROT	EXTRAV
N1	5.0435	- -
N2	4.7517	- -
N3	5.3993	- -
N4	5.0272	- -
E1	- -	4.8111
E2	- -	5.1693
E3	- -	4.4999
E4	- -	3.9146

PHI

	NEUROT	EXTRAV
NEUROT	1.0000	
EXTRAV	-0.4350	1.0000

Completely Standardized Solution

LAMBDA-X

	NEUROT	EXTRAV
N1	0.8848	- -
N2	0.8485	- -
N3	0.8436	- -
N4	0.8820	- -
E1	- -	0.8018
E2	- -	0.8338
E3	- -	0.7895
E4	- -	0.6990

(continued)

TABLE 4.4. *(continued)*

PHI

	NEUROT	EXTRAV
NEUROT	1.0000	
EXTRAV	-0.4350	1.0000

THETA-DELTA

N1	N2	N3	N4	E1	E2
0.2171	0.2800	0.2883	0.2221	0.3570	0.3048

THETA-DELTA

E3	E4
0.3768	0.5113

Note. LISREL estimates = unstandardized parameter estimates. LISREL 9.1 output.

tion" section of Table 4.4). However, the sign of this relationship is in accord with theory and the scaling of the latent variables; that is, Neuroticism is inversely related to Extraversion. Moreover, the unstandardized factor loadings of N1 and E1 are 1.0 because these observed measures are used as marker indicators; their loadings are fixed to 1.0 in order to pass the metric of N1 and E1 on to the Neuroticism and Extraversion factors, respectively. Consequently, the standard errors of these estimates are 0.0, and thus no significance test (i.e., z value) is available for the marker indicators (see the "LISREL Estimates" section of Table 4.4).

From a substantive perspective, the *directions of the parameter estimates* provided in Table 4.4 are in accord with prediction. The N1, N2, N3, and N4 indicators are positively related to the latent construct of Neuroticism, and the E1, E2, E3, and E4 indicators are positively related to the latent construct of Extraversion. For example, the unstandardized factor loading for N2 (hostility) is 0.942, which can be interpreted as indicating that a 1-unit increase in the latent dimension of Neuroticism is associated with a 0.942-unit increase in the observed measure of hostility (see Table 4.4, under the "LISREL Estimates" heading).

The results provided in Table 4.4 also indicate that every freely estimated parameter is *statistically significant*. As noted earlier, statistical significance is determined by dividing the parameter estimate by its standard error (standard errors are provided in parentheses under the unstandardized parameter estimates in the "LISREL Estimates" section of Table 4.4); for a discussion of how standard errors are calculated in ML estimation, see Kaplan (2000). Because this ratio can be interpreted as a z score, ±1.96 would be the critical value at an alpha level of .05 (two-tailed). For example, the z value for the unstandardized E2 factor loading is 13.64 (i.e., 1.074 / .079 = 13.66; see the "LIS-

TABLE 4.5. Selected Results from the Two-Factor CFA Model of Neuroticism and Extraversion (Mplus 7.11)

MODEL RESULTS

	Estimate	S.E.	Est./S.E.	Two-Tailed P-Value
NEUROT BY				
N1	1.000	0.000	999.000	999.000
N2	0.942	0.052	17.981	0.000
N3	1.071	0.060	17.791	0.000
N4	0.997	0.052	19.312	0.000
EXTRAV BY				
E1	1.000	0.000	999.000	999.000
E2	1.074	0.079	13.636	0.000
E3	0.935	0.072	12.927	0.000
E4	0.814	0.072	11.247	0.000
EXTRAV WITH				
NEUROT	-10.512	1.920	-5.476	0.000
Variances				
NEUROT	25.335	2.900	8.736	0.000
EXTRAV	23.054	3.188	7.232	0.000
Residual Variances				
N1	7.025	0.907	7.746	0.000
N2	8.747	0.999	8.755	0.000
N3	11.760	1.328	8.855	0.000
N4	7.188	0.916	7.846	0.000
E1	12.802	1.581	8.100	0.000
E2	11.671	1.602	7.285	0.000
E3	12.192	1.458	8.361	0.000
E4	15.972	1.666	9.587	0.000

STANDARDIZED MODEL RESULTS

STDYX Standardization

	Estimate	S.E.	Est./S.E.	Two-Tailed P-Value
NEUROT BY				
N1	0.885	0.018	49.547	0.000
N2	0.849	0.021	39.952	0.000
N3	0.844	0.022	38.868	0.000
N4	0.882	0.018	48.691	0.000
EXTRAV BY				
E1	0.802	0.029	27.203	0.000
E2	0.834	0.027	30.772	0.000
E3	0.789	0.030	25.931	0.000
E4	0.699	0.038	18.478	0.000

(continued)

TABLE 4.5. *(continued)*

	Estimate	S.E.	Est./S.E.	P-Value
EXTRAV WITH				
NEUROT	-0.435	0.059	-7.410	0.000
Variances				
NEUROT	1.000	0.000	999.000	999.000
EXTRAV	1.000	0.000	999.000	999.000
Residual Variances				
N1	0.217	0.032	6.869	0.000
N2	0.280	0.036	7.770	0.000
N3	0.288	0.037	7.871	0.000
N4	0.222	0.032	6.952	0.000
E1	0.357	0.047	7.553	0.000
E2	0.305	0.045	6.747	0.000
E3	0.377	0.048	7.838	0.000
E4	0.511	0.053	9.668	0.000

STD Standardization

	Estimate	S.E.	Est./S.E.	Two-Tailed P-Value
NEUROT BY				
N1	5.033	0.288	17.472	0.000
N2	4.742	0.290	16.337	0.000
N3	5.389	0.333	16.190	0.000
N4	5.017	0.289	17.381	0.000
EXTRAV BY				
E1	4.801	0.332	14.465	0.000
E2	5.159	0.337	15.294	0.000
E3	4.491	0.317	14.150	0.000
E4	3.907	0.326	11.974	0.000
EXTRAV WITH				
NEUROT	-0.435	0.059	-7.410	0.000
Variances				
NEUROT	1.000	0.000	999.000	999.000
EXTRAV	1.000	0.000	999.000	999.000
Residual Variances				
N1	7.025	0.907	7.746	0.000
N2	8.747	0.999	8.755	0.000
N3	11.760	1.328	8.855	0.000
N4	7.188	0.916	7.846	0.000
E1	12.802	1.581	8.100	0.000
E2	11.671	1.602	7.285	0.000
E3	12.192	1.458	8.361	0.000
E4	15.972	1.666	9.587	0.000

(continued)

TABLE 4.5. *(continued)*

R-SQUARE

Observed Variable	Estimate	S.E.	Est./S.E.	Two-Tailed P-Value
N1	0.783	0.032	24.774	0.000
N2	0.720	0.036	19.976	0.000
N3	0.712	0.037	19.434	0.000
N4	0.778	0.032	24.345	0.000
E1	0.643	0.047	13.602	0.000
E2	0.695	0.045	15.386	0.000
E3	0.623	0.048	12.966	0.000
E4	0.489	0.053	9.239	0.000

Note. First set of results under the "Model Results" heading = unstandardized parameter estimates; STDYX, completely standardized estimates; STDX, partially standardized estimates (latent variable is standardized, indicator is unstandardized, STDY solution has been omitted); S.E., standard error; Est./S.E., test statistic (*z* value). Mplus 7.11 output.

REL Estimates" section of Table 4.4), which can thus be interpreted as indicating that the gregariousness indicator loads significantly on the latent variable of Extraversion. In addition to the factor loadings, the factor covariance (Extraversion with Neuroticism), factor variances, and indicator error variances (provided in the Phi and Theta-Delta matrices in the "LISREL Estimates" section of Table 4.4) all differ significantly from zero (but see Appendix 4.1).[5]

The *z* tests of the statistical significance of the unstandardized parameter estimates cannot be relied upon to determine the statistical significance of corresponding (completely) standardized estimates (see Bollen, 1989, for discussion). Although not identical, the significance test results for standardized and unstandardized parameter estimates are usually close; that is, especially in the case of strong effects, these tests will more often than not yield the same conclusion regarding statistical significance. If significance tests of standardized coefficients are desired, standard errors of these estimates must be estimated (from which *z*-test statistics can be computed). Most of the major latent variable software programs do not readily provide the standard errors of partially or completely standardized coefficients, although an exception is the Mplus program (see Table 4.5).

In addition, it is important to evaluate the *standard errors of the parameter estimates* to determine if their magnitude is appropriate, or problematically too large or too small. Standard errors represent estimates of how much sampling error is operating in the model's parameter estimates (i.e., how closely the model's parameter estimates approximate the true population parameters). Stated another way, standard errors provide an estimate of how stable the model's parameter estimates would be if we were able to fit the model repeatedly by taking multiple samples from the population of interest. The 95% confidence interval of an unstandardized parameter estimate can be calculated by adding and subtracting the estimate by the product of 1.96 times the standard error. For example, the 95% confidence interval of the E2 factor loading is .919–1.229—that is, 1.074 ± 1.96 (.079) (computed from values obtained in Table 4.4). This is interpreted as indicating we can be 95% certain that the true population value of this parameter is between 0.919 and 1.229.[6]

Although small standard errors may imply considerable precision in the estimate of the parameter, the significance test of the parameter (i.e., z statistic) cannot be calculated if the standard error approximates zero. Conversely, excessively large standard errors indicate problematically imprecise parameter estimates (i.e., very wide confidence intervals) and hence are associated with low power to detect the parameter as statistically significant from zero. Problematic standard errors may stem from a variety of difficulties, such as a misspecified model, small sample size, or use of non-normal data; an improper estimator or matrix type; or some combination. Unfortunately, there are no specific guidelines available to assist the researcher in determining if the magnitude of standard errors is problematic in a given data set. This is because the size of standard errors is determined in part by the metric of the indicators and latent variables (cf. Appendix 4.1), and in part by the size of the actual parameter estimate, which can vary from data set to data set. However, keeping the metric of the variables in mind, the researcher should be concerned about standard errors that have standout values or approach zero, as well as parameter estimates that appear reasonably large but are not statistically significant.

Assuming that the problems related to insufficient sample size and inappropriate standard errors can be ruled out, the researcher must consider parameter estimates in the model that fail to reach statistical significance (further reflected by Wald tests; see preceding section). For example, a nonsignificant factor loading in a congeneric CFA solution indicates that an observed measure is not related to its purported latent dimension, and typically suggests that the indicator should be eliminated from the measurement model. In a noncongeneric solution, a nonsignificant cross-loading suggests that this parameter is not important to the model and can be dropped. Likewise, a nonsignificant error covariance suggests that the parameter does not assist in accounting for the relationship between two indicators (beyond the covariance explained by the factors). As in model respecifications where parameters are added to the initial solution, nonsignificant parameters should be deleted from the model one at a time as guided by a conceptual or empirical rationale. A factor variance that does not significantly differ from zero typically signals significant problems in the solution, such as the use of a marker variable that does not have a relationship with the factor (another potential consequence of this is empirical underidentification; see Chapter 3), substantial non-normality in the input matrix, or use of a sample that is too small. Whether or not nonsignificant factor covariances (correlations) should be of concern depends on the theoretical context of the CFA solution (e.g., is the lack of overlap among factors in accord with prediction?). Error variances are inversely related to the size of their respective factor loadings; that is, the more indicator variance explained by a factor, the smaller the error variance will be. Thus, assuming no other problems with the solution (e.g., a leptokurtotic indicator, small N), nonsignificant error variances should not prompt remedial action and in fact may signify that an indicator is very strongly related to its purported factor. In applied research, however, indicator error variances almost always differ significantly from zero because an appreciable portion of an indicator's variance is usually not explained by the factor, in tandem with the use of large sample sizes (see below).

The acceptability of parameter estimates should not be determined solely on the basis of their direction and statistical significance. Because CFA is typically conducted in large samples, the analysis is often highly powered to detect rather trivial effects as statistically significant. Thus it is important not only to demonstrate that the specified model reproduces the relationships in the input data well, but that the resulting parameter estimates are of a *magnitude that is substantively meaningful*. For an illustration of how goodness of model fit may be unrelated to the reasonability of the CFA parameter estimates, see Appendix 4.2. For example, the size of the factor loadings should be considered closely to determine if all indicators can be regarded as reasonable measures of their latent constructs. The issue of what constitutes a sufficiently large parameter estimate varies across empirical contexts. For example, in applied factor analytic research of questionnaires, completely standardized factor loadings of .30 (or .40) and above are commonly used to operationally define a "salient" factor loading or cross-loading. However, such guidelines may be viewed as too liberal in many forms of CFA research, such as construct validation studies where scale composite scores, rather than individual items, are used as indicators.

In CFA models where there are no cross-loading indicators, the completely standardized factor loading can be interpreted as the correlation between the indicator and the factor.[7] Accordingly, squaring the completely standardized factor loading provides the proportion of variance of the indicator that is explained by the factor—that is, a *communality* (Eq. 3.5, Chapter 3). For example, the completely standardized loading of N4 (self-consciousness) is estimated to be .882, which can be interpreted as indicating that a standard score increase in Neuroticism is associated with a .882 standard score increase in self-consciousness. The squared factor loading is .778, indicating that 77.8% of the variance in the observed measure of self-consciousness is accounted for by the latent variable of Neuroticism. These values are provided in the "Squared Multiple Correlations" section of the LISREL output (Table 4.4), and under the "R-Square" heading in the Mplus output (Table 4.5). Especially in the context of psychometric research (e.g., evaluating the latent structure of a test instrument), these squared factor loadings can be considered as estimates of the indicator's *reliability*—that is, the proportion of the indicator's variance that is estimated to be *true-score variance* (see Chapter 8 for further discussion). Accordingly, the proportion of explained variance in the indicators can be quite useful in formulating conclusions about whether the measures are meaningfully related to their purported latent dimensions.

As noted earlier, the interpretability of the size and statistical significance of factor intercorrelations depends on the specific research context. For instance, in the Figure 4.1 solution, the correlation between the latent variables of Neuroticism and Extraversion is estimated to be −.435. The corresponding factor covariance is statistically significant ($p < .05$), as reflected by an absolute value of z greater than 1.96 (i.e., $z = −5.49$; see Table 4.4). The substantive sensibility of this relationship is evaluated by considering prior evidence and theory that bear on this analysis. In general, the size of the factor correlations in multifactorial CFA solutions should also be interpreted with regard to the *discriminant validity* of the latent constructs. Small, or statistically nonsignificant, factor covariances are usually not considered problematic and are typically retained in the solution (i.e., they provide evidence that the discriminant validity of the factors is good). However, if factor correlations approach 1.0, there is strong evidence to question the notion that the factors represent distinct constructs. In applied research, a factor correlation that exceeds .80 or .85 is often used as the criterion to define poor discriminant validity. When two factors are highly overlapping, a common research strategy is to respecify the model by collapsing the dimensions into a single factor, and determine whether this modification results in a significant degradation in model fit (see Chapter 5). If the respecified model provides an acceptable fit to the data, it is usually favored because of its superior parsimony, although substantive considerations must be brought to bear (i.e., the researcher must consider the implications in regard to the literature that has guided the formation of the initial, less parsimonious model).

INTERPRETATION AND CALCULATION OF CFA MODEL PARAMETER ESTIMATES

This section reviews and expands on how the various parameter estimates in the two-factor CFA model of Neuroticism and Extraversion are calculated and interpreted. Figure 4.2 presents this model with the unstandardized and completely standardized parameter estimates. Recall that in an *unstandardized solution*, all parameter estimates (factor loadings, factor variances and covariances, indicator error variances and covariances) are based on the original metrics of the indicators and factors when the marker indicator method is used to define the scales of the latent variables (i.e., the metric of the marker indicator is passed on to the factor). For example, in unstandardized solutions, factor loadings can be interpreted as unstandardized regression coefficients; for instance, a 1-unit increase in Neuroticism is associated with a 1.07 increase in depression (N3; see Figure 4.2).

In *completely standardized solutions*, the metrics of both the indicators and latent variables are standardized (i.e., $M = 0.0$, $SD = 1.0$). Thus factor loadings in a completely standardized solution can be interpreted as standardized regression coefficients; for instance, a 1-standardized score increase in Neuroticism is associated with a .844 standardized score increase in depression (N3; see Figure 4.2). However, when the measurement model contains no double-loading indicators, a completely standardized factor loading can also be interpreted as the correlation of the indicator with the factor because the factor is the sole predictor of the indicator. Accordingly, squaring a completely standardized factor loading provides the proportion of variance in the indicator that is explained by the factor; for example, Neuroticism accounts for 71.2% of the variance in the indicator (N3) of depression ($.8436^2 = .712$). Thus it follows that the basic equation of "1.0 minus the squared factor loading" (Eq. 3.6, Chapter 3) provides the proportion of variance in an indicator that is not explained by the factor; for instance, $1 - .8436^2 =$

.288, indicating that 28.8% of the observed variance in the depression indicator (N3) is estimated to be unique or error variance (see Figure 4.2).

By means of simple calculations, it is straightforward to transform a completely standardized CFA solution into an unstandardized solution, and vice versa. The following calculations transform a completely standardized solution into an unstandardized solution. When the marker indicator approach to model identification is employed, the variance of the factor is calculated by squaring the completely standardized factor loading of the marker indicator and multiplying this result by the observed variance of the marker indicator. To illustrate, the variance of the latent variable of Neuroticism is estimated to be 25.44 (see Figure 4.2). In Chapter 3 (Eq. 3.10), it has been shown that this unstandardized parameter estimate can be calculated by squaring the completely standardized factor loading of N1 (anxiety; λ_{11} = .8848) and multiplying this result by the observed variance of this indicator (σ_1^2 = 32.49; see Table 4.2):

$$\phi_{11} = .8848^2(32.49) = 25.44 \tag{4.1}$$

The factor variance of Extraversion (ϕ_{22} = 23.15; Figure 4.2) can be computed in the same fashion by multiplying the squared completely standardized factor loading of E1 (warmth; λ_{52} = .8018) by the observed variance of E1 (σ_5^2 = 36.00; see Table 4.2):

$$\phi_{22} = .8018^2(36.00) = 23.15 \tag{4.2}$$

The standard deviations of the factors are calculated by simply taking the square roots of the factor variances; that is, Neuroticism *SD* = SQRT(25.44) = 5.04, Extraversion *SD* = SQRT(23.15) = 4.81. (Some readers may notice that these values equal the *partially standardized* factor loadings for the marker indicators N1 and E1; see the "Standardized Solution" section of Table 4.4.)

The unstandardized error variances of the indicators can be calculated by multiplying the completely standardized residuals (δ) by the observed variance of the indicators. For example, the error variance of N2 (hostility) is estimated by the two-factor model to be 8.78 (see Figure 4.2). This estimate can be reproduced by multiplying N2's completely standardized residual (δ_2 = .28; Figure 4.2) by its observed variance (σ_2^2 = 31.36; Table 4.2):

$$\delta_2 = .28(31.36) = 8.78 \tag{4.3}$$

Alternatively, error variances can be computed by using the squared completely standardized factor loadings and the observed variances of indicators:

$$\delta_2 = \sigma_2^2 - \sigma_2^2(\lambda_{21}^2) \tag{4.4}$$
$$= 31.36 - (.8485^2)(31.36)$$
$$= 8.78$$

Factor covariances can be hand-calculated by using the SDs and correlations of the factors, in the same equation used for calculating covariances among the input indicators:

$$\phi_{21} = r_{21}(SD_1)(SD_2) \tag{4.5}$$

In the two-factor model, the correlation between Neuroticism and Extraversion is estimated to be −.435 (see Figure 4.2), and the SDs of these factors are estimated to be 5.04 and 4.81, respectively. Thus the factor covariance of −10.55 (see Figure 4.2) can be calculated by multiplying these three estimates:

$$\phi_{21} = -.435(5.04)(4.81) = -10.55 \tag{4.6}$$

Although not relevant to the current two-factor solution, indicator error covariances can be calculated in the same fashion (i.e., multiplying the error correlation by the indicator error standard deviations).

As noted earlier, factor loadings are regression coefficients. Therefore, the equations used for calculating unstandardized and standardized regression coefficients can be applied to the transformation of a completely standardized factor loading into an unstandardized loading, and vice versa. In a regression equation involving a single predictor (x) and criterion variable (y), the unstandardized regression coefficient (b) can be computed by the following formula:

$$b = (r_{yx}SD_y) / (SD_x) \tag{4.7}$$

This equation readily generalizes to the CFA of congeneric indicator sets, in which b is the unstandardized factor loading; r_{yx} is the completely standardized factor loading; SD_y is the standard deviation of the indicator; and SD_x is the SD of the factor. For example, the completely standardized factor loading of the E2 indicator (gregariousness; $\lambda_{62} = .8338$) can be converted into an unstandardized loading ($\lambda_{62} = 1.075$; see Figure 4.2) as follows:

$$\lambda_{62} = .8338(6.2) / 4.81 = 1.075 \tag{4.8}$$

(The observed standard deviation of E2 is 6.2, and the standard deviation of the Extraversion factor is 4.81.)

It is also straightforward to transform an unstandardized solution into a completely standardized solution. As a function of standardization, the factor variance in a completely standardized solution must be 1.0 (see Table 4.4). A completely standardized indicator error can be readily calculated by dividing the model-estimated error variance (δ) by the observed variance (σ^2) of the indicator. For example, the completely standardized error estimate of .28 for N2 (hostility) is reproduced by dividing its unstandardized error variance ($\delta_2 = 8.78$; Figure 4.2) by its observed variance ($\sigma^2_2 = 31.36$; Table 4.2):

$$\delta_2 = 8.78 / 31.36 = .280 \tag{4.9}$$

Again, this value reflects the proportion of the variance in N2 that is not explained by the latent variable of Neuroticism.

A factor intercorrelation can be calculated by dividing a factor covariance by the product of the *SD*s of the factors. Indeed, the factor correlation between Neuroticism and Extraversion was estimated to be −.435. This coefficient can be computed by dividing the factor covariance (ϕ_{21} = −10.55; Figure 4.2) by the product of the *SD*s of Neuroticism and Extraversion (5.04 and 4.81, respectively):

$$\phi_{21} = -10.55 / [(5.04)(4.81)] = -.435 \tag{4.10}$$

(This formula generalizes to the calculation of the correlations between indicator errors.) Squaring a factor correlation provides the proportion of overlapping variance between two factors; for example, Neuroticism and Extraversion share 18.9% of their variance ($-.435^2 = .189$).

As is the case for unstandardized factor loadings, an equation from multiple regression can be readily applied to the computation of completely standardized loadings from unstandardized estimates. In multiple regression, a standardized regression coefficient (b^*) can be computed by the following formula:

$$b^* = (bSD_x) / (SD_y) \tag{4.11}$$

where b is the unstandardized regression coefficient; SD_x is the standard deviation of the predictor; and SD_y is the standard deviation of the criterion variable. The parameters of a CFA solution can be substituted in this equation: b^* is the completely standardized factor loading; b is the unstandardized factor loading; SD_x is the standard deviation of the factor; and SD_y is the standard deviation of the indicator. For example, the unstandardized factor loading of the E2 indicator (gregariousness; λ_{62} = 1.075) can be converted into a completely standardized factor loading (λ_{62} = .8338; see Figure 4.2) as follows:

$$\lambda_{62} = 1.075 (4.81) / 6.2 = .8338 \tag{4.12}$$

(The *SD* of the Extraversion factor is 4.81, the observed *SD* of E2 is 6.2.)

In addition to unstandardized and completely standardized solutions, most latent variable software packages (e.g., LISREL, Mplus) provide *partially standardized solutions*, where indicators are unstandardized and the latent variables are standardized, or vice versa. (Mplus provides both types of partially standardized solutions; LISREL provides the type of partially standardized solution where the indicators are unstandardized and the factors are standardized.) Table 4.4 provides the partially standardized parameter estimates in the "Standardized Solution" section of the LISREL output (in Mplus, these estimates are provided under the STD heading; the other form of partially standardized estimates have been deleted from the Mplus output). For example, the standardized factor loading of N2 (hostility) is estimated to be 4.75. This estimate is interpreted as indicating that for every 1-unit standardized score unit increase in Neuroticism, N2 is predicted to increase by 4.75 unstandardized units. This parameter is calculated by multiplying the unstandardized factor loading of N2 (λ_{21} = 0.9421) by the standard deviation of the Neuroticism factor (5.04):

$$\lambda_{21} = 0.9421(5.04) = 4.75 \qquad (4.13)$$

The term *standardized solution* is frequently used in applied and even basic quantitative research in reference to a completely standardized solution. Although partially standardized solutions are rarely reported in applied SEM and CFA research, the reader must take care to determine whether a completely standardized solution or a partially standardized solution is being presented by the study's authors as the "standardized solution."

Although partially standardized estimates often do not provide useful additional information for common forms of CFA analyses, they are helpful to the interpretation of models where latent variables are regressed onto categorical background variables (e.g., a dummy code representing Gender: 0 = female, 1 = male), as is done in MIMIC models (see Chapter 7). In these instances, it does not make interpretative sense to use completely standardized estimates of the relationships between the factors and the observed score covariates. Consider a scenario where the latent variable of Neuroticism is regressed onto a dummy code for gender (Gender → Neuroticism). Because gender is a binary observed variable (0 / 1), it makes little substantive sense to consider the completely standardized regressive path that reflects the predicted standardized score change in Neuroticism for every *standardized score increase in gender*. Instead, it is better to interpret the path by using estimates that involve the original metric of the dummy code for gender. This standardized parameter estimate would indicate how many standardized scores Neuroticism is predicted to change as a function of an *unstandardized unit increase in gender*; in other words, it reflects the standardized score difference between males and females on the latent dimension of Neuroticism (cf. Cohen's *d*; Cohen, 1988).

Thus far, parameter estimates have been considered under the assumption that marker indicators are used to define the metric of the latent variables. However, as discussed in Chapter 3, the metric of latent variables can be defined by fixing the factor variance to a specific value. When this method is used, factor variances are usually fixed to 1.0, consistent with traditional forms of standardization (i.e., σ^2 = *SD* = 1.0). When factor variances are fixed to 1.0, many aspects of the CFA solution do not change (relative to the marker indicator approach), including overall model fit indices, fit diagnostic information (e.g., modification indices, standardized residuals), standardized and completely standardized parameter estimates, and the unstandardized estimates of indicator error variances and their corresponding standard errors and significance tests (*z* statistics). The primary impact that this approach will have is on the unstandardized estimates of the factor loadings, factor variances, and factor covariances. Factor variances will equal 1.0 as the result of the method used to define the latent variable metric, and there will be no corresponding standard errors and z statistics for these fixed parameters (as in the marker indicator approach for indicators N1 and E1; see Table 4.4). Because all factor variances are standardized, the factor covariances will reflect completely standardized estimates (e.g., it would equal −.435 in the Table 4.4 solution). In addition, the "unstandardized" factor loadings will take on the same values of the partially standardized estimates that were produced by a solution using marker indicators; that is, the "unstandardized" and partially standardized parameter estimates of the CFA model will be identical (for the type of partially standardized solution provided by LISREL where only the latent variables are standardized). Accordingly, the standard errors and z statistics associated with the factor loadings (and factor covariances) will also change.

A few additional comparisons between the outputs provided by LISREL (Table 4.4) and Mplus (Table 4.5) are worth noting. In the Mplus output, the unstandardized, partially standardized, and completely standardized solutions are provided under the "Model Results," "STD Standardization," and "STDYX Standardization" headings, respectively. As with all latent variable software programs other than LISREL, Mplus does not arrange the results according to matrices (e.g., the corresponding theta-delta estimates in LISREL are presented under the "Residual Variances" heading in the Mplus output). The reader will note that many of the estimates produced by LISREL and Mplus are the same (e.g., unstandardized factor loadings and their standard errors, and all completely standardized estimates). Unlike Mplus, where indicator error variances are presented in each type of solution, LISREL does not provide theta-delta estimates in the "Standardized Solution" section because these error variances are identical to those provided in the LISREL Estimates section (i.e., because the indicators are unstandardized in both solutions). Mplus provides standard errors (and test ratios and p values) for all solutions, not just the unstandardized solution as in LISREL.

Another key difference between the two programs pertains to parameter estimates of unstandardized variances and covariances, and estimates that rely on unstandardized variances in their computation (e.g., partially standardized factor loadings). For instance, the unstandardized variance estimate for the Neuroticism latent variable is 25.335 in Mplus, whereas the corresponding estimate in LISREL is 25.44. This is because Mplus uses N instead of $N − 1$ in the calculation of variance estimates (and model χ^2; see Chapter 3 for discussion). Thus, to convert a LISREL estimate into an Mplus estimate, the LISREL variance estimate should be multiplied by $N − 1$ and then divided by N; for example, (25.44 * 249) / 250 = 25.34 (the same transformation applies to unstandardized indicator error variances). The estimates for factor variances (and covariances) and indicator error variances (and covariances) are thus affected, as well as the partially standardized factor loadings (cf. Eq. 4.13). It is important for the reader to be mindful of the differences between LISREL and Mplus (as well as between these and other software programs) when interpreting the results presented in the applied research literature and later in this book (although the differences are trivial in large samples).

CFA MODELS WITH SINGLE INDICATORS

Another advantage of CFA over EFA is the ability to include single indicator variables in the analysis. It may be necessary to use single indicators in a CFA or an SEM analysis, given the nature of the variable (e.g., sex, age) or the unavailability of multiple measures of a construct. Variables assessed by a single measure should not be interpreted as factors (i.e., a factor accounts for shared variance of multiple observed measures). However, the inclusion of single indicators in CFA is very useful in many situations. For instance, single indicators are used in MIMIC models where latent variables and the indicators of the factors are regressed onto covariates (see Chapter 7). Moreover, CFA is routinely used as a precursor to SEM because the measurement model must be worked out before a more parsimonious structural solution can be evaluated (cf. Figure 3.2, Chapter 3). If the structural model will contain a mix of single indicators and latent variables, it is important to include the single indicators in the measurement model. If not, specification error may occur when the single indicators are added to the structural model; for instance, the measure may have poor discriminant validity with a latent variable or may operate better as an indicator of a latent variable (see Chapter 5 for a detailed discussion of specification errors). Including single indicators in the CFA allows the researcher to examine the correlations among the latent variables and single indicators before a structural model is imposed on these relationships. These estimates will provide important information about the viability of the structural model; for example, if a single indicator is weakly correlated with a latent variable, this bodes poorly for a proposed structural model in which the single indicator is specified to have a significant indirect effect on the latent variable.

When observed measures are used as indicators of latent variables, the CFA provides the estimate of their measurement error—that is, the amount of variance in the indicator that is not explained by the factor. Obviously, this cannot be done for variables assessed by single indicators. However, an error theory can be invoked for these variables by fixing the unstandardized error of the indicator to some predetermined value. If the error variance is fixed to zero, then it is assumed that the indicator is perfectly reliable (i.e., entirely free of measurement error). This assumption may be reasonable for some variables such as age, height, or weight. If the single indicator is an indirect measure of a construct (e.g., a self-esteem questionnaire), the assumption of perfect reliability is less tenable. Fortunately, measurement error can be readily incorporated into a dimensional indicator by fixing its unstandardized error to some nonzero value, calculated on the basis of the measure's sample variance estimate and known psychometric information (e.g., internal consistency estimate):

$$\delta_x = VAR(X)(1 - \rho) \tag{4.14}$$

where $VAR(X)$ is the sample variance of the single indicator, and ρ is the reliability estimate of the indicator. Ideally, ρ is derived from prior psychometric research that generalizes well to the sample for which the current analysis is being undertaken.[8]

To illustrate, if the sample variance of an indicator is 42.00 and the best estimate of its scale reliability is .85, its error variance would be calculated as

$$\delta_x = (42)(1 - .85) = 6.3 \tag{4.15}$$

The value of 6.3 can then be used to fix the error variance of this indicator in the CFA model.

A CFA model entailing a combination of latent variables and single indicators is presented in Figure 4.3. In this example, an epidemiological researcher is interested in examining whether dimensions of health status are predictive of various public health outcomes (e.g., quality of life, use of medical services) in a sample of 500 participants; the example is loosely based on the SF-36 Health Survey (e.g., Ware & Sherbourne, 1992). For the sake of illustration, only a portion of the variable set is included in the example, but in practice, all predictor and outcome variables to be used in the SEM analysis should be included in the measurement model. There are two latent variables, Physical Functioning and Mental Functioning, defined by three observed measures each. A third health status variable, General Well-Being, is assessed by one self-report measure of general health. Prior population-based studies indicate that the best reliability estimate (ρ) of this measure is .89; thus 11% of the total variance of this measure is estimated to be error (i.e., $1 - .89$). As shown in Figure 4.3, the standard deviation of the general health measure is 8.462, and thus its sample variance is 71.605 (i.e., 8.462^2). Because this example uses the Mplus program, the sample variance estimate is recomputed with N instead of $N - 1$; that is, $(71.605 * 499) / 500 = 71.462$. Participant age is an important covariate in the SEM analysis, so it is included in the measurement model.

In Table 4.6, Mplus syntax and selected output are presented. The correlation matrix and indicator standard deviations are presented in Figure 4.3, but a raw data file is used as input in this syntax example. The syntax programming for Physical Functioning and Mental Functioning is conducted in the usual manner. Because two indicators of Physical Functioning are subscales from the same measure, a correlated error is specified (i.e., ACTIV WITH SOMA). Although it may be substantively plausible to expect that a method effect exists for the indicator of general health (e.g., it is also a subscale from the measure used to obtain indicators of physical functioning), correlated errors cannot be specified for single indicators because their error variances are fixed (see Chapter 5). Although General Well-Being and Age are not latent variables, they are programmed as such (e.g., GWB BY GENHLTH, Table 4.6). The unstandardized factor loading of the single indicator on its "pseudofactor" is fixed to 1.0 (in Mplus, this is a default). Next, the appropriate error variance constraints are imposed. It is assumed that age has been measured without measurement error, and thus the error variance of this indicator is fixed to zero (i.e., AGE@0). With Equation 4.14, the error variance of the general health indicator (GENHLTH) is calculated to be (71.462) $(1 - .89) = 7.861$; this value is added to the model (i.e., GENHLTH@7.861). Programming will follow similar lines in other software packages—for example, LISREL: VA 7.88 TD(7,7); EQS: /VARIANCES E7 = 7.88.

The Figure 4.3 model fits the data reasonably well: $\chi^2(15) = 45.01$, $p < .001$, SRMR = .026, RMSEA = 0.063 (CFit = .137), TLI = 0.971, CFI = .984. To ascertain that the error variance constraints have been modeled as intended, one should first inspect the appropriate section of the output; in Mplus, this section is under the heading "Residual

Variances." As can be seen in Table 4.6, the unstandardized and completely standardized error variances of GENHLTH are 7.861 and .11, respectively, in accord with the reliability information used to specify the model (e.g., proportion of error variance = 1 – .89). Similarly, the error variance of age is 0.00, as intended. Observe that factor loadings are estimated by Mplus for the single indicator and their "pseudofactor." These loadings are byproducts of the error variance constraints. For instance, the completely standardized loading of GENHLTH on the General Well-Being factor (GWB) is .943. Squaring this loading produces the reliability coefficient used to calculate the error variance of GENHLTH—that is, $.943^2$ = .89 (cf. Eq. 3.5, Chapter 3). Correlations involving the GWB factor reflect the relationship of the GENHLTH indicator with other variables in the model, adjusted for measurement error; for instance, the correlation of GENHLTH with Physical Functioning is .642. The factor loading of Age on the Age "factor" is 1.00, reflecting a perfect relationship between the observed measure and its underlying "true" score.

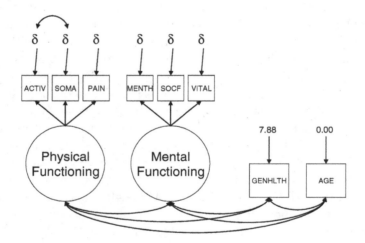

Correlations/Standard Deviations (SD):

	ACTIV	SOMA	PAIN	MENTH	SOCF	VITAL	GENHLTH	AGE
ACTIV	1.000							
SOMA	0.779	1.000						
PAIN	0.412	0.463	1.000					
MENTH	0.366	0.311	0.285	1.000				
SOCF	0.280	0.273	0.269	0.784	1.000			
VITAL	0.227	0.193	0.218	0.696	0.748	1.000		
GENHLTH	0.417	0.452	0.365	0.539	0.605	0.456	1.000	
AGE	-0.118	-0.122	-0.146	-0.082	-0.073	-0.119	-0.096	1.000
SD:	19.961	15.851	3.270	5.545	9.366	13.838	8.462	11.865

FIGURE 4.3. CFA model of health status containing latent variables and single indicators. ACTIV, physical activity; SOMA, somatic complaints; PAIN, bodily pain; MENTH, mental health; SOCF, social functioning; VITAL, vitality; GENHLTH, general health; N = 500.

TABLE 4.6. Mplus Syntax and Selected Results for a Measurement Model of Health Status Involving Latent Variables and Single Indicators

Mplus syntax

```
TITLE: MPLUS 7.11 CFA MODEL WITH TWO FACTORS, TWO SINGLE INDICATORS
DATA:
       FILE IS FIG4.3.DAT;
       FORMAT IS F3,8F3;
VARIABLE:
       NAMES ARE SUBJID ACTIV SOMA PAIN MENTH SOCF VITAL GENHLTH AGE;
       USEV = ACTIV SOMA PAIN MENTH SOCF VITAL GENHLTH AGE;
ANALYSIS: ESTIMATOR=ML;
MODEL:
       PHYSF BY ACTIV SOMA PAIN;
       MENTF BY MENTH SOCF VITAL;
       GWB BY GENHLTH;
       AGEF BY AGE;
       GENHLTH@7.861; AGE@0;
       ACTIV WITH SOMA;
OUTPUT:  SAMPSTAT STANDARDIZED MODINDICES(5);
```

Selected results

MODEL RESULTS

		Estimate	S.E.	Est./S.E.	Two-Tailed P-Value
PHYSF	BY				
ACTIV		1.000	0.000	999.000	999.000
SOMA		0.868	0.050	17.527	0.000
PAIN		0.150	0.018	8.480	0.000
MENTF	BY				
MENTH		1.000	0.000	999.000	999.000
SOCF		1.850	0.074	25.077	0.000
VITAL		2.357	0.109	21.659	0.000
GWB	BY				
GENHLTH		1.000	0.000	999.000	999.000
AGEF	BY				
AGE		1.000	0.000	999.000	999.000
MENTF	WITH				
PHYSF		28.371	4.307	6.587	0.000
GWB	WITH				
PHYSF		69.058	7.958	8.677	0.000
MENTF		25.179	2.280	11.044	0.000
AGEF	WITH				
PHYSF		-30.303	9.003	-3.366	0.001
MENTF		-5.290	2.627	-2.014	0.044
GWB		-9.664	4.502	-2.147	0.032

(continued)

TABLE 4.6. *(continued)*

ACTIV WITH				
SOMA	87.988	17.805	4.942	0.000
Variances				
PHYSF	182.151	29.415	6.192	0.000
MENTF	22.140	1.948	11.367	0.000
GWB	63.601	4.520	14.072	0.000
AGEF	140.490	8.885	15.811	0.000
Residual Variances				
ACTIV	215.498	24.582	8.766	0.000
SOMA	113.524	15.800	7.185	0.000
PAIN	6.567	0.583	11.260	0.000
MENTH	8.548	0.782	10.935	0.000
SOCF	11.780	1.949	6.045	0.000
VITAL	68.147	5.272	12.926	0.000
GENHLTH	7.861	0.000	999.000	999.000
AGE	0.000	0.000	999.000	999.000

STANDARDIZED MODEL RESULTS

STDYX Standardization

	Estimate	S.E.	Est./S.E.	Two-Tailed P-Value
PHYSF BY				
ACTIV	0.677	0.045	15.015	0.000
SOMA	0.740	0.043	17.204	0.000
PAIN	0.620	0.041	14.989	0.000
MENTF BY				
MENTH	0.849	0.017	51.457	0.000
SOCF	0.930	0.013	73.919	0.000
VITAL	0.802	0.019	42.728	0.000
GWB BY				
GENHLTH	0.943	0.004	255.852	0.000
AGEF BY				
AGE	1.000	0.000	999.000	999.000
MENTF WITH				
PHYSF	0.447	0.051	8.734	0.000
GWB WITH				
PHYSF	0.642	0.043	14.778	0.000
MENTF	0.671	0.030	22.644	0.000
AGEF WITH				
PHYSF	-0.189	0.054	-3.509	0.000
MENTF	-0.095	0.046	-2.044	0.041
GWB	-0.102	0.047	-2.178	0.029

(continued)

TABLE 4.6. *(continued)*

	Estimate	S.E.	Est./S.E.	P-Value
ACTIV WITH				
SOMA	0.563	0.054	10.461	0.000
Variances				
PHYSF	1.000	0.000	999.000	999.000
MENTF	1.000	0.000	999.000	999.000
GWB	1.000	0.000	999.000	999.000
AGEF	1.000	0.000	999.000	999.000
Residual Variances				
ACTIV	0.542	0.061	8.882	0.000
SOMA	0.453	0.064	7.116	0.000
PAIN	0.615	0.051	11.982	0.000
MENTH	0.279	0.028	9.933	0.000
SOCF	0.135	0.023	5.746	0.000
VITAL	0.357	0.030	11.840	0.000
GENHLTH	0.110	0.007	15.811	0.000
AGE	0.000	999.000	999.000	999.000

R-SQUARE

Observed Variable	Estimate	S.E.	Est./S.E.	Two-Tailed P-Value
ACTIV	0.458	0.061	7.507	0.000
SOMA	0.547	0.064	8.602	0.000
PAIN	0.385	0.051	7.495	0.000
MENTH	0.721	0.028	25.729	0.000
SOCF	0.865	0.023	36.960	0.000
VITAL	0.643	0.030	21.364	0.000
GENHLTH	0.890	0.007	127.926	0.000
AGE	1.000	999.000	999.000	999.000

REPORTING A CFA STUDY

Good-fitting CFA solutions have been presented in this chapter to illustrate the specification and interpretation of latent variable measurement models. In addition to understanding how to conduct and interpret these analyses properly, applied researchers must be aware of what information should be presented when they are reporting the results of a CFA study. Indeed, although many excellent guides have been published for reporting SEM and CFA results (e.g., Hoyle & Panter, 1995; McDonald & Ho, 2002; Raykov, Tomer, & Nesselroade, 1991), recent surveys have indicated that applied research articles continue to often omit key aspects of the analyses—such as the type of input matrix used, the identifiability and exact specification of the model, and the resulting parameter estimates (MacCallum & Austin, 2000; McDonald & Ho, 2002). Recommended information to include in a CFA research report is listed in Table 4.7. Appendix 4.3 provides a sample write-up of some of this suggested information, using the example of the two-factor CFA model of Neuroticism and Extraversion.

Although Table 4.7 and Appendix 4.3 are fairly self-explanatory, a few elaborations are warranted. The parameter estimates from the two-factor model used in this example lend themselves to presentation in a path diagram (Figure 4.2), but many applied CFA models are too complex to be presented in this fashion (e.g., models often contain a large number of indicators and factors). Thus applied CFA findings are frequently presented in tabular formats—specifically, a p (indicator) $\times$ m (factor) matrix of factor loadings. While a tabular approach may be preferred for large CFA models (cf. McDonald & Ho, 2002), the researcher must be sure to provide the remaining parameter estimates (e.g., factor and error correlations) in the text, a table note, or a companion figure. Although methodologists have underscored the importance of providing standard errors (and/or confidence intervals) of parameter estimates, this information is rarely reported in CFA research (MacCallum & Austin, 2000; McDonald & Ho, 2002). These data could also be readily presented in a table or figure note, or in the body of the table or figure itself when only unstandardized estimates are provided.

It should be noted that there exists no "gold standard" for how a path diagram should be prepared. Although some constants exist (e.g., representing observed and latent variables by rectangles and circles, respectively; depicting a direct effect by a unidirectional arrow), the reader will encounter many variations in the applied and quantitative literature; for instance, an indicator error variance may be represented by e, θ, δ, ε, a circle, a double-headed curved arrow, or some combination. No approach is considered to be more correct than another. One particular method is used throughout this book, but readers are encouraged to peruse other sources to decide which approach best suits their own tastes and purposes.

Another consideration is whether to present an unstandardized solution, a completely standardized solution, or both. (Also, as noted in the previous section, partially standardized estimates may be relevant for some parameters, such as in MIMIC models.) In applied CFA research, the convention has been to report completely standardized parameter estimates. This may be due in part to the fact that CFA is frequently employed after EFA in the psychometric analysis of test instruments (e.g., multiple-item questionnaires). As noted in Chapter 2, the tradition of EFA is to standardize both the observed and latent variables. Often neither the observed nor latent variables are standardized in CFA (e.g., a variance–covariance matrix is inputted, a marker indicator is specified). Although the completely standardized CFA solution can be informative (e.g., a completely standardized error indicates the proportion of variance in the indicator that is not accounted for by the factor), the unstandardized solution (or both) may be preferred in some instances. These include measurement invariance evaluations, where constraints are placed on the unstandardized parameter estimates (Chapter 7); construct validity studies, where the indicators are composite measures with readily interpretable metrics; and analyses using item parcels (Chapter 9), where information regarding the relative magnitudes of the relationships to the factors has little substantive importance or does not convey information about the original items. As discussed in Chapter 3, there are some CFA and SEM scenarios where exclusive reliance on completely standardized solutions can result in misleading or erroneous conclusions (Bollen, 1989; Willett et al., 1998).

TABLE 4.7. Information to Report in a CFA Study

Model specification

- Conceptual/empirical justification for the hypothesized model
- Complete description of the parameter specification of the model
  - List the indicators for each factor
  - Indicate how the metric of the factors was defined (e.g., specify which observed variables were used as marker indicators)
  - Describe all freely estimated, fixed, and constrained parameters (e.g., factor loadings and cross-loadings, random and correlated indicator errors, factor correlations, intercepts, and factor means[a])
- Demonstrate that the model is identified (e.g., positive model df, scaling of latent variables, absence of empirical underidentification)

Input data

- Description of sample characteristics, sample size, and sampling method
- Description of the type of data used (e.g., nominal, interval; scale range of indicators)
- Tests of estimator assumptions (e.g., multivariate normality of input indicators)
- Extent and nature of missing data, and the method of missing data management (e.g., direct ML, multiple imputation[b])
- Provide sample correlation matrix and indicator SDs (and means, if applicable[a]), or make such data available on request

Model estimation

- Indicate the software and version used (e.g., LISREL 9.1)
- Indicate the type of data/matrices analyzed (e.g., variance–covariance, tetrachoric correlations/asymptotic covariances[b])
- Indicate the estimator used (e.g., ML, WLS[b]), as justified by properties of the input data

Model evaluation

- Overall goodness of fit
  - Report model χ^2 along with its df and p
  - Report multiple fit indices (e.g., SRMR, RMSEA, CFI), and indicate cutoffs used (e.g., RMSEA ≤ 0.06); provide confidence intervals, if applicable (e.g., RMSEA)
- Localized areas of ill fit
  - Report strategies used to assess for focal strains in the solution (e.g., modification indices/Lagrange multipliers, standardized residuals, Wald tests, EPC values)
  - Report absence of areas of ill fit (e.g., largest modification index) or indicate the areas of strain in the model (e.g., modification index, EPC value)
- If model is respecified, provide a compelling substantive rationale for the added or removed parameters, and clearly document (improvement in) fit of the modified models
- Parameter estimates
  - Provide all parameter estimates (e.g., factor loadings, error variances, factor variances), including any nonsignificant estimates
  - Consider the clinical as well as the statistical significance of the parameter estimates (e.g., are all indicators meaningfully related to the factors?)
  - Ideally, include the standard errors or confidence intervals of the parameter estimates
- If necessary (e.g., suitability of N could be questioned), report steps taken to verify the power and precision of the model estimates (e.g., Monte Carlo evaluation using the model estimates as population values[c])

Substantive conclusions

- Discuss CFA results in regard to their substantive implications, directions for future research, and so on
- Interpret the findings in context of study limitations (e.g., range and properties of the indicators and sample) and other important considerations (e.g., equivalent CFA models[d])

[a]See Chapter 7. [b]See Chapter 9. [c]See Chapter 10. [d]See Chapter 5.

If possible, the sample input data used in the CFA should be published in the research report (e.g., variance–covariance matrix if ML estimation is used). In instances where this is not feasible (e.g., a large set of indicators is used), it is helpful for authors to make these data available upon request (e.g., to provide them in a supplement to the research paper or to post the data on a website). Inclusion of these data provides a wealth of information (e.g., magnitudes and patterns of relationships among variables), and it allows a reader to replicate the study's models and to explore possible conceptually viable alternative models (equivalent or better-fitting; see Chapter 5). In general, the sample correlation matrix (accompanied with SDs and Ms, if applicable) should be provided rather than a variance–covariance matrix. This is because the reader will be able to analyze a variance–covariance matrix that contains less rounding error (than the typical variance–covariance matrix published in research reports) by creating it directly from the sample correlations and SDs. Of course, there are some situations where a CFA study cannot be reproduced from a published table of data. Whereas tabled data work fine for ML analysis of data sets that contain no missing data (or where pairwise or listwise deletion was used to manage missing data), this is not the case for analyses that are conducted on raw data (e.g., as in analyses that use direct ML for missing data; see Chapter 9) or that require companion matrices (e.g., asymptotic covariance as well as tetrachoric correlation matrices in WLS-estimated models; see Chapter 9). In these cases, it may be better to make the data available by request or possibly by download from the author's or journal's website (e.g., as a supplement).

Finally, it should be emphasized that the suggestions provided in Table 4.7 are most germane to a measurement model conducted in a single group, and thus must be adapted to the nature of the particular CFA study. For example, many CFA investigations entail the comparison of a target model to a substantively viable competing solution (see Chapter 5). In such studies, the recommendations listed in Table 4.7 should be extended to the alternative models as well (e.g., the author should also provide conceptual/empirical justification for the competing models). In multiple-groups CFA studies (see Chapter 7), it is important to evaluate the CFA solutions separately in each group before conducting the simultaneous analysis. In addition, the multiple-groups analysis typically entails invariance evaluation of the CFA parameters (e.g., are the factor loadings invariant, consistent with the notion that the indicators measure the latent construct in comparable ways in all groups?). Thus, although all of the steps listed in Table 4.7 are relevant, the report of a multiple-groups CFA study is considerably more extensive than the sample write-up provided in Appendix 4.3 (e.g., data normality screening and model fit estimation/evaluation in each group, nested invariance evaluation; see Chapter 7).

SUMMARY

The fundamental concepts and procedures of CFA have been illustrated in this chapter with a full example. As the chapter has shown, the proper conduct of CFA requires a series of steps and decisions, including the specification of the measurement model (based on prior evidence and theory); selection of a statistical estimator appropriate for the type and distributional properties of the data (e.g., ML); choice of a latent variable software program (e.g., EQS, LISREL); evaluation of the acceptability of the model (e.g., overall goodness of fit, focal areas of strain in the solution, interpretability/strength of parameter estimates); and the interpretation and presentation of results. Although this material has been presented in context of a well-specified measurement model, some of the complications and issues often encountered in applied CFA research have been introduced (e.g., potential sources of ill fit, Heywood cases, significance testing of parameter estimates). These issues are considered at much greater length in the next chapter, which focuses on the respecification and comparison of CFA models.

NOTES

1. In the case of the Figure 4.1 model, however, specification of a direct effect between Neuroticism and Extraversion will produce the same fit as a solution that simply allows these factors to be intercorrelated. This is because the potential structural component of the model (e.g., Neuroticism $\rightarrow$ Extraversion, or Extraversion $\rightarrow$ Neuroticism) is just-identified.

2. However, in EQS, the term *standardized residual* is used differently. In EQS, a standardized residual reflects the difference between the observed correlation and the model-implied correlation; for example, for the N1 and N2 relationship in the Figure 4.1 model, the EQS standardized residual is .016 (i.e., .767 − .751).

3. Like Lagrange multipliers, Wald tests can be used as multivariate statistics that estimate the change in model χ^2 if *sets* of freed parameters are fixed.

4. Although completely standardized loadings > 1.0 are generally considered to be Heywood cases, Jöreskog (1999) has demonstrated instances where such estimates are valid (i.e., models that contain double-loading indicators).

5. One-tailed (directional) tests are appropriate for parameter estimates involving variances (i.e., indicator error variances, factor variances) because these parameters cannot have values below zero (z_{crit} = 1.645, α = .05, one-tailed).

6. For a technically more accurate interpretation, see the section of Chapter 11 on Bayesian analysis.

7. For indicators that load on more than one factor, factor loadings should be interpreted as partial regression coefficients; for example, a given factor loading for a double-loading indicator would be interpreted as how much the indicator is predicted to change, given a unit increase in one factor while the other factor is held constant.

8. The question often arises as to which reliability estimate should be selected for the error variance constraint in situations where the measure in question has an extensive psychometric history or the quality of the extant psychometric evidence is poor. Although qualitative considerations are important (e.g., selection should be guided by the quality and generalizability of psychometric studies), it is often useful to conduct a sensitivity analysis in which the stability of the results is examined with a range of viable reliability estimates.

Model Identification Affects the Standard Errors of the Parameter Estimates

Chapter 9 discusses the fact that the standard errors of the parameter estimates of the CFA solution should not be trusted when the ML estimator is used if the data are non-normal (i.e., an estimator robust to non-normality should be employed). If the standard errors are incorrect, significance testing and confidence interval estimation of the parameter estimates are undermined. However, even when the assumptions of the fitting function hold, usually a problem still exists in the estimation of standard errors and their corresponding z tests, confidence intervals, and Wald tests. Specifically, *standard errors are not invariant to the method used to define the scale of the latent variable*. In other words, the magnitude of the standard error and corresponding z test of a parameter estimate tend to vary based on the selection of the marker indicator, or when the scale of the latent variable is defined by fixing its variance to 1.0.

To illustrate, the factor covariance from the two-factor model of Neuroticism and Extraversion (Figure 4.1) is used as the example (although the illustration applies to any freely estimated parameter in the solution). The model is estimated three times, with differing approaches to scaling the factors: (1) using N1 and E1 as the marker indicators (as in Figure 4.2); (2) using N4 and E4 as the marker indicators; and (3) fixing the variances of Neuroticism and Extraversion to 1.0.

It is important to note that the manner in which the scale of the latent variables is identified has no impact on overall goodness of fit. Indeed, the three models produce identical goodness of fit indices—for example, $\chi^2(19) = 13.285$, $p = .82$. However, each model produces a different standard error (SE) and test statistic (z):

	Factor covariance estimate	SE	z
Model 1 (N1, E1 as markers)	−10.554 (−.453)	1.924	5.487
Model 2 (N4, E4 as markers)	−8.560 (−.453)	1.612	5.309
Model 3 ($\phi_{11} = \phi_{22} = 1.0$)	−.453 (−.453)	0.059	7.430

(Completely standardized estimates are in parentheses; i.e., these estimates are not affected by the scaling approach.)

In this example, the results are consistent in their indication that the null hypothesis is false (i.e., the factor covariance significantly differs from zero). However, this will not always be the case in applied data sets; that is, *different approaches to scaling the latent variable may yield different conclusions regarding the statistical significance of freely estimated parameters*. Because standard errors are used in the computation of confidence intervals and Wald tests, the same problem applies to these statistics as well.

The explanation for why standard errors vary across identification methods is very technical (e.g., standard errors are determined on the basis of the curvature of fit function when the maximum height is reached; curvature at the maximum differs depending on the identification method). Interested readers are referred to Bollen (1989) and Gonzalez and Griffin (2001) to learn more about this issue. However, a more general question is this: *How can one test the statistical significance of CFA parameters if standard errors are not invariant to the latent variable scaling method?* Fortunately, the standard χ^2 difference procedure can be implemented for this purpose.

As noted in Chapter 3, a *nested model* has a subset of the free parameters of another model (often referred to as the *parent model*). Gonzalez and Griffin (2001) note that instead of approximating the value of the fitting function by a measure of curvature that is sensitive to the scaling of the latent variable (i.e., as in typical standard errors and the corresponding z tests), it is better to evaluate statistical significance by comparing model fit when the parameter of interest is freely estimated (i.e., maximum height) to when the parameter is constrained to the value of the null hypothesis. In the current example, this entails comparing the fit of the model where the factor covariance is freely estimated (the parent model) to the model that constrains the factor variance to zero (the nested model). The resulting χ^2 difference test (also referred to as the *likelihood ratio test*) is not affected by the manner in which the metric of the latent variable is set (cf. Figure 3 and associated text in Gonzalez & Griffin, 2001). Using Model 1 as the parent model (Model P), the χ^2 difference test would be as follows:

	χ^2	df
Model N (ϕ_{21} fixed to 0)	54.416	20
Model P (ϕ_{21} free)	13.285	19
χ^2 difference	41.131	1

As the critical value of χ^2 with $df = 1$ is 3.84 ($\alpha = .05$), it can be concluded that the parent model provides a significantly better fit to the data than the nested model. This is because the difference in model χ^2 follows a χ^2 distribution with one degree of freedom: $\chi^2_{\text{diff}}(1) = 41.131$, $p < .001$. More germane to the present discussion is the relationship between the z and χ^2 distributions that has been noted in Chapter 3 and this chapter: That is, at $df = 1$, the square root of the χ^2 is normally distributed. Thus, in the current example, the z test of the null hypothesis (that the factor covariance of Neuroticism and Extraversion does not differ from zero) can be obtained by taking the square root of the χ^2 difference value:

$$z = \text{SQRT}(41.131) = 6.41 \qquad (p < .001)$$

Observe that the z value of 6.41 does not correspond to any of the z values derived from the three models estimated above. Although this example uses Model 1 as the parent model, the same result will be obtained with either Model 2 or Model 3 instead.

Although this approach is recommended over the typical z tests provided by software program outputs for determining the statistical significance of parameter estimates, a few limitations should be noted. First, the method is not feasible in situations where the null (nested) model will be underidentified. For instance, we cannot employ this method to test the significance of the factor covariance if Neuroticism and Extraversion are measured by two indicators each (cf. Model D, Figure 3.7 in Chapter 3). Second, this procedure is not appropriate when the null hypothesis entails fixing a parameter estimate to a value that is on the border of inadmissibility (e.g., setting a factor variance, indicator error variance, or disturbance to a value of 0). As discussed in later chapters (e.g., Chapter 5), models that contain borderline values may not yield proper χ^2 distributions (cf. Self & Liang, 1987). Third, this approach only addresses significance testing and does not address calculation of confidence intervals (but see Neale & Miller, 1997). There are alternative ways of obtaining standard errors in each of the three scenarios above, but these methods are somewhat complex and cumbersome. Finally, applied users may balk at this approach because performing nested χ^2 tests for each parameter is time-consuming. Gonzalez and Griffin (2001) recommend that at the very least, researchers perform these tests on the parameters of the model that are most important to the research question (e.g., structural relationships among latent variables).

Goodness of Model Fit Does Not Ensure Meaningful Parameter Estimates

Novice CFA researchers occasionally confuse the notions of goodness of model fit and the meaningfulness of the model's parameter estimates. Even if a model is very successful at reproducing the observed relationships in the input matrix (S), this does not ensure that the latent variables are substantively interrelated or account for meaningful variance in the indicators. Thus it is just as important to consider the size of the model's parameter estimates as it is to consider the model's goodness of fit in determining the acceptability of the solution. This point is illustrated below with a somewhat exaggerated example.

In keeping with the example presented in Figure 4.1, a researcher has developed a brief questionnaire designed to measure the constructs of Neuroticism and Extraversion (three items each). Thus a two-factor model is anticipated in which the first three items (X1–X3) load onto a latent variable of Neuroticism, and the remaining three items (X4–X6) load onto a distinct factor of Extraversion. The questionnaire is administered to 300 participants, and the two-factor model is fitted to the following sample data (for ease of interpretation, a completely standardized matrix is presented):

	X1	X2	X3	X4	X5	X6
X1	1.0000					
X2	0.0500	1.0000				
X3	0.0600	0.0750	1.0000			
X4	−0.0100	−0.0125	−0.0150	1.0000		
X5	−0.0120	−0.0150	−0.0180	0.0750	1.0000	
X6	−0.0100	−0.0125	−0.0150	0.0875	0.0750	1.0000

The following completely standardized solution results from this sample matrix and two-factor model specification:

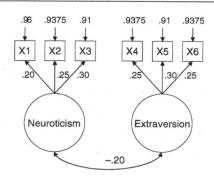

In fact, this solution provides a perfect fit to the data, $\chi^2(8) = 0.00$. For example, multiplying the factor loadings of X1 and X2 perfectly reproduces their observed relationship (.20 × .25 = .05), and multiplying the factor loading of X1 and X5 with the factor correlation perfectly reproduces the observed relationship between X1 and X5 (.20 × .30 × –.20 = –.012). However, goodness of fit of the two-factor solution is not determined by the *absolute magnitude* of the sample correlations, but by whether the *differential magnitude* of the correlations can be reproduced by the specified model and its resulting parameter estimates. Accordingly, because the X1–X3 indicators are more strongly intercorrelated with each other than they are with the X4–X6 indicators (and vice versa), the two-factor model specification fits the data well. Nevertheless, the size of the resulting parameter estimates will lead the sensible researcher to reject the two-factor model and the questionnaire on which it is based. For example, the latent variable of Neuroticism accounts for only 4% of the variance in the X1 indicator (.20² = .04); that is, the vast majority of its variance (96%) is not explained by the latent variable. Indeed, the largest percentage of variance explained in the items is only 9% (X3 and X5). Thus the questionnaire items are very weakly related to the factors and should not be considered to be reasonable indicators of their purported constructs.

Example Report of the Two-Factor CFA Model of Neuroticism and Extraversion

Based on prior evidence and theory bearing on the Big Five model of personality, a two-factor model was specified in which anxiety (N1), hostility (N2), depression (N3), and self-consciousness (N4) loaded onto the latent variable of Neuroticism, and in which warmth (E1), gregariousness (E2), assertiveness (E3), and positive emotions (E4) loaded onto the latent variable of Extraversion. The indicators were subscales of the NEO Personality Inventory and had a range of scores from 0 to 32, with higher scores reflecting higher levels of the personality dimension. Figure 4.1 depicts the complete specification of the two-factor model. Anxiety (N1) and warmth (E1) were used as marker indicators for Neuroticism and Extraversion, respectively. The measurement model contained no double-loading indicators, and all measurement error was presumed to be uncorrelated. The latent variables of Neuroticism and Extraversion were permitted to be correlated, based on prior evidence of a moderate inverse relationship between these dimensions. Accordingly, the model was overidentified with 19 *df*.

As noted in the "Method" section, the NEO was administered to 250 college undergraduates who participated in the study for course credit (see "Method" for description of sample demographics). All 250 cases had complete NEO data. Prior to the CFA analysis, the data were evaluated for univariate and multivariate outliers by examining leverage indices for each participant. An outlier was defined as a leverage score that was five times greater than the sample average leverage value. No univariate or multivariate outliers were detected. Normality of the indicators was examined by using PRELIS 2.30 (Jöreskog & Sörbom, 1996b). The standardized skewness score was 1.17, and the standardized kurtosis score was 1.32 (*p*s > .05). The joint test of nonnormality in terms of skewness and kurtosis was not significant, $\chi^2 = 8.35$, $p = .38$.

Thus the sample variance–covariance matrix was analyzed with LISREL 9.1, and a maximum likelihood minimization function (sample correlations and *SD*s are provided in Figure 4.1). Goodness of fit was evaluated by using the standardized root mean square residual (SRMR), root mean square error of approximation (RMSEA) and its 90% confidence interval (90% CI) and test of close fit (CFit), comparative fit index (CFI), and the Tucker–Lewis index (TLI). Guided by suggestions provided in Hu and Bentler (1999), acceptable model fit was defined by the following criteria: RMSEA (≤ 0.06, 90% CI ≤ 0.06, CFit *ns*), SRMR (≤ 0.08), CFI (≥ 0.95), and TLI (≥ 0.95). Multiple indices were used because they provided different information about model fit (i.e., absolute fit, fit adjusting for model parsimony, fit relative to a null model); used together, these indices provided a more conservative and reliable evaluation of the solution.

Each of the overall goodness-of-fit indices suggested that the two-factor model fit the data well: $\chi^2(19)$ = 13.285, p = .82, SRMR = .019, RMSEA = 0.00 (90% CI = 0.00–.034, CFit = .99), TLI = 1.005, CFI = 1.00. Inspection of standardized residuals and modification indices indicated no localized points of ill fit in the solution (e.g., largest modification index = 3.48, largest standardized residual = 1.66). Unstandardized and completely standardized parameter estimates from this solution are presented in Figure 4.2 (standard errors of the estimates are provided in Table 4.4). All freely estimated unstandardized parameters were statistically significant (ps < .001). Factor loading estimates revealed that the indicators were strongly related to their purported factors (range of R^2s = .49–.78), consistent with the position that the NEO scales are reliable indicators of the constructs of Neuroticism and Extraversion. Moreover, estimates from the two-factor solution indicate a moderate relationship between the dimensions of Neuroticism and Extraversion (–.435), in accord with previous evidence and theory.

(*Discussion of other implications, limitations, and future directions would follow.*)

5

Model Revision and Comparison

In Chapter 4, many of the procedures and issues associated with model specification have been introduced in context of a good-fitting, two-factor CFA solution of Neuroticism and Extraversion. In this chapter, these concepts are further illustrated and extended by using initially poor-fitting solutions. Although poor-fitting models are frequently encountered in applied research, SEM sourcebooks rarely deal with this topic in detail because of the numerous potential sources of ill fit, and the fact that proper model (re)specification hinges directly on the substantive context of the analysis. Thus, although some examples are provided, readers must adapt these general guidelines and principles to the specific aspects of their own data sets and models. This chapter discusses the fact that model respecification can also be carried out to improve the parsimony and interpretability of a CFA solution. The sources of improper solutions, and ways to deal with them, are also discussed. In addition, the techniques of EFA within the CFA framework (E/CFA) and exploratory SEM (ESEM) are presented as intermediate steps between EFA and CFA. This methodology allows a researcher to explore measurement structures more fully in order to develop more realistic (i.e., better-fitting) CFA solutions. The chapter concludes with a discussion of equivalent CFA models—an issue that is germane to virtually all measurement models but is largely unrecognized in applied research.

GOALS OF MODEL RESPECIFICATION

Often a CFA model will need to be revised. The most common reason for respecification is to improve the fit of the model. In this case, the results of an initial CFA indicate that one or more of the three major criteria used to evaluate the acceptability of the model are not satisfied; that is, the model does not fit well on the whole, does not reproduce some indicator relationships well, or does not produce uniformly interpretable parameter estimates (see Chapter 4). On the basis of fit diagnostic information (e.g., modification indices) and substantive justification, the model is revised and fitted to the data again in the hopes of improving its goodness of fit. The sources of CFA model misspecification, and the methods of detecting and rectifying them, are discussed in the next section of this chapter.

In addition, respecification is often conducted to improve the parsimony and interpretability of the CFA model. Rarely do these forms of respecification improve the fit of the solution; in fact, they may worsen overall fit to some degree. For example, the results of an initial CFA may indicate that some factors have poor discriminant validity (e.g., two factors are so highly correlated that the notion that they represent distinct constructs is untenable). Based on this outcome, the model may be respecified by collapsing the highly overlapping factors; that is, the indicators that loaded on separate, overlapping factors are respecified to load on a single factor. Although this respecification may foster the parsimony and interpretability of the measurement model, it will lead to some decrement in goodness of fit relative to the more complex initial solution.

Two other types of model respecification frequently used to improve parsimony are multiple-groups solutions and higher-order factor models. These respecifications are conducted after an initial CFA model has been found to fit the data well. For instance, equality constraints are placed on parameters in multiple-groups CFA solutions (e.g., factor loadings) to determine the equivalence of a measurement model across groups (e.g., do test items show the same relationships to the underlying construct of cognitive ability in men and women?). With the possible exception of parsimony fit indices (e.g., RMSEA, TLI), these constraints will not improve the fit of the model compared to a baseline solution where the parameters are freely estimated in all groups. In higher-order CFA models, the goal is to reproduce the correlations among the factors of an initial CFA solution with a more parsimonious higher-order factor structure; for example, can the six correlations among the four factors of an initial CFA model be reproduced by a single higher-order factor? As with the previous examples, this respecification cannot improve model fit because the number of parameters in the higher-order factor structure is less than the number of freely estimated factor correlations in the initial CFA model. Multiple-groups solutions and higher-order factor models are discussed in Chapters 7 and 8, respectively.

SOURCES OF POOR-FITTING CFA SOLUTIONS

In a CFA model, the main potential sources of misspecification are the number of factors (too few or too many), the indicators (e.g., selection of indicators, patterning of indicator–factor loadings), and the error theory (e.g., uncorrelated vs. correlated measurement errors). As discussed in Chapter 4, a misspecified CFA solution may be evidenced by several aspects of the results: (1) overall goodness-of-fit indices that fall below accepted thresholds (e.g., CFI, TLI < .95); (2) large standardized residuals or modification indices; and (3) unexpectedly large or small parameter estimates or *Heywood cases*, which are estimates with out-of-range values. Standardized residuals and modification indices are often useful for determining the particular sources of strain in the solution. However, these statistics are most apt to be helpful when the solution contains minor misspecifications. When the initial model is grossly misspecified, specification searches are not nearly as likely to be successful (MacCallum, 1986).

Number of Factors

In practice, misspecifications resulting from an improper number of factors should occur rarely. When this occurs, it is likely that the researcher has moved into the CFA framework prematurely. CFA hinges on a strong conceptual and empirical basis. Thus, in addition to a compelling substantive justification, CFA model specification is usually supported by prior (but less restrictive) exploratory analyses (i.e., EFA) that have established the appropriate number of factors, and pattern of indicator–factor relationships. Accordingly, gross misspecifications (e.g., specifying too many or too few factors) should be unlikely when the proper groundwork for CFA has been conducted.

However, there are some instances where EFA has the potential to provide misleading information regarding the appropriate number of factors in CFA. This is particularly evident in scenarios where the relationships among indicators are better accounted for by correlated errors than separate factors. A limitation of EFA is that its identification restrictions preclude the specification of correlated indicator errors (see Chapters 2 and 3). Thus the results of EFA may suggest additional factors when in fact the relationships among some indicators are better explained by correlated errors from *method effects*. A method effect exists when some of the differential covariance among items is due to the measurement approach rather than the substantive latent variables. Method effects may stem from the general assessment modality (e.g., questionnaires, behavioral observation ratings, clinical interview ratings). Or more specifically, these effects may be due to similarly or reverse-worded assessment items, or to other sources such as items with differential proneness to response set, acquiescence, or social desirability.[1] A comprehensive review of the potential sources of method effects is provided in Podsakoff, MacKenzie, Lee, and Podsakoff (2003).

The influence of method effects has been illustrated in factor analyses of the Self-Esteem Scale (SES; Rosenberg, 1965) (e.g., Marsh, 1996; Tomás & Oliver, 1999; Wang, Siegal, Falck, & Carlson, 2001) and the Penn State Worry Questionnaire (PSWQ; Meyer, Miller, Metzger, & Borkovec, 1990) (e.g., Brown, 2003; Fresco, Heimberg, Mennin, & Turk, 2002). Specifically, this research has shown the impact of method effects in questionnaires composed of some mixture of positively and negatively worded items. (The SES contains 4 positively worded items, e.g., "I feel good about myself," and 3 negatively worded items, e.g., "At times I think I am no good at all"; the PSWQ contains 11 items worded in the symptomatic direction, e.g., "I worry all the time," and 5 items worded in the nonsymptomatic direction, e.g., "I never worry about anything.") In other words, the differential covariance among these items is not based on the influence of distinct, substantively important latent dimensions. Rather, this covariation reflects an artifact of response styles associated with the wording of the items (cf. Marsh, 1996).

Nevertheless, studies that conducted EFAs with these measures routinely reported two-factor solutions, with one factor composed of positively worded items (SES: Positive Self-Evaluation, PSWQ: Absence of Worry) and the second factor composed of negatively worded items (SES: Negative Self-Evaluation, PSWQ: Worry Engagement). However, subsequent CFA studies challenged the conceptual utility of these two-factor models (e.g., Brown, 2003; Hazlett-Stevens, Ullman, & Craske, 2004; Marsh, 1996). For example, because the PSWQ was designed to measure the trait of pathological worry, what is the practical and conceptual importance of a dimension of Absence of Worry (Brown, 2003)? In both lines of research, the CFA studies (e.g., Brown, 2003; Marsh, 1996) demonstrated the substantive (i.e., interpretability) and empirical (i.e., goodness of fit) superiority of single-factor solutions where the additional covariance stemming from the directionality of item wording was accounted for by correlated measurement errors (for a detailed discussion of this approach, see Chapter 6). This also highlights the importance of keeping substantive issues firmly in mind when researchers are formulating and interpreting EFA and CFA solutions. In the examples above, two-factor EFA solutions for SES and PSWQ items are apt to provide a better fit (in terms of χ^2, RMSEA, etc.) than a one-factor model. Although the viability of a one-factor model with correlated errors could not be explored within the traditional EFA framework, the acceptability of these two-factor solutions could be challenged on substantive grounds, despite their superior fit.

In the case of a typical measurement model of congeneric indicator sets (a model in which there are no double-loading items and no correlated errors), a CFA solution with too few factors will fail to adequately reproduce the observed relationships among several indicators. For instance, consider the scenario where the two-factor model of Neuroticism and Extraversion in Chapter 4 is specified as a one-factor solution with no correlated indicator errors (ML estimation). The overall fit of this solution is poor: $\chi^2(20) = 375.33$, $p < .001$, SRMR = .187, RMSEA = 0.267 (90% CI = 0.243–0.290, CFit < .001), TLI = 0.71, CFI = .79 (LISREL 9.1). As seen in Table 5.1, both the standardized residuals and modification indices indicate that as a consequence of forcing the four Extraversion indicators (E1–E4) to load on the same latent variable as the four indicators of Neuroticism (N1–N4), the parameter estimates of the solution markedly underestimate the observed relationships among the E1–E4 measures (see Chapter 4 for guidelines on interpreting modification indices and standardized residuals). Specifically, the standardized residuals for the Extraversion indicators range from 7.17 to 9.76, and modification indices range from 51.43 to 95.08. It is noteworthy that in the case of a one-factor solution, modification indices can only appear in sections of the results that pertain to indicator measurement errors (e.g., the theta-delta matrix in LISREL; see Table 5.1). Although the "true" model in this instance is a two-factor solution (see Figure 4.2, Chapter 4), the fact that fit diagnostics only appear in this fashion might lead novice CFA researchers to conclude that correlated measurement errors are required. Modification indices can point to problems with the model that are not the real source of misfit. Again, this underscores the importance of an explicit substantive basis (both conceptual and empirical) for model (re)specification; for example, specifying a model with

TABLE 5.1. Standardized Residuals and Modification for a One-Factor CFA Model of Indicators of Neuroticism and Extraversion

Standardized residuals

	N1	N2	N3	N4	E1	E2	E3	E4
N1	0.0000							
N2	2.2795	0.0000						
N3	-1.0160	-0.0931	0.0000					
N4	1.0853	0.2913	2.2968	0.0000				
E1	1.5625	2.4952	0.5194	2.7257	0.0000			
E2	1.4925	1.9975	1.2917	3.3007	9.7596	0.0000		
E3	2.3872	1.7009	1.4268	2.1411	8.9562	9.3893	0.0000	
E4	1.6416	1.8726	0.5404	2.9372	7.1785	8.4024	7.8935	0.0000

Modification indices for theta–delta

	N1	N2	N3	N4	E1	E2	E3	E4
N1	– –							
N2	6.4304	– –						
N3	1.2797	0.0099	– –					
N4	1.6175	0.1024	6.3720	– –				
E1	2.5111	6.3311	0.2743	7.6157	– –			
E2	2.2852	4.0507	1.6942	11.1424	95.0844	– –		
E3	5.8465	2.9372	2.0670	4.6888	80.0749	87.9790	– –	
E4	2.7598	3.5557	0.2961	8.8096	51.4303	70.4408	62.1663	– –

Note. Results obtained from LISREL 9.1.

correlated errors among the Extraversion indicators is not well founded in relation to a model entailing two distinct factors. In this example, the pattern of relationships in the input matrix (see Figure 4.1, Chapter 4), in addition to the aggregation of standardized residuals and modification indices associated with a set of indicators (E1–E4; see Table 5.1), provides clear empirical evidence against a simple one-factor solution. However, the determination of whether such outcomes signify distinct factors vs. method effects (or *minor factors*) is not always clear.

The nested model comparison methodology is often used in the applied literature to statistically compare the fit of CFA models that differ in their number of factors (e.g., does a two-factor model of Neuroticism and Extraversion provide a better fit to the data than a one-factor model?). Recall that a *nested model* contains a subset of the freed parameters of another solution (Chapters 3 and 4). Consider the following one-factor models involving the same set of 5 indicators; number of elements of the input variance–covariance matrix = 5(6)/2 = 15 = 5 variances, 10 covariances. Model N (the nested model) is a simple one-factor solution with no correlated measurement errors; thus it consists of 10 freed parameters (5 factor loadings, 5 indicator errors; factor variance is fixed to 1.0 to define latent variable metric), and the model has 5 df (15 – 10). Model P (the parent model) is identical to Model N, with the exception that a correlated error is specified for the fourth and fifth indicators; thus it consists of 11 freely estimated parameters (5 factor loadings, 5 indicator errors, 1 correlated error), and this model's df = 4 (15 – 11). In this scenario, Model N is nested under Model P; if a path is dropped from Model P—the correlated residual for indicators 4 and 5—Model N is formed. In other words, Model N contains a subset of the freed parameters of the parent model, Model P. (A nested model will possess a larger number of dfs than the parent model; in this case, the df difference is 1, i.e., 5 – 4.) In Model N, the correlations among all indicator errors are fixed to zero. In Model P, this is not the case because the correlation between the residuals of two indicators is freely estimated.

Models that differ in the number of latent variables are considered nested models. For example, consider the one- and two-factor solutions for the 8 indicators of Neuroticism and Extraversion (a = 36 = 8 variances and 28 covariances). The two-factor solution discussed in Chapter 4 contains 17 freely estimated parameters: 6 loadings, 8 indicator errors, 2 factor variances, and 1 factor covariance (the loadings of N1 and E1 are fixed as marker indicators). Thus, this two-factor model has df = 19 (36 – 17). In contrast, the one-factor solution presented above contains 16 freed parameters: 7 loadings, 8 errors, and 1 factor variance (the loading of N1 is fixed as the marker indicator). The one-factor solution has df = 20 (36 – 16). Although structurally more discrepant than the Model N vs. Model P example described previously, the one-factor model could be construed as nested under the two-factor solution, again with a difference of df = 1 (i.e., 20 – 19). This single df difference relates to factor correlation in the two-factor solution. Specifically, the two-factor solution will provide a model fit identical to the one-factor model if the correlation between Neuroticism and Extraversion is fixed to 1.0. Under this specification, the factors will be identical (i.e., 100% shared variance), and thus the indicators will relate to the factors in the same fashion as in the one-factor solution. Accordingly, the one-factor solution can be viewed as a more constrained version of the two-factor model; that is, in the two-factor model the factor correlation is freely estimated, and thus has one fewer *df*. This principle generalizes to solutions with larger numbers of factors (e.g., a four-factor solution vs. a three-factor solution in which two of the factors from the former model are collapsed).

When models are nested, the χ^2 statistic can be used to statistically compare the fit of the solutions. Used in this fashion, χ^2 is often referred to as the χ^2 difference test (χ^2_{diff}) or the *nested χ^2 test* (an equally valid but lesser used term in this context is the *likelihood ratio test*). If a model is nested under a parent model, the simple difference in the model χ^2s is also distributed as χ^2 in many circumstances (for exceptions to this guideline, see Chapter 9). For example, in the case of the one- vs. two-factor model of Neuroticism and Extraversion, the χ^2 difference test is calculated as follows:

	df	χ^2
One-factor model	20	375.33
Two-factor model	19	13.29 (see Chapter 4)
χ^2 difference (χ^2_{diff})	1	362.04

Thus $\chi^2_{diff}(1) = 362.04$. In this example, the χ^2 difference test has 1 *df*, which reflects the difference in model *df*s for the one- and two-factor solutions (20 –19). Therefore, the critical value for χ^2_{diff} in this example is 3.84 ($\alpha = .05$, *df* = 1). Because the χ^2_{diff} test value exceeds 3.84 (362.04), it can be concluded that the two-factor model provides a significantly better fit to the data than the one-factor model. It is also important that the two-factor model fits the data well. Use of the χ^2 difference test to compare models is not justified when neither solution provides an acceptable fit to the data.

However, some methodologists would argue that models differing in regard to the number of factors are not nested. This is because the restriction required to make a two-factor model equivalent to a one-factor model (or a three-factor model equivalent to a two-factor model, etc.) entails a fixed parameter that is on the border of admissible parameter space: a factor correlation of 1.0 (i.e., a factor correlation > 1.0 constitutes an out-of-range parameter). In other words, nested models that contain such border-line values (e.g., unit factor correlations, indicator error variances fixed to zero) may not yield proper χ^2 distributions, and thus the χ^2 difference test may be compromised. Although it is not clear how often and to what degree this statistical technicality results in improper conclusions in applied research, the researcher should also consider and report information criterion indices (e.g., Akaike information criterion, Bayesian information criterion) in instances involving the comparison of CFA models that differ in the number of factors. These indices are discussed later in this chapter.

From the illustration above, the relationship between χ^2_{diff} and modification indices should be more apparent. As discussed in Chapter 4, modification indices represent the predicted decrease in model χ^2 if a fixed or constrained parameter is freely estimated. Accordingly, modification indices (and univariate Lagrange multipliers; cf.

EQS) reflect expected χ^2 changes associated with a single *df*. However, nested models often involve solutions that differ by more than a single *df*. For example, a two-factor and a three-factor measurement model of the same data set will differ by two degrees of freedom; that is, the two-factor model contains one factor covariance, whereas the three-factor model contains three factor covariances. Measurement invariance evaluation (Chapter 7) typically entails simultaneously placing constraints on multiple parameters (e.g., constraining factor loadings to equality in two or more groups). Such models are also nested, but differ by more than a single *df*. Moreover, modification indices differ from χ^2_{diff} test values in the fact that they represent *estimates* of how much model χ^2 will decrease after freeing a fixed or constrained parameter. Quite often, the *actual* difference in model χ^2 (reflected by χ^2_{diff}) produced by a single-*df* model modification differs from the estimate of χ^2 change provided by the modification index (this is also true for expected parameter change values; see Chapter 4).

Although applied researchers often compare CFA measurement models that differ in number of factors, it is important that a strong conceptual rationale exists for doing so. Occasionally such analyses are "straw man" comparisons, where the models specified as competing solutions to the hypothesized model have dubious substantive bases and little likelihood of providing an equivalent or superior fit to the data (e.g., in the psychometric evaluation of a questionnaire, comparing a three-factor model to a one-factor model, when in fact the measure was designed to be multifactorial and prior EFA research has supported this structure). Although the one- vs. two-factor models of Neuroticism and Extraversion have been presented in this chapter to discuss some of the concepts and issues of model comparison, this criticism will apply; that is, there is no basis for the one-factor model, in view of compelling theory and scientific evidence for the distinctiveness of Neuroticism and Extraversion. Thus the researcher should provide justification for both the hypothesized and alternative CFA models.

If too many factors have been specified in the CFA model, this is likely to be detected by correlations between factors that approximate ±1.0, and so the latent dimensions have poor *discriminant validity*. In applied research, a factor correlation that equals or exceeds .85 is often used as the cutoff criterion for problematic discriminant validity (cf. guidelines for multicollinearity in regression; Cohen et al., 2003; Tabachnick & Fidell, 2013). When factors overlap to this degree, it may be possible to combine factors to acquire a more parsimonious solution. The goal of such respecification is not to improve the overall fit of the model (e.g., a model with fewer factors entails a smaller number of freely estimated parameters), but ideally the fit of the more parsimonious solution will be similar to that of the initial model (assuming that overall fit of the initial model is satisfactory, except for the excessive correlation between two factors). Again, revisions of this nature require a clear rationale.

Earlier in this chapter, studies on method effects arising from reverse-worded items have been briefly discussed. A paper from this literature provides an applied example of excessive overlap between latent variables. Recall that in the Brown (2003) study, a one-factor model of the PSWQ was hypothesized in which all 16 items were specified to load on a single dimension of Pathological Worry, with correlated errors to account for method effects from reversed items. Because a two-factor solution (Worry Engagement, Absence of Worry; cf. Fresco et al., 2002) had prevailed in prior EFA and CFA studies of the PSWQ, it was also fitted to the data in Brown (2003) to serve as a competing solution to the hypothesized one-factor model. Results of this CFA indicated poor discriminant validity of the Worry Engagement and Absence of Worry latent dimensions; that is, these factors were highly correlated (e.g., $r = .87$; Brown, 2003). This finding, along with other evidence and considerations (e.g., superior fit of the hypothesized one-factor model, no substantive basis for an Absence of Worry dimension), strongly challenged the acceptability of a two-factor solution.

An alternative to combining two highly correlated factors is to drop one of the factors and its constituent indicators. Although the best remedial approach depends on the specific scientific context, dropping a factor may be favored if one of the factors is defined by only a few indicators or has limited variance, or if substantive and practical considerations support this strategy (e.g., there are clear advantages to developing a briefer questionnaire).

Indicators and Factor Loadings

Another potential source of CFA model misspecification is an incorrect designation of the relationships between indicators and the factors. This can occur in the following manners (assuming that the correct number of factors has been specified): (1) The indicator is specified to load on one factor, but actually has salient loadings on two or more factors; (2) the indicator is specified to load on the wrong factor; or (3) the indicator is specified to load on a factor, but actually has no salient relationship to any factor. Depending on the problem, the remedy will be either to respecify the pattern of relationships between the indicator and the factors, or to eliminate the indicator from the model altogether.

Fit diagnostics for these forms of misspecifications are presented in context of the CFA measurement model presented in Figure 5.1. The path diagram in this figure represents the latent structure of the population measurement model; a sample variance–covariance matrix for this model ($N = 500$) has been generated by using the Monte Carlo utility in Mplus (see Chapter 10). In this example, a researcher has developed a 12-item questionnaire (items are rated on 0–8 scales) designed to assess young adults' motives to consume alcoholic beverages (cf. Cooper, 1994). The measure is intended to assess three facets of this construct (4 items each): (1) Coping Motives (items 1–4), (2) Social Motives (items 5–8), and (3) Enhancement Motives (items 9–12). All of the items are phrased in the positive direction (e.g., item 5: "Because you feel more self-confident about yourself"), with the exception of items 11 and 12, which are reverse-worded and scored. Although the three-factor model is intended to possess congeneric item sets, the "true" model contains one double-loading item (item 4) and one correlated error resulting from the reverse wording (cf. Marsh, 1996). Figure 5.1 presents the completely standardized parameter estimates and overall fit of this model; LISREL syntax and the input matrix are provided in Table 5.2. As seen in this figure, fit indices are consistent

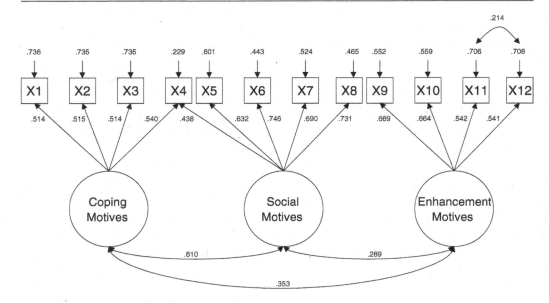

FIGURE 5.1. Completely standardized parameter estimates from the three-factor CFA model of a 12-item questionnaire about alcohol-drinking motives (*N* = 500). Overall fit of the model: $\chi^2(49)$ = 44.955, *p* = .638, SRMR = .025, RMSEA = 0.00 (90% CI = 0.00–0.025, CFit *p* = 1.00), TLI = 1.002, CFI = 1.00. All freely estimated unstandardized parameter estimates are statistically significant (*p* < .001) (LISREL 9.1).

with good model fit; this outcome is supported by standardized residuals below 2.00 and modification indices below 4.00. All parameter estimates are statistically significant (*p*s < .001) and of a magnitude in accord with expectation.

Now consider the various forms of indicator–factor misspecifications previously mentioned. For instance, a poor-fitting model may result from specification of congeneric indicator sets, when in fact some indicators load on more than one factor. In the current example, the researcher is predicting such a model in which four items each load on the three latent dimensions of drinking motives (including item 4, which is expected to load only on the Social Motives factor). When this model is fitted to the data (the correlated error of items 11 and 12 is included), the following fit indices result: $\chi^2(50)$ = 61.658, *p* = .125, SRMR = .032, RMSEA = 0.022 (90% CI = 0.00–0.038, CFit = .999, TLI = 0.994, CFI = .996. Note that this solution has one more *df* than the "true" model because the relationship between item 4 and Social Motives has been fixed to zero; hence this solution is nested under the "true" model.[2]

Selected results of this solution are presented in Table 5.3: standardized residuals, modification indices, completely standardized expected parameter change (EPC) values, and completely standardized estimates of factor loadings and factor correlations. These data exemplify several points made earlier in this chapter and in Chapter 4. For instance, researchers (e.g., MacCallum, 1986) have found that specification searches based on modification indices are more likely to be successful when the model contains only minor misspecifications. Thus, by following the recommendation (cf. Jöreskog,

TABLE 5.2. LISREL Syntax for a Three-Factor Model of a Questionnaire about Alcohol-Drinking Motives

```
TITLE THREE FACTOR MODEL FOR DRINKING MOTIVES
DA NI=12 NO=500 MA=CM
LA
X1 X2 X3 X4 X5 X6 X7 X8 X9 X10 X11 X12
KM
1.000
0.300 1.000
0.229 0.261 1.000
0.411 0.406 0.429 1.000
0.172 0.252 0.218 0.481 1.000
0.214 0.268 0.267 0.579 0.484 1.000
0.200 0.214 0.241 0.543 0.426 0.492 1.000
0.185 0.230 0.185 0.545 0.463 0.548 0.522 1.000
0.134 0.146 0.108 0.186 0.122 0.131 0.108 0.151 1.000
0.134 0.099 0.061 0.223 0.133 0.188 0.105 0.170 0.448 1.000
0.160 0.131 0.158 0.161 0.044 0.124 0.066 0.061 0.370 0.350 1.000
0.087 0.088 0.101 0.198 0.077 0.177 0.128 0.112 0.356 0.359 0.507 1.000
SD
2.06 1.52 1.92 1.41 1.73 1.77 2.49 2.27 2.68 1.75 2.57 2.66
MO NX=12 NK=3 PH=SY,FR LX=FU,FR TD=SY,FR
LK
Coping Social Enhance
PA LX
0 0 0
1 0 0
1 0 0
1 1 0     !double loading item
0 0 0
0 1 0
0 1 0
0 1 0
0 0 0
0 0 1
0 0 1
0 0 1
VA 1.0 LX(1,1) LX(5,2) LX(9,3)
PA TD
1
0 1
0 0 1
0 0 0 1
0 0 0 0 1
0 0 0 0 0 1
0 0 0 0 0 0 1
0 0 0 0 0 0 0 1
0 0 0 0 0 0 0 0 1
0 0 0 0 0 0 0 0 0 1
0 0 0 0 0 0 0 0 0 0 1
0 0 0 0 0 0 0 0 0 0 1 1   !correlated residual
PA PH
1
1 1
1 1 1
OU ME=ML RS MI SC AD=OFF IT=100 ND=4
```

TABLE 5.3. Selected Results for the Drinking Motives Three-Factor Model When a Double-Loading Item Has Not Been Specified

Standardized Residuals

X1	0.00										
X2	3.22	0.00									
X3	1.01	1.89	0.00								
X4	-0.14	-2.26	-0.22	0.00							
X5	-1.36	0.96	-0.29	-0.08	0.00						
X6	-1.44	0.29	-0.05	0.79	0.60	0.00					
X7	-1.16	-0.79	-0.21	1.23	-0.53	-1.55	0.00				
X8	-2.15	-0.75	-2.56	-0.81	0.07	0.21	1.19	0.00			
X9	0.94	1.22	0.19	-1.14	-0.03	-0.45	-0.77	0.29	0.00		
X10	0.94	0.04	-1.00	0.39	0.27	1.34	-0.88	0.86	0.34	0.00	
X11	1.99	1.27	1.87	-0.35	-1.44	0.19	-1.13	-1.46	0.85	-0.64	0.00
X12	0.21	0.22	0.47	0.81	-0.60	1.64	0.51	-0.09	-0.33	-0.22	0.00	0.00

Modification Indices for LAMBDA-X

	Coping	Social	Enhance
X1	- -	6.57	1.89
X2	- -	0.03	0.89
X3	- -	1.90	0.02
X4	- -	**16.16**	2.15
X5	0.03	- -	0.19
X6	0.44	- -	1.37
X7	0.80	- -	1.06
X8	1.92	- -	0.01
X9	0.76	0.35	- -
X10	0.18	0.53	- -
X11	0.16	1.53	- -
X12	0.77	1.16	- -

Completely Standardized Expected Change for LAMBDA-X

	Coping	Social	Enhance
X1	- -	-0.27	0.07
X2	- -	-0.02	0.05
X3	- -	-0.15	0.01
X4	- -	**0.90**	-0.10
X5	-0.01	- -	-0.02
X6	0.06	- -	0.05
X7	0.08	- -	-0.05
X8	-0.12	- -	0.01
X9	-0.04	-0.03	- -
X10	0.02	0.04	- -
X11	-0.02	-0.06	- -
X12	0.04	0.05	- -

(continued)

TABLE 5.3. *(continued)*

Modification Indices for THETA-DELTA

	X1	X2	X3	X4	X5	X6	X7	X8	X9	X10	X11	X12
	----	----	----	----	----	----	----	----	----	----	----	----
X1	- -											
X2	10.37	- -										
X3	1.02	3.58	- -									
X4	0.03	7.02	0.07	- -								
X5	0.45	1.45	0.03	0.10	- -							
X6	0.47	0.11	0.23	0.08	0.37	- -						
X7	0.28	1.12	0.01	2.22	0.28	2.45	- -					
X8	1.53	0.36	5.21	0.14	0.00	0.04	1.43	- -				
X9	0.50	2.18	0.27	2.59	0.45	1.28	0.01	0.93	- -			
X10	0.04	0.70	3.39	0.49	0.10	0.83	1.71	0.68	0.12	- -		
X11	4.98	2.06	5.92	1.37	0.72	0.12	0.41	1.58	0.67	0.24	- -	
X12	1.93	0.87	0.32	1.14	0.74	1.53	0.68	0.23	0.26	0.00	- -	- -

Completely Standardized Solution

    LAMBDA-X

	Coping	Social	Enhance
	------	------	-------
X1	0.4311	- -	- -
X2	0.4349	- -	- -
X3	0.4502	- -	- -
X4	0.9548	- -	- -
X5	- -	0.6333	- -
X6	- -	0.7478	- -
X7	- -	0.6895	- -
X8	- -	0.7291	- -
X9	- -	- -	0.6663
X10	- -	- -	0.6688
X11	- -	- -	0.5370
X12	- -	- -	0.5414

    PHI

	Coping	Social	Enhance
	------	------	-------
Coping	1.0000		
Social	0.7978	1.0000	
Enhance	0.3348	0.2917	1.0000

1993) of freeing a fixed or constrained parameter with the largest modification index (provided that this parameter can be interpreted substantively), the correct model can be readily obtained. As seen in Table 5.3, the fixed parameter with by far the highest modification index is λ_{42} (16.16), corresponding to a double loading of item 4 with the Social Motives factor (cf. Figure 5.1).

Second, these results demonstrate that the acceptability of the model should not be based solely on indices of overall model fit, although this practice is still somewhat common in the applied literature. Note that all descriptive fit indices are consistent with good model fit. However, these global indices mask the fact that at least one relationship is not well represented by the model (i.e., the double loading of item 4).

Third, the modification index (16.16) and completely standardized EPC values (0.90) do not correspond exactly to the actual change in model χ^2 and parameter estimate, respectively, when the relationship between item 4 and the Social Motives factor is freely estimated. As seen in Figure 5.1, the completely standardized parameter estimate of this double loading is .438. This revision produces a significant improvement in model fit, $\chi^2_{diff}(1) = 16.70$, $p < .01$ (i.e., $61.658 - 44.955 = 16.70$). However, the χ^2 difference is slightly larger than the modification index associated with this parameter (16.16). Again, this underscores the fact that modification indices and EPC values are approximations of model change if a fixed or constrained parameter is freed.

Fourth, aspects of these results illustrate that parameter estimates should not be interpreted when the model is poor-fitting.[3] For instance, it can be seen in Table 5.3 that some of the consequences of failing to specify the item 4 cross-loading include an inflated estimate of the factor correlation between Coping Motives and Social Motives (.798), an inflated estimate of the loading of item 4 on Coping Motives (.955), and underestimates of the loadings of items 1–3 on Coping Motives. Because item 4 is not specified to load on Social Motives, its moderate relationships with the Social Motives indicators (items 5–8, $rs = .48–.58$; see input correlation matrix in Table 5.2) have to be reproduced primarily through its factor loading with Coping Motives ($\lambda_{41} = .955$) and the factor correlation of Coping Motives and Social Motives ($\phi_{21} = .798$); for example, the model-predicted relationship of X4 and X5 (completely standardized) $= \lambda_{41}\phi_{21}\lambda_{52} = .955(.798)(.633) = .482$ (observed r of X4 and X5 is .481; see Table 5.2). Thus, as the result of the iterative process of minimizing F_{ML}, these parameters (λ_{41}, ϕ_{21}) are inflated (and λ_{11}, λ_{21}, and λ_{31} are underestimated) in order to acquire a model that best reproduces the observed relationships ($S - \Sigma$). This also exemplifies why Heywood cases (offending estimates) may arise from a misspecified solution: Through the iterative process, the parameters may take on out-of-range values to minimize F_{ML}. In this example, a solution with Heywood cases can be produced by fitting the Figure 5.1 model, but specifying X7 to load on Coping Motives instead of Social Motives. Table 5.3 also shows that the standardized residuals are not clearly diagnostic of the source of misspecification in this instance (e.g., these values suggest that the relationships between item 4 and the Social Motives indicators have been reproduced adequately). This is because the ML iterations have resulted in parameter estimates that are able to reproduce most relationships reasonably well. Focal areas of strains are nonetheless evident; for example, some relationships are overestimated (e.g., standardized residual for X3, X8 = −2.56), and one relationship is underestimated (standardized residual for X1, X2 = 3.22), each relating to strains in the model for reproducing the relationship of the Coping Motives and Social Motives indicators. In addition to emphasizing the need to examine multiple aspects of fit (i.e., overall fit, standardized residuals, modification indices, parameter estimate values), this illustrates that respecifying one parameter may successfully eliminate what seem to be multiple strains because all relationships in the model are interdependent.

Another way an indicator–factor relationship may be misspecified is when an indicator loads on the wrong factor. Table 5.4 presents selected results of an analysis of the drinking motives model in which item 12 is specified to load on the Social Motives factor (its relationship to Enhancement Motives is fixed to 0, but its correlated error with item 11 is freely estimated). As in the prior example, overall fit statistics suggest acceptable model fit: $\chi^2(49) = 126.590$, $p < .001$, SRMR = .063, RMSEA = 0.056 (90% CI = 0.044–0.068, CFit = .186), TLI = 0.961, CFI = .971, based on the guidelines recommended by Hu and Bentler (1999). However, as shown in Table 5.4, modification indices and standardized residuals indicate localized points of strain in the solution. As before, simply freeing the fixed parameter ($\lambda_{12,3}$) with the largest modification index (76.93) will result in the proper measurement model. This misspecification is evident in other ways, such as (1) large standardized residuals indicating that the observed relationships among the Enhancement Motives indicators (e.g., item 12 with items 9–11) are being underestimated by the model parameter estimates; and (2) a factor loading of item 12 with Social Motives ($\lambda_{12,2} = .20$) that, while statistically significant ($z = 4.46$, not shown in Table 5.4), is well below conventional guidelines for a "salient" indicator–factor relationship. Note that another impact of this misspecification is an elevated model estimate of the correlated error of items 11 and 12 ($\delta_{12,11} = .40$ vs. $\delta_{12,11} = .21$ in the properly specified solution). This inflated estimate reflects the attempts of the iterations to reproduce the observed correlation between items 11 and 12 ($r = .507$), although the model still underestimates this relationship (standardized residual = 7.31). In the correct solution, the sample correlation is reproduced by the sum of the product of the item 11 and item 12 factor loadings [$\lambda_{11,3}\lambda_{12,3} = .542 (.541) = .293$] and the correlated error of these indicators (i.e., .293 + .214 = .507). Because the misspecified model cannot use the product of the item 11 and item 12 factor loadings (because item 12 loads on a different factor than item 11 does), the solution must rely more on the correlated error ($\delta_{12,11}$) to reproduce this observed relationship (a very small portion of this relationship is also estimated in the solution by $\lambda_{12,2}\phi_{32}\lambda_{11,3}$).

Unlike some previous examples, the Figure 5.1 model and the current model are not nested. Both models entail 29 freely estimated parameters ($df = 49$), and thus one does not contain a subset of the freed parameters of the other. Therefore, the χ^2_{diff} test cannot be used to statistically compare these two solutions. In this scenario, a strategy that can be used to compare solutions is to qualitatively evaluate each with regard to the three major aspects of model acceptability: overall goodness of fit, focal areas of ill fit, and interpretability/strength of parameter estimates. For instance, if one model satisfies each of these criteria and the other does not, the former model would be favored.

In addition, methodologists have developed procedures for using χ^2 in the comparison of non-nested models. Two popular methods are the Akaike information criterion (AIC; Akaike, 1987) and the Bayesian information criterion (BIC; Schwarz, 1978). These indices are closely related, in that they both take into account model fit (as reflected by the estimated model's –2log-likelihood, or –2LL; provided by most software programs in the fit statistics section of the output)[4] and model complexity–parsimony (as reflected by model df or the number of freely estimated parameters; cf. the RMSEA), although the penalty function is more severe in the BIC.

TABLE 5.4. Selected Results for the Drinking Motives Three-Factor Model When Item 12 Has Been Specified to Load on the Wrong Factor

Standardized Residuals

	X1	X2	X3	X4	X5	X6	X7	X8	X9	X10	X11	X12
	----	----	----	----	----	----	----	----	----	----	----	----
X1	0.00											
X2	1.31	0.00										
X3	-1.49	-0.24	0.00									
X4	-0.35	-1.10	1.83	0.00								
X5	-0.76	1.57	0.60	-0.21	0.00							
X6	-0.68	1.08	1.07	0.37	0.77	0.00						
X7	-0.50	-0.09	0.77	0.93	-0.37	-1.45	0.00					
X8	-1.42	0.01	-1.42	-1.28	0.20	0.23	1.21	0.00				
X9	0.35	0.68	-0.38	-1.60	-0.40	-1.01	-1.26	-0.20	0.00			
X10	0.39	-0.60	-1.64	0.21	-0.03	0.97	-1.28	0.49	1.39	0.00		
X11	2.36	1.62	2.31	1.15	-0.77	0.96	-0.40	-0.66	6.05	5.17	8.46	
X12	0.55	0.57	0.89	1.48	-1.54	0.85	-0.41	-1.26	7.33	7.40	7.31	0.38

Modification Indices for LAMBDA-X

	Coping	Social	Enhance
	------	------	-------
X1	- -	1.97	1.03
X2	- -	0.67	0.29
X3	- -	0.34	0.28
X4	- -	- -	0.64
X5	0.13	- -	0.09
X6	0.77	- -	0.09
X7	0.18	- -	2.60
X8	3.60	- -	0.00
X9	0.98	1.21	- -
X10	0.34	0.67	- -
X11	4.59	0.17	- -
X12	1.11	- -	**76.93**

Completely Standardized Expected Change for LAMBDA-X

	Coping	Social	Enhance
	------	------	-------
X1	- -	-0.12	0.06
X2	- -	0.07	0.03
X3	- -	0.05	-0.03
X4	- -	- -	-0.04
X5	0.03	- -	-0.02
X6	0.06	- -	0.01
X7	0.03	- -	-0.08
X8	-0.14	- -	0.00
X9	-0.07	-0.07	- -
X10	-0.04	0.05	- -
X11	0.12	0.02	- -
X12	0.08	- -	**0.53**

Modification Indices for THETA-DELTA

	X1	X2	X3	X4	X5	X6	X7	X8	X9	X10	X11	X12
	----	----	----	----	----	----	----	----	----	----	----	----
X1	- -											
X2	1.73	- -										

(continued)

TABLE 5.4. (continued)

X3	2.23	0.06	- -									
X4	0.80	3.15	1.56	- -								
X5	0.26	2.44	0.17	0.39	- -							
X6	0.15	0.60	0.83	0.03	0.60	- -						
X7	0.01	0.43	0.37	1.15	0.14	2.15	- -					
X8	0.37	0.02	2.95	0.12	0.04	0.05	1.48	- -				
X9	0.03	1.05	0.02	1.78	0.03	1.31	0.05	0.26	- -			
X10	0.01	1.19	3.84	0.69	0.07	0.72	2.30	0.07	2.29	- -		
X11	2.77	0.76	3.97	0.20	0.30	0.66	0.04	0.68	0.00	1.33	- -	
X12	0.59	0.38	0.53	0.36	1.85	0.12	0.20	1.15	20.95	22.71	- -	- -

Completely Standardized Solution

LAMBDA-X

	Coping	Social	Enhance
X1	0.5163	- -	- -
X2	0.5178	- -	- -
X3	0.5158	- -	- -
X4	0.5301	0.4468	- -
X5	- -	0.6297	- -
X6	- -	0.7471	- -
X7	- -	0.6886	- -
X8	- -	0.7293	- -
X9	- -	- -	0.6683
X10	- -	- -	0.6599
X11	- -	- -	0.3816
X12	- -	0.2048	- -

PHI

	Coping	Social	Enhance
Coping	1.0000		
Social	0.6084	1.0000	
Enhance	0.3524	0.3222	1.0000

THETA-DELTA

	X1	X2	X3	X4	X5	X6	X7	X8	X9	X10	X11	X12
X1	0.73											
X2	- -	0.73										
X3	- -	- -	0.73									
X4	- -	- -	- -	0.23								
X5	- -	- -	- -	- -	0.60							
X6	- -	- -	- -	- -	- -	0.44						
X7	- -	- -	- -	- -	- -	- -	0.53					
X8	- -	- -	- -	- -	- -	- -	- -	0.47				
X9	- -	- -	- -	- -	- -	- -	- -	- -	0.55			
X10	- -	- -	- -	- -	- -	- -	- -	- -	- -	0.56		
X11	- -	- -	- -	- -	- -	- -	- -	- -	- -	- -	0.85	
X12	- -	- -	- -	- -	- -	- -	- -	- -	- -	- -	0.40	0.96

Although there is some variation across latent variable software programs in how the AIC and BIC are calculated, the AIC is computed as follows in LISREL and Mplus (the formula below is equivalent to the one shown in LISREL output):

$$AIC = -2LL + 2b \qquad (5.1)$$

where b is the number of freely estimated parameters in the model (in Figure 5.1, $b = 29$). Thus, for the Figure 5.1 model, AIC would be 12,895.089; that is, 12837.089 + 2(29). The BIC is calculated with the equation below:

$$BIC = -2LL + [b * \ln(N)] \qquad (5.2)$$

Thus the BIC for the Figure 5.1 solution is 13,017.313; that is, 12837.089 + [29 * ln(500]. For the current misspecified model involving the item 12 loading, AIC and BIC are 12,976.724 and 13,098.948, respectively.

Whereas the AIC and BIC can be used in tandem with χ^2_{diff} for the comparison of nested solutions, these indices are more often considered in the evaluation of competing, non-nested models. Generally, models with the lowest AIC and BIC values are judged to fit the data better in relation to alternative solutions (regardless of the method used in their calculation). From this standpoint, the Figure 5.1 solution will be favored over the current model because it is associated with lower AIC and BIC values (although in the present case, this can be determined simply by comparing model χ^2s because the two solutions do not differ in df). It is important to emphasize that, unlike χ^2_{diff}, the AIC and BIC do not provide a *statistical* comparison of competing models. Rather, these indices foster the comparison of the overall fit of models, with adjustments for the complexity of each.[5]

Another possible problematic outcome of a CFA solution is that an indicator does not load on any factor. This problem is readily diagnosed by results showing that the indicator has a nonsignificant or nonsalient loading on the conjectured factor, as well as modification indices (and EPC values) suggesting that the fit of the model cannot be improved by allowing the indicator to load on a different factor. This conclusion should be further supported by inspection of standardized residuals (<2.00) and sample correlations pointing to the fact that the indicator is weakly related to other indicators in the model. Unlike the prior examples, this scenario does not substantially degrade the fit of the model (assuming that the model is well specified otherwise). Thus, although the proper remedial action is to eliminate the problematic indicator, the overall fit of the model will usually not improve. In addition, the revised model is not nested with the initial solution because the input variance–covariance matrix has changed.

Although most of this section has described various ways a model may be respecified while retaining the original set of indicators, it is often the case that a much better-fitting solution can be obtained by simply dropping bad indicators from the model. For example, an indicator may be associated with several large modification indices and standardized residuals, reflecting that the indicator is rather nonspecific in that it evidences similar relationships to all latent variables in the solution. Simply dropping this indicator from the model will eliminate multiple strains in the solution.

Correlated Errors

A CFA solution can also be misspecified with respect to the relationships among the indicator error variances. When no correlated errors (error covariances) are specified, the researcher is asserting that all of the covariation among indicators loading on a given factor is due to that latent dimension (and that all measurement error is random). Correlated errors between indicators are specified on the basis of the notion that some of the covariance in the indicators not explained by the latent variable is due to another exogenous common cause (i.e., some of the shared variance is due to the factor, while some of this variance is due to an outside cause). It is also possible that correlated errors exist for indicators that load on separate factors. In this case, most of the shared variance may be due to an outside cause (some of the observed covariation may also be reproduced by the product of the indicators' factor loadings and the factor correlation). Occasionally variables measured by single indicators are included in a CFA model (see Chapters 4 and 7). Because such models would be underidentified, it is not possible to correlate the measurement errors of single indicators with the errors of other indicators in the solution.

As discussed earlier in this chapter, in the case of the analysis of multiple questionnaire items, correlated errors may arise from items that are very similarly worded, reverse-worded, differentially prone to social desirability, or the like. In CFA construct validation studies, correlated errors may be needed to account for method covariance, such as in the analysis of indicators collected from different assessment modalities (e.g., self-report, behavioral observation, interview rating; cf. the multitrait–multimethod (MTMM) approach, discussed in Chapter 6).

Unnecessary correlated errors can be readily detected by results indicating their statistical or clinical nonsignificance (e.g., z values below 1.96, or very small parameter estimates that reflect trivial shared variance of the errors). The next step is simply to refit the model by fixing these covariances to zero, and verify that this respecification does not result in a significant decrease in model fit. The χ^2_{diff} test can be used in this situation (but see below). The more common difficulty is the failure to include salient correlated errors in the solution. As in the prior examples, the omission of these parameters is typically manifested by large standardized residuals, modification indices, and EPC values.

Table 5.5 presents selected results of the drinking motives CFA solution where the correlated error between items 11 and 12 has not been specified (all other aspects of the solution have been properly specified). Although the overall fit of the model is good—for example, $\chi^2(50) = 69.44$, $p = .036$, TLI = 0.99, CFI = .99, RMSEA = 0.028, SRMR = .031—standardized residuals (5.04) and modification indices ($\delta_{12,11} = 25.94$) indicate that the relationship between these items has not been adequately reproduced by the model's parameter estimates. The need for a correlated error for these items is further

TABLE 5.5. Selected Results for the Drinking Motives Three-Factor Model without the Correlated Error of Item 11 and Item 12

Standardized Residuals

	X1	X2	X3	X4	X5	X6	X7	X8	X9	X10	X11	X12
X1	0.00											
X2	1.40	0.00										
X3	-1.43	-0.19	0.00									
X4	-0.33	-1.04	1.56	0.00								
X5	-0.76	1.58	0.57	-0.22	0.00							
X6	-0.65	1.11	1.06	0.55	0.70	0.00						
X7	-0.50	-0.09	0.74	0.94	-0.51	-1.54	0.00					
X8	-1.42	0.02	-1.45	-1.25	0.06	0.18	1.07	0.00				
X9	0.67	0.97	-0.02	0.08	0.56	0.34	-0.06	0.97	0.00			
X10	0.70	-0.21	-1.20	1.36	0.88	2.00	-0.10	1.54	4.46	0.00		
X11	1.09	0.30	1.02	-1.51	-1.84	-0.22	-1.59	-2.01	-1.90	-2.74	0.00	
X12	-0.87	-0.85	-0.50	-0.14	-0.93	1.39	0.20	-0.48	-2.67	-2.11	5.04	0.00

Modification Indices for LAMBDA-X

	Coping	Social	Enhance
X1	--	1.93	0.31
X2	--	0.71	0.00
X3	--	0.30	0.04
X4	--	--	0.11
X5	0.08	--	0.46
X6	1.40	--	2.70
X7	0.24	--	0.47
X8	3.64	--	0.09
X9	0.24	0.31	--
X10	0.56	3.16	--
X11	0.12	4.91	--
X12	0.60	0.02	--

Completely Standardized Expected Change for LAMBDA-X

	Coping	Social	Enhance
X1	--	-0.12	0.03
X2	--	0.07	0.00
X3	--	0.05	-0.01
X4	--	--	-0.01
X5	0.02	--	-0.03
X6	0.09	--	0.07
X7	0.04	--	-0.03
X8	-0.14	--	-0.01
X9	0.03	0.03	--
X10	0.04	0.08	--
X11	-0.02	-0.10	--
X12	-0.04	0.01	--

Modification Indices for THETA-DELTA

	X1	X2	X3	X4	X5	X6	X7	X8	X9	X10	X11	X12
X1	--											
X2	1.96	--										
X3	2.04	0.03	--									
X4	0.77	3.02	1.09	--								
X5	0.24	2.50	0.17	0.34	--							
X6	0.19	0.51	0.68	0.03	0.50	--						
X7	0.02	0.47	0.31	1.05	0.26	2.40	--					
X8	0.36	0.01	3.07	0.07	0.00	0.03	1.17	--				
X9	0.15	1.29	0.02	0.54	0.85	0.98	0.02	1.80	--			
X10	0.10	0.47	3.82	1.13	0.29	0.52	1.59	1.26	20.04	--		
X11	2.02	0.34	3.49	1.31	1.02	0.34	0.17	1.79	3.64	7.59	--	
X12	1.91	0.97	0.08	0.32	1.22	1.52	0.58	0.59	7.20	4.49	**25.94**	--

Completely Standardized Expected Change for THETA-DELTA

	X1	X2	X3	X4	X5	X6	X7	X8	X9	X10	X11	X12
X1	--											
X2	0.06	--										
X3	-0.06	-0.01	--									
X4	0.03	-0.07	0.04	--								
X5	-0.02	0.05	0.01	-0.02	--							
X6	-0.01	0.02	0.03	0.00	0.02	--						
X7	0.00	-0.02	0.02	0.03	-0.02	-0.05	--					
X8	-0.02	0.00	-0.05	-0.01	0.00	0.01	0.03	--				
X9	0.01	0.04	-0.01	-0.02	0.03	-0.03	0.00	0.04	--			
X10	0.01	-0.02	-0.07	0.03	0.02	0.02	-0.04	0.03	0.18	--		
X11	0.05	0.02	0.06	-0.03	-0.03	0.02	-0.01	-0.04	-0.08	-0.12	--	
X12	-0.05	-0.03	-0.01	0.01	-0.03	0.03	0.02	-0.02	-0.12	-0.09	**0.25**	--

(continued)

TABLE 5.5. (continued)

Completely Standardized Solution

LAMBDA-X

	Coping	Social	Enhance
X1	0.5137	- -	- -
X2	0.5149	- -	- -
X3	0.5160	- -	- -
X4	0.5380	0.4390	- -
X5	- -	0.6319	- -
X6	- -	0.7465	- -
X7	- -	0.6906	- -
X8	- -	0.7309	- -
X9	- -	- -	0.6026
X10	- -	- -	0.5947
X11	- -	- -	0.6649
X12	- -	- -	0.6629

PHI

	Coping	Social	Enhance
Coping	1.0000		
Social	0.6096	1.0000	
Enhance	0.3500	0.2645	1.0000

THETA-DELTA

X1	X2	X3	X4	X5	X6
0.74	0.73	0.73	0.23	0.60	0.44

X7	X8	X9	X10	X11	X12
0.52	0.47	0.64	0.65	0.56	0.56

evidenced by a rather large completely standardized EPC value ($\delta_{12,11}$ = .25). Because this correlated error can be defended substantively (i.e., items 11 and 12 are the only reverse-worded items in this questionnaire), this parameter is freed in the respecified solution. This modification significantly improves model fit, $\chi^2_{\text{diff}}(1)$ = 24.48, p < .001, and the completely standardized estimate of this correlated error is .214 (see Figure 5.1).

As can be seen in Table 5.5, consequences of this misspecification include higher factor loadings for items 11 and 12, and lower loadings for items 9 and 10, on the Enhancement Motives factor. The factor loadings of items 11 and 12 are inflated in the iterative process because a considerable portion of the observed correlation between these indicators cannot be reproduced by the correlated error (e.g., $\lambda_{x11,3}$ = .665 vs. .542 in the Table 5.5 and Figure 5.1 solutions, respectively). However, in an attempt to avoid marked overestimation of the relationships between items 9 and 10 with items 11 and 12, the magnitude of the item 9 and item 10 factor loadings is attenuated (e.g., $\lambda_{x10,3}$ = .595 vs. .664 in the Table 5.5 and Figure 5.1 solutions, respectively). In other words, because the factor loadings for items 11 and 12 have to be increased to better approximate the observed relationship of these indicators, the factor loadings of items 9 and 10 have been lowered in the iterations to better approximate the relationships of item 9 with items 11 and 12, and item 10 with items 11 and 12. However, these relationships are still overestimated by the parameter estimates (e.g., standardized residual for items 10 and 11 = −2.74, Table 5.5), and the reduction in the magnitude of the item 9 and item 10 factor loadings results in a model underestimate of the observed relationship of these indicators (i.e., standardized residual for items 9 and 10 = 4.46, Table 5.5). This again illustrates how correcting one misspecified parameter ($\delta_{12,11}$) may resolve several strains in the solution.

Because of the large sample sizes typically involved in CFA, the researcher will often encounter "borderline" modification indices (e.g., larger than 3.84, but not of particularly strong magnitude), which suggest that the fit of the model could be improved if correlated errors were added to the model. As with any type of parameter specification in CFA, correlated errors must be supported by a substantive rationale and should not be freely estimated simply to improve model fit. In addition, the magnitude of EPC values should also contribute to the decision about whether to free these parameters. As discussed in Chapter 4, the researcher should resist any temptation to use borderline modification indices to overfit the model. These trivial additional estimates usually have minimal impact on the key parameters of the CFA solution (e.g., factor loadings) and are apt to be highly unstable (i.e., to reflect sampling error rather than an important relationship; cf. MacCallum, 1986).

It is also important to be consistent in the decision rules used to specify correlated errors; that is, if there is a plausible reason for correlating the errors of two indicators, then all pairs of indicators for which this reasoning applies should also be specified with correlated errors. For instance, if it is believed that method effects exist for questionnaire items that are reverse-worded (e.g., Marsh, 1996), correlated errors should be freely estimated for all such indicators, not just a subset of them. Similarly, if the errors of indicators X1 and X2 and indicators X2 and X3 are correlated for the same reason, then the errors of X1 and X3 should also be estimated. Earlier in this chapter, where the one- vs. two-factor solution of Neuroticism and Extraversion has been considered, it has been shown that the patterning of standardized residuals and modification indices may suggest the existence of a distinct factor (see Table 5.1). However, in some situations, such as in the analysis of questionnaires with reverse-worded items (e.g., Brown, 2003; Marsh, 1996), this patterning may not reflect important latent dimensions but rather the impact of substantively irrelevant method effects. Theoretical considerations must be brought to bear strongly in this determination, as this decision may have far-reaching implications to the future measurement and conceptualization of the construct.

Improper Solutions and Nonpositive Definite Matrices

A measurement model should not be deemed acceptable if the solution contains one or more parameter estimates that have out-of-range values. As noted earlier, such estimates are usually referred to as *Heywood cases* or *offending estimates*. The most common type of Heywood case in a CFA model is a negative error variance. Moreover, a completely standardized factor loading with a value greater than 1.0 is problematic if the CFA consists of congeneric indicator sets (i.e., each indicator loads on one factor only). In CFA models with indicators that load on more than one factor (and the factors are specified to be intercorrelated), the factor loadings of such indicators are regression coefficients, *not* correlations between the indicators and the factors. A completely standardized factor loading above 1.0 may be admissible in such models, although this result might be indicative of multicollinearity in the sample data (see below). This section discusses the various sources of improper solutions and their remedies.

A necessary condition for obtaining a proper CFA solution is that both the input variance–covariance matrix and the model-implied variance–covariance matrix are *positive definite*. Appendix 3.3 (Chapter 3) has introduced the concept of positive definiteness in context of the calculation of determinants and F_{ML}. As noted in Appendix 3.3, a determinant is a single number (*scalar*) that conveys the amount of nonredundant variance in a matrix (i.e., the extent to which variables in the matrix are free to vary). When a determinant equals 0, the matrix is said to be *singular*, meaning that one or more rows or columns in the matrix are linearly dependent on other rows and columns. An example of a singular matrix is one composed of three test scores: Subscale A, Subscale B, and Total Score (sum of Subscale A and Subscale B). The resulting matrix is singular because the third variable (Total Score) is redundant; that is, it is linearly dependent on the other two variables and thus has no freedom to vary. A singular matrix is problematic because it has no inverse. Consequently, multivariate statistics that require the inverse of a matrix cannot be computed (e.g., F_{ML}; cf. Appendix 3.3). Singularity is one reason why a matrix will not be positive definite. For the condition of positive definiteness to hold, the input and model-implied matrices, and every principal submatrix of these matrices, must have determinants greater than zero (*principal submatrices* are all possible subsets of the original matrix created by removing variables from the original matrix).[6]

The condition of positive definiteness can be evaluated by submitting the variance–covariance matrix in question to principal components analysis (PCA; see Chapter 2). PCA will produce as many eigenvalues as the number of variables (p) in the input matrix. If all eigenvalues are greater than zero, the matrix is positive definite. If one or more eigenvalues is less than zero, the matrix is *indefinite*. The term *semidefinite* is used in reference to matrices that produce at least one eigenvalue equaling zero, but no negative eigenvalues (Wothke, 1993).

There are several potential causes of a nonpositive definite input variance–covariance matrix. As noted above, this problem may stem from high multicollinearities or linear dependencies in the sample data. This problem is usually addressed by eliminating collinear variables from the input matrix or combining them (cf. parceling, Chapter 9). Often a nonpositive definite input matrix is due to a minor data entry problem, such as typographical errors in preparing the input matrix (e.g., a negative or zero sample variance, a correlation > 1.0) or errors reading the data into the analysis (e.g., formatting errors in the syntax file). Large amounts of missing data, in tandem with use of an inferior approach to missing data management, can create a nonpositive definite input matrix. In Chapter 9, it is demonstrated that the range of possible values that a correlation (or covariance) may possess is dependent on all other relationships in the input matrix. For example, if $r_{x,z} = .80$ and $r_{y,z} = .80$, then $r_{x,y}$ must not be below .28 (see Eq. 9.1, Chapter 9), or this submatrix would not be positive definite. Input matrices created from complete data of a large sample are usually positive definite. However, pairwise deletion of missing data can cause definiteness problems (out-of-bound correlations) because the input matrix is computed on different subsets of the sample. Listwise deletion can produce a nonpositive definite matrix by decreasing sample size (see below) or causing other problems (e.g., creating constants in the sample data that have zero variance). Here the best solution would be to use a state-of-the-art approach to missing data (i.e., direct ML, multiple imputation; see Chapter 9).

There are several other problems that may lead to improper solutions (i.e., nonpositive definite model matrices, Heywood cases). Perhaps the most common cause is a misspecified model: Improper solutions frequently occur when the specified model is very different from models that the data would support. Structurally and empirically underidentified models will also lead to nonconverging or improper solutions (e.g., see Figure 3.7, Chapter 3; see also the "Model Identification Revisited" section of this chapter, below). In these situations, it is often possible to revise the model by using the fit diagnostic procedures described in this chapter. If the model is grossly misspecified, the researcher may need to move back into a purely exploratory analytic framework (EFA) to revamp the measurement model.[7] Bad starting values can be the root of improper solutions (see Chapter 3). However, with the possible exception of complex models (e.g., models involving the specification of nonlinear constraints), starting values are not often the cause of improper solutions in today's latent variable software programs. This is because these programs have become very sophisticated in the automatic generation of such numbers.

Problems often arise with the use of small samples. For instance, an input matrix may not be positive definite due to sampling error. Small samples often work in concert with other problems to create model estimation difficulties. For instance, small samples are more prone to the influence of outliers (i.e., cases with aberrant values on one or more variables). Outliers can cause collinearities and non-normality in the sample data, and can lead to Heywood cases such as negative variance estimates (e.g., an indicator error less than zero). Moreover, some non-normal theory estimators (e.g., weighted least squares) perform very poorly with small or moderate-sized samples because their associated weight matrices cannot be inverted (see Chapter 9). Anderson and Gerbing (1984) reported how sample size may interact with other aspects of a model to affect the risk for improper solutions. Specifically, these authors found that the risk of negative variance estimates is highest in small samples when there are only two or three indicators per latent variable and when the communalities of the indicators are low (see also Chen, Bollen, Paxton, Curran, & Kirby, 2001). As discussed in Chapter 10, each of these aspects also contributes to statistical power. In addition to reducing the risk of empirical underidentification, having multiple indicators per factor also decreases the likelihood of improper solutions. If an improper solution is caused by these issues, additional cases or data must be obtained; for example, the researcher must collect a larger sample, or obtain a larger set of indicators with stronger relationships to the latent variables.

An estimator appropriate to the sample data must be used. For example, if the indicators are binary (e.g., yes–no items), it will be inappropriate to conduct the CFA on a matrix of Pearson correlations or covariances with the ML estimator. This analysis will provide incorrect results. The proper procedures for conducting CFA with non-normal and categorical data are presented in Chapter 9.

In addition, the risk of nonconvergence and improper solutions is positively related to model complexity (i.e., models that have a large number of freely estimated parameters). A very good example of an overparameterized model is the correlated methods CFA of MTMM data (see Figure 6.1, Chapter 6). As discussed in Chapter 6, these models produce improper solutions most of the time. In other types of models, it may be possible to rectify the inadmissibility problem by removing some freely estimated parameters (e.g., if substantively justified, by dropping nonsignificant parameters or placing other restrictions on the model such as equality constraints; see Chapter 7). However, the estimation problems associated with model complexity are magnified by small sample size.

Some examples of nonpositive definite matrices and improper solutions are presented in Figure 5.2. In Model A, the input matrix is semidefinite because of a linear dependency in the sample data (i.e., Y1 is the sum of the other indicators). The fact that this matrix is not positive definite is upheld by the results of a PCA, which shows that one eigenvalue equals zero. Model B analyzes an indefinite input matrix (i.e., one of its associated eigenvalues is negative). In this example, the correlation of Y1 and Y2 cannot be −.20, given the relationships of Y1 and Y2 with other indicators in the sample data. This problem may have been produced by a data entry error or the use of pairwise deletion as the missing data strategy. Depending on the latent variable software program, Models A and B either will not be executed (i.e., the program will issue the warning that the matrix is not positive definite and will stop) or will produce a factor loading above 1.0 and a negative indicator error variance.

Model A: Semidefinite Input Matrix
   (Linear dependency: Y1 is the sum of Y2, Y3, and Y4)

Input Matrix   Eigenvalues

	Y1	Y2	Y3	Y4	
Y1	1.0000				2.877
Y2	0.8035	1.0000			0.689
Y3	0.6303	0.4379	1.0000		0.435
Y4	0.8993	0.5572	0.3395	1.0000	0.000

Consequence: Factor loading > 1.0, negative error variance

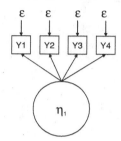

Model B: Indefinite Input Matrix
   (Correlation of Y1–Y2 is out of possible range, given other relationships)

Input Matrix   Eigenvalues

	Y1	Y2	Y3	Y4	
Y1	1.0000				2.432
Y2	-0.2000	1.0000			1.209
Y3	0.7000	0.7000	1.0000		0.513
Y4	0.4500	0.6000	0.5000	1.0000	-0.155

Consequence: Factor loading > 1.0, negative error variance

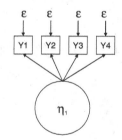

(continued)

FIGURE 5.2. Examples of nonpositive definite matrices and improper solutions.

Model C: Positive Definite Input Matrix, Indefinite Model-Implied Matrix
(Specification problem: Incorrect specification of indicator–factor relationships)

Input Matrix   Eigenvalues

	Y1	Y2	Y3	Y4	
Y1	1.0000				2.501
Y2	0.4000	1.0000			0.851
Y3	0.7000	0.4000	1.0000		0.360
Y4	0.4500	0.6500	0.4000	1.0000	0.289

Consequence: Factor correlation > 1.0

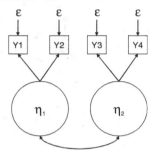

Model D: Positive Definite Input and Model-Implied Matrices
(Correctly Specified Model)

Input Matrix   Eigenvalues

	Y1	Y2	Y3	Y4	
Y1	1.0000				2.501
Y2	0.4000	1.0000			0.851
Y3	0.7000	0.4000	1.0000		0.360
Y4	0.4500	0.6500	0.4000	1.0000	0.289

Consequence: Good-fitting model, $\chi^2(1, N = 500) = 2.78$, reasonable parameter estimates

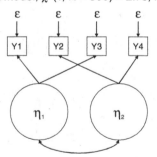

FIGURE 5.2. (*continued*)

In Model C, the input matrix is positive definite: PCA indicates that all eigenvalues are positive. However, the two-factor measurement model is misspecified. The proposed model will not support the pattern of relationships in the sample data (e.g., Y1 and Y2 are specified as indicators of η_1, yet Y1 is much more strongly related to Y3 than Y2). In an effort to reproduce the sample relationships, given the model that is specified, the ML estimation process is "forced to push" one or more parameter estimates out of the range of admissibility. In this case, ML estimation produces an indefinite factor correlation matrix (i.e., the correlation of η_1 and η_2 is estimated to be 1.368). When the model is properly specified (Model D), the results indicate a good-fitting model—for example, $\chi^2(1, N = 500) = 2.77, p = .096$—and reasonable parameter estimates (e.g., factor correlation = .610, range of factor loadings = .778–.867).

In practice, the problem of improper solutions is often circumvented by a "quick fix" method. For example, a negative indicator error variance may be addressed by respecifying the model with additional constraints (e.g., fixing the error variance to zero or a small positive value). Although overall fit will worsen somewhat, a proper solution may be obtained.[8] In fact, a default in the EQS program prevents negative variances by setting the lower bound of these estimates to zero. In some versions of LISREL, when the program encounters an indefinite or semidefinite matrix, it invokes a *ridge option* (a smoothing function to eliminate negative or zero eigenvalues) to make the input data suitable for the analysis. These remedial strategies are not recommended. As noted in this section, a negative indicator error variance may be due to a variety of problems, such as sampling fluctuation (small sample size), non-normality (e.g., outliers), multicollinearity, or model misspecification. In addition to causing the Heywood case, these issues signal other problems with the analysis and the sample data (e.g., low statistical power, poorly screened data, redundant indicators). Thus it is better to diagnose and correct the true source of the problem rather than sidestep it with one of these "quick fix" remedies.

INTERMEDIATE STEPS FOR FURTHER DEVELOPING A MEASUREMENT MODEL FOR CFA

A common sequence in scale development and construct validation is to conduct CFA as the next step after latent structure has been explored by using EFA. However, the researcher frequently encounters a poor-fitting CFA solution because of the potential sources of misfit that are not present in EFA. For example, unlike in EFA models, indicator cross-loadings and residual covariances are usually fixed to zero in initial CFA models. The researcher is then faced with potentially extensive post hoc model testing subject to the criticisms of specification searches in a single data set (MacCallum, 1986).

There are two analytic procedures that allow the researcher to explore measurement structures more fully before moving into a confirmatory framework: EFA within the CFA framework, or E/CFA (Jöreskog, 1969; Jöreskog & Sörbom, 1979; Muthén & Muthén, 1998–2012); and exploratory SEM, or ESEM (Asparouhov & Muthén, 2009). Although underutilized in applied research, both methods represent an intermediate step between EFA and CFA that provides substantial information important in the development of realistic confirmatory models.

EFA in the CFA Framework

In the E/CFA strategy, the CFA applies the same number of identifying restrictions used in EFA (m^2) by fixing factor variances to unity, by freely estimating the factor covariances, and by selecting an anchor item for each factor whose cross-loadings are fixed to zero (the loadings of nonanchor items are freely estimated on each factor). Whereas this specification produces the same model fit as ML EFA, the CFA estimation provides considerably more information, including the statistical significance of cross-loadings and the potential presence of salient error covariances. Thus the researcher can develop a realistic measurement structure prior to moving into the more restrictive CFA framework (for applied examples of this approach, see Brown, White, & Barlow, 2005; Brown, White, Forsyth, & Barlow, 2004; Campbell-Sills, Liverant, & Brown, 2004). In addition, E/CFA can be used to bring other variables (i.e., predictors or distal outcomes of the factors) into an EFA-type solution, eliminating the need for factor scores (see Chapters 2 and 3 for an overview of the limitations of factor scores).

To illustrate this strategy, the data from the drinking motives questionnaire are again used. However, this time let us suppose that the psychometric development of this measure is in the early stages, and the researcher only has a sense of the correct number of common factors and the hypothesized pattern of item–factor relationships, based on theory-driven item development and preliminary exploratory research (which may have led to the elimination of some poorly behaved items). As the next exploratory step, the researcher conducts an EFA in a larger sample (i.e., the current sample with $N = 500$) with the intent of verifying that a three-factor solution provides acceptable fit, and that the primary loadings of the items are generally in accord with prediction (e.g., items 1–4 have their highest loadings on the latent dimension of Coping Motives). At this stage of psychometric evaluation, use of CFA is premature. Although the researcher has a firm conceptual sense of this measure (i.e., the number of factors and the conjectured pattern of item–factor relationships, as supported by preliminary research), the initial EFA findings are limited in their ability to fully guide the CFA specification (e.g., reasonability of fixing all cross-loadings and error covariances to zero).

Table 5.6 presents Mplus syntax and selected results of an EFA using the $N = 500$ sample. As can be seen in this table, the three-factor EFA solution provides a good fit to the data, $\chi^2(33) = 55.546$, $p = .008$, RMSEA = 0.037 (90% CI = 0.019–0.053, CFit = .898). Although these results support the viability of a three-factor model, the researcher wishes to explore the latent structure of this measure further before specifying a CFA solution in an independent sample.

Table 5.7 provides the syntax from several programs (LISREL, Mplus, EQS, SAS/CALIS) for specifying an E/CFA in the $N = 500$ data set. The programs appear very similar to the CFAs presented in earlier examples (e.g., see Table 4.1, Chapter 4), except for two major differences: (1) All factor loadings and cross-loadings are freely estimated, with the

**TABLE 5.6. Mplus Syntax and Selected Results for an EFA
of a Drinking Motives Questionnaire**

```
TITLE:   DRINKING MOTIVES EFA
DATA:
  FILE IS "C:\efa.sav";  ! the correlation matrix in Table 5.2 could be
VARIABLE:                ! used as input here
  NAMES ARE X1-X12 GP;
  USEV ARE X1-X12;
ANALYSIS:
  TYPE IS EFA 3 3;
  ESTIMATOR IS ML;
  ITERATIONS = 1000;
  CONVERGENCE = 0.00005;
OUTPUT: SAMPSTAT;
```

EIGENVALUES FOR SAMPLE CORRELATION MATRIX

	1	2	3	4	5
1	3.876	1.906	1.150	0.837	0.722

	6	7	8	9	10
1	0.669	0.576	0.557	0.487	0.471

	11	12
1	0.426	0.323

EXPLORATORY FACTOR ANALYSIS WITH 3 FACTOR(S):

Chi-Square Test of Model Fit
Value	55.546
Degrees of Freedom	33
P-Value	0.0083

RMSEA (Root Mean Square Error Of Approximation)
Estimate	0.037	
90 Percent C.I.	0.019	0.053
Probability RMSEA <= .05	0.898	

GEOMIN ROTATED LOADINGS (* significant at 5% level)

	1	2	3	
X1	0.564*	-0.003	0.020	to be used as anchor item
X2	0.480*	0.097	-0.008	
X3	0.454*	0.103	0.001	
X4	0.434*	0.554*	0.000	
X5	0.010	0.626*	-0.035	
X6	0.015	0.726*	0.041	
X7	0.011	0.681*	-0.031	
X8	-0.114	0.806*	0.004	to be used as anchor item
X9	0.026	0.145	0.550*	
X10	-0.023	0.209*	0.544*	
X11	0.166*	-0.019	0.656*	
X12	-0.003	0.147	0.651*	to be used as anchor item

TABLE 5.7. Syntax for Conducting an EFA within the CFA Framework for the Drinking Motives Questionnaire

LISREL 9.1

```
TITLE E/CFA OF DRINKING MOTIVES ITEMS
DA NI=13 NO=500 MA=CM
LA
X1 X2 X3 X4 X5 X6 X7 X8 X9 X10 X11 X12 GP
RA FI = C:\EFA.SAV
SE
X1 X2 X3 X4 X5 X6 X7 X8 X9 X10 X11 X12 /
MO NX=12 NK=3 PH=SY,FR LX=FU,FR TD=DI
LK
COPING SOCIAL ENHANCE
PA LX
1 0 0                        ! ITEM 1 IS AN ANCHOR ITEM FOR COPING
1 1 1
1 1 1
1 1 1
1 1 1
1 1 1
1 1 1
0 1 0                        ! ITEM 8 IS AN ANCHOR ITEM FOR SOCIAL
1 1 1
1 1 1
1 1 1
0 0 1                        ! ITEM 12 IS AN ANCHOR ITEM FOR ENHANCE
PA PH
0
1 0
1 1 0                        ! FACTOR COVARIANCES FREELY ESTIMATED
VA 1.0 PH(1,1) PH(2,2) PH(3,3)   ! FACTOR VARIANCES FIXED TO 1.0
OU ME=ML RS MI SC ND=4
```

Mplus 7.11

```
TITLE:  DRINKING MOTIVES ECFA
DATA:
  FILE IS "C:\efa.sav";
VARIABLE:
  NAMES ARE X1-X12 GP;
  USEV ARE X1-X12;
ANALYSIS: ESTIMATOR IS ML;
          ITERATIONS=1000;
MODEL:
  COPING BY  X1-X12*.5 X8@0 X12@0;  !X1 is ANCHOR ITEM
  SOCIAL BY  X1-X12*.5 X1@0 X12@0;  !X8 is ANCHOR ITEM
  ENHANCE BY X1-X12*.5 X1@0 X8@0;   !X12 is ANCHOR ITEM
  COPING-ENHANCE@1;                 !FACTOR VARIANCES FIXED TO 1.0
OUTPUT:  STANDARDIZED MODINDICES(10.00) SAMPSTAT;
```

EQS 6.2

```
/TITLE
 E/CFA OF DRINKING MOTIVES ITEMS
/SPECIFICATIONS
```

(continued)

TABLE 5.7. *(continued)*

```
CASES=500; VAR=13; ME=ML; MA=RAW;
DA = 'EFA.SAV';
/LABELS
v1=item1; v2= item2; v3= item3; v4= item4; v5= item5; v6= item6;
v7= item7; v8= item8; v9= item9; v10=item10; v11=item11; v12=item12; v13=gp;
f1 = coping; f2 = social; f3 = enhance;
/EQUATIONS
V1 =   *F1+E1;              ! ITEM 1 IS ANCHOR ITEM FOR COPING
V2 =   *F1+*F2+*F3+E2;
V3 =   *F1+*F2+*F3+E3;
V4 =   *F1+*F2+*F3+E4;
V5 =   *F1+*F2+*F3+E5;
V6 =   *F1+*F2+*F3+E6;
V7 =   *F1+*F2+*F3+E7;
V8 =   *F2+E8;             ! ITEM 8 IS ANCHOR ITEM FOR SOCIAL
V9 =   *F1+*F2+*F3+E9;
V10 = *F1+*F2+*F3+E10;
V11 = *F1+*F2+*F3+E11;
V12 = *F3+E12;             ! ITEM 12 IS ANCHOR ITEM FOR ENHANCE
/VARIANCES
F1 TO F3 = 1.0;           ! FACTOR VARIANCES FIXED TO 1.0
E1 TO E12 = *;
/COVARIANCES
F1 TO F3 = *;             ! FACTOR COVARIANCES FREELY ESTIMATED
/PRINT
 fit=all;
/LMTEST
/END
```

SAS PROC CALIS 9.4

```
proc calis data=EFADATA cov method=ml pall pcoves;
var = X1-X12;
lineqs
 X1 =  lm11 f1 + e1,
 X2 =  lm21 f1 + lm22 f2 + lm23 f3 + e2,
 X3 =  lm31 f1 + lm32 f2 + lm33 f3 + e3,
 X4 =  lm41 f1 + lm42 f2 + lm43 f3 + e4,
 X5 =  lm51 f1 + lm52 f2 + lm53 f3 + e5,
 X6 =  lm61 f1 + lm62 f2 + lm63 f3 + e6,
 X7 =  lm71 f1 + lm72 f2 + lm73 f3 + e7,
 X8 =  lm82 f2 + e8,
 X9 =  lm91 f1 + lm92 f2 + lm93 f3 + e9,
 X10 = lm101 f1 + lm102 f2 + lm103 f3 + e10,
 X11 = lm111 f1 + lm112 f2 + lm113 f3 + e11,
 X12 = lm123 f3 + e12;
std
  f1-f3 = 1.0,
  e1-e12 = td1-td12;
cov
  f1-f3 = ph21 ph31 ph32;
run;
```

Note. N = 500.

exception of the cross-loadings of the items selected as anchor indicators (X1, X6, X8); and (2) as in EFA, the metric of the factors is specified by fixing the factor variances to 1.0 (but the three factor correlations are freely estimated). Note that items 1 (X1), 6 (X6), and 8 (X8) have been selected as anchor indicators for Coping Motives, Social Motives, and Enhancement Motives, respectively. Anchor items are selected on the basis of EFA results (Table 5.6), and entail one item from each factor that has a high (or the highest) primary loading on the factor and low (or the lowest) cross-loadings on the remaining factors. For example, item 8 (X8) has been chosen as the anchor indicator for Social Motives because it has the highest loading on this factor (.759) and the lowest cross-loadings on Coping Motives and Enhancement Motives (.011 and .001, respectively; see Table 5.6).

Selected Mplus output of the E/CFA is provided in Table 5.8. First, note that although this analysis is conducted in the CFA framework, the degrees of freedom and overall fit are the same as in the EFA solution presented in Table 5.6: $\chi^2(33) = 55.546$, $p = .008$, RMSEA = 0.037 (90% CI = 0.019–0.053, CFit = .898). Although current versions of Mplus provide standard errors and statistical significance testing for EFA estimates (see Table 5.6), this is not the case for many software programs that conduct EFA (e.g., SPSS). In these instances, this analysis provides considerably more information than the EFA. The E/CFA provides z tests (under the "Est./S.E." column in Table 5.8) to determine the statistical significance of primary and secondary loadings (except for the cross-loadings of the three anchor items). For example, items 5–8 (X5–X8) have statistically significant ($ps < .001$) loadings on Social Motives (range of $zs = 10.23–17.09$), but none of these items have statistically significant cross-loadings on Coping Motives or Enhancement Motives (range of $zs = 0.66–1.64$). Conversely, whereas items 1–4 have statistically significant ($ps < .001$) loadings on Coping Motives, the E/CFA results indicate that item 4 also has a large ($\lambda_{42} = .546$) and statistically significant ($z = 7.79$, $p < .001$) loading on Social Motives. This suggests that a CFA model of congeneric indicator sets may not be viable; that is, item 4 should be specified to load on both Coping Motives and Social Motives, although its loading on Enhancement Motives can be fixed to zero (i.e., $z = 0.35$, Table 5.8). Other than item 4, the E/CFA results are generally supportive of fixing all other cross-loadings at zero in a subsequent CFA model. In addition, the E/CFA provides significance tests of the factor covariances; as seen in Table 5.8, the three drinking motives factors are significantly interrelated (range of $zs = 2.82–4.04$).

Although EFA may also furnish evidence of the presence of double-loading items (see promax rotated loadings of item 4 in Table 5.6), EFA does not provide any direct indications of the potential existence of salient correlated errors (as noted in Chapters 2 and 3, EFA identification restrictions prevent the specification of correlated indicator errors). This is another area where the results of E/CFA can be quite valuable. Because the analysis is conducted within the CFA framework, modification indices and other fit diagnostic information (e.g., standardized residuals) are available to allow the researcher to examine whether the observed correlations among indicators can be adequately reproduced by the factors alone. Modification indices are provided only for the measurement error portion of the model (e.g., the theta-delta matrix in LISREL) because all other portions of the solution are saturated (e.g., factor loadings and cross-loadings).

TABLE 5.8. Selected Mplus Output of E/CFA of the Drinking Motives Questionnaire Items

```
MODEL FIT INFORMATION

Number of Free Parameters                    57

Chi-Square Test of Model Fit

        Value                           55.546
        Degrees of Freedom                  33
        P-Value                         0.0083

RMSEA (Root Mean Square Error Of Approximation)

        Estimate                         0.037
        90 Percent C.I.                  0.019   0.053
        Probability RMSEA <= .05         0.898

CFI/TLI

        CFI                              0.986
        TLI                              0.972

SRMR (Standardized Root Mean Square Residual)

        Value                            0.021


MODEL RESULTS

                                                Two-Tailed
                  Estimate     S.E.   Est./S.E.  P-Value

COPING    BY
   X1             0.613       0.062      9.839     0.000
   X2             0.514       0.074      6.985     0.000
   X3             0.487       0.083      5.849     0.000
   X4             0.536       0.074      7.262     0.000
   X5             0.097       0.070      1.386     0.166
   X6             0.114       0.070      1.636     0.102
   X7             0.110       0.072      1.523     0.128
   X8             0.000       0.000    999.000   999.000
   X9             0.032       0.083      0.388     0.698
   X10           -0.008       0.083     -0.102     0.919
   X11            0.150       0.075      2.008     0.045
   X12            0.000       0.000    999.000   999.000

SOCIAL    BY
   X1             0.000       0.000    999.000   999.000
   X2             0.101       0.074      1.378     0.168
   X3             0.105       0.083      1.275     0.202
   X4             0.546       0.070      7.790     0.000
   X5             0.574       0.056     10.228     0.000
```

(continued)

TABLE 5.8. (continued)

X6	0.651	0.055	11.923	0.000
X7	0.649	0.057	11.315	0.000
X8	0.791	0.046	17.086	0.000
X9	0.020	0.068	0.294	0.769
X10	0.083	0.068	1.216	0.224
X11	-0.158	0.066	-2.388	0.017
X12	0.000	0.000	999.000	999.000
ENHANCE BY				
X1	0.000	0.000	999.000	999.000
X2	-0.028	0.066	-0.416	0.677
X3	-0.018	0.070	-0.251	0.802
X4	-0.022	0.064	-0.345	0.730
X5	-0.042	0.053	-0.783	0.434
X6	0.034	0.052	0.659	0.510
X7	-0.039	0.055	-0.717	0.473
X8	0.000	0.000	999.000	999.000
X9	0.582	0.059	9.786	0.000
X10	0.580	0.060	9.632	0.000
X11	0.693	0.057	12.203	0.000
X12	0.671	0.049	13.675	0.000
SOCIAL WITH				
COPING	0.400	0.099	4.039	0.000
ENHANCE WITH				
COPING	0.313	0.111	2.818	0.005
SOCIAL	0.255	0.081	3.143	0.002

MODEL MODIFICATION INDICES

Minimum M.I. value for printing the modification index   10.000

		M.I.	E.P.C.	Std E.P.C.	StdYX E.P.C.
WITH Statements					
X10	WITH X9	24.856	0.213	0.213	0.309
X12	WITH X11	23.393	0.283	0.283	0.527

As shown in Table 5.8, large modification indices are obtained in regard to the correlated errors of items 11 and 12, and items 9 and 10 (standardized EPCs = .28 and .21, respectively). The correlated error between items 11 and 12 may be grounded substantively by the fact that these are the only reverse-worded items in the questionnaire. It has also been described earlier in this chapter why the failure to specify this correlated error may have a negative impact on the solution's ability to reproduce the observed relationship between items 9 and 10. Accordingly, the researcher is likely to take these results as evidence for the need to freely estimate an error covariance between items 11 and 12 in a subsequent CFA solution. The absence of other large modification indices suggests that other measurement errors can be presumed to be random. Collectively, these findings may foster the refinement of the solution initially suggested by EFA—namely, that the CFA model should not be specified as congeneric (i.e., item 4 should load on two factors), and that all error covariances may be fixed to zero with the exception of items 11 and 12.

Exploratory SEM

The second method, ESEM (Asparouhov & Muthén, 2009), is a newer approach that is only available in the Mplus software program (beginning with Version 5.21). ESEM allows for the integration of EFA and CFA measurement models within the same solution. That is, within a given measurement model, some factors can be specified according to the conventions of CFA (i.e., zero cross-loadings), whereas other factors can be specified according to those of EFA (i.e., rotation of a full factor loading matrix). Unlike traditional EFA, the EFA measurement model in ESEM provides the same information as ML CFA, such as multiple indices of goodness of fit, standard errors for all rotated parameters, and modification indices (i.e., highlighting possible correlated residuals among indicators). Moreover, most of the modeling possibilities of CFA are available in ESEM, including correlated residuals, regressions of factors on covariates, regression among factors (among different EFA factor blocks or between EFA and CFA factors), multiple-groups solutions, mean structure analysis, and measurement invariance examination across groups or across time. At this writing, ESEM possesses some relatively minor practical and analytic limitations, including the inability to read in summary data (i.e., raw data must be used as input) and certain restrictions in the specification of structural parameters (e.g., exploratory factors from the same block cannot be regressed on each other and cannot be used as lower-order factors in a hierarchical factor model; a structural path linking an exploratory factor to another variable must be specified for all factors within the same exploratory block). A technical description of ESEM can be found in Asparouhov and Muthén (2009); see Marsh et al. (2009, 2010) and Rosellini and Brown (2011) for initial applied studies.

Table 5.9 presents Mplus syntax and selected output for the three-factor model of drinking motives in ESEM. In the MODEL command, the BY statement specifies that the three named factors (Coping, Social, and Enhancement Motives) are measured by the indicators X1–X12. The label (*1) after the BY statement is used to indicate that Coping, Social, and Enhancement Motives are a block of EFA factors. Because a rotation option has not been specified, the Mplus default of oblique geomin rotation is used. As seen in Table 5.9, by Mplus default the variances of the factors are fixed to one; in addition, the residual variances (and intercepts, not shown) of the indicators are freely estimated, and the residuals are not correlated as the default. However, the correlated residual default can be overridden by using the same Mplus command language used in CFA (e.g., X11 WITH X12 will freely estimate this error covariance). Inspection of the geomin-rotated factor matrix shows that all primary loadings are statistically significant and three cross-loadings are significant, including the cross-loading of X4 on the Social Motives factor.

TABLE 5.9. Mplus Syntax and Selected Output for ESEM of the Drinking Motives Questionnaire Items

```
TITLE:  DRINKING MOTIVES ESEM
DATA:
FILE IS EFA.DAT;
VARIABLE:
  NAMES ARE X1-X12 GP;
  USEV ARE X1-X12;
ANALYSIS: ESTIMATOR IS ML;
MODEL:
  COPING SOCIAL ENHANCE BY  X1-X12 (*1);
  ! X11 WITH X12;  this command could be included to estimate the X11-X12
  !                error covariance
OUTPUT: MODINDICES(10.00);
```

MODEL FIT INFORMATION

Chi-Square Test of Model Fit

Value	55.546
Degrees of Freedom	33
P-Value	0.0083

RMSEA (Root Mean Square Error Of Approximation)

Estimate	0.037	
90 Percent C.I.	0.019	0.053
Probability RMSEA <= .05	0.898	

CFI/TLI

CFI	0.986
TLI	0.972

SRMR (Standardized Root Mean Square Residual)

Value	0.021

MODEL RESULTS

	Estimate	S.E.	Est./S.E.	Two-Tailed P-Value
COPING BY				
X1	0.610	0.065	9.439	0.000
X2	0.496	0.082	6.033	0.000
X3	0.469	0.090	5.222	0.000
X4	0.451	0.069	6.578	0.000
X5	0.010	0.047	0.205	0.838
X6	0.014	0.044	0.326	0.744
X7	0.011	0.049	0.231	0.817

(continued)

TABLE 5.9. *(continued)*

X8	-0.120	0.077	-1.568	0.117
X9	0.026	0.069	0.380	0.704
X10	-0.024	0.064	-0.373	0.709
X11	0.170	0.072	2.369	0.018
X12	-0.003	0.032	-0.095	0.924
SOCIAL BY				
X1	-0.003	0.017	-0.180	0.857
X2	0.100	0.086	1.165	0.244
X3	0.106	0.094	1.126	0.260
X4	0.577	0.073	7.930	0.000
X5	0.604	0.051	11.738	0.000
X6	0.703	0.049	14.360	0.000
X7	0.686	0.053	12.916	0.000
X8	0.847	0.069	12.218	0.000
X9	0.148	0.083	1.781	0.075
X10	0.215	0.080	2.686	0.007
X11	-0.019	·0.012	-1.538	0.124
X12	0.146	0.075	1.938	0.053
ENHANCE BY				
X1	0.022	0.055	0.399	0.690
X2	-0.008	0.044	-0.177	0.860
X3	0.001	0.043	0.023	0.982
X4	0.000	0.016	0.027	0.978
X5	-0.034	0.046	-0.743	0.458
X6	0.040	0.043	0.927	0.354
X7	-0.031	0.045	-0.684	0.494
X8	0.004	0.018	0.221	0.825
X9	0.562	0.055	10.273	0.000
X10	0.560	0.055	10.153	0.000
X11	0.673	0.052	12.838	0.000
X12	0.647	0.051	12.714	0.000
SOCIAL WITH				
COPING	0.520	0.096	5.427	0.000
ENHANCE WITH				
COPING	0.179	0.098	1.824	0.068
SOCIAL	0.061	0.108	0.566	0.571
Variances				
COPING	1.000	0.000	999.000	999.000
SOCIAL	1.000	0.000	999.000	999.000
ENHANCE	1.000	0.000	999.000	999.000
Residual Variances				
X1	0.792	0.073	10.790	0.000
X2	0.762	0.061	12.516	0.000
X3	0.787	0.061	12.982	0.000
X4	0.276	0.036	7.555	0.000
X5	0.562	0.041	13.690	0.000

(continued)

TABLE 5.9. *(continued)*

X6	0.428	0.035	12.070	0.000
X7	0.536	0.041	12.960	0.000
X8	0.477	0.048	9.889	0.000
X9	0.685	0.056	12.172	0.000
X10	0.694	0.058	12.044	0.000
X11	0.534	0.056	9.615	0.000
X12	0.539	0.052	10.281	0.000

```
MODEL MODIFICATION INDICES

Minimum M.I. value for printing the modification index    10.000
```

		M.I.	E.P.C.	Std E.P.C.	StdYX E.P.C.
WITH Statements					
X10	WITH X9	24.869	0.213	0.213	0.309
X12	WITH X11	23.351	0.283	0.283	0.527

Although goodness-of-fit statistics are identical, the reader will note that the parameter estimates (e.g., factor loadings, factor correlations) differ somewhat between E/CFA and ESEM. Primarily, this is due to the use of factor rotation (geomin) in ESEM, which also eliminates the need to specify anchor items (unlike in E/CFA, all factor loadings are freely estimated in ESEM). It is also noteworthy that the factor correlation estimates produced by E/CFA and ESEM are of smaller magnitude than the corresponding estimates from CFA; for example, the correlations between Coping Motives and Social Motives are .61, .40, and .52 in CFA, E/CFA, and ESEM, respectively. This is for the same reasons previously discussed in the comparison of traditional EFA and CFA results in Chapter 3 (e.g., Appendix 3.1). Specifically, in CFA, where virtually all cross-loadings are fixed to zero, there is more burden on the factor correlations to reproduce the correlations among indicators loading on different factors because there are no cross-loadings to assist in these model-implied estimates. Indeed, the requirement of fixing cross-loadings to zero in traditional CFA has been criticized by some researchers (e.g., Marsh et al., 2009). First, this practice may be unrealistic for some types of measurement models where it would be very difficult to attain acceptable levels of model fit by conventional standards (i.e., fixing several small, yet statistically significant, cross-loadings to zero may result in poor fit). Second, as illustrated in this example, CFA may overstate the size of the relationships among factors because the majority of cross-loadings are fixed to zero. For these reasons, a method such as ESEM, although less parsimonious than CFA, may be more appropriate than CFA in some contexts (e.g., factor analysis of a multiple-item, multidimensional measurement scale in a large sample; cf. Marsh et al., 2010; Rosellini & Brown, 2011). Nonetheless, given the recent advent of ESEM, best-practice guidelines for this procedure await future research and application.

MODEL IDENTIFICATION REVISITED

General rules and guidelines for CFA model identification have been discussed in Chapter 3. This discussion includes basic principles such as the need to define a metric for the latent variables, the need for model *df* to be positive (i.e., the number of elements of the input matrix should equal or exceed the number of freely estimated parameters), and the issue of empirical underidentification (e.g., the marker indicator must be significantly related to its latent variable). However, in the current chapter, more complex measurement models have been considered that entail double-loading indicators and correlated indicator errors. Although these general rules still apply, researchers are more likely to encounter identification problems with more complex solutions. Thus identification issues with such models are now discussed briefly.

Because latent variable software programs are capable of evaluating whether a given model is identified, it is often most practical to simply try to estimate the solution and let the computer determine the model's identification status. Nevertheless, it is helpful for the researcher to be aware of general model identification guidelines in order to avoid pursuing structurally or empirically underidentified solutions (such as proposing them in research grant applications). In addition to the guidelines presented in Chapter 3, the researcher should be mindful of the fact that specification of a large number of correlated errors may produce an underidentified model (even when model *df* is positive). As a general rule, for every indicator there should be at least one other indicator in the solution (which may or may not load on the same factor), with which it does not share an error covariance (although in some cases, models that satisfy this rule may still be underidentified; cf. Kenny, Kashy, & Bolger, 1998). In addition, models specified with double-loading indicators may be more prone to underidentification. As discussed in Chapter 6, empirical underidentification is a serious problem in *correlated methods* approaches to CFA MTMM analyses, where each indicator loads on both a trait factor and a methods factor. The risk for underidentification is increased in models that contain some mixture of double-loading indicators *and* correlated errors. For instance, a solution will be underidentified if an indicator, X1, is specified to load on Factors A and B, but is also specified to have correlated errors with each of the Factor B indicators.

EQUIVALENT CFA SOLUTIONS

Another important consideration in model specification and evaluation is the issue of *equivalent solutions*. Equivalent solutions exist when different model specifications produce identical goodness of fit (with the same number of *df*) and predicted covariance matrices (Σ) in any given data set. Consider the two mediation models depicted below. Both models are overidentified with 1 *df* corresponding to the nontautological relationship between A and C. Although these models differ greatly in terms of their substantive meaning (Model 1: A is a direct cause of B and an indirect cause of C; Model 2: C is a direct cause of B and an indirect cause of A), they generate the same predicted covariance matrix; that is, in both solutions, the model-implied relationship of A and C is .30.[9]

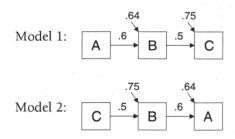

Interested readers are referred to the papers by Stelzl (1986), Lee and Hershberger (1994), and Hershberger (1994), who have developed rules for explicating equivalent solutions for various types of structural equation models. A noteworthy aspect of this work is that the number of equivalent solutions is related to the saturation of the model. Stated another way, models with fewer *df*s have a greater number of alternative, equivalent solutions than more parsimonious models.

Figure 5.3 presents four equivalent CFA solutions. In this example, the researcher is interested in examining the latent dimensionality of situational social anxiety (i.e., anxiety in social situations due to concerns about being negatively evaluated by others, embarrassment, etc.). Of the six indicators, four could be construed as measuring anxiety in one-on-one social interactions (S1–S4), and the remaining two pertain to anxiety in public speaking situations (S5, S6). Each of the four CFA models in Figure 5.3 is overidentified with *df* = 8 and fits the data equally well.

Models A and B exemplify a quandary often faced by applied CFA researchers: In this case, should the additional covariance existing between the two public speaking indicators be accounted for as a distinct factor (Model A), or within a unidimensional solution containing an error covariance between S5 and S6 (Model B)? These alternative specifications may have substantial conceptual implications in the researcher's applied area of study. For instance, the Model A specification will forward the conceptual notion that the construct of social anxiety is multidimensional. Model B will assert that social anxiety is a broader unitary construct and that the differential covariance of indicators S5 and S6 represents a correlated error, perhaps due to a substantively trivial method effect (unlike S1–S4, these two indicators are more specifically worded to assess anxiety in large groups). Because Models A and B provide the same fit to the data, the procedures of CFA cannot be employed to resolve the question of which model is more acceptable. This is particularly problematic in instances where two or more equivalent models are substantively plausible.

This example also illustrates how model fit and equivalent solutions can be affected by the composition of the indicator set. The latent dimensionality of a collection of indicators may be strongly influenced by potentially artifactual issues, such as inclusion of individual items with very similar or reverse wordings, or use of highly overlapping (multicollinear) measures. In the present case, it may be decided prior to specifying a unidimensional CFA model of social anxiety (along the lines of Model B) that the S5 and S6 indicators are overly redundant (i.e., both assess anxiety about speaking in front of large groups). If one of these indicators is omitted from the analyses (or if S5 and S6 are combined; cf. parceling in Chapter 9), the researcher can avert the problems of poor model fit (if a correlated error has not been specified between S5 and S6 in an initial one-factor model) and equivalence with a two-factor specification (Model A).

Figure 5.3 presents two additional equivalent CFA models of social anxiety (it is assumed that a two-factor solution is conceptually viable). Model C depicts a higher-

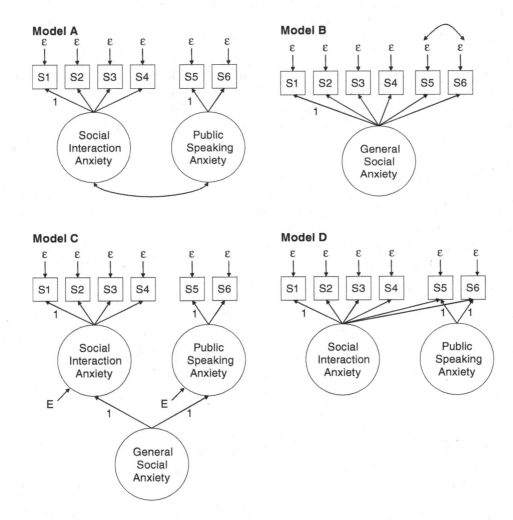

FIGURE 5.3. Examples of equivalent CFA models of social anxiety. S1, making eye contact; S2, maintaining a conversation; S3, meeting strangers; S4, speaking on the telephone; S5, giving a speech to a large group; S6, introducing yourself to large groups. All path diagrams use latent Y notation. In order for Model C to be identified, the two higher-order factor loadings (i.e., General Social Anxiety → Social Interaction Anxiety, General Social Anxiety → Public Speaking Anxiety) must be constrained to equality. To identify Model D, the factor loadings of S5 and S6 on the latent variable of Public Speaking Anxiety must be constrained to equality.

order factor model (see Chapter 8) in which the correlation between the lower-order factors of Social Interaction Anxiety and Public Speaking Anxiety (Model A) is accounted for by a second-order factor of General Social Anxiety. This specification implies that Social Interaction Anxiety and Public Speaking Anxiety represent distinct subdimensions influenced by the broader construct of General Social Anxiety. It should be noted that because there are only two lower-order factors, the higher-order portion of Model C will be underidentified if the loadings of General Social Anxiety → Social Interaction Anxiety and General Social Anxiety → Public Speaking Anxiety are freely estimated (i.e., it is analogous to the underidentified model depicted by Model A in Figure 3.6, Chapter 3). Thus, to identify Model C, the two higher-order factor loadings must be constrained to be equal (equality constraints are discussed in detail in Chapter 7).[10]

Methodologists have noted that some equivalent solutions can be readily dismissed on the basis of logic or theory (e.g., MacCallum, Wegener, Uchino, & Fabrigar, 1993). Model D can be regarded as one such example. In this model, S5 and S6 are purported to be indicators of both Social Interaction Anxiety and Public Speaking Anxiety (this specification requires an equality constraint for the S5 and S6 loadings on Public Speaking Anxiety). However, Social Interaction Anxiety and Public Speaking Anxiety are assumed to be unrelated (i.e., the factor covariance is fixed to zero). It is likely that both of these assertions (S5 and S6 as indicators of anxiety in one-on-one social interactions; orthogonal nature of Social Interaction Anxiety and Public Speaking Anxiety) can be quickly rejected on conceptual grounds. Other examples of nonsensical equivalent solutions include longitudinal models with directional relationships among temporal variables (e.g., regressing a Time 1 variable onto a Time 2 variable; i.e., Time 1 → Time 2 vs. Time 2 → Time 1) or models with direct effects of predetermined exogenous indicators such as gender or age (e.g., Gender → Job Satisfaction vs. Job Satisfaction → Gender).

Figure 5.4 presents three equivalent solutions that entail a single factor. In this example, the four indicators are observed measures related to the construct of Depres-

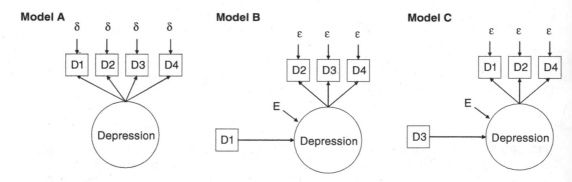

FIGURE 5.4. Examples of equivalent CFA models of depression. D1, hopelessness; D2, depressed mood; D3, guilt; D4, loss of interest in activities.

sion: hopelessness (D1), depressed mood (D2), feelings of guilt (D3), and loss of interest in usual activities (D4). In Model A, a simple one-factor measurement model is specified in which the four indicators are presumed to be interrelated because they are each influenced (caused) by the underlying dimension of Depression. In Model B, the latent variable of Depression (now defined by indicators D2–D4) is regressed onto the single indicator of hopelessness (D1). Although providing the same fit to the data, Models A and B are profoundly different in terms of their conceptual implications. Unlike Model A, which presumes that hopelessness is just another manifest symptom of Depression, Model B is in accord with a conceptualization that a sense of hopelessness is a cause of Depression. Note that Model A can be respecified so that any of the four indicators are designated as having direct effects on the Depression latent variable (e.g., Model C is similar to Model B except that guilt, D3, is specified as the cause of Depression). Models B and C can be regarded as examples of CFA with *covariates* (covariates are sometimes referred to as *background variables*; cf. Chapter 7). In such models, the factors are endogenous (i.e., latent Y variables) because the solution attempts to explain variance in them with an exogenous indicator (e.g., D1 in Model B). Accordingly, in Models B and C, the residual variance in Depression (often referred to as a *disturbance*; see Chapters 3 and 7) is freely estimated in the psi matrix (cf. Model A, where the factor variance is estimated in the phi matrix). The residual variance in Depression is depicted as "E" in the path diagrams of Models B and C (i.e., "E" = φ_1). As will be seen in Chapter 9, the issue of equivalent models is also germane to solutions that contain formative constructs (i.e., latent variables "caused" by a composite set of indicators) because such models can often be alternatively parameterized as MIMIC models (Chapter 7).

Perhaps the most widely cited paper on the issue of equivalent structural equation models was published by MacCallum and colleagues in 1993. These authors found that although equivalent models were quite common in applied SEM research, this issue was virtually ignored by investigators. For example, of the 20 SEM papers published in the *Journal of Personality and Social Psychology* between 1988 and 1991, 100% reported a model where there existed 3 or more equivalent models. The median number of equivalent models in these studies was 12 (range of equivalent models = 3 to 33,925!). However, none of these studies contained an acknowledgment of the existence of equivalent solutions. In the 20+ years since the publication of the MacCallum et al. (1993) paper, little has changed in regard to applied researchers' explicit recognition of equivalent models. This may be due in part to the unavailability of utilities in latent variable software packages to generate the equivalent models associated with the models specified by the researcher. If such utilities were available, the equivalent models revealed by the software could be evaluated by the researcher in terms of their substantive plausibility. In some cases, this process may lend further support for the hypothesized model (i.e., in situations where there exist no zero equivalent models, or where each of the equivalent models is conceptually not viable). In other instances, the process may have considerable heuristic value (e.g., it may reveal theoretically plausible alternative models that were not recognized by the researcher).

SUMMARY

Poor-fitting models are common in applied data sets. It is important to have a sound knowledge of the sample data and model before proceeding with a CFA. Data-screening procedures (e.g., normality, outliers), in tandem with a proper strategy for missing data (see Chapter 9), should be conducted to help ensure that the sample data are appropriate for CFA and the statistical estimator (e.g., ML). Principal components analysis is another helpful procedure in this phase to verify that the input matrix is positive definite (i.e., all resulting eigenvalues are positive). EFA, E/CFA, and ESEM are important precursors to CFA for developing and refining a measurement model. Although use of these methods will foster the chances of success in the restrictive CFA framework (e.g., most or all cross-loadings and error covariances are fixed to zero), the solution still may not satisfy one or more of the three major criteria for model acceptability (i.e., overall goodness of fit, areas of localized strain in the solution, interpretability/strength or parameter estimates). Thus the researcher must be adept at diagnosing and rectifying the sources of ill fit (interpretation of modification indices, standardized residuals, and reasonability of parameter estimates), while keeping in mind both substantive issues (i.e., conceptual justification for the respecified model) and methodological considerations pertaining to the principles of model identification, equivalent models, model comparison, and improper solutions (e.g., knowing when additional parameters will lead to underidentification).

The first five chapters of this book have covered what can be considered the "fundamentals" of CFA—that is, assuming no complications such as non-normality or missing data, what every researcher "needs to know" to conduct CFA properly with applied data sets. Beginning with Chapter 6, which focuses on the topic of CFA with MTMM matrices, the remaining chapters of this book address more specialized applications of CFA (e.g., higher-order factor models, multiple-groups solutions, MIMIC models, scale reliability evaluation, Bayesian factor analysis) and other issues that frequently arise in the analysis of real data (e.g., missing data, non-normality, categorical indicators, statistical power analysis).

NOTES

1. Correlated errors are sometimes referred to as *minor factors*, especially in instances where three or more indicators loading on a broader factor have intercorrelated measurement errors. This terminology is most appropriate when the error covariances are presumed to reflect a substantively salient dimension (rather than a byproduct of the measurement approach such as reverse-worded items) that is subsumed by a broader latent construct.

2. Unlike factor correlations = 1.0 or error variances = 0.0 (see discussion of comparing one- and two-factor models), factor loadings, factor covariances, and indicator error covariances that are fixed to 0.0 are not restrictions on the edge of admissible parameter spaces (i.e., theoretically, these unstandardized parameters can have values of $\pm \infty$).

3. Interested readers are referred to Yuan, Marshall, and Bentler (2003) for a detailed demonstration of how model misspecification affects parameter estimate biases.

4. In fact, the model χ^2 is equal to the difference in the –2LL values produced by a saturated model (where all parameters are freely estimated) and by the model that was fitted to the data (the latter –2LL value is used in the computation of the AIC and BIC).

5. An alternative, yet less frequently used, approach to comparing non-nested models involves tests based on *nested tetrads* (Bollen & Ting, 1993, 2000). *Tetrads* are differences in the products of pairs of covariances (e.g., $\tau_{1234} = \sigma_{12}\sigma_{34} - \sigma_{13}\sigma_{24}$). Depending on the model specification, some tetrads will equal zero (these are referred to as *vanishing tetrads*). Bollen (1990) derived a χ^2 statistic to perform a simultaneous test of vanishing tetrads of a model to assess its fit to data (df = the number of vanishing tetrads). Bollen and Ting have demonstrated scenarios where two models are not structurally nested (i.e., one model does not contain a subset of the freed parameters of the other) but are nested in their vanishing tetrads (i.e., the vanishing tetrads of one model are a subset of the other's). In these instances, two models can be compared by taking the difference in their χ^2s.

6. This rule also applies to weight matrices associated with non-normal theory estimators (e.g., WLS; see Chapter 9) and variance–covariance matrices within the model (i.e., Θ_δ, Θ_ε, Ψ, and Φ).

7. On a related note, Heywood cases may occur in EFA when too many factors have been extracted from the data.

8. However, constraining a variance to zero places the estimate on the border of inadmissibility (see discussion in this chapter on the use of χ^2 difference testing to compare CFA models that differ in number of factors).

9. On rare occasions, equal goodness of fit (in terms of χ^2, CFI, etc.) may occur by chance for two models that do not produce the same model-implied covariance matrix. This is not an instance of equivalent models.

10. As shown in Chapter 8, a hierarchical model involving a single higher-order factor and three lower-order factors produces the same goodness of fit as a three-factor model where the factors are freely intercorrelated. This is because the higher-order portion of the hierarchical model is just-identified (i.e., analogous to Model B in Figure 3.6, Chapter 3).

6

CFA of Multitrait–Multimethod Matrices

As seen in prior chapters, CFA provides an elegant analytic framework for evaluating the validity of constructs, and for examining the interrelations of constructs while adjusting for measurement error and an error theory. The application of CFA to multitrait–multimethod (MTMM) matrices offers an even more sophisticated methodology for estimating convergent validity, discriminant validity, and method effects in evaluating the construct validity of concepts in the social and behavioral sciences. This chapter discusses the various ways an MTMM CFA model can be parameterized and extended, and the strengths and drawbacks of each approach. In addition to bolstering the concepts of convergent and discriminant validity in the context of CFA, the chapter illustrates the deleterious consequences that may result from failing to account for measurement error and method effects.

CORRELATED VERSUS RANDOM MEASUREMENT ERROR REVISITED

In Chapter 5, the need for specifying correlated errors has been discussed in instances where some of the observed covariance of two or more indicators is believed to be the effects of the measurement approach (i.e., method covariance), over and above covariance explained by the substantive latent variables. Thus the error theory of a CFA measurement model can entail some combination of *random measurement error* (i.e., the unexplained variance of one indicator does not covary with the unexplained variance of another) and *correlated measurement error* or uniqueness (i.e., the unexplained variance of one indicator covaries with the unexplained variance of another), as guided by the conceptual or empirical basis of the model specification. As discussed in Chapter 5, correlated errors should not be specified solely for the purpose of improving model fit. In the majority of models involving latent variables defined by multiple indicators, measurement errors are freely estimated (and correlated errors, if applicable). However, in some instances it is justified to impose constraints on these parameters (e.g., constraining error variances to equality, as in the evaluation of parallel tests), or to fix these estimates to predetermined values (e.g., prespecifying the amount of measurement error into a variable measured by a single indicator; Chapter 4). These alternative specifications for measurement errors are discussed in other chapters (see Chapters 4 and 7).

THE MULTITRAIT–MULTIMETHOD MATRIX

A common limitation of applied research is that the dimensionality and validity of constructs are evaluated in a cross-sectional fashion with a single measurement scale. For example, a researcher may hypothesize that the negative symptoms of schizophrenia consist of three components: flat affect, alogia (poverty of speech), and social amotivation. To examine this notion, he or she develops a multiple-item clinical observation rating system that is subsequently used to rate the behavior of patients admitted to a state hospital. After a sufficient sample has been collected, the ratings are submitted to factor analysis. The results indicate a three-factor solution, which the researcher interprets as supporting the conjectured tripartite model of negative symptoms. He or she concludes that the findings attest to favorable convergent and discriminant validity, in that features of flat affect, alogia, and social amotivation are differentially intercorrelated to such an extent that they load on distinct factors. Further support for validity is obtained by results showing that the three latent variables are more strongly related to measures of schizophrenia severity than to indicators of other disorders (e.g., bipolar disorder).

The aforementioned scenario is a common empirical sequence in scale development and construct validation. Although a useful part of such endeavors, this sequence provides an incomplete evaluation of construct validity. It is not clear to what extent the multidimensionality of the negative symptom rating scale could be attributed to artifacts of the indicator set. For example, the chances of obtaining a distinct factor of Flat Affect may be fostered by creation of a rating scale containing several similarly worded items that assess this feature. This issue is particularly salient in instances where factor analysis uncovers more latent dimensions than initially predicted; that is, are these additional dimensions conceptually and practically useful, or do they reflect artifacts of scale development (cf. Models A and B in Figure 5.3, Chapter 5)? In addition, method effects may obscure the discriminant validity of the constructs. That is, when each construct is assessed by the same measurement approach (e.g., observer rating), it cannot be determined how much of the observed overlap (i.e., factor correlations) is due to method effects as opposed to "true" covariance of the traits. In sum, construct validation is limited in instances where a single assessment method is employed.

Campbell and Fiske (1959) developed the multitrait–multimethod (MTMM) matrix as a method for establishing the construct validity of psychological measures. This methodology entails a matrix of correlations arranged in a manner that fosters the evaluation of construct validity. *Construct validity* is the overarching principle of validity, referring to the extent to which a psychological measure in fact measures the concept it purports to measure. This approach requires that several *traits* (*T*; e.g., attitudes, personality characteristics, behaviors) are each assessed by several *methods* (*M*; e.g., alternative test forms, alternative assessment modalities such as questionnaires and observer ratings, or separate testing occasions). The result is a $T \times M$ correlation matrix that is interpreted with respect to convergent validity, discriminant validity, and method effects.

An example of an MTMM matrix is presented in Table 6.1. In this illustration, the researcher wishes to examine the construct validity of the DSM-5 Cluster A personality disorders, which are enduring patterns of symptoms characterized by odd or eccentric behaviors (American Psychiatric Association, 2013). Cluster A is composed of three personality disorder constructs: (1) Paranoid (an enduring pattern of distrust and suspicion, such that others' motives are interpreted as malevolent); (2) Schizoid (an enduring pattern of detachment from social relationships and restricted range of emotional expression); and (3) Schizotypal (an enduring pattern of acute discomfort in social relationships, cognitive and perceptual distortions, and behavioral eccentricities). In a sample of 500 patients, each of these three traits is measured by three assessment methods: (1) a self-report inventory of personality disorders; (2) dimensional ratings from a structured clinical interview of personality disorders; and (3) observational ratings made by paraprofessional staff. Thus Table 6.1 is a 3 (*T*) × 3 (*M*) matrix, arranged such that the correlations among the different traits (personality disorders: Paranoid, Schizotypal, Schizoid) are nested within each method (assessment type: inventory, clinical interview, observer ratings). The MTMM matrix is a symmetric correlation matrix with one exception: Reliability estimates (e.g., Cronbach's alphas) of the measures are inserted in the diagonal in place of ones (e.g., in Table 6.1, the internal consistency estimate of the inventory measure of Paranoid personality is .93). As Campbell and Fiske (1959) note, ideally the reliability diagonal should contain the largest coefficients in the matrix; that is, the measure should be more strongly correlated with itself than with any other indicator in the MTMM matrix.

The MTMM matrix consists of two general types of blocks of coefficients: (1) *monomethod blocks*, which contain correlations among indicators derived from the same assessment method; and (2) *heteromethod blocks*, which contain correlations among indicators assessed by different methods (see Table 6.1). Of central interest is the *validity diagonal*, which corresponds to the diagonal within each heteromethod block—the boldfaced correlations in Table 6.1. Correlations on the validity diagonal represent estimates of *convergent validity*: Different measures of theoretically similar or overlapping constructs should be strongly interrelated. In the MTMM matrix, convergent validity is evidenced by strong correlations among methods measuring the same trait (i.e., monotrait–heteromethod coefficients). For example, the findings in Table 6.1 indicate that the three different measures of Schizotypal personality are strongly interrelated (range of *r*s = .676–.749). The off-diagonal elements of the heteromethod blocks reveal *discriminant validity*: Measures of theoretically distinct constructs should not be highly intercorrelated. Discriminant validity in the MTMM matrix is evidenced by weaker correlations between different traits measured by different methods (i.e., heterotrait–heteromethod coefficients) in relation to correlations on the validity diagonal (monotrait–heteromethod coefficients). In Table 6.1, support for discriminant validity is obtained by the finding

TABLE 6.1. MTMM Matrix of Cluster A Personality Disorder Constructs

	Inventory			Clinical interview			Observer ratings		
	PAR	SZT	SZD	PAR	SZT	SZD	PAR	SZT	SZD
Inventory									
Paranoid (PAR)	(.930)								
Schizotypal (SZT)	.290	(.900)							
Schizoid (SZD)	.372	.478	(.910)						
Clinical interview									
Paranoid (PAR)	**.587**	.238	.209	(.800)					
Schizotypal (SZT)	.201	**.586**	.126	.213	(.870)				
Schizoid (SZD)	.218	.281	**.681**	.195	.096	(.830)			
Observer ratings									
Paranoid (PAR)	**.557**	.228	.195	**.664**	.242	.232	(.820)		
Schizotypal (SZT)	.196	**.644**	.146	.261	**.641**	.248	.383	(.790)	
Schizoid (SZD)	.219	.241	**.676**	.290	.168	**.749**	.361	.342	(.800)
Standard deviations:	3.610	3.660	3.590	2.940	3.030	2.850	2.220	2.420	2.040

Note. Values on the diagonal in parentheses are indicator reliabilities; correlations in boldface type represent convergent validities; single-line boxes are heteromethod blocks; double-line boxes are monomethod blocks; correlations in monomethod blocks represent heterotrait–monomethod coefficients; nonboldfaced correlations in heteromethod blocks represent heterotrait–heteromethod coefficients.

that correlations in the off-diagonal elements of the heteromethod blocks are uniformly lower (range of rs = .126–.290) than the validity coefficients (range of rs = .557–.749).

Finally, evidence of *method effects* is obtained by an examination of the off-diagonal elements of the monomethod blocks. The extent of method effects is reflected by the differential magnitude of correlations between different traits measured by the same method (heterotrait–monomethod coefficients) relative to the correlations between the same two traits measured by different methods. As shown in Table 6.1, although not extreme, some method variance is evident, especially for the inventory and observer rating measures. For example, the observer ratings of the traits of Paranoid and Schizotypal personality are more highly correlated (r = .383) than heteromethod measures of these traits (e.g., the correlation between Paranoid and Schizotypal personality traits measured by inventory and observer ratings, respectively, is .196; see Table 6.1). As in the present example, when the collective results indicate that convergent and discriminant validity are high and method effects are negligible, construct validity is supported.

CFA APPROACHES TO ANALYZING THE MTMM MATRIX

Despite the fact that the Campbell and Fiske (1959) methodology represented a significant advance in the conceptualization and evaluation of construct validity, the MTMM approach was not widely used over the years immediately following its inception. In addition, several limitations of the MTMM methodology were noted, including the subjective nature of its interpretation (e.g., ambiguity in terms of what patterns of correlations reflect satisfactory convergent and discriminant validity), its reliance on correlations among fallible observed measures to draw inferences about trait and method factors (cf. Schmitt & Stults, 1986), and the failure of EFA to obtain meaningful solutions of MTMM data (e.g., identification restrictions of EFA prevent specification of correlated errors; see Chapter 2).

Interest in the MTMM matrix increased with the realization that the procedures of CFA could be readily applied to its analysis (cf. Cole, 1987; Flamer, 1983; Marsh & Hocevar, 1983; Widaman, 1985). In other words, MTMM matrices, like any other form of correlation or covariance matrix, can be analyzed by CFA to permit inferences about potential underlying dimensions such as trait and method factors. Although several different types of CFA models can be applied to MTMM data (cf. Marsh & Grayson, 1995; Widaman, 1985), two forms of CFA specification have been predominant. To use more contemporary terminology, these two types of solutions have been referred to as *correlated methods* and *correlated uniqueness* models (Marsh & Grayson, 1995).

Correlated Methods Models

The correlated methods parameterization reflects the traditional CFA approach to analyzing MTMM matrices. There are five major aspects of correlated methods model specifications (Widaman, 1985): (1) To be identified, there must be at least three traits (T) and three methods (M); (2) $T \times M$ indicators are used to define $T + M$ factors (i.e., the number of trait factors = T; the number of method factors = M); (3) each indicator is specified to load on two factors—its trait factor and its method factor (all other cross-loadings are fixed to zero); (4) correlations among trait factors and among method factors are freely estimated, but the correlations between trait and method factors are usually fixed to zero; and (5) indicator uniquenesses (i.e., variance in the indicators not explained by the trait and method factors) are freely estimated but cannot be correlated with the uniquenesses of other indicators. Accordingly, in this specification, each indicator is considered to be a function of trait, method, and unique factors.

Figure 6.1 depicts the path diagram of the correlated methods CFA specification for the MTMM matrix of Cluster A personality disorders (Table 6.1). Table 6.2 provides the LISREL syntax for this model. For reasons discussed later in this chapter (e.g., propensity for improper solutions), results of the correlated methods CFA specification for the Cluster A personality disorders example are not provided.

An important special case of correlated methods solutions is the *uncorrelated methods* model. Its specification is identical to that of correlated methods, except that the covariances of the methods factors are fixed to zero (Widaman, 1985); that is, in the LISREL syntax provided in Table 6.2, this would simply entail fixing the PH(5,4), PH(6,4), and PH(6,5) elements to zero instead of freely estimating these parameters. Because these two models are nested, this comparison provides a statistical evaluation of whether the effects associated with the different assessment methods are correlated; for example, a lack of correlated method effects would be indicated by a nonsignificant χ^2 difference test.

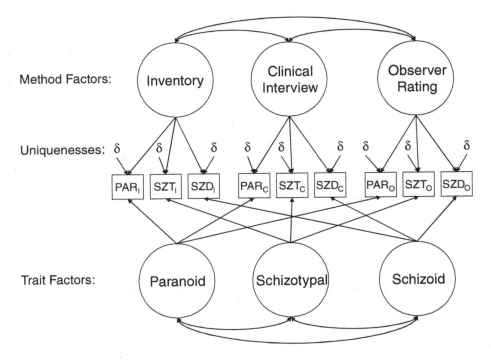

FIGURE 6.1. Correlated methods CFA specification of the MTMM matrix of Cluster A personality disorders.

TABLE 6.2. LISREL Syntax for Specification of a Correlated Methods CFA of the MTMM Matrix of Cluster A Personality Disorders

```
TITLE LISREL SYNTAX FOR CORRELATED METHODS MTMM SPECIFICATION
DA NI=9 NO=500 MA=CM
LA
PARI SZTI SZDI PARC SZTC SZDC PARO SZTO SZDO
KM
1.000
0.290  1.000
0.372  0.478  1.000
0.587  0.238  0.209  1.000
0.201  0.586  0.126  0.213  1.000
0.218  0.281  0.681  0.195  0.096  1.000
0.557  0.228  0.195  0.664  0.242  0.232  1.000
0.196  0.644  0.146  0.261  0.641  0.248  0.383  1.000
0.219  0.241  0.676  0.290  0.168  0.749  0.361  0.342  1.000
SD
3.61 3.66 3.59 2.94 3.03 2.85 2.22 2.42 2.04
MO NX=9 NK=6 PH=SY,FR LX=FU,FR TD=SY,FR
LK
PARANOID SCHIZOTYP SCHIZOID INVENTRY INTERVW OBSERVE
PA LX
1 0 0 1 0 0
0 1 0 1 0 0
0 0 1 1 0 0
1 0 0 0 1 0
0 1 0 0 1 0
0 0 1 0 1 0
1 0 0 0 0 1
0 1 0 0 0 1
0 0 1 0 0 1
PA TD
1
0 1
0 0 1
0 0 0 1
0 0 0 0 1
0 0 0 0 0 1
0 0 0 0 0 0 1
0 0 0 0 0 0 0 1
0 0 0 0 0 0 0 0 1
PA PH
0
1 0
1 1 0
0 0 0 0
0 0 0 1 0
0 0 0 1 1 0
VA 1.0 PH(1,1) PH(2,2) PH(3,3) PH(4,4) PH(5,5) PH(6,6)
OU ME=ML RS MI SC AD=OFF IT=500 ND=4
```

Correlated Uniqueness Models

The correlated uniqueness CFA model (Kenny, 1979; Marsh, 1989) was introduced as an alternative approach to analyzing MTMM data. Figure 6.2 depicts the path diagram of the correlated uniqueness CFA specification for the MTMM matrix of Cluster A personality disorders (Table 6.1). A recent applied example of this approach can be found in Meyer, Frost, Brown, Steketee, and Tolin (2013). In order for the correlated uniqueness model to be identified, there must be at least two traits (T) and three methods (M) (although a $2T \times 2M$ model can be fitted to the data if the factor loadings of indicators loading on the same trait factor are constrained to equality). Specification of the trait portion of the correlated uniqueness model is the same as that of the correlated methods approach: (1) Each indicator is specified to load on one trait factor (all other cross-loadings are fixed to zero); and (2) correlations among trait factors are freely estimated. Thus the key difference between these parameterizations is the manner in which method effects are estimated. In the correlated uniqueness model, method effects are estimated by specifying correlated uniquenesses (errors) among indicators based on the same assessment method rather than by method factors. Table 6.3 provides the programming syntax for this model in the LISREL, Mplus, EQS, and SAS/CALIS languages.[1]

A strong practical advantage of the correlated uniqueness approach is that, unlike correlated methods models, this parameterization rarely results in improper solutions (Kenny & Kashy, 1992; Marsh & Bailey, 1991; Tomás, Hontangas, & Oliver, 2000). Indeed, the correlated uniqueness MTMM model of Cluster A personality disorders converges and produces a proper solution that provides an acceptable fit to the data: $\chi^2(15)$ = 14.37, p = .50, SRMR = .025, RMSEA = 0.00 (90% CI = 0.00–0.04, CFit = .99), TLI =

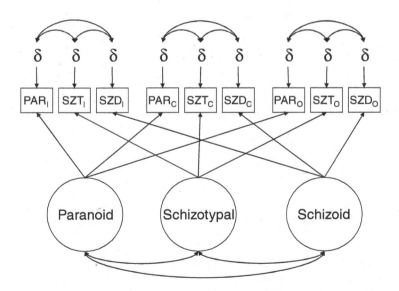

FIGURE 6.2. Correlated uniqueness CFA specification of the MTMM matrix of Cluster A personality disorders.

TABLE 6.3. Computer Syntax (LISREL, Mplus, EQS, SAS/CALIS) for Specification of a Correlated Uniquenesses CFA of the MTMM Matrix of Cluster A Personality Disorders

LISREL 9.1

```
TITLE LISREL SYNTAX FOR CORRELATED UNIQUENESS MTMM SPECIFICATION
DA NI=9 NO=500 MA=CM
LA
PARI SZTI SZDI PARC SZTC SZDC PARO SZTO SZDO
KM
1.000
0.290  1.000
0.372  0.478  1.000
0.587  0.238  0.209  1.000
0.201  0.586  0.126  0.213  1.000
0.218  0.281  0.681  0.195  0.096  1.000
0.557  0.228  0.195  0.664  0.242  0.232  1.000
0.196  0.644  0.146  0.261  0.641  0.248  0.383  1.000
0.219  0.241  0.676  0.290  0.168  0.749  0.361  0.342  1.000
SD
3.61 3.66 3.59 2.94 3.03 2.85 2.22 2.42 2.04
MO NX=9 NK=3 PH=SY,FR LX=FU,FR TD=SY,FR
LK
PARANOID SCHIZOTYP SCHIZOID
PA LX
1 0 0
0 1 0
0 0 1
1 0 0
0 1 0
0 0 1
1 0 0
0 1 0
0 0 1
PA TD
1
1 1
1 1 1
0 0 0 1
0 0 0 1 1
0 0 0 1 1 1
0 0 0 0 0 0 1
0 0 0 0 0 0 1 1
0 0 0 0 0 0 1 1 1
PA PH
0
1 0
1 1 0
VA 1.0 PH(1,1) PH(2,2) PH(3,3)
OU ME=ML RS MI SC AD=OFF IT=500 ND=4
```

Mplus 7.11

```
TITLE:     MPLUS PROGRAM FOR CORRELATED UNIQUENESS MTMM MODEL
DATA:      FILE IS "C:\INPUT5.DAT";
           TYPE IS STD CORR;
           NOBS ARE 500;
```

(continued)

TABLE 6.3. *(continued)*

```
VARIABLE:     NAMES ARE PARI SZTI SZDI PARC SZTC SZDC PARO SZTO SZDO;
ANALYSIS:     ESTIMATOR=ML;
MODEL:        PARANOID BY PARI* PARC PARO;
              SCHZOTYP BY SZTI* SZTC SZTO;
              SCHIZOID BY SZDI* SZDC SZDO;
              PARANOID@1.0; SCHZOTYP@1.0; SCHIZOID@1.0;
              PARI SZTI WITH SZDI; PARI WITH SZTI; PARC SZTC WITH SZDC;
              PARC WITH SZTC; PARO SZTO WITH SZDO; PARO WITH SZTO;
              PARANOID SCHZOTYP WITH SCHIZOID; PARANOID WITH SCHZOTYP;
OUTPUT:       SAMPSTAT MODINDICES(3.84) STAND RESIDUAL;
```

EQS 6.2

```
/TITLE
 EQS SYNTAX FOR CORRELATED UNIQUENESS MTMM SPECIFICATION
/SPECIFICATIONS
 CASES=500; VARIABLES=9; METHODS=ML; MATRIX=COR; ANALYSIS=COV;
/LABELS
 v1=PARI; v2=SZTI; v3=SZDI; v4=PARC;
 v5=SZTC; v6=SZDC; v7=PARO; v8=SZTO; v9=SZDO;
 f1 = paranoid; f2 = schizotyp; f3 = schizoid;
/EQUATIONS
 V1 =   *F1+E1;
 V2 =   *F2+E2;
 V3 =   *F3+E3;
 V4 =   *F1+E4;
 V5 =   *F2+E5;
 V6 =   *F3+E6;
 V7 =   *F1+E7;
 V8 =   *F2+E8;
 V9 =   *F3+E9;
/VARIANCES
 F1 TO F3 = 1.0;
 E1 TO E9= *;
/COVARIANCES
 F1 TO F3 = *;
 E1 TO E3= *; E4 TO E6= *; E7 TO E9= *;
/MATRIX
1.000
0.290  1.000
0.372  0.478  1.000
0.587  0.238  0.209  1.000
0.201  0.586  0.126  0.213  1.000
0.218  0.281  0.681  0.195  0.096  1.000
0.557  0.228  0.195  0.664  0.242  0.232  1.000
0.196  0.644  0.146  0.261  0.641  0.248  0.383  1.000
0.219  0.241  0.676  0.290  0.168  0.749  0.361  0.342  1.000
/STANDARD DEVIATIONS
3.61 3.66 3.59 2.94 3.03 2.85 2.22 2.42 2.04
/PRINT
 fit=all;
/LMTEST
/WTEST
/END
```

(continued)

TABLE 6.3. *(continued)*

SAS 9.4 PROC CALIS

```
Title "CALIS SYNTAX FOR CORRELATED UNIQUENESS MTMM SPECIFICATION";
Data CLUSTA (type=CORR);
  input _TYPE_ $ _NAME_ $ V1-V9;
  label V1 = 'pari'
        V2 = 'szti'
        V3 = 'szdi'
        V4 = 'parc'
        V5 = 'sztc'
        V6 = 'szdc'
        V7 = 'paro'
        V8 = 'szto'
        V9 = 'szdo';
cards;
mean  .     0     0     0     0     0     0     0     0     0
 std  .  3.61  3.66  3.59  2.94  3.03  2.85  2.22  2.42  2.04
   N  .   500   500   500   500   500   500   500   500   500
corr V1 1.000    .     .     .     .     .     .     .     .
corr V2 0.290 1.000    .     .     .     .     .     .     .
corr V3 0.372 0.478 1.000    .     .     .     .     .     .
corr V4 0.587 0.238 0.209 1.000    .     .     .     .     .
corr V5 0.201 0.586 0.126 0.213 1.000    .     .     .     .
corr V6 0.218 0.281 0.681 0.195 0.096 1.000    .     .     .
corr V7 0.557 0.228 0.195 0.664 0.242 0.232 1.000    .     .
corr V8 0.196 0.644 0.146 0.261 0.641 0.248 0.383 1.000    .
corr V9 0.219 0.241 0.676 0.290 0.168 0.749 0.361 0.342 1.000
;
run;

proc calis data= CLUSTA cov method=ml pall pcoves;
var = V1-V9;
lineqs
  V1 = lam1 f1 + e1,
  V2 = lam2 f2 + e2,
  V3 = lam3 f3 + e3,
  V4 = lam4 f1 + e4,
  V5 = lam5 f2 + e5,
  V6 = lam6 f3 + e6,
  V7 = lam7 f1 + e7,
  V8 = lam8 f2 + e8,
  V9 = lam9 f3 + e9;
std
  f1-f3 = 1.0,
  e1-e9 = td1-td9;
cov
  f1-f3 = ph1-ph3,
  e1-e3 = td10-td12,
  e4-e6 = td13-td15,
  e7-e9 = td16-td18;
run;
```

1.00, CFI = 1.00. Inspection of standardized residuals and modification indices indicates a reasonable solution. The completely standardized parameter estimates of this solution are presented in Table 6.4. With the exception of one correlated uniqueness (δ_{54}), all unstandardized parameter estimates are significantly different from zero ($p < .05$).

In regard to the parameter estimates produced by a correlated uniqueness CFA model specification, trait factor loadings that are large and statistically significant can be viewed as supporting convergent validity. On the other hand, large correlations among the trait factors can be viewed as indicative of poor discriminant validity. The presence of appreciable method effects is reflected by correlated uniquenesses among indicators assessed by the same method that are moderate or greater in magnitude. The results in Table 6.4 support the construct validity of the Cluster A personality disorders. The trait factor loadings are consistently large (range = .712–.872), providing evidence that the indicators are strongly related to their purported latent constructs (convergent validity), while adjusting for the effects of assessment method. Adequate discriminant validity is evidenced by the modest correlations among trait factors (range = .310–.381). Although significant method effects are obtained in all but one instance (SZT_C with PAR_C), the size of these effects is modest (standardized values range = −.037–.293). As in the MTMM matrix presented in Table 6.1, method effects estimated by the CFA model are smallest for the clinical ratings indicators.

ADVANTAGES AND DISADVANTAGES OF CORRELATED METHODS AND CORRELATED UNIQUENESS MODELS

As Kenny and Kashy (1992) note, an appealing feature of the correlated methods model is that it corresponds directly to Campbell and Fiske's (1959) original conceptualization of the MTMM matrix. Under this specification, each indicator is considered to be a function of trait, method, and unique variance. For instance, with the completely standardized solution, the squared trait factor loading (squared multiple correlations or communalities), squared method factor loading, and uniqueness for any given indicator sum to 1.0. Thus these estimates can be interpreted as the proportions of trait, method, and unique variance of each indicator. Moreover, the parameter estimates produced by correlated methods solutions provide seemingly straightforward interpretations with regard to construct validity. For example, large trait factor loadings suggest favorable convergent validity; small or nonsignificant method factor loadings imply an absence of method effects; and modest trait factor intercorrelations suggest favorable discriminant validity. The specification of method factors fosters the substantive interpretation of method effects. Because the covariance associated with a given method is (ideally) accounted for by a single factor (e.g., a method factor for Observer Ratings; see Figure 6.1), method effects are assumed to be unidimensional (although, as discussed later, this assumption does not always hold). Moreover, unlike the correlated uniqueness model, the correlated methods approach allows for the evaluation of the extent to which method factors are intercorrelated.

TABLE 6.4. Completely Standardized Estimates for the Correlated Uniqueness Model of the MTMM Matrix of Cluster A Personality Disorders

	Trait factor loadings			SMC	Uniqueness	Correlated uniquenesses		
	Paranoid	Schizotypal	Schizoid					
PAR$_I$	.712			.507	.493	1.000		
SZT$_I$		.788		.621	.379	.094	1.000	
SZD$_I$			.769	.592	.408	.198	.293	1.000
PAR$_C$	.841			.707	.293	1.000		
SZT$_C$		.768		.589	.411	−.037	1.000	
SZD$_C$			.860	.740	.260	−.073	−.108	1.000
PAR$_O$	.788			.620	.380	1.000		
SZT$_O$		.843		.711	.289	.137	1.000	
SZD$_O$			.872	.760	.240	.111	.126	1.000

	Trait factor correlations		
	Paranoid	Schizotypal	Schizoid
Paranoid	1.000		
Schizotypal	.381	1.000	
Schizoid	.359	.310	1.000

Note. SMC, squared multiple correlation (i.e., λ^2). All unstandardized parameters are significantly different ($p < .05$) from zero except for the correlation between the unique variances of SZT$_C$ and PAR$_C$ (LISREL 9.1 output).

However, an overriding drawback of the correlated methods model is that it is usually empirically underidentified. Consequently, a correlated methods solution will typically fail to converge. If it does converge, the solution will usually be associated with Heywood cases (Chapter 5) and large standard errors. For instance, using over 400 MTMM matrices derived from real and simulated data, Marsh and Bailey (1991) found that the correlated methods model resulted in improper solutions 77% of the time. These authors noted that improper solutions were most probable when the MTMM design was small (e.g., $3T \times 3M$), when sample size was small, and when the assumption of unidimensional method effects was untenable. Moreover, Kenny and Kashy (1992) have demonstrated that the correlated methods model was empirically underidentified in two special cases: (1) when the loadings on a trait or methods factor are equal; and (2) when there is no discriminant validity between two or more factors. Although these conditions are never perfectly realized, Kenny and Kashy (1992) have shown that research data frequently approximate these cases (e.g., factor loading estimates that are roughly equal in magnitude). When factor loadings are roughly equal or when discriminant validity is poor, this empirical underidentification results in severe estimation difficulties. Because of these problems, Kenny and Kashy (1992) and other methodologists (e.g., Marsh & Grayson, 1995) have recommended the use of the correlated uniqueness model over the correlated methods approach for the analysis of MTMM data. However, other researchers (Lance, Noble, & Scullen, 2002) have concluded that, given the substantive strengths of the correlated methods model, the correlated uniqueness should be used only if the correlated methods model fails. These authors underscore the various design features that may foster the chances of obtaining an admissible correlated methods solution (e.g., increased sample size, larger MTMM designs).[2]

Although its propensity for estimation problems is the primary disadvantage of correlated methods models, researchers have noted another limitation with this specification approach. In particular, these models do not allow for multidimensional method effects. A method effect is multidimensional if there exist two or more systematic sources of variability (aside from the underlying trait) that affect some or all of the indicators in the model (e.g., response set, reverse-worded questionnaire items, indicators differentially affected by social desirability or the tendency for over- or underreporting). As noted earlier, correlated methods models attempt to explain all the covariance associated with a given assessment method by a single method factor; hence the method effects are assumed to be unidimensional. Although these models do not permit correlations between trait and method factors, researchers have demonstrated instances where this assumption is unrealistic (e.g., Kumar & Dillon, 1992). Moreover, methodologists have illustrated scenarios in which the partitioning of observed measure variance into trait and method components does not produce trait-free and method-free interpretations (e.g., Bagozzi, 1993).

Unlike the correlated methods model, correlated uniqueness models rarely pose estimation problems. For instance, in the Marsh and Bailey (1991) study, correlated uniqueness model specifications resulted in proper solutions 98% of the time. In addition, these models can accommodate both unidimensional and multidimensional method effects because method covariance is reproduced by freely estimated correlations (which may differ greatly in magnitude) among indicators based on the same assessment approach. However, the interpretation of correlated uniquenesses as method effects is not always straightforward (e.g., Bagozzi, 1993). Although this parameterization allows for multidimensional method effects, the resulting solution does not provide information on the interpretative nature of these effects (e.g., does not foster a substantive explanation for why the magnitude of the correlated errors may vary widely), and so, for instance, it can be difficult to determine which method has the most method variance.

Another potential drawback of the correlated uniqueness model is its assumption that the correlations among methods, and the correlations between traits and methods, are zero. In fact, when the number of traits (T) is three, the correlated uniqueness model and the uncorrelated methods model virtually always produce equivalent solutions (e.g., model df and goodness of fit are identical; the parameter estimates from one solution can be transformed into the other). Methodologists (e.g., Byrne & Goffin, 1993; Kenny & Kashy, 1992) have shown how the parameter estimates of a correlated uniqueness solution may be biased when the assumption of zero correlations among methods and between methods and traits does not hold. If these zero correlation constraints are not tenable, the amount of trait variance and covariance between trait factors will be overestimated, resulting in inflated estimates of convergent validity and lower estimates of discriminant validity, respectively (Kenny & Kashy, 1992; Marsh & Bailey, 1991). Kenny and Kashy (1992) have illustrated this biasing effect in situations where the covariances among methods are mistakenly assumed to be zero and the true factor loadings are all equal:

> The average method–method covariance is added to each element of the trait–trait covariance matrix. So, if the methods are similar to one another, resulting in positive method–method covariances, the amount of trait variance will be overestimated as will be the amount of trait–trait covariance. (pp. 169–170)

Although the size of these biases is usually trivial (Marsh & Bailey, 1991), Kenny and Kashy (1992) nonetheless recommend that researchers using correlated uniqueness models to analyze MTMM data should try to employ assessment methods that are as independent as possible.

OTHER CFA PARAMETERIZATIONS OF MTMM DATA

A number of other CFA-based strategies for analyzing MTMM data have been developed. One of the more prominent alternative approaches is the *direct product model* (Browne, 1984a; Cudeck, 1988; Wothke & Browne, 1990). Unlike the preceding CFA approaches, the direct product model addresses the possibility that method factors interact with trait factors in a multiplicative manner rather than additively (cf. Campbell & O'Connell, 1967). In other words, method effects may augment the correlations of strongly related traits more than they augment the correlations of weakly related constructs; for example, the higher the correlation between traits, the greater the effects of methods. When the data conform to the direct product model, this parameterization provides an elegant test of the criteria outlined by Campbell and Fiske (1959) for evaluating convergent and discriminant validity (for applied examples, see Bagozzi & Yi, 1990; Coovert, Craiger, & Teachout, 1997; and Lievens & Conway, 2001). Another advantage of this method is that it estimates a correlation matrix for the methods. This matrix can be inspected to evaluate the similarity of methods; for instance, high correlations between purportedly distinct methods would challenge their discriminant validity (i.e., the methods are in effect the same). However, the direct product model is relatively difficult to program and interpret (the reader is referred to Bagozzi & Yi, 1990, and Wothke, 1996, for fully worked-through LISREL parameterizations of direct product solutions). The direct product model often produces improper solutions, although much less so than correlated methods approaches (e.g., Lievens & Conway, 2001). A number of other disadvantages have been noted. For instance, some researchers (e.g., Podsakoff et al., 2003) have concluded that the extant evidence indicates that trait × methods interactions are not very common or potent and thus a simpler CFA parameterization will usually be just as good. Nevertheless, more extensive evaluation of the direct product model is needed.

A newer alternative is the *correlated trait–correlated method minus one model*, or CT-C($M - 1$) (Eid, 2000; Eid, Lischetzke, Nussbeck, & Trierweiler, 2003). The CT-C($M - 1$) model is very similar to the correlated methods model (Figure 6.1) except that it contains one method factor less than the methods included ($M - 1$); e.g., the path diagram in Figure 6.1 could be converted into CT-C($M - 1$) model by eliminating the Inventory method factor.[3] This parameterization resolves some identification problems of the correlated methods model. The omitted method factor becomes the "comparison standard." The principal notion behind the CT-C($M - 1$) model is that for each trait, the true-score variables of indicators of the comparison standard are regressors in a latent regression analysis in which the dependent measures are the true-score variables of the "nonstandard" methods ("nonstandard" = methods other than the comparison standard). Thus a method factor is a residual factor common to all variables measured by the same method. In other words, it represents the portion of a trait measured by a nonstandard method that cannot be predicted by the true-score variable of the indicators measured by the comparison standard method. A step-by-step illustration of the CT-C($M - 1$) model is presented in Eid et al. (2003); for applied examples, see Lischetzke and Eid (2003), and Kollman, Brown, and Barlow (2009). The CT-C($M - 1$) model has many of the strengths of the correlated methods model (e.g., straightforward decomposition of trait, method, and error effects), but it avoids the serious problems of underidentification and improper solutions. The CT-C($M - 1$) model has several other advantages. When multiple indicators are used (see note 3), this model can test for trait-specific method effects (i.e., there exists a source of variance specific to each trait–method combination), allowing the researcher to examine the generalizability of method effects across traits. Moreover, the model also provides correlations among method factors, and to some extent, information on the relationships between method factors and trait factors. One limitation is the asymmetry of the CT-C(M − 1) model because a comparison standard method must be selected. Choice of the comparison standard may be clear-cut in some situations (e.g., when a "gold standard" exists), but not others (e.g., when all methods are somewhat similar or randomly chosen). Like the direct product model, the performance of the CT-C(M − 1) model in applied and simulated data sets could benefit from more extensive study (e.g., Eid et al., 2008).

CONSEQUENCES OF NOT MODELING METHOD VARIANCE AND MEASUREMENT ERROR

It has been noted in earlier chapters (e.g., Chapters 3 and 5) that the ability to model method effects is an important advantage of CFA over EFA. For instance, EFAs of questionnaires that have been designed to measure unidimensional constructs (e.g., self-esteem) tend to produce two-factor solutions (e.g., Positive Self-Esteem, Negative Self-Esteem). These multidimensional latent structures stem from method effects introduced by including reverse-worded items. A conceptually more viable unidimensional structure can be upheld by CFA through specification of a single trait factor (e.g., Self-Esteem) and method effects (e.g., correlated errors) among the reverse-worded items (e.g., Marsh, 1996). In CFA, specification of a one-factor model without method effects will result in a poor-fitting solution. The correlated errors are needed to account for the additional covariance among items that is due to nonrandom measurement error (e.g., acquiescent response style). In addition, this specification yields better estimates of the relationships between the indicators and the latent construct (i.e., reduces bias in the factor loadings).

These points also apply to the CFA of MTMM data. For instance, when the data in Table 6.1 are reanalyzed without modeling method variance (i.e., three trait factors are specified, and all measurement error is presumed to be random), a poor-fitting solution results: $\chi^2(24) = 455.32$, $p < .001$, SRMR = .063, RMSEA = 0.190 (90% CI = 0.175–0.205, CFit < .001), TLI = 0.765, CFI = .843. Inspection of standardized residuals and modification indices reveal that relationships among indicators obtained from the same assessment method are not well explained by the model—for example, standardized residuals (LISREL) for observer ratings:

	PARO	SZTO	SZDO
PARO	– –		
SZTO	4.3763	– –	
SZDO	3.9064	3.6103	– –

The failure to account for method variance may lead to the false conclusion that the constructs under study have poor discriminant validity. This is because the correlations among the factors (traits) are apt to be inflated by the CFA estimation process. Because method effects have not been taken into account in the model specification, the estimation process attempts to reproduce the additional covariance of indicators sharing the same assessment method by increasing the magnitude of the factor correlations (cf. Eq. 3.8, Chapter 3). This is exemplified in the current data set, where each of the correlations among the Cluster A personality traits are inflated when method effects are not taken into account; for example, correlation of Schizotypal and Schizoid = .31 and .36 in CFA solutions with and without method effects, respectively. This is less problematic in the Table 6.1 data because the parameter estimates will not be interpreted in context of a poor-fitting model (e.g., RMSEA = 0.183), and the factor correlations in both solutions are not large (e.g., .31 vs. .36). However, this will not always be the case in applied data sets (e.g., inflation of factor correlations can be more extreme).

Another good example in the applied literature on the effects of (not) modeling measurement error is in the area of emotion. Investigators have long debated whether positive and negative mood (e.g., happiness and sadness) are largely independent (i.e., separate constructs) or bipolar (i.e., represent opposite ends of a single dimension). Although original theories of emotion assumed bipolarity, much of the early research did not support this position. For instance, the initial evidence showed that single indicators of happy and sad mood were not strongly correlated. Consequently, EFAs typically revealed two-factor solutions (along the lines of EFAs of questionnaires containing reversed items, described above). However, researchers later came to learn that the low correlations between indicators of positive and negative moods might be due to the failure to consider and model the effects of random and nonrandom response error (e.g., Green, Goldman, & Salovey, 1993; Russell, 1979). A thorough description of the sources of systematic error in the measurement of affect can be found in Russell (1979) and Green et al. (1993).[4] But when random and systematic error is adjusted for, the correlation between positive and negative emotions increases to an extent that severely challenges the differentiation (discriminant validity) of these dimensions.

To illustrate, a sample correlation matrix from Green et al. (1993) is presented in Figure 6.3 (N = 304 undergraduates). The constructs of Happy and Sad mood were assessed by four different types of self-report formats: adjective checklist; item response options ranging from strong agreement to strong disagreement; self-description response options ranging from very well to not at all; and a semantic differential Likert scale. Some of the correlations seen in Figure 6.3 might be mistakenly interpreted in support of the notion that Happy and Sad mood are distinct dimensions; for example, the correlation between adjective checklist indicators of Happy and Sad mood is −.10. When a CFA model with the appropriate error theory is fitted to these data (see path diagram in Figure 6.3), a different picture emerges. The two-factor measurement model, which takes random and nonrandom error into account, provides a reasonable fit to the data: $\chi^2(15)$ = 26.21, p = .036, SRMR = .026, RMSEA = 0.050 (90% CI = 0.01–0.08, CFit = .468), TLI = 0.993, CFI = .996. However, the factor correlation of Happy and Sad mood is found to be −.87; this result seriously challenges the discriminant validity of these constructs and upholds the contention of the bipolarity of emotion. These findings strongly underscore the importance of multimethod research designs and analytic approaches (i.e., CFA) that model measurement error in construct validation (cf. Castro-Schilo, Widaman, & Grimm, 2013; Cole, Ciesla, & Steiger, 2007).

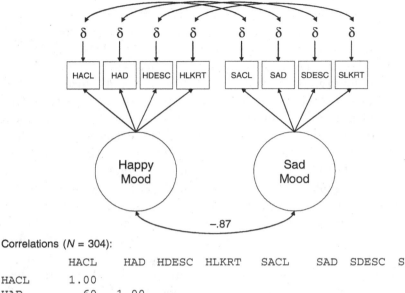

Correlations (N = 304):

	HACL	HAD	HDESC	HLKRT	SACL	SAD	SDESC	SLKRT
HACL	1.00							
HAD	.60	1.00						
HDESC	.59	.69	1.00					
HLKRT	.59	.67	.74	1.00				
SACL	-.10	-.46	-.53	-.50	1.00			
SAD	-.44	-.53	-.63	-.64	.68	1.00		
SDESC	-.44	-.48	-.59	-.59	.52	.60	1.00	
SLKRT	-.47	-.59	-.65	-.66	.60	.69	.61	1.00

FIGURE 6.3. Path diagram and input data for CFA model of positive and negative mood. Correlation matrix is taken from Green et al. (1993, Table 7). The first letter of each indicator refers to the construct it purportedly measures: H, Happy; S, Sad. The remaining letters of each indicator pertain to the self-report assessment format: ACL, adjective checklist; AD, agree–disagree format; DESC, self-descriptive format; LKRT, semantic differential Likert scale.

SUMMARY

CFA approaches to MTMM matrices (and the MTMM approach in general) continue to be somewhat underutilized in many areas of the applied literature. This is presumably due in part to the fact that in many cases, multiple measures are not available for a given construct (e.g., the correlated uniqueness approach ordinarily requires at least three assessment methods). Although the examples used in this chapter involved relatively disparate assessment methods, these approaches can be employed for constructs assessed by a single assessment modality—for example, DSM-5 Cluster A personality dimensions assessed by three different self-report inventories (also see Green et al., 1993).

MTMM models can be extended in useful ways. First, an MTMM model can be embedded in a larger structural equation model, such as one that relates the trait factors to background variables or distal outcomes that represent external validators;

for example, are the Cluster A personality dimensions differentially related to salient clinical variables such as comorbidity, overall functioning, or long-term adjustment? Such analyses can strongly bolster the importance of the substantive constructs of the MTMM model (e.g., can establish their predictive validity). A second possible extension is analyzing the MTMM model within a multiple-groups solution to determine whether salient parameters (e.g., trait factor loadings) are invariant across salient subgroups; for example, is each observed measure of Cluster A personality related to the substantive trait in an equivalent manner for men and women? Approaches to evaluating invariance of CFA models are presented in Chapter 7, along with the analysis of mean structures (i.e., indicator intercepts, latent variable means).

NOTES

1. Readers who run this model in Mplus will encounter larger standardized error covariances than the estimates presented in Table 6.4. In later releases, Mplus began to compute standardized error covariances differently than other leading software programs (e.g., LISREL). The following formula converts a covariance into a correlation: $CORR_{1,2} = COV_{1,2} / SQRT(VAR_1 * VAR_2)$. While the unstandardized error covariances ($COV_{1,2}$) are virtually the same across software programs, Mplus computes the error correlation by using the indicator residual variances (VAR_1, VAR_2) rather than the sample variances of the indicators (the unstandardized residual variances and covariances will also differ slightly in Mplus, given the use of N instead of $N - 1$ in these calculations; see Chapter 3). Consequently, the standardized error covariances reported by Mplus are larger than those derived from the same unstandardized factor solution in other software programs. If the user wishes to obtain the correct model-implied correlations among indicators from a measurement model solution involving indicator error covariances, the traditional computational method should be used (e.g., the estimates provided by LISREL).

2. For instance, Tomás et al. (2000) found that under some conditions, the correlated methods model works reasonably well (and perhaps better than the correlated uniqueness approach) when more than two indicators per trait–method combination are available. However, because the typical MTMM study is a $3T \times 3M$ design with a single indicator per each trait–method combination (Marsh & Grayson, 1995), evaluation of the correlated methods model under the conditions studied by Tomás et al. (2000) may not be practical in many applied research scenarios.

3. The CT-C($M - 1$) model can be identified with as few as two traits and two methods. However, due to separate trait, method, and error effects in the model, at least two indicators per each trait–method combination are required (e.g., for each trait in Figure 6.1, at least two inventory indicators, two clinical interview indicators, and two observer rating indicators are needed; $p = 18$).

4. In addition, a comprehensive review of the most common types of method effects found in social and behavioral sciences research is provided by Podsakoff et al. (2003).

7

CFA with Equality Constraints, Multiple Groups, and Mean Structures

The previous CFA examples presented in this book have been estimated within a single group, have used a variance–covariance matrix as input, and have entailed model parameters that are either freely estimated or fixed. In this chapter, these analyses are extended in several ways. For instance, some CFA specifications place equality constraints on selected parameters of the measurement model. Such constraints may be applicable in CFA analyses involving a single group (e.g., do the items of a questionnaire assess the same latent construct in equivalent units of measurement?) or two or more groups (e.g., do males and females respond to items of a measuring instrument in a similar manner?). In addition, two different approaches to CFA with multiple groups are presented (multiple-groups CFA, MIMIC models) in context of the analysis of measurement invariance and population heterogeneity. Finally, CFA is extended to the analysis of mean structures, involving the estimation of indicator intercepts and factor means and the evaluation of their equivalence in multiple groups. Consequently, the input matrix must include the sample means of the indicators, in addition to their variances and covariances. The substantive applications of each approach are discussed.

OVERVIEW OF EQUALITY CONSTRAINTS

As first noted in Chapter 3, parameters in a CFA solution can be freely estimated, fixed, or constrained. A free parameter is unknown, and the researcher allows the analysis to find its optimal value that, in tandem with other model estimates, minimizes the differences between the observed and predicted variance–covariance matrices (e.g., in a one-factor CFA model, to obtain the set of factor loadings that best reproduces the observed correlations among four input indicators). A fixed parameter is prespecified by the researcher to be a specific value, most commonly either 1.0 (e.g., in the case of marker indicators or factor variances to define the metric of a latent variable) or 0 (e.g., the absence of cross-loadings or error covariances). Like a free parameter, a constrained parameter is unknown. However, the parameter is not free to be any value; rather, the specification places restrictions on the values it may assume. The most common forms of constrained parameters are *equality constraints*, in which unstandardized parameters are restricted to be equal in value (other types of constrained parameters are discussed later in this chapter in context of latent variable scaling, and in the section in Chapter 8 on scale reliability evaluation). Consider a CFA model in which four indicators are specified to load on a single factor. If the CFA specification entails an equality constraint on the four factor loadings, the specific value of these loadings is unknown a priori (as in models with freely estimated factor loadings), but the analysis must find a single estimate (applied to all four loadings) that best reproduces the observed relationships among the four indicators. This is unlike the models with freely estimated parameters, where the factor loadings are free to take on any set of values that maximize the fit of the solution.

Two important principles apply to the various examples of equality constraints discussed in this chapter. First, as with fixed parameters, these constraints are placed on the unstandardized solution. Accordingly, the indicators whose parameters are to be held equal should have the same metric; for example, the unstandardized loading of an indicator defined by a 0–100 scale will inherently differ from the unstandardized loading of an indicator measured on a 1–8 scale. Second, because a CFA model with equality constraints is nested under the measurement model without these constraints (i.e., it entails a subset of the parent model's freely estimated parameters), χ^2 difference testing can be employed as a statistical comparative evaluation of the constrained solution. For example, if the CFA model where the four factor loadings are held to equality does not produce a significant reduction in fit relative to the corresponding unconstrained solution, it can be concluded that the four indicators have equivalent relationships to the latent variable.

EQUALITY CONSTRAINTS WITHIN A SINGLE GROUP

Congeneric, Tau-Equivalent, and Parallel Indicators

As noted in Chapter 3, most CFA specifications in applied research entail *congeneric* indicator sets. Congeneric indicators are presumed to measure the same construct, and the size of their factor loadings and measurement errors are free to vary; in addition, the assumption of independent measurement errors must hold. Figure 7.1 provides an example of a congeneric model in which the first three observed measures (X1, X2, X3) are indicators of one latent construct (Auditory Memory), and the second set of measures (X4, X5, X6) are indicators of another latent construct (Visual Memory). In addition to congeneric models, the psychometric literature distinguishes more restrictive solutions that test for the conditions of *tau-equivalent* and *parallel* indicators. A tau-equivalent model entails a congeneric solution in which the indicators of a given factor have equal

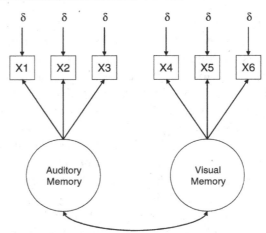

Sample Correlations and Standard Deviations (*SD*s); *N* = 200

	X1	X2	X3	X4	X5	X6
X1	1.000					
X2	0.661	1.000				
X3	0.630	0.643	1.000			
X4	0.270	0.300	0.268	1.000		
X5	0.297	0.265	0.225	0.805	1.000	
X6	0.290	0.287	0.248	0.796	0.779	1.000
SD:	2.610	2.660	2.590	1.940	2.030	2.050

FIGURE 7.1. Two-factor model of memory. X1, logical memory; X2, verbal paired association; X3, word list; X4, faces; X5, family pictures; X6, visual reproduction.

loadings but differing error variances. As noted previously, when the condition of equal factor loadings holds, it can be asserted that the indicators have equivalent relationships with the underlying construct they measure. Stated another way, a 1-unit change in the latent variable is associated with the same amount of change in each indicator that loads on that factor. The most restrictive solution treats indicators as parallel, in which the observed measures are posited to have equal factor loadings and equal error variances. Thus, in addition to the assumption that indicators measure the latent construct in the same units of measurement (tau equivalence), parallel indicators are assumed to measure the latent construct with the same level of precision (i.e., reflected by equivalent error variances).[1] These distinctions have psychometric implications. For instance, Raykov (2001a, 2012) has shown that Cronbach's coefficient alpha is a misestimator of the scale reliability of a multicomponent measuring instrument (e.g., a multiple-item questionnaire) when the assumption of tau equivalence does not hold (in Chapter 8, a CFA-based approach to estimating scale reliability is presented that does not rest on this assumption). If the conditions of parallel indicators are met, this lends support for the notion that the measures in question are psychometrically interchangeable—a finding that is germane to the endeavor of establishing parallel test forms or justifying the practice of operationalizing a latent construct by summation of its indicators' observed scores (McDonald, 1999).

The test for tau-equivalent and parallel indicators begins with the evaluation of the congeneric measurement model, in which the factor loadings and residual variances are free to vary. If the conditions of a congeneric solution are not met (e.g., an indicator loads on more than one factor), the analysis will not proceed to the evaluation of tau equivalence unless, perhaps, substantive considerations allow for the measurement model to be revised to conform to a congeneric model (e.g., removal of a cross-loading indicator). In context of a congeneric measurement model, the test for tau equivalence is conducted by placing the appropriate restrictions on the solution (i.e., equality constraints on indicators that load on the same factor) and evaluating the resulting change in model χ^2, using the less restricted congeneric model as the baseline solution. If the results are in accord with tau equivalence (i.e., constraining the factor loadings to equality does not result in a significant increase in χ^2), the analysis can proceed to the evaluation of parallel indicators.

To illustrate this procedure, a two-factor model of memory is used in which three observed measures are posited to be indicators of the construct of Auditory Memory (X1, logical memory; X2, verbal paired association; X3, word list), and three additional measures are posited to be indicators of the construct of Visual Memory (X4, faces; X5, family pictures; X6, visual reproduction). The example is loosely based on the structure of the Wechsler Memory Scale—Third Edition (Wechsler, 1997); for applied CFA examples in this research domain, see Price, Tulsky, Millis, and Weiss (2002) and Tulsky and Price (2003). The measurement model is presented in Figure 7.1, along with the sample (N = 200) correlations and standard deviations of the six indicators.

The congeneric, two-factor solution provides a good fit to the data: $\chi^2(8) = 4.88$, $p = .77$, SRMR = .012, RMSEA = 0.00 (90% CI = 0.00–0.06, CFit = .93), TLI = 1.01, CFI = 1.00 (Mplus 7.11 output). The inspection of modification indices and standardized residuals reveals no areas of strain in the solution. The model parameter estimates are presented in Table 7.1. All six factor loadings are statistically significant ($ps < .001$) and sufficiently large (range of completely standardized estimates = .78–.91) (in this example, the metric of the latent variables is defined by fixing factor variances to 1.0). As expected, the constructs of Auditory Memory and Visual Memory are significantly correlated ($p < .001$), $\phi_{21} = .38$.

First, the tau equivalence of the indicators that load on Auditory Memory is examined. This is performed by placing an equality constraint on the factor loadings of the X1, X2, and X3 indicators. Table 7.2 provides the programming syntax for this model in the LISREL, Mplus, EQS, and SAS/CALIS languages. As shown in this table, the software programs differ in their syntax for imposing equality constraints on the model parameters. In LISREL, this restriction is accomplished by adding the line

EQ LX(1,1) LX(2,1) LX(3,1)

where EQ is the keyword for holding the parameters that follow to equality. The syntax also illustrates a programming shortcut in LISREL, TD = DI, where the theta-delta

TABLE 7.1. Mplus Results of the Two-Factor Model of Memory

MODEL FIT INFORMATION

Chi-Square Test of Model Fit

Value	4.877
Degrees of Freedom	8
P-Value	0.7706

RMSEA (Root Mean Square Error Of Approximation)

Estimate	0.000	
90 Percent C.I.	0.000	0.057
Probability RMSEA <= .05	0.929	

CFI/TLI

CFI	1.000
TLI	1.008

SRMR (Standardized Root Mean Square Residual)

Value	0.012

STANDARDIZED MODEL RESULTS

STDYX Standardization

	Estimate	S.E.	Est./S.E.	Two-Tailed P-Value
AUDITORY BY				
X1	0.807	0.036	22.410	0.000
X2	0.823	0.035	23.431	0.000
X3	0.779	0.038	20.674	0.000
VISUAL BY				
X4	0.907	0.019	47.999	0.000
X5	0.887	0.021	43.188	0.000
X6	0.878	0.021	41.340	0.000
VISUAL WITH				
AUDITORY	0.382	0.070	5.464	0.000
Variances				
AUDITORY	1.000	0.000	999.000	999.000
VISUAL	1.000	0.000	999.000	999.000
Residual Variances				
X1	0.349	0.058	6.008	0.000
X2	0.323	0.058	5.601	0.000
X3	0.393	0.059	6.682	0.000
X4	0.177	0.034	5.158	0.000
X5	0.214	0.036	5.872	0.000
X6	0.229	0.037	6.127	0.000

(continued)

TABLE 7.1. *(continued)*

R-SQUARE

Observed Variable	Estimate	S.E.	Est./S.E.	Two-Tailed P-Value
X1	0.651	0.058	11.205	0.000
X2	0.677	0.058	11.716	0.000
X3	0.607	0.059	10.337	0.000
X4	0.823	0.034	24.000	0.000
X5	0.786	0.036	21.594	0.000
X6	0.771	0.037	20.670	0.000

matrix is specified to be diagonal (DI), meaning that the indicator error variances are freely estimated and the error covariances are fixed to 0 (i.e., the only freed parameters within TD are on the diagonal). In Mplus, the equality constraints are reflected in the line

AUDITORY BY X1* X2 X3 (1);

where parameters are constrained to be equal by placing the same number in parentheses following the parameters that are to be held equal. An * is placed directly after the X1 indicator to override the Mplus default of setting the first indicator in the list as the marker indicator. This is done because the metric of Auditory Memory is defined by fixing its variance to 1.0. In EQS, the same constraints are imposed by adding a /CONSTRAINT section to the command file as follows:

/CONSTRAINT
(V1, F1) = (V2,F1) = (V3,F1);

In SAS/CALIS, these constraints are made in the following lines:

V1 = lam1 f1 + e1,
V2 = lam1 f1 + e2,
V3 = lam1 f1 + e3,

in which the factor loadings for the first factor (f1; Auditory Memory) are each given the same parameter name (lam1; the user may provide any name for these or other model parameters).

As shown in Table 7.3, this restriction results in an increase in χ^2 to 5.66 with *df* = 10 (*p* = .84); for a more in-depth discussion of model estimation under this equality constraint, see Appendix 7.1. This solution has 2 more *df*s than the congeneric model (*df* = 8) because of the equality constraint on the factor loadings of Auditory Memory; that is, the congeneric model contains three separate estimates of the X1, X2, and X3 factor

TABLE 7.2. Computer Syntax (LISREL, Mplus, EQS, SAS/CALIS) for Specification of Tau-Equivalent Indicators of Auditory Memory within a Two-Factor Measurement Model of Memory

LISREL 9.10

```
TITLE LISREL PROGRAM FOR TAU EQUIVALENT AUDITORY INDICATORS
DA NI=6 NO=200 MA=CM
LA
X1 X2 X3 X4 X5 X6
KM
1.000
0.661  1.000
0.630  0.643  1.000
0.270  0.300  0.268  1.000
0.297  0.265  0.225  0.805  1.000
0.290  0.287  0.248  0.796  0.779  1.000
SD
2.61 2.66 2.59 1.94 2.03 2.05
MO NX=6 NK=2 PH=SY,FR LX=FU,FR TD=DI
LK
AUDITORY VISUAL
PA LX
1 0
1 0
1 0
0 1
0 1
0 1
PA PH
0
1 0
VA 1.0 PH(1,1) PH(2,2)
EQ LX(1,1) LX(2,1) LX(3,1)                ! EQUALITY CONSTRAINT
OU ME=ML RS MI SC AD=OFF IT=100 ND=4
```

Mplus 7.11

```
TITLE:      MPLUS PROGRAM FOR TAU EQUIVALENT AUDITORY INDICATORS
DATA:       FILE IS FIG7.1.DAT;
            TYPE IS STD CORR;
            NOBS ARE 200;
VARIABLE:   NAMES ARE X1-X6;
ANALYSIS:   ESTIMATOR=ML;
MODEL:      AUDITORY BY X1* X2 X3 (1);  ! EQUALITY CONSTRAINT
            VISUAL BY X4* X5 X6;
            AUDITORY@1.0; VISUAL@1.0;
OUTPUT:     SAMPSTAT MODINDICES(4.00) STAND RESIDUAL;
```

EQS 6.2

```
/TITLE
 EQS SYNTAX FOR TAU EQUIVALENT AUDITORY INDICATORS
/SPECIFICATIONS
 CASES=200; VARIABLES=6; METHODS=ML; MATRIX=COR; ANALYSIS=COV;
```

(continued)

TABLE 7.2. *(continued)*

```
/LABELS
 v1=logical; v2=verbal; v3=word; v4=faces; v5=family; v6=visrep;
 f1 = auditory; f2 = visual;
/EQUATIONS
 V1 =   *F1+E1;
 V2 =   *F1+E2;
 V3 =   *F1+E3;
 V4 =   *F2+E4;
 V5 =   *F2+E5;
 V6 =   *F2+E6;
/VARIANCES
 F1 TO F2 = 1.0;
 E1 TO E6= *;
/COVARIANCES
 F1 TO F2 = *;
/CONSTRAINT
 (V1,F1) = (V2,F1) = (V3,F1);
/MATRIX
1.000
0.661  1.000
0.630  0.643  1.000
0.270  0.300  0.268  1.000
0.297  0.265  0.225  0.805  1.000
0.290  0.287  0.248  0.796  0.779  1.000
/STANDARD DEVIATIONS
2.61 2.66 2.59 1.94 2.03 2.05
/PRINT
 fit=all;
/LMTEST
/END
```

SAS 9.4 PROC CALIS

```
Title "CALIS SYNTAX FOR TAU EQUIVALENT AUDITORY INDICATORS";
Data WMS (type=CORR);
 input _TYPE_ $ _NAME_ $ V1-V6;
 label V1 = 'logical'
       V2 = 'verbal'
       V3 = 'word'
       V4 = 'faces'
       V5 = 'family'
       V6 = 'visrep';
cards;
mean  .    0      0      0      0      0      0
 std  .  2.61   2.66   2.59   1.94   2.03   2.05
  N   .   200    200    200    200    200    200
corr V1  1.000    .      .      .      .      .
corr V2  0.661  1.000    .      .      .      .
corr V3  0.630  0.643  1.000    .      .      .
corr V4  0.270  0.300  0.268  1.000    .      .
corr V5  0.297  0.265  0.225  0.805  1.000    .
corr V6  0.290  0.287  0.248  0.796  0.779  1.000
;
run;
```

(continued)

TABLE 7.2. (continued)

```
proc calis data=WMS cov method=ml pall pcoves;
var = V1-V6;
lineqs
 V1 = lam1 f1 + e1,
 V2 = lam1 f1 + e2,
 V3 = lam1 f1 + e3,
 V4 = lam4 f2 + e4,
 V5 = lam5 f2 + e5,
 V6 = lam6 f2 + e6;
std
  f1-f2 = 1.0,
  e1-e6 = td1-td6;
cov
  f1-f2 = ph3;
run;
```

loadings, and the tau-equivalent model contains one factor loading estimate applied to each of these three indicators. The difference in χ^2 of the current solution and the congeneric solution is 0.78, with $df = 2$. Because this χ^2 difference is below the critical value of the χ^2 distribution at $df = 2$ (i.e., $\chi^2_{crit} = 5.99$, at $\alpha = .05$), it can be concluded that the three indicators of Auditory Memory are tau-equivalent; that is, a unit increase in the latent construct of Auditory Memory is associated with the same amount of change in each of the X1, X2, and X3 indicators. The selected Mplus output provided below shows that these equality constraints apply to both the unstandardized factor loadings and their standard errors (and hence the z tests of significance as well).

```
MODEL RESULTS

                                                          Two-Tailed
                    Estimate        S.E.    Est./S.E.     P-Value

 AUDITORY BY
    X1               2.100         0.125     16.800        0.000
    X2               2.100         0.125     16.800        0.000
    X3               2.100         0.125     16.800        0.000

 VISUAL   BY
    X4               1.756         0.108     16.183        0.000
    X5               1.795         0.115     15.608        0.000
    X6               1.796         0.117     15.378        0.000
```

Because this restriction of factor loading equality holds, it is retained in subsequent tests of the two-factor measurement model of memory.

Next, the model is respecified with the added constraint of holding the factor loadings of Visual Memory to equality, to evaluate the tau equivalence of these indicators. As seen in Table 7.3, this restriction also does not significantly degrade the fit of the solution, $\chi^2_{diff}(2) = 0.22$, ns; therefore, the indicators of Visual Memory can also be con-

TABLE 7.3. Model Fit of Congeneric, Tau-Equivalent, and Parallel Solutions of a Two-Factor Model of Memory

	χ^2	df	χ^2_{diff}	Δdf	RMSEA (90% CI)	CFit	SRMR	CFI	TLI
Congeneric solution	4.88	8			.000 (.000–.057)	.93	.012	1.00	1.01
Tau-equivalent: Auditory Memory	5.66	10	0.78	2	.000 (.000–.044)	.96	.021	1.00	1.01
Tau-equivalent: Auditory and Visual Memory	5.88	12	0.22	2	.000 (.000–.025)	.99	.022	1.00	1.01
Parallel: Auditory Memory	5.97	14	0.09	2	.000 (.000–.000)	1.00	.021	1.00	1.01
Parallel: Auditory and Visual Memory	9.28	16	3.31	2	.000 (.000–.028)	.99	.027	1.00	1.01

Note. $N = 200$. χ^2_{diff}, nested χ^2 difference; RMSEA, root mean square error of approximation; 90% CI, 90% confidence interval for RMSEA; CFit, test of close fit (probability RMSEA ≤ .05); SRMR, standardized root mean square residual; CFI, comparative fit index; TLI, Tucker–Lewis index. All χ^2 and χ^2_{diff} values nonsignificant (*ps* > .05). Based on Mplus 7.11 output.

sidered to be tau-equivalent. Because tau equivalence has been established, the analysis can proceed to evaluating the condition of parallel indicators.

The current example has defined the metric of Auditory Memory and Visual Memory by fixing their variances to 1.0. It is noteworthy that the same results are obtained if the unstandardized factor loadings of all indicators are set to 1.0 (with the factor variances freely estimated). This is because when an indicator is selected to be a marker, its unstandardized loading is fixed to 1.0. Thus, if all factor loadings are to be tested for equality (tau equivalence), they must also equal the value of the marker (1.0).

In addition to tau equivalence, the condition of parallelism requires that the error variances of indicators loading on the same factor are the same. This added restriction is tested for Auditory Memory by placing equality constraints on the measurement errors of X1, X2, and X3. The findings presented in Table 7.3 show that this restriction does not result in a significant increase in χ^2, $\chi^2_{\text{diff}}(2) = 0.09$, *ns*. Finally, the results in this table also indicate that the indicators of Visual Memory can be regarded as parallel, $\chi^2_{\text{diff}}(2) = 3.31$, *ns*. The collective findings can be interpreted as being consistent with the notion that X1–X3 and X4–X6 are interchangeable indicators of the latent constructs of Auditory Memory and Visual Memory, respectively.

For the reader's information, the model estimates for the final solution are presented in Table 7.4. Table 7.5 provides the programming syntax for the final model in LISREL, Mplus, EQS, and SAS/CALIS. Before the discussion continues in the next section, a couple of additional points are made about the preceding analyses. First, the nested χ^2 procedure has been employed by comparing the model in question to the previous, slightly less restricted solution (e.g., a model in which the indicators of both Auditory Memory and Visual Memory are held to be parallel vs. the model in which only the indicators of Auditory Memory are held parallel). It should be noted that any two models presented in Table 7.3 are nested, and thus any less restricted solution could serve as the comparison model. For instance, the final solution (in which both the indicators of Auditory Memory and Visual Memory are parallel) can be tested against the congeneric solution. In this case, the χ^2_{diff} value is 4.40 with *df* = 8; that is, $\chi^2 = 9.28 - 4.88 = 4.40$, *df* = 16 – 8 = 8. This comparison produces the equivalent result that the fully parallel model does not significantly degrade the fit of the solution, because the χ^2 difference of 4.40 is less than the critical value of the χ^2 distribution at *df* = 8 (i.e., χ^2_{crit} = 15.51, at α = .05). It might appear that it would be more efficient to move straight from the congeneric model to the most restrictive solution. However, the conditions of tau-equivalent and parallel indicators often do not hold in applied data sets (in particular, the condition of parallel indicators is quite restrictive). Thus it is usually better to employ an incremental strategy that will allow one to more readily detect the sources of noninvariance if significant degradations in model fit are encountered because the restrictions are placed on a single set of parameters at a time rather than all at once.

Second, in the case of a multifactorial, congeneric solution, the evaluation of tau equivalent and parallel indicators of one factor does not rely on the respective findings for indicators loading on different factors. For instance, if the condition of tau equivalence for the indicators of Auditory Memory is not met (i.e., these equality constraints

TABLE 7.4. Selected Mplus Results of the Final Two-Factor Measurement Model of Memory

```
MODEL FIT INFORMATION

Chi-Square Test of Model Fit

          Value                              9.277
          Degrees of Freedom                    16
          P-Value                           0.9016

RMSEA (Root Mean Square Error Of Approximation)

          Estimate                           0.000
          90 Percent C.I.                    0.000   0.028
          Probability RMSEA <= .05           0.989

CFI/TLI

          CFI                                1.000
          TLI                                1.009

SRMR (Standardized Root Mean Square Residual)

          Value                              0.027
```

```
MODEL RESULTS

                                                     Two-Tailed
                    Estimate     S.E.   Est./S.E.     P-Value

AUDITORY BY
   X1                2.099       0.125    16.795        0.000
   X2                2.099       0.125    16.795        0.000
   X3                2.099       0.125    16.795        0.000

VISUAL   BY
   X4                1.782       0.097    18.364        0.000
   X5                1.782       0.097    18.364        0.000
   X6                1.782       0.097    18.364        0.000

VISUAL   WITH
   AUDITORY          0.381       0.070     5.431        0.000

Variances
   AUDITORY          1.000       0.000   999.000      999.000
   VISUAL            1.000       0.000   999.000      999.000

Residual Variances
   X1                2.427       0.172    14.142        0.000
   X2                2.427       0.172    14.142        0.000
   X3                2.427       0.172    14.142        0.000
   X4                0.832       0.059    14.142        0.000
```

(continued)

TABLE 7.4. *(continued)*

X5	0.832	0.059	14.142	0.000
X6	0.832	0.059	14.142	0.000

STANDARDIZED MODEL RESULTS

STDYX Standardization

	Estimate	S.E.	Est./S.E.	Two-Tailed P-Value
AUDITORY BY				
X1	0.803	0.021	38.825	0.000
X2	0.803	0.021	38.825	0.000
X3	0.803	0.021	38.825	0.000
VISUAL BY				
X4	0.890	0.012	72.344	0.000
X5	0.890	0.012	72.344	0.000
X6	0.890	0.012	72.344	0.000
VISUAL WITH				
AUDITORY	0.381	0.070	5.431	0.000
Variances				
AUDITORY	1.000	0.000	999.000	999.000
VISUAL	1.000	0.000	999.000	999.000
Residual Variances				
X1	0.355	0.033	10.699	0.000
X2	0.355	0.033	10.699	0.000
X3	0.355	0.033	10.699	0.000
X4	0.208	0.022	9.476	0.000
X5	0.208	0.022	9.476	0.000
X6	0.208	0.022	9.476	0.000

R-SQUARE

Observed Variable	Estimate	S.E.	Est./S.E.	Two-Tailed P-Value
X1	0.645	0.033	19.413	0.000
X2	0.645	0.033	19.413	0.000
X3	0.645	0.033	19.413	0.000
X4	0.792	0.022	36.172	0.000
X5	0.792	0.022	36.172	0.000
X6	0.792	0.022	36.172	0.000

TABLE 7.5. Computer Syntax (LISREL, Mplus, EQS, SAS/CALIS) for Specification of Parallel Indicators of Auditory Memory and Visual Memory within a Two-Factor Measurement Model of Memory

LISREL 9.10

```
TITLE LISREL PROGRAM FOR PARALLEL INDICATORS
DA NI=6 NO=200 MA=CM
LA
X1 X2 X3 X4 X5 X6
KM
<insert correlation matrix from Figure 7.1 here>
SD
2.61 2.66 2.59 1.94 2.03 2.05
MO NX=6 NK=2 PH=SY,FR LX=FU,FR TD=DI
LK
AUDITORY VISUAL
PA LX
1 0
1 0
1 0
0 1
0 1
0 1
PA PH
0
1 0
VA 1.0 PH(1,1) PH(2,2)
EQ LX(1,1) LX(2,1) LX(3,1)
EQ LX(4,2) LX(5,2) LX(6,2)
EQ TD(1,1) TD(2,2) TD(3,3)
EQ TD(4,4) TD(5,5) TD(6,6)
OU ME=ML RS MI SC AD=OFF IT=100 ND=4
```

Mplus 7.11

```
TITLE:      MPLUS PROGRAM FOR PARALLEL INDICATORS
DATA:       FILE IS "C:\input6.dat";
            TYPE IS STD CORR;
            NOBS ARE 200;
VARIABLE:   NAMES ARE X1-X6;
ANALYSIS:   ESTIMATOR=ML;
MODEL:      AUDITORY BY X1* X2 X3 (1);
            VISUAL   BY X4* X5 X6 (2);
            AUDITORY@1.0; VISUAL@1.0;
            X1 X2 X3 (3);
            X4 X5 X6 (4);
OUTPUT:     SAMPSTAT MODINDICES(4.00) STAND RESIDUAL;
```

EQS 6.2

```
/TITLE
 EQS SYNTAX FOR PARALLEL INDICATORS
/SPECIFICATIONS
 CASES=200; VARIABLES=6; METHODS=ML; MATRIX=COR; ANALYSIS=COV;
```

(continued)

TABLE 7.5. *(continued)*

```
/LABELS
 v1=logical; v2=verbal; v3=word; v4=faces; v5=family; v6=visrep;
 f1 = auditory; f2 = visual;
/EQUATIONS
 V1 =   *F1+E1;
 V2 =   *F1+E2;
 V3 =   *F1+E3;
 V4 =   *F2+E4;
 V5 =   *F2+E5;
 V6 =   *F2+E6;
/VARIANCES
 F1 TO F2 = 1.0;
 E1 TO E6= *;
/COVARIANCES
 F1 TO F2 = *;
/CONSTRAINT
 (V1,F1) = (V2,F1) = (V3,F1);
 (V4,F2) = (V5,F2) = (V6,F2);
 (E1, E1) = (E2, E2) = (E3, E3);
 (E4, E4) = (E5, E5) = (E6, E6);
/MATRIX
<insert correlation matrix from Figure 7.1 here>
/STANDARD DEVIATIONS
2.61 2.66 2.59 1.94 2.03 2.05
/PRINT
 fit=all;
/LMTEST
/END
```

SAS 9.4 PROC CALIS

```
Title "CALIS SYNTAX FOR PARALLEL INDICATORS";
Data WMS (type=CORR);
 input _TYPE_ $ _NAME_ $ V1-V6;
 label V1 = 'logical'
       V2 = 'verbal'
       V3 = 'word'
       V4 = 'faces'
       V5 = 'family'
       V6 = 'visrep';
cards;
mean  .   0      0      0      0      0      0
 std  .  2.61   2.66   2.59   1.94   2.03   2.05
   N  .  200    200    200    200    200    200
corr V1  1.000   .      .      .      .      .
corr V2  0.661  1.000   .      .      .      .
corr V3  0.630  0.643  1.000   .      .      .
corr V4  0.270  0.300  0.268  1.000   .      .
corr V5  0.297  0.265  0.225  0.805  1.000   .
corr V6  0.290  0.287  0.248  0.796  0.779  1.000
;
run;
```

(continued)

TABLE 7.5. *(continued)*

```
proc calis data=WMS cov method=ml pall pcoves;
var = V1-V6;
lineqs
 V1 = lam1 f1 + e1,
 V2 = lam1 f1 + e2,
 V3 = lam1 f1 + e3,
 V4 = lam2 f2 + e4,
 V5 = lam2 f2 + e5,
 V6 = lam2 f2 + e6;
std
  f1-f2 = 1.0,
  e1-e3 = td1,
  e4-e6 = td2;
cov
  f1-f2 = ph3;
run;
```

lead to a significant increase in model χ^2), the researcher can still proceed to evaluating whether the indicators of Visual Memory are tau-equivalent and parallel.

Longitudinal Measurement Invariance

Another type of invariance evaluation that can be conducted on CFA models within a single group concerns the equality of construct measurement over time. Although rarely addressed in the applied literature, longitudinal measurement invariance is a fundamental aspect of evaluating temporal change in a construct. In the absence of such evaluation, it cannot be determined whether temporal change observed in a construct is due to true change or to changes in the structure or measurement of the construct over time. Drawing on the work of Golembiewski, Billingsley, and Yeager (1976), Chan (1998) outlined three types of change that may be encountered in repeated measurements: *alpha*, *beta*, and *gamma* change (these terms do not correspond to parameters of structural equation models). Alpha change refers to true-score change in a construct, given a constant conceptual domain and constant measurement. Alpha change (true-score change) can only be said to occur in the context of longitudinal measurement invariance (i.e., evidence that the measurement of the construct does not change over time). Chan (1998) notes that longitudinal measurement invariance can be construed as an absence of beta and gamma change. Beta change occurs in instances where the construct of interest remains constant, but the measurement properties of the indicators of the construct are temporally inconsistent (e.g., numerical values across assessment points are not on the same measurement scale). Gamma change occurs when the meaning of the construct changes over time (e.g., the number of factors that represent the construct vary across assessment waves). In applied longitudinal research, measurement invariance is often simply (implicitly) assumed and not examined. However, when measurement is not invariant over time, it is misleading to analyze and interpret the temporal change in observed measures or latent constructs; that is, change may be misinterpreted as alpha change when in fact the precision of measurement of the construct, or the construct itself, varies across time. Thus the examination of measurement invariance should precede applications of SEM procedures with longitudinal data (e.g., latent growth curve models, autoregressive/cross-lagged panel models; Bollen & Curran, 2006; Curran & Hussong, 2003; Duncan, Duncan, & Strycker, 2006).

These procedures are illustrated with the longitudinal measurement model presented in Figure 7.2. In this example, the researcher wishes to evaluate whether an intervention has been successful in improving employees' job satisfaction. Employees ($N = 250$) of a large company are administered four measures of job satisfaction (Measures A through D, which vary in assessment modality; e.g., questionnaires, supervisor ratings) immediately before and after the intervention (the pre- to posttest interval is 4 weeks). For each mea-

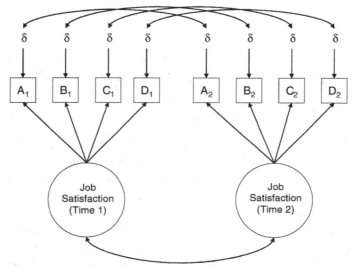

Sample Correlations, Means (*M*), and Standard Deviations (*SD*); *N* = 250

	A1	B1	C1	D1	A2	B2	C2	D2
A1	1.000							
B1	0.736	1.000						
C1	0.731	0.648	1.000					
D1	0.771	0.694	0.700	1.000				
A2	0.685	0.512	0.496	0.508	1.000			
B2	0.481	0.638	0.431	0.449	0.726	1.000		
C2	0.485	0.442	0.635	0.456	0.743	0.672	1.000	
D2	0.508	0.469	0.453	0.627	0.759	0.689	0.695	1.000
M:	1.500	1.320	1.450	1.410	6.600	6.420	6.560	6.310
SD:	1.940	2.030	2.050	1.990	2.610	2.660	2.590	2.550

FIGURE 7.2. Longitudinal measurement model of Job Satisfaction. Indicators A through D are various measures (e.g., questionnaires, supervisor ratings) of job satisfaction. Rectangles denote within-time portions of the input correlation matrix.

sure, higher scores reflect higher job satisfaction. Prior to examining whether the intervention has resulted in an overall increase in job satisfaction, the researcher wishes to verify that the construct of Job Satisfaction, and its measurement, remain stable over time.

Figure 7.2 presents the hypothesized path model in which the construct of Job Satisfaction is posited to be structurally the same (i.e., unidimensional) at both assessment points (cf. gamma change). If the factor structure is temporally equivalent, additional tests can be performed to examine the other aspects of measurement invariance (e.g., equality of factor loadings) in a manner similar to the procedures used for evaluating tau-equivalent and parallel indicators. In addition, note that correlated errors have been specified for each pair of repeated measures (e.g., the residual of indicator A1 is allowed to covary freely with the residual of indicator A2). Recall from earlier chapters (e.g., Chapter 2) that the uniqueness of an indicator consists of some combination of random measurement error and indicator-specific variance. Specification of correlated errors is based on the premise that these indicator-specific variances are temporally stable. For instance, such parameters may be posited to reflect method effects associated with repeated administrations of the same measure (e.g., in addition to the underlying dimension of Job Satisfaction, some of the variance in the A indicator is due to the influence of social desirability, present at each testing occasion). Depending on the research scenario, it may or may not be necessary to specify autocorrelations among indicators' error variances; for example, these effects may be less likely evident in designs employing wide assessment intervals or monomethod indicators.

It should be noted that longitudinal measurement invariance can also be evaluated by using the multiple-groups approach discussed later in this chapter (i.e., each "group" is represented by a different wave of assessment—e.g., Group 1 = Time 1; Group 2 = Time 2). In reviewing the measurement invariance literature, Vandenberg and Lance (2000) have discussed the advantages and disadvantages of assessing longitudinal measurement invariance by using a single sample (all assessment waves combined in a single input matrix) versus a multiple-groups approach (assessment waves represented by separate input matrices for each "group"). The one-sample approach takes into account the complete data structure—that is, the lagged relationships among indicators in addition to the within-time covariances. The entire matrix presented in Figure 7.2 is used in the analysis, including the between-time correlations (such as A1 with A2, B2, C2, and D2). In the multiple-groups approach, only the within-time portions of the matrix (denoted by boxes in Figure 7.2) are input to the analysis (each box represents a different "group"). Accordingly, a primary advantage of employing a one-sample approach is that correlated errors of the repeated measurements can be estimated and controlled for in the estimation of other model parameters. The main disadvantage of the single-sample method pertains to use of an input matrix larger than that used in the multiple-groups approach (cf. Figure 7.2). Because a larger matrix is used as input, the single-sample approach may be more prone to poor model fit or possibly improper solutions resulting from the greater complexity of the model (cf. Chapter 5). Nevertheless, given the ability to model the full temporal measurement structure (e.g., correlated measurement errors over time), the one-sample approach is strongly favored over the multiple-groups method for evaluating the longitudinal invariance of measurement models.

The longitudinal measurement example is also used to introduce the reader to the analysis of mean structures. All examples presented in this book thus far have entailed the analysis of covariance structures. That is, only variances and covariances have been included in the input matrix, and thus the indicators and resulting model parameters are deviations from their means (i.e., their means are presumed to be zero). For example, as with coefficients in multiple regression, a factor loading can be interpreted as the amount of predicted change in an indicator (the Y variable) given a unit change in the factor (the X variable). These coefficients do not reflect the exact predicted score of the Y variable, but estimate how much this variable is predicted to change, given a unit change in X. However, a typical application of multiple regression involves the prediction of specific scores of the Y variable, as shown in the simple regression equation below:

$$\hat{Y} = a + bX \tag{7.1}$$

where $\hat{Y}$ is the predicted Y score; a is the intercept; b is the unstandardized regression coefficient; and X is a given score of the predictor variable. The actual (observed) value of Y can be reproduced by adding the residual (e) to the sum of $a + bX$:

$$Y = a + bX + e \tag{7.2}$$

A basic equation in regression solves for the intercept (a) on the basis of knowledge of the sample means of X and Y (M_y, M_x) and the regression coefficient (b):

$$a = M_y - bM_x \tag{7.3}$$

Through a simple algebraic manipulation, this formula can be reexpressed as

$$M_y = a + bM_x \tag{7.4}$$

A fundamental difference between these equations and the equations presented in context of CFA thus far is its inclusion of the intercept parameter. In the multiple regression framework, the intercept represents the position where the least squares regression line crosses the Y axis (i.e., the predicted value of Y when all X variables are zero). In Chapter 3 (Eq. 3.4), it has been shown that the observed variance of an indicator could be calculated from the model estimates of a covariance structure CFA by using this formula (latent X notation):

$$VAR(X) = \lambda_x^2 \phi + \delta \tag{7.5}$$

where λ_x is the unstandardized factor loading; ϕ is the factor variance; and δ is the indicator's error variance. With the exception of the intercept parameter, this equation is very similar to the previous regression equation (e.g., $\lambda_x = b$; $\delta = e$).

The equations of CFA can be expanded to include intercept (and latent mean) parameters; for example,

$$X = \tau_x + \Lambda_x \xi + \theta_\delta \qquad (7.6)$$

where τ_x is the indicator intercept (τ_x = tau X; cf. Figure 3.3, Chapter 3). Similarly, the CFA equivalent to the equation $M_y = a + bM_x$ is

$$M_x = \tau_x + \lambda_x \kappa \qquad (7.7)$$

where κ is the mean of the latent exogenous variable (κ = kappa; cf. Figure 3.3, Chapter 3). Thus the mean of a given indicator (M_x) can be reproduced by the CFA model's parameter estimates of the indicator's intercept (τ_x), factor loading (λ_x), and latent variable mean (κ).

Later in this chapter, it is shown that, like marker indicators and factor variances, indicator intercepts (means) and factor means are closely related. In addition, the analysis of mean structure poses new identification issues that can be addressed in different fashions in the single-sample and multiple-groups approaches.

To return to the example of longitudinal measurement invariance, this analysis can be conducted with or without the inclusion of mean structures. For instance, if the ultimate goal is to examine models based on a covariance structure (e.g., as in autoregressive/cross-lagged panel modeling), the analysis of mean structures is less relevant. However, if the goal is to examine the trajectory of change in the level of a given construct (e.g., as in latent growth curve modeling), the measurement invariance evaluation should include the analysis of indicator means; that is, the comparison of means is meaningful only if the factor loadings and measurement intercepts are found to be invariant.

The first step of the longitudinal analysis of Job Satisfaction is to establish that the same factor structure is present at both testing occasions (equal form). As shown in Figure 7.2, it is predicted that a unidimensional measurement model of job satisfaction is viable at both assessment points. Correlated errors are specified in anticipation that additional covariance will exist between repeated measures due to temporally stable indicator-specific variance (method effects). In this first example, the metric of the latent variables is defined by using the same observed measure at both testing occasions (A1 and A2) as the marker indicator. Later in this chapter, this model is respecified to illustrate another method of latent variable scaling that can be particularly useful in tests of measurement invariance.

The estimation of indicator intercepts and factor means requires that the observed means of the indicators be included as input data. Thus the number of elements of the input matrix increases by the number of indicator means that are included in the analysis; that is,

$$a = [p(p + 1) / 2] + p \qquad (7.8)$$

where a is the number of elements of the input matrix, and p is the number of indicators. Thus, in the current example ($p = 8$ indicators), there are 36 variances and covariances [$p(p + 1) / 2$] and 8 means (p), totaling 44 elements of the input matrix; Eq. 7.8 can be alternatively written as $a = [p(p + 3)] / 2$. Although this exceeds the number of freely estimated parameters of the equal form solution, the mean structure portion of the solution is underidentified in the absence of other model restrictions. In this example, there are 8 observed means (knowns) but 10 unknown parameters in the CFA mean structure (8 indicator intercepts, 2 latent variable means). Hence this aspect of the model is under-identified because the number of unknowns exceeds the number of knowns (cf. Chapter 3). Moreover, just as latent variables must be provided with a metric (e.g., by fixing the loading of an indicator to 1.0), they must also be assigned an origin (i.e., a mean). In covariance structure analysis, the latent variable means are assumed to be zero. In mean structure analysis, where the latent variables may take on mean values other than zero, origins must be assigned. In a single-sample analysis, the mean structure aspect of the measurement model may be identified in one of three ways: (1) fixing the latent mean to zero; (2) assigning the factor mean to take on the mean as one of its indicators (by fixing the intercept of one indicator to zero); or (3) using the effects coding approach, introduced in Chapter 3 and illustrated later in this example. In the first method, all the indicator intercepts are freely estimated but will equal the observed means of the indicators. In the second strategy, all but one intercept is freely estimated, and these freed intercepts will take on values different from their sample means; the factor mean is freely estimated but will equal the observed mean of the indicator whose intercept is fixed to zero. In the third approach, constraints are placed on the measurement model such that the indicator intercepts for each latent construct sum to zero. The restrictions associated with each approach will result in just-identification of the mean structure portion of the solution. In all approaches, additional restrictions can be placed on the model to overidentify its mean structure component (e.g., equality constraints on indicator intercepts). Later in this chapter, a slightly different method is used to identify the mean structure of a multiple-groups CFA solution.

In this initial example, Indicator A's intercept is fixed to zero at both testing occasions. Consequently, in the initial, less constrained solutions (e.g., the test of equal factor structure), the mean of the latent variable of Job Satisfaction is equal to the observed mean of Indicator A. In addition, these restrictions lead to just-identification of the mean structure aspect of the solution (i.e., 8 observed means; 8 freely estimated mean parameters = 6 intercepts + 2 latent variable means). Thus, although the input matrix is expanded to include indicator means, the overall model df is the same as in covariance structure analysis ($df = 15$) because the mean structure component does not provide additional degrees of freedom to the model ($df = 8 - 8 = 0$).

Selected results of the equal form (factor structure) solution are presented in Table 7.6. The overall fit of this solution is presented in Table 7.7. All results are in accord with the conclusion that a unidimensional measurement model of Job Satisfaction is viable at both testing occasions. Each of the overall fit statistics is consistent with good model fit—for example, $\chi^2(15) = 2.09$, $p = 1.0$—and fit diagnostics indicate the absence

TABLE 7.6. Mplus Results of the Equal Form Longitudinal Model of Job Satisfaction

MODEL RESULTS

		Estimate	S.E.	Est./S.E.	Two-Tailed P-Value
SATIS1	BY				
A1		1.000	0.000	999.000	999.000
B1		0.951	0.050	19.212	0.000
C1		0.970	0.049	19.616	0.000
D1		0.990	0.046	21.394	0.000
SATIS2	BY				
A2		1.000	0.000	999.000	999.000
B2		0.916	0.048	18.973	0.000
C2		0.922	0.046	20.085	0.000
D2		0.934	0.045	20.770	0.000
SATIS2	WITH				
SATIS1		2.680	0.353	7.586	0.000
A1	WITH				
A2		0.701	0.119	5.915	0.000
B1	WITH				
B2		1.043	0.163	6.407	0.000
C1	WITH				
C2		1.042	0.158	6.576	0.000
D1	WITH				
D2		0.770	0.134	5.735	0.000
Means					
SATIS1		1.500	0.122	12.343	0.000
SATIS2		6.600	0.164	40.183	0.000
Intercepts					
A1		0.000	0.000	999.000	999.000
B1		-0.107	0.117	-0.910	0.363
C1		-0.005	0.118	-0.042	0.967
D1		-0.075	0.108	-0.692	0.489
A2		0.000	0.000	999.000	999.000
B2		0.375	0.340	1.102	0.270
C2		0.477	0.324	1.475	0.140
D2		0.146	0.316	0.462	0.644
Variances					
SATIS1		2.985	0.321	9.297	0.000
SATIS2		5.414	0.586	9.244	0.000

(continued)

TABLE 7.6. *(continued)*

Residual Variances

A1	0.707	0.105	6.722	0.000
B1	1.409	0.153	9.178	0.000
C1	1.434	0.157	9.137	0.000
D1	1.038	0.127	8.165	0.000
A2	1.330	0.195	6.815	0.000
B2	2.418	0.263	9.183	0.000
C2	2.113	0.239	8.856	0.000
D2	1.836	0.219	8.390	0.000

STANDARDIZED MODEL RESULTS

STDYX Standardization

	Estimate	S.E.	Est./S.E.	Two-Tailed P-Value
SATIS1　BY				
A1	0.899	0.017	52.913	0.000
B1	0.811	0.024	33.162	0.000
C1	0.814	0.024	33.776	0.000
D1	0.859	0.020	42.264	0.000
SATIS2　BY				
A2	0.896	0.017	51.746	0.000
B2	0.808	0.025	32.617	0.000
C2	0.828	0.023	36.109	0.000
D2	0.849	0.021	39.785	0.000
SATIS2　WITH				
SATIS1	0.667	0.039	17.272	0.000
A1　WITH				
A2	0.723	0.052	13.947	0.000
B1　WITH				
B2	0.565	0.051	11.085	0.000
C1　WITH				
C2	0.598	0.049	12.159	0.000
D1　WITH				
D2	0.558	0.056	9.927	0.000
Means				
SATIS1	0.868	0.084	10.283	0.000
SATIS2	2.836	0.169	16.795	0.000

(continued)

TABLE 7.6. *(continued)*

Intercepts				
A1	0.000	0.000	999.000	999.000
B1	-0.053	0.057	-0.920	0.357
C1	-0.002	0.057	-0.042	0.967
D1	-0.038	0.054	-0.697	0.486
A2	0.000	0.000	999.000	999.000
B2	0.142	0.132	1.078	0.281
C2	0.184	0.128	1.435	0.151
D2	0.057	0.125	0.458	0.647
Variances				
SATIS1	1.000	0.000	999.000	999.000
SATIS2	1.000	0.000	999.000	999.000
Residual Variances				
A1	0.192	0.031	6.271	0.000
B1	0.343	0.040	8.652	0.000
C1	0.338	0.039	8.626	0.000
D1	0.262	0.035	7.501	0.000
A2	0.197	0.031	6.356	0.000
B2	0.347	0.040	8.682	0.000
C2	0.315	0.038	8.297	0.000
D2	0.280	0.036	7.735	0.000

of significant areas of strain in the solution (e.g., all modification indices < 4.0). At both assessments, the indicators are found to be significantly ($ps < .001$) and strongly related to the latent construct of Job Satisfaction (range of completely standardized factor loadings = .81–.90). In addition, the four error covariances are statistically significant ($ps < .001$). The test–retest covariance of the latent construct of Job Satisfaction is statistically significant ($\phi_{21} = 2.68$, $p < .001$). As noted above, the intercept of Indicator A is fixed to zero at both assessments in order to identify the mean structure component of the solution. This is reflected by the values of zero for all estimates of A1 and A2 in the Intercepts portion of the model results. As the result of these restrictions, the latent means of Job Satisfaction equal the observed means of the A1 and A2 indicators (1.5 and 6.6 for SATIS1 and SATIS2, respectively; cf. Table 7.6 and Figure 7.2). To illustrate an equation presented earlier in this chapter, the sample mean of the indicators can be reproduced by inserting the appropriate model estimates into Eq. 7.7. For instance, the sample mean of B1 = 1.32 and can be reproduced by using the model estimates of its intercept ($\tau_1 = -0.107$), unstandardized factor loading ($\lambda_{21} = 0.951$), and the mean of its factor ($\kappa = 1.5$):

$$-0.107 + 0.951(1.5) = 1.32 \tag{7.9}$$

Given evidence of equal form (cf. gamma change), additional tests of longitudinal measurement invariance may proceed. The next analysis tests for the equality of factor

TABLE 7.7. Longitudinal Invariance of a Measurement Model of Job Satisfaction

	χ^2	df	χ^2_{diff}	Δdf	RMSEA (90% CI)	CFit	SRMR	CFI	TLI
Equal form	2.09	15			.000 (.000–.000)	1.00	.010	1.00	1.01
Equal factor loadings	3.88	18	1.79	3	.000 (.000–.000)	1.00	.014	1.00	1.01
Equal indicator intercepts	7.25	21	3.37	3	.000 (.000–.000)	1.00	.026	1.00	1.01
Equal indicator error variances	90.73***	25	83.48***	4	.103 (.080–.126)	0.00	.037	0.96	0.96

Note. $N = 250$. χ^2_{diff}, nested χ^2 difference; RMSEA, root mean square error of approximation; 90% CI, 90% confidence interval for RMSEA; CFit, test of close fit (probability RMSEA $\leq$ .05); SRMR, standardized root mean square residual; CFI, comparative fit index; TLI, Tucker–Lewis index.

***p < .001.

loadings over the two assessment points. Unlike the evaluation of tau equivalence, this analysis does not constrain the indicators that load on the same factor to equality (e.g., $\lambda_{21} = \lambda_{31} = \lambda_{41}$). Rather, the equality constraint pertains to the factor loadings of indicators administered repeatedly across testing occasions (e.g., $\lambda_{21} = \lambda_{62}$). As in the analysis of tau equivalence, the nested χ^2 test can be employed to determine whether these constraints significantly degrade model fit. As shown in Table 7.7, the model χ^2 of the equal loadings solution is 3.88 ($df = 18$, $p = 1.0$), resulting in a nonsignificant χ^2 difference test, $\chi^2_{diff}(3) = 1.79$, ns [critical value of $\chi^2(3) = 7.81$, $\alpha = .05$]. The df of this nested model comparison is equal to 3 (and the equal factor loadings model $df = 18$) because only three pairs of factor loadings (B1 = B2; C1 = C2; D1 = D2) are involved in this equality constraint. This constraint does not apply to the loadings of A1 and A2 because these parameters have been previously fixed to 1.0 to set the scale of the latent variables.

On the basis of the results of the equal factor loading analysis, it can be concluded that the indicators evidence equivalent relationships to the latent construct of Job Satisfaction over time. Keeping the equality constraints of the factor loadings in place, the next model places additional equality constraints on the indicators' intercepts, except for indicators A1 and A2, whose intercepts have been previously fixed to zero for the purposes of model identification. These restrictions also do not lead to a significant reduction in model fit, $\chi^2_{diff}(3) = 3.37$, ns, suggesting that the indicator's intercepts are invariant between the two testing occasions. Recall that in multiple regression, an intercept can be regarded as the predicted value of Y when X is at zero. Measurement intercepts in CFA can be interpreted in an analogous manner; that is, τ_x = the predicted value of indicator X when $\kappa = 0$. Thus, if the intercept of an indicator is found to be temporally *noninvariant*, the predicted score of the indicator will vary across time at a constant level of the latent construct (i.e., $\kappa_1 = \kappa_2$). Stated another way, even when the "true score" (latent variable) remains unchanged, the observed scores of the indicators will vary over time. Thus it is erroneous to interpret changes in observed scores as true change (alpha change) because the observed change is due in some part to temporal variation in the measurement properties of the indicator. For example, although factor loading equivalence will suggest that the indicator possesses a temporally stable relationship to the underlying construct (i.e., a unit increase in the latent construct is associated with comparable changes in the indicator at all assessment points), noninvariant indicator intercepts will suggest inequality of the indicator's location parameters over time (a spurious shift from using one portion of the indicator's response scale at Time 1 to another portion of the response scale at Time 2, as may occur in various forms of rater drift such as leniency bias; cf. Vandenberg & Lance, 2000). However, in the present illustration both the factor loadings and indicator intercepts are found to be invariant, suggesting that the analysis of mean change over time can be attributed to true change in the construct (cf. Eq. 7.7).

The final analysis tests for the equality of the indicator's error variances. Table 7.8 provides Mplus and LISREL syntax for this analysis. This restriction results in a significant decrease in model fit, $\chi^2_{diff}(4) = 83.48$, $p < .001$ (critical χ^2 value at $df = 4$, $\alpha = .05$,

TABLE 7.8. Computer Syntax (LISREL, Mplus) for Testing a Fully Invariant Longitudinal Measurement Model of Job Satisfaction (Equal Form, Equal Factor Loadings, Equal Intercepts, Equal Residual Variances)

Mplus 7.11

```
TITLE:       MPLUS PROGRAM FOR TIME1-TIME2 MSMT MODEL OF JOB SATISFACTION
DATA:        FILE IS FIG7.2.DAT;
             TYPE IS MEANS STD CORR; ! INDICATOR MEANS ALSO INPUTTED
             NOBS ARE 250;
VARIABLE:    NAMES ARE A1 B1 C1 D1 A2 B2 C2 D2;
ANALYSIS:    ESTIMATOR=ML;
MODEL:       SATIS1 BY A1 B1 (1)
             C1 (2)
             D1 (3);
             SATIS2 BY A2 B2 (1)
             C2 (2)
             D2 (3);
             A1 WITH A2; B1 WITH B2; C1 WITH C2; D1 WITH D2;
             [A1@0]; [A2@0];          ! FIXES THE A INDICATOR INTERCEPTS TO ZERO
             [SATIS1*]; [SATIS2*];  ! FREELY ESTIMATES FACTOR MEANS
             [B1 B2] (4); [C1 C2] (5); [D1 D2] (6);
             A1 A2 (7); B1 B2 (8); C1 C2 (9); D1 D2 (10);
OUTPUT:      SAMPSTAT MODINDICES(4.00) STAND RESIDUAL;
```

LISREL 9.10

```
TITLE LISREL PROGRAM FOR TIME1-TIME2 MSMT MODEL OF JOB SATISFACTION
DA NI=8 NO=250 MA=CM
LA
A1 B1 C1 D1 A2 B2 C2 D2
KM
1.000
0.736  1.000
0.731  0.648  1.000
0.771  0.694  0.700  1.000
0.685  0.512  0.496  0.508  1.000
0.481  0.638  0.431  0.449  0.726  1.000
0.485  0.442  0.635  0.456  0.743  0.672  1.000
0.508  0.469  0.453  0.627  0.759  0.689  0.695  1.000
ME                                           ! MEANS INCLUDED AS INPUT
1.50 1.32 1.45 1.41 6.60 6.42 6.56 6.31
SD
1.94 2.03 2.05 1.99 2.61 2.66 2.59 2.55
MO NX=8 NK=2 PH=SY,FR LX=FU,FR TD=SY,FR TX=FR KA=FR       ! TAU-X AND KAPPA
LK
SATIS1 SATIS2
PA LX
0 0
1 0
1 0
1 0
0 0
0 1
0 1
0 1
VA 1.0 LX(1,1) LX(5,2)                 ! SET THE METRIC OF THE LATENT VARIABLES
```

(continued)

TABLE 7.8. *(continued)*

```
PA TD
1
0 1
0 0 1
0 0 0 1
1 0 0 0 1                        ! OFF-DIAGONAL 1s ARE CORRELATED ERRORS
0 1 0 0 0 1
0 0 1 0 0 0 1
0 0 0 1 0 0 0 1
PA PH
1
1 1
FI TX(1) TX(5)
VA 0.0 TX(1) TX(5)              ! FIX INDICATOR A INTERCEPTS TO ZERO
EQ LX(2,1) LX(6,2)             ! FACTOR LOADING EQUALITY CONSTRAINT
EQ LX(3,1) LX(7,2)             ! FACTOR LOADING EQUALITY CONSTRAINT
EQ LX(4,1) LX(8,2)             ! FACTOR LOADING EQUALITY CONSTRAINT
EQ TX(2) TX(6)                 ! INTERCEPT EQUALITY CONSTRAINT
EQ TX(3) TX(7)                 ! INTERCEPT EQUALITY CONSTRAINT
EQ TX(4) TX(8)                 ! INTERCEPT EQUALITY CONSTRAINT
EQ TD(1,1) TD(5,5)             ! ERROR VARIANCE EQUALITY CONSTRAINT
EQ TD(2,2) TD(6,6)             ! ERROR VARIANCE EQUALITY CONSTRAINT
EQ TD(3,3) TD(7,7)             ! ERROR VARIANCE EQUALITY CONSTRAINT
EQ TD(4,4) TD(8,8)             ! ERROR VARIANCE EQUALITY CONSTRAINT
OU ME=ML RS MI SC AD=OFF IT=100 ND=4
```

Note. In Mplus, there can be only one number in parentheses on each line. Recall that parameters that are followed by the same number in parentheses are constrained to be equal. Thus the syntax (1), (2), and (3) holds the factor loadings of B1 and B2, C1 and C2, and D1 and D3 to equality (although A1 and A2 are on the same line as B1 and B2, these parameters are not constrained to equality because they have been fixed to 1.0 by Mplus default). The Mplus language uses brackets, [ ], to represent indicator intercepts and factor means. Thus the command [B1 B2] (4) holds the intercepts of indicators B1 and B2 to equality. The last line on the MODEL: section of the syntax—for instance, A1 A2 (7)—instructs the analysis to constrain the indicator error variances to equality.

is 9.49). Fit diagnostics suggest that the indicator error variances are temporally noninvariant:

```
MODEL MODIFICATION INDICES

Variances/Residual Variances

A1                    35.186    -0.421    -0.421    -0.108
B1                    24.357    -0.579    -0.579    -0.125
C1                    15.381    -0.421    -0.421    -0.093
D1                    23.307    -0.459    -0.459    -0.112
A2                    35.185     0.474     0.474     0.076
B2                    24.357     0.602     0.602     0.087
C2                    15.381     0.442     0.442     0.065
D2                    23.307     0.483     0.483     0.078
```

Heterogeneity of variance is a common outcome in repeated measures designs, such as the current example. Thus the test of equal residual variances usually fails in actual data sets because of the temporal fanspread of indicator variances. In the present context, this could be reflective of individual differences in response to the intervention to improve job satisfaction. That is, at Time 1 the variances were more homogeneous because individuals were more similar with regard to their level of job satisfaction. By Time 2, individual differences were more pronounced because some participants responded favorably to the intervention, whereas others did not. This is reflected in the input matrix where it can be seen that the *SD*s increase in magnitude from Time 1 to Time 2 (in addition to the *M*s, which reflect overall improvement in satisfaction). Accordingly, methodologists concede that the test of equal indicator residual variances is highly stringent and will rarely hold in realistic data sets (e.g., Chan, 1998). As will be discussed in more detail later in this chapter in the context of multiple-groups solutions, fortunately the test of the equality of indicator error variances is not germane to invariance evaluation in most applied research scenarios.

The Effects Coding Approach to Scaling Latent Variables

As first noted in Chapter 3, in addition to the marker indicator approach or the strategy of fixing the factor variances to unity, there exists a third but infrequently employed method of defining the metrics of latent variables. In this *effects coding approach* (Little et al., 2006), a priori constraints are placed on the solution so that the set of factor loadings for a given construct average to 1.0 and the corresponding indicator intercepts sum to zero. As a result, the variance of the latent variables reflects the average of the indicators' variances explained by the construct, and the mean of the latent variable is the optimally weighted average of the means for the indicators of that construct. Thus, unlike the marker indicator approach (where the variances and means of the latent variables will vary depending on which indicator is selected as the marker indicator), the effects coding approach has been termed *nonarbitrary* because the latent variable will have the same unstandardized metric as the average of all its manifest indicators. Especially in measurement invariance evaluation (i.e., multiple occasions or multiple groups), and more generally in situations where the indicators have interpretable metrics, the effects coding approach may be viewed as superior to the more commonly employed marker indicator method of latent variable scaling. This is because the unstandardized parameter estimates produced by the effects coding approach are more interpretable since they reflect a nonarbitrary scale of measurement.

This approach is now illustrated by using the two-occasion measurement model of Job Satisfaction. Table 7.9 provides the LISREL, Mplus, and EQS syntax for specifying the equal form model. Note that in each program, the factor loadings and intercepts for the A1 and A2 indicators are initially specified to be freely estimated (e.g., A1* in the Mplus program). However, the estimation of these parameters is superseded by the model constraints that follow. For instance, in LISREL and Mplus, the A1 factor loading is estimated to solve the constraint $\lambda_{1,1} = 4 - \lambda_{2,1} - \lambda_{3,1} - \lambda_{4,1}$. Note that the value of 4.0 is used in this constraint because there are four indicators of the Job Satisfaction factor. This constant should always equal the number of indicators that load on a given factor, and all loadings for the factor other than the one being constrained should be subtracted from this constant. Similarly, in LISREL and Mplus, a constraint is placed on the A1

TABLE 7.9. Computer Syntax (LISREL, Mplus, EQS) for Specification of the Equal Form Longitudinal Measurement Model, Using the Effects Coding Approach to Latent Variable Scaling

LISREL 9.10

```
TITLE LISREL PROGRAM FOR TIME1-TIME2 MSMT MODEL OF JOB SATISFACTION:
      EFFECTS CODING APPROACH
DA NI=8 NO=250 MA=CM
LA
A1 B1 C1 D1 A2 B2 C2 D2
KM
1.000
0.736  1.000
0.731  0.648  1.000
0.771  0.694  0.700  1.000
0.685  0.512  0.496  0.508  1.000
0.481  0.638  0.431  0.449  0.726  1.000
0.485  0.442  0.635  0.456  0.743  0.672  1.000
0.508  0.469  0.453  0.627  0.759  0.689  0.695  1.000
ME
1.50 1.32 1.45 1.41 6.60 6.42 6.56 6.31
SD
1.94 2.03 2.05 1.99 2.61 2.66 2.59 2.55
MO NX=8 NK=2 PH=SY,FR LX=FU,FR TD=SY,FR TX=FR KA=FR
LK
SATIS1 SATIS2
PA LX
1 0
1 0
1 0
1 0
0 1
0 1
0 1
0 1
CO LX(1,1) = 4 - LX(2,1) - LX(3,1) - LX(4,1)
CO TX(1) = 0 - TX(2) - TX(3) - TX(4)
CO LX(5,2) = 4 - LX(6,2) - LX(7,2) - LX(8,2)
CO TX(5) = 0 - TX(6) - TX(7) - TX(8)
PA TD
1
0 1
0 0 1
0 0 0 1
1 0 0 0 1                            ! OFF-DIAGONAL 1s ARE CORRELATED ERRORS
0 1 0 0 0 1
0 0 1 0 0 0 1
0 0 0 1 0 0 0 1
PA PH
1
1 1
ST 0.97 LX(1,1) LX(2,1) LX(3,1) LX(4,1)
ST 0.92 LX(5,2) LX(6,2) LX(7,2) LX(8,2)
ST 2.30 PH(1,1)
ST 5.44 PH(2,2)
OU ME=ML RS MI SC AD=OFF IT=100 ND=4
```

(*continued*)

TABLE 7.9. *(continued)*

Mplus 7.11

```
TITLE:        MPLUS PROGRAM FOR TIME1-TIME2 MSMT MODEL OF JOB SATISFACTION:
              EFFECTS CODING APPROACH TO LATENT VARIABLE SCALING
DATA:         FILE IS FIG7.2.DAT;
              TYPE IS MEANS STD CORR;
              NOBS ARE 250;
VARIABLE:     NAMES ARE A1 B1 C1 D1 A2 B2 C2 D2;
ANALYSIS:     ESTIMATOR=ML;
MODEL:        SATIS1 BY A1* (Lam11)
                        B1 (Lam21)
                        C1 (Lam31)
                        D1 (Lam41);
              SATIS2 BY A2* (Lam52)
                        B2 (Lam62)
                        C2 (Lam72)
                        D2 (Lam82);
              [A1-D2] (TX1-TX8);
              [SATIS1 SATIS2];
              A1 WITH A2; B1 WITH B2; C1 WITH C2; D1 WITH D2;
MODEL CONSTRAINT:
              Lam11 = 4 - Lam21 - Lam31 - Lam41;
              Lam52 = 4 - Lam62 - Lam72 - Lam82;
              TX1 = 0 - TX2 - TX3 - TX4;
              TX5 = 0 - TX6 - TX7 - TX8;
OUTPUT:       SAMPSTAT STAND RESIDUAL;
```

EQS 6.2

```
/TITLE
 EQS SYNTAX FOR EFFECTS CODING IN MODEL OF JOB SATISFACTION
/SPECIFICATIONS
 CASES=250; VAR=8; ME=ML; MA=COR; ANALYSIS=MOM;
/MATRIX
1.000
0.736  1.000
0.731  0.648  1.000
0.771  0.694  0.700  1.000
0.685  0.512  0.496  0.508  1.000
0.481  0.638  0.431  0.449  0.726  1.000
0.485  0.442  0.635  0.456  0.743  0.672  1.000
0.508  0.469  0.453  0.627  0.759  0.689  0.695  1.000
/MEANS
1.50 1.32 1.45 1.41 6.60 6.42 6.56 6.31
/STANDARD DEVIATIONS
1.94 2.03 2.05 1.99 2.61 2.66 2.59 2.55
/LABELS
 V1=A1; V2=B1; V3=C1; V4=DD1; V5=A2; V6=B2;
 V7=C2; V8=DD2; F1 = SATIS1; F2 = SATIS2;
/EQUATIONS
 V1 = *V999 + .97*F1 + E1;
 V2 = *V999 + .97*F1 + E2;
 V3 = *V999 + .97*F1 + E3;
```

(continued)

TABLE 7.9. *(continued)*

```
V4  =  *V999 +  .97*F1 +  E4;
V5  =  *V999 +  .92*F2 +  E5;
V6  =  *V999 +  .92*F2 +  E6;
V7  =  *V999 +  .92*F2 +  E7;
V8  =  *V999 +  .92*F2 +  E8;
F1  =  *V999 +  D1;
F2  =  *V999 +  D2;
/VARIANCES
 E1  TO  E8  =  *;
 D1  =  2.3*;
 D2  =  5.44*;
/COVARIANCES
 E1,E5  =  *;
 E2,E6  =  *;
 E3,E7  =  *;
 E4,E8  =  *;
 D1,D2  =  *;
/CONSTRAINTS
(V1,F1)  +  (V2,F1)  +  (V3,F1)  +  (V4,F1)  =  4;
(V1,V999)  +  (V2,V999)  +  (V3,V999)  +  (V4,V999)  =  0;
(V5,F2)  +  (V6,F2)  +  (V7,F2)  +  (V8,F2)  =  4;
(V5,V999)  +  (V6,V999)  +  (V7,V999)  +  (V8,V999)  =  0;          !
/PRINT
 fit=all;
/LMTEST
/END
```

intercept to solve the equation $\tau_1 = 0 - \tau_2 - \tau_3 - \tau_4$. Constraints are also placed on the estimation of the factor loading and intercept for one indicator (A2) of the Job Satisfaction factor at Time 2. In this example, the constraints are placed on the variables that have been marker indicators in the previous example (A1 and A2). However, any one indicator per factor can be selected for this purpose; this selection is less important than in the marker indicator approach because no single indicator is used to define the mean and variance of the latent variable.

These constraints are specified in LISREL by using the CONSTRAINT (CO) command (see Table 7.9). Occasionally the effects coding method requires the use of starting values to achieve a valid solution (see Chapter 3 for a discussion of starting values). This in fact is the case for the current example when estimated by LISREL 9.10 (also used in EQS but not needed in Mplus 7.11 for this example). As shown in Table 7.9, starting values (ST) are provided for the unstandardized factor loadings and factor variances. The values used for these initial estimates are obtained from the unstandardized solution from LISREL when the marker indicator approach is used to scale the Job Satisfaction latent variables (e.g., 0.97 is the rough average unstandardized factor loading for Job Satisfaction at Time 1).

In Mplus, these constraints are specified by using the MODEL CONSTRAINT command. In the current example, MODEL CONSTRAINT specifies parameter constraints by using labels previously defined for parameters in the MODEL command. These labels are defined in parentheses that follow the parameter; for instance, the $\lambda_{1,1}$ loading is given the label "Lam11" by the statement SATIS1 BY A1* (Lam11) (see Table 7.9).

In EQS, these constraints are imposed in a slightly different fashion because in this program, the constant (e.g., 4.0 for factor loadings) must stand alone on the right-hand side of the general constraint equation. Thus, as shown in Table 7.9, the equations under the /CONSTRAINTS command specify that the four factor loadings on each factor must sum to 4.0 (thus the average factor loading estimate is 1.0). Additional equations are specified in this section of the syntax to impose the constraint that the four indicator intercepts per factor must sum to zero.

Selected output from the LISREL program is presented in Table 7.10. Although not shown in this table, it is important to note that the effect size coding approach does not alter the fit of the solution—for example, χ^2 (15) = 2.093—or the completely standardized parameter estimates. What are affected by the scaling approach are the unstandardized parameter estimates and their standard errors (shown in Table 7.10). For instance, as a result of the constraint imposed on the factor loading of the A1 indicator, the unstandardized factor loadings for the four indicators of Job Satisfaction at Time 1 average to 1.0 (range = 0.973–1.023). Thus the factor loading parameters have been estimated to have an optimal balance around 1.0, but no specific factor loading is fixed to a value of 1.0 (unlike in the marker indicator approach). Moreover, due to the constraint placed on the intercept of the A1 indicator, the unstandardized intercepts for the four indicators of Job Satisfaction at Time 1 average to zero (range = −0.0612–0.048); unlike in the marker indicator approach, however, no individual intercept is fixed to zero.

The consequences of these constraints can also be seen in the estimates of the factor means and variances. For instance, the latent means of Job Satisfaction now reflect the average of the means of the four indicators used to define these factors; for example, κ_1 = 1.42 = (1.50 + 1.32 + 1.45 + 1.41) / 4 (values are the observed indicator means presented in Figure 7.2). Strictly speaking, the estimate of the factor mean reflects the average of the observed indicator means weighted by their unstandardized factor loadings. In this example, taking the unweighted average of the indicator means reproduces the latent variable means (within rounding error) because the factor loadings are of roughly the same magnitude (i.e., the indicators evidence very comparable relationships to the underlying dimension). The variances of the factors (e.g., $\phi_{1,1}$ = 2.865) reflect the average of the indicators' variances explained by the latent variables. These estimates can be hand-calculated by multiplying the observed variance of each indicator by its communality (squared multiple correlations in LISREL output), and then taking the average of this product for the four indicators loading on the Job Satisfaction factor.

These results should convey a key difference between the marker indicator and effects coding approaches to defining the metric of latent variables. When the marker indicator method is used, the metric of the factor is defined primarily by only one measure of the construct, and thus the resulting estimates are not necessarily interpretable with regard to the construct's true mean and variance. This is better represented by the effects coding approach, in which all measures of the construct contribute to the metric

TABLE 7.10. Selected LISREL Results for the Equal Form Longitudinal Measurement Model, Using the Effects Coding Approach to Latent Variable Scaling

LISREL Estimates (Maximum Likelihood)

     LAMBDA-X

	SATIS1	SATIS2
	--------	--------
A1	1.0228	- -
	(0.0283)	
	36.1295	
B1	0.9727	- -
	(0.0353)	
	27.5480	
C1	0.9920	- -
	(0.0350)	
	28.3288	
D1	1.0125	- -
	(0.0323)	
	31.3504	
A2	- -	1.0606
		(0.0293)
		36.2254
B2	- -	0.9714
		(0.0356)
		27.3235
C2	- -	0.9775
		(0.0336)
		29.1132
D2	- -	0.9905
		(0.0327)
		30.3255

     PHI

	SATIS1	SATIS2
	--------	--------
SATIS1	2.8650	
	(0.2815)	
	10.1767	
SATIS2	2.4803	4.8326
	(0.3087)	(0.4747)
	8.0353	10.1804

(continued)

TABLE 7.10. (continued)

Squared Multiple Correlations for X - Variables

A1	B1	C1	D1	A2	B2
0.8084	0.6571	0.6619	0.7380	0.8028	0.6526

Squared Multiple Correlations for X - Variables

C2	D2
0.6851	0.7200

TAU-X

A1	B1	C1	D1	A2	B2
0.0477	-0.0612	0.0413	-0.0278	-0.2647	0.1324
(0.0648)	(0.0806)	(0.0806)	(0.0730)	(0.2014)	(0.2444)
0.7358	-0.7594	0.5127	-0.3807	-1.3139	0.5418

TAU-X

C2	D2
0.2334	-0.1011
(0.2311)	(0.2243)
1.0101	-0.4510

KAPPA

SATIS1	SATIS2
1.4200	6.4725
(0.1121)	(0.1455)
12.6697	44.4767

of the factor, and the factor mean and variance estimates reflect the average mean and "true-score" variances of the indicators of the factor. These estimates are thus more informative for measurement invariance evaluation (e.g., comparing latent means that reflect the average of all indicators of the construct) or in any scenario where the factors are defined by indicators possessing interpretable metrics. When the metrics are not important or meaningful, then the researcher may wish to estimate a standardized solution only by fixing the factor variances to 1.0.

Although not shown in this chapter, longitudinal measurement invariance evaluation using the effects coding approach will proceed in exactly the same fashion as previously demonstrated for the marker indicator method. Indeed, the fit of the models under the various levels of invariance evaluation (e.g., equal factor loadings, equal indicator intercepts) will be identical regardless of the scaling approach. Thus substantive conclusions about the equivalence of measurement properties over time will be the same. The reader is referred to T. D. Little (2013) for fully worked examples of longitudinal measurement invariance evaluation using the effects coding method.

CFA IN MULTIPLE GROUPS

Overview of Multiple-Groups Solutions

The themes introduced in this chapter are now extended to the simultaneous analysis of more than one group. As noted in previous chapters, one of the major advantages of CFA over EFA is its capability to examine the equivalence of all measurement and structural parameters of the factor model across multiple groups. The measurement model pertains to the measurement characteristics of the indicators (observed measures) and thus consists of the factor loadings (lambda), intercepts (tau), and residual variances (theta). Hence the evaluation of across-groups equivalence of these parameters reflects tests of *measurement invariance*. The structural parameters of the CFA model involve evaluation of the latent variables themselves, and thus consist of the factor variances (phi), covariances (phi), and latent means (kappa). If latent Y terminology is used, the corresponding matrices are psi and alpha (cf. Figure 3.4, Chapter 3). These parameters describe characteristics of the population from which the sample was drawn. Thus the examination of the group concordance of structural parameters can be considered tests of *population heterogeneity*; that is, do the dispersion, interrelationships, and levels of the factors vary across groups?

CFA with multiple groups has many potential practical applications. For instance, the issues addressed by measurement invariance evaluation are key to the psychometric development of psychological tests; for example, do the items of a questionnaire measure the same constructs (same factor structure) and evidence equivalent relationships to these constructs (equal factor loadings) in all subgroups of the population for whom the measure will be used? Or are there sex, ethnic/racial, age, or other subgroup differences that preclude responding to the questionnaire in comparable ways? Does the questionnaire contain items that are biased against a particular subgroup; that is, does it yield substantially higher or lower observed scores in a group at equivalent levels of the latent or "true" score? The evaluation of measurement invariance is also important to determining the generalizability of psychological constructs across groups; for instance, does the construct underlying the formal definition of a given psychiatric diagnosis operate equivalently across cultures, sexes, age groups, and so forth? Tests of structural parameters reveal potential group differences adjusting for measurement error and an error theory. For example, tests of equality of factor covariances can be construed as the CFA counterparts to inferential evaluation of the differential magnitude of independent correlations; that is, are two constructs more strongly correlated in one group than in another? Tests of the equality of latent means are analogous to the comparison of observed group means via *t* test or ANOVA. However, the major strength of the CFA-based approach is that such comparisons are made in the context of a latent variable measurement model, which hence adjusts for measurement errors, correlated residuals, and so forth.

Two methods can be used to evaluate CFA solutions in multiple groups: (1) *multiple-groups CFA* and (2) *MIMIC modeling*. Multiple-groups CFA entails the simultaneous analysis of CFA in more than one group. For instance, if the analysis involves two groups (e.g., males and females), two *separate* input matrices are analyzed, and constraints can be placed on like parameters (e.g., factor loadings) in both groups to examine the equivalence of the measurement (measurement invariance) and structural solution (population heterogeneity). Although somewhat underutilized in applied research (cf. Vandenberg & Lance, 2000), multiple-groups CFA can entail the analysis of mean structures to evaluate the equality of indicator intercepts (measurement invariance) and latent means (population heterogeneity). A key advantage of multiple-groups CFA is that all aspects of measurement invariance and population heterogeneity can be examined (i.e., factor loadings, intercepts, residual variances, factor variances, factor covariances, latent means).

Conversely, MIMIC modeling entails the analysis of a single covariance matrix that, in addition to the indicators, includes the dummy code(s) conveying group membership. MIMIC, an acronym for "multiple indicators, multiple causes," has also been referred to as *CFA with covariates*. In this approach, both the latent variables and indicators are regressed onto dummy code(s) denoting group membership. A significant direct effect of the dummy code (covariate) on the latent variable indicates population heterogeneity (group differences on latent means), and a significant direct effect of the dummy code on an indicator is evidence of measurement noninvariance (group differences on the indicator's intercept—i.e., *differential item functioning*). Because a single input matrix is used, advantages of MIMIC modeling over multiple-groups CFA include its greater parsimony (MIMIC entails fewer freely estimated parameters); its relatively greater ease of implementation when several groups are involved (i.e., depending on the complexity of the measurement model, multiple-groups CFA may be cumbersome when the number of groups exceeds two); and its less restrictive sample size requirements (i.e., multiple-groups CFA requires a sufficiently large sample size for each group). However, the key limitation of MIMIC modeling relative to multiple-groups CFA is its ability to examine just two potential sources of invariance (indicator intercepts, factor means).

Multiple-Groups CFA

Before the implementation of multiple-groups CFA is illustrated, it is important to acknowledge the variability that exists in the methodology literature in its terminologies and procedures. As noted above, an advantage of multiple-groups CFA is that all potential aspects of invariance across groups can be examined. Different terminologies exist in the literature for these various tests of invariance (cf. Horn & McArdle, 1992; Meredith, 1993). For example, the test of equal factor structures (*equal form*, meaning that the number of factors and pattern of indicator–factor loadings are identical across groups) has been referred to as *configural invariance*. Equality of factor loadings has been referred to as *metric invariance* or *weak factorial invariance*. The equality of indicator intercepts has been alternatively termed *scalar invariance* or *strong factorial invariance*. Finally, evaluation of the equality of indicator residuals has also been referred to as a test of *strict factorial invariance* (Meredith, 1993). Some confusion surrounds the use of these varying terminologies (cf. Vandenberg & Lance, 2000). For this reason, a more descriptive and pedagogically useful terminology is encouraged and used in this book (e.g., *equal form*, *equal factor loadings*, *equal intercepts*).

Second, some discrepancies exist in the order in which the model restrictions within multiple-groups CFA are evaluated. Most commonly, the stepwise procedures previously illustrated in the tests of longitudinal invariance are employed, in which the analysis begins with the least restricted solution (equal form), and subsequent models are evaluated (using nested χ^2 methods) that entail increasingly restrictive constraints; that is, equal factor loadings → equal intercepts → equal residual variances, and so on. However, some methodologists (e.g., Horn & McArdle, 1992) have proffered a "step-down" strategy in which the starting model contains all the pertinent invariance restrictions, and subsequent models are then evaluated that sequentially relax these constraints. The former approach is recommended for several reasons. Especially in context of a complex CFA solution (i.e., multiple factors and indicators; >2 groups), it may be difficult to determine the (multiple) sources of noninvariance when a model held to full invariance is poor-fitting. Any aspects of ill fit that are encountered are more easily identified and adjusted for in a model-building approach, where new restrictions are placed on the solution at each step. In addition, tests of some aspects of invariance rest on the assumption that other aspects of invariance hold. Group comparisons of latent means are meaningful only if the factor loadings and indicator intercepts have been found to be invariant (see the "Longitudinal Measurement Invariance" section). Group comparisons of factor variances and covariances are meaningful only when the factor loadings are invariant. The viability of the fully constrained model rests on the results of the less restricted solutions. Thus it is more prudent for model evaluation to work upward from the least restricted solution (equal form) to determine if further tests of measurement invariance and population heterogeneity are warranted. Of relevance here is the issue of *partial invariance*. As discussed later in this chapter, it may be possible for the evaluation of group equivalence to proceed even in instances where some noninvariant parameters have been encountered (cf. Byrne et al., 1989). For instance, it may be possible to compare groups on latent means if some (but not all) of the factor loadings and intercepts are invariant. Again, a "step-up" approach to invariance evaluation would foster this endeavor.

For the aforementioned reasons, the recommended sequence of multiple-groups CFA invariance evaluation is as follows: (1) Test the CFA model separately in each group. (2) Conduct the simultaneous test of equal form (identical factor structure). (3) Test the equality of factor loadings. (4) Test the equality of indicator intercepts. (5) Test the equality of indicator residual variances (optional). Then, if substantively meaningful: (6) Test the equality of factor variances. (7) Test the equality of factor covariances (if applicable, i.e., >1 factor). (8) Test the equality of latent means. Steps 1–5 are tests of measurement invariance; Steps 6–8 are tests of population heterogeneity.

Finally, some debate continues as to whether multiple-groups CFA should be prefaced by an overall test of covariance matrices across groups. This procedure was introduced by Jöreskog (1971b); its rationale is that if group differences exist in the parameters of the CFA model, then some values within the covariance matrix should also differ across groups. If the overall test of the equality of covariances fails to reject the null hypothesis, no further analyses are conducted, and it is concluded that the groups are invariant. Rejection of the null hypothesis (e.g., $\Sigma_1 \neq \Sigma_2$) is interpreted as justification for conducting multiple-groups CFA to identify the source(s) of noninvariance (for an illustration of how to employ this test, see Vandenberg & Lance, 2000). Although some researchers have supported its continued use (e.g., Vandenberg & Lance, 2000), many methodologists have questioned the rationale and utility of the omnibus test of equal covariance matrices (e.g., Byrne, 2014; Byrne et al., 1989; Jaccard & Wan, 1996). For instance, Byrne (2014) notes that this test often produces contradictory findings with respect to equivalencies across groups (i.e., occasions where the omnibus test indicates $\Sigma_1 = \Sigma_2$ but subsequent hypothesis tests of the invariance of specific CFA measurement or structural parameters must be rejected, and vice versa). Jaccard and Wan (1996) add that if the researcher has a specific hypothesis regarding group differences on selected parameters, it is better to proceed directly to the multiple-groups CFA framework because this more focused test will have greater statistical power than the omnibus comparison of covariance matrices. Accordingly, these methodologists have concluded that the omnibus test of equal covariance matrices provides little guidance for testing the equivalence of CFA parameters and thus should not be regarded as a prerequisite to multiple-groups CFA.

The multiple-groups CFA methodology is now illustrated by using an actual data set of 750 adult outpatients (375 men, 375 women) with current mood disorders.[2] In this example, the researcher is interested in examining the generalizability of the DSM-5 (American Psychiatric Association, 2013) criteria for the construct of Major Depressive Disorder (MDD) between sexes. The analysis is motivated by questions that have arisen in the field regarding the possibility of salient sex differences in the expression of mood disorders (e.g., somatic symptoms such as appetite/weight change may be more strongly related to depression in women). Patients were rated by experienced clinicians on the severity of the nine symptoms constituting the diagnostic criteria for MDD on 0–8 scales (0 = none, 8 = very severely disturbing/disabling; see Figure 7.3 for description of the nine symptoms). Prior to the CFAs, the data were screened to ensure their suitability for the ML estimator (i.e., normality, absence of multivariate outliers). In accord with its DSM-5 conceptualization, a unidimensional model of MDD has been posited (see Figure 7.3). For substantive reasons (cf. DSM-5), a correlated residual is specified between the first two diagnostic criteria (i.e., depressed mood, loss of interest in usual activities). The first criterion (M1, depressed mood) is used as a marker indicator to define the metric of the latent variable. Accordingly, the covariance structure aspect of the model is overidentified in both groups with $df = 26$ (45 variances/covariances, 19 freely estimated parameters).

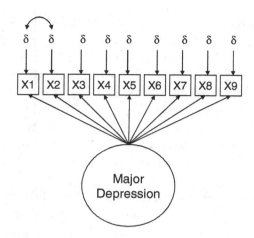

Females: Sample Correlations, Means (*M*), and Standard Deviations (*SD*); *N* = 375

	MDD1	MDD2	MDD3	MDD4	MDD5	MDD6	MDD7	MDD8	MDD9
MDD1	1.000								
MDD2	0.616	1.000							
MDD3	0.315	0.313	1.000						
MDD4	0.349	0.332	0.261	1.000					
MDD5	0.314	0.250	0.270	0.327	1.000				
MDD6	0.418	0.416	0.298	0.328	0.317	1.000			
MDD7	0.322	0.313	0.096	0.117	0.130	0.140	1.000		
MDD8	0.409	0.415	0.189	0.314	0.303	0.281	0.233	1.000	
MDD9	0.318	0.222	0.051	0.115	0.140	0.150	0.217	0.222	1.000
M:	4.184	3.725	1.952	3.589	2.256	3.955	3.869	3.595	1.205
SD:	1.717	2.015	2.096	2.212	2.132	2.005	2.062	2.156	1.791

Males: Sample Correlations, Means (*M*), and Standard Deviations (*SD*); *N* = 375

	MDD1	MDD2	MDD3	MDD4	MDD5	MDD6	MDD7	MDD8	MDD9
MDD1	1.000								
MDD2	0.689	1.000							
MDD3	0.204	0.218	1.000						
MDD4	0.335	0.284	0.315	1.000					
MDD5	0.274	0.320	0.153	0.265	1.000				
MDD6	0.333	0.333	0.221	0.364	0.268	1.000			
MDD7	0.258	0.211	0.114	0.139	0.185	0.132	1.000		
MDD8	0.319	0.346	0.176	0.207	0.231	0.279	0.146	1.000	
MDD9	0.316	0.269	0.111	0.140	0.117	0.131	0.263	0.163	1.000
M:	4.171	3.685	1.739	3.357	2.235	3.661	3.421	3.517	1.259
SD:	1.598	2.018	2.094	2.232	2.108	2.113	2.286	2.174	1.788

FIGURE 7.3. Measurement model of DSM-5 Major Depressive Disorder (MDD). MDD1, depressed mood; MDD2, loss of interest in usual activities; MDD3, weight/appetite change; MDD4, sleep disturbance; MDD5, psychomotor agitation/retardation; MDD6, fatigue/loss of energy; MDD7, feelings of worthlessness/guilt; MDD8, concentration difficulties; MDD9, thoughts of death/suicidality.

Prior to conducting any multiple-groups CFA, it is important to ensure that the posited one-factor model is acceptable in both groups. If markedly disparate measurement models are obtained between groups, this outcome will contraindicate further invariance evaluation. As shown in Table 7.11, in both men and women in this data set, overall fit statistics for the one-factor solution are consistent with good model fit. In both groups, all freely estimated factor loadings are statistically significant (all $ps < .001$) and salient (completely standardized factor loadings range from .34 to .73). No remarkable points of strain are noted in either solution, as reflected by small modification indices, expected parameter change values, and standardized residuals.

Next, the simultaneous analysis of equal form is conducted. Table 7.12 provides the LISREL, Mplus, and EQS syntax for the equal form analysis. In LISREL and EQS, the programming essentially entails "stacking" the CFA of one group on top of the other. In LISREL and EQS, the programs are alerted to the multiple-groups analysis by an additional command on the second line of syntax (NG = 2 on the DATA line of LISREL; GROUPS = 2 on the /SPECIFICATIONS line of EQS). What follows in the EQS programming is stacking the two CFAs for men and women (i.e., the programming is identical to the syntax that would be written for conducting the CFA in a single group). The LISREL programming also follows this logic, but some programming shortcuts are implemented. Note that on the MODEL (MO) line for men, the keyword value PS is used in the specification of the lambda-X (LX), phi (PH), and theta-delta (TD) matrices. In LISREL, PS is used to signify the same *pattern* and *starting* values of fixed and freed parameters of the corresponding matrix of the prior group (not to be confused with *keyword* PS, which appears to the left of the equal sign in the model specification of latent Y variable variances and covariances). Thus, because no invariance constraints are placed in the equal form solution, the PS keyword value allows the user to inform LISREL that the parameter specifications for the measurement model for men and women are identical, without having to repeat the syntax previously written for women (i.e., pattern matrices, marker indicators).

The Mplus example in Table 7.12 runs the analysis on a raw data file instead of variance–covariance matrices (although multiple-groups CFA can also be conducted in Mplus by reading separate input matrices). In this case, the data file contains a dummy code for sex (SEX: 0 = female, 1 = male), in addition to the nine clinical ratings of MDD (M1–M9). The USEVARIABLE command (USEVAR) selects out the nine indicators that will be used in the CFA. The GROUPING command identifies the variable in the data set denoting the levels of groups that will be used in the multiple-groups CFA. The default in later versions of Mplus is to include the means and intercepts in the analysis model. Because these estimates are not required at this stage, this default is overridden by the MODEL = NOMEANSTRUCTURE option and the accompanying INFORMATION = EXPECTED command, which selects the expected information matrix (as opposed to the observed information matrix, the Mplus default) when computing standard errors in ML, according to the conventions of covariance structure analysis. The first MODEL command is the same as would be used to specify the Figure 7.3 measurement model in a single group—in this instance, the first level of GROUP (females) identified by the

TABLE 7.11. Tests of Measurement Invariance and Population Heterogeneity of DSM-5 Major Depressive Disorder (MDD) in Men and Women

	χ^2	df	χ^2_{diff}	Δdf	RMSEA (90% CI)	CFit	SRMR	CFI	TLI
Single-group solutions									
Men ($n = 375$)	45.96**	26			.045 (.022–.066)	.62	.038	.97	.96
Women ($n = 375$)	52.95**	26			.053 (.032–.073)	.39	.040	.96	.94
Measurement invariance									
Equal form	98.91***	52			.049 (.034–.064)	.52	.043	.96	.95
Equal factor loadings	102.84***	60	3.93	8	.044 (.029–.058)	.76	.046	.97	.96
Equal indicator intercepts	115.31***	68	12.47	8	.046 (.029–.056)	.79	.046	.96	.96
Equal indicator error variances	125.02***	77	9.71	9	.041 (.027–.053)	.88	.056	.96	.96
Population heterogeneity									
Equal factor variance	116.44***	69	1.13	1	.043 (.029–.056)	.80	.049	.96	.96
Equal latent mean	118.35***	70	1.91	1	.043 (.029–.056)	.80	.048	.96	.96

Note. $N = 750$. χ^2_{diff}, nested χ^2 difference; RMSEA, root mean square error of approximation; 90% CI, 90% confidence interval for RMSEA; CFit, test of close fit (probability RMSEA ≤ .05); SRMR, standardized root mean square residual; CFI, comparative fit index; TLI, Tucker–Lewis index.

** $p < .01$, *** $p < .001$.

Mplus 7.11 output.

TABLE 7.12. Computer Syntax (LISREL, Mplus, EQS) for Equal Form Multiple-Groups Model of MDD

LISREL 9.10

```
TITLE LISREL PROGRAM FOR EQUAL FORM OF MAJOR DEPRESSION (FEMALES)
DA NG=2 NI=9 NO=375 MA=CM                                        ! NOTE: NG = 2
LA
M1 M2 M3 M4 M5 M6 M7 M8 M9
KM
1.000
0.616  1.000
0.315  0.313  1.000
0.349  0.332  0.261  1.000
0.314  0.250  0.270  0.327  1.000
0.418  0.416  0.298  0.328  0.317  1.000
0.322  0.313  0.096  0.117  0.130  0.140  1.000
0.409  0.415  0.189  0.314  0.303  0.281  0.233  1.000
0.318  0.222  0.051  0.115  0.140  0.150  0.217  0.222  1.000
SD
1.717  2.015  2.096  2.212  2.132  2.005  2.062  2.156  1.791
MO NX=9 NK=1 PH=SY,FR LX=FU,FR TD=SY,FR
LK
DEPRESS
PA LX
0
1
1
1
1
1
1
1
1
VA 1.0 LX(1,1)                          ! SET THE METRIC OF THE LATENT VARIABLE
PA TD
1
1 1
0 0 1
0 0 0 1
0 0 0 0 1
0 0 0 0 0 1
0 0 0 0 0 0 1
0 0 0 0 0 0 0 1
0 0 0 0 0 0 0 0 1
OU ME=ML RS MI SC AD=OFF IT=100 ND=4
EQUAL FORM OF MAJOR DEPRESSION (MALES)              ! SEPARATE TITLE FOR MALES
DA NI=9 NO=375 MA=CM
LA
M1 M2 M3 M4 M5 M6 M7 M8 M9
KM
1.000
0.689  1.000
0.204  0.218  1.000
0.335  0.284  0.315  1.000
0.274  0.320  0.153  0.265  1.000
0.333  0.333  0.221  0.364  0.268  1.000
0.258  0.211  0.114  0.139  0.185  0.132  1.000
```

(continued)

TABLE 7.12. *(continued)*

```
0.319  0.346  0.176  0.207  0.231  0.279  0.146  1.000
0.316  0.269  0.111  0.140  0.117  0.131  0.263  0.163  1.000
SD
1.598  2.018  2.094  2.232  2.108  2.113  2.286  2.174  1.788
MO NX=9 NK=1 PH=PS LX=PS TD=PS                    ! PROGRAMMING SHORTCUTS
OU ME=ML RS MI SC AD=OFF IT=100 ND=4
```

Mplus 7.11

```
TITLE:      MPLUS PROGRAM FOR EQUAL FORM OF MAJOR DEPRESSION
DATA:       FILE IS MDDALL.DAT;            ! DATA READ FROM RAW DATA FILE
VARIABLE:   NAMES ARE SEX M1-M9;
            USEVAR ARE M1-M9;
            GROUPING IS SEX (0=FEMALE 1=MALE); ! SPECIFY GROUPING FACTOR
ANALYSIS:   ESTIMATOR=ML;
            MODEL=NOMEANSTRUCTURE;        ! OVERRIDE DEFAULT TO INCLUDE MEANS AND
            INFORMATION = EXPECTED;       ! INTERCEPTS IN THE ANALYSIS MODEL
MODEL:      DEPRESS BY M1-M9;
            M1 WITH M2;
MODEL MALE: DEPRESS BY M2-M9;             ! FREELY ESTIMATE PARAMETERS
            M1 WITH M2;                   ! IN MEN AS WELL AS IN WOMEN
OUTPUT:     SAMPSTAT MODINDICES(10.00) STAND RESIDUAL;
```

EQS 6.2

```
/TITLE
 EQS SYNTAX FOR EQUAL FORM OF MAJOR DEPRESSION (FEMALES)
/SPECIFICATIONS
 CASES=375; VAR=9; ME=ML; MA=COR; ANALYSIS=COV; GROUPS=2;  ! NOTE: GROUPS=2
/MATRIX
1.000
0.616  1.000
0.315  0.313  1.000
0.349  0.332  0.261  1.000
0.314  0.250  0.270  0.327  1.000
0.418  0.416  0.298  0.328  0.317  1.000
0.322  0.313  0.096  0.117  0.130  0.140  1.000
0.409  0.415  0.189  0.314  0.303  0.281  0.233  1.000
0.318  0.222  0.051  0.115  0.140  0.150  0.217  0.222  1.000
/STANDARD DEVIATIONS
1.717  2.015  2.096  2.212  2.132  2.005  2.062  2.156  1.791
/LABELS
 V1=depmood; V2=anhedon; V3=weight; V4=sleep; V5=motor; V6=fatigue;
 V7=guilt; V8=concent; V9=suicide;
 F1 = DEPRESS;
/EQUATIONS
 V1 =  F1+E1;
 V2 = *F1+E2;
 V3 = *F1+E3;
 V4 = *F1+E4;
 V5 = *F1+E5;
 V6 = *F1+E6;
 V7 = *F1+E7;
 V8 = *F1+E8;
 V9 = *F1+E9;
/VARIANCES
 E1 TO E9= *;
 F1 = *;
```

(continued)

TABLE 7.12. (continued)

```
/COVARIANCES
 E1,E2 = *;
/END
/TITLE
 EQUAL FORM OF MAJOR DEPRESSION (MALES)
/SPECIFICATIONS
 CASES=375; VAR=9; ME=ML; MA=COR; ANALYSIS=COV;
/MATRIX
1.000
0.689  1.000
0.204  0.218  1.000
0.335  0.284  0.315  1.000
0.274  0.320  0.153  0.265  1.000
0.333  0.333  0.221  0.364  0.268  1.000
0.258  0.211  0.114  0.139  0.185  0.132  1.000
0.319  0.346  0.176  0.207  0.231  0.279  0.146  1.000
0.316  0.269  0.111  0.140  0.117  0.131  0.263  0.163  1.000
/STANDARD DEVIATIONS
1.598  2.018  2.094  2.232  2.108  2.113  2.286  2.174  1.788
/LABELS
 V1=depmood; V2=anhedon; V3=weight; V4=sleep; V5=motor; V6=fatigue;
 V7=guilt; V8=concent; V9=suicide;
 F1 = DEPRESS;
/EQUATIONS
 V1 =  F1+E1;
 V2 = *F1+E2;
 V3 = *F1+E3;
 V4 = *F1+E4;
 V5 = *F1+E5;
 V6 = *F1+E6;
 V7 = *F1+E7;
 V8 = *F1+E8;
 V9 = *F1+E9;
/VARIANCES
 E1 TO E9= *;
 F1 = *;
/COVARIANCES
 E1,E2 = *;
/PRINT
 fit=all;
/LMTEST
/END
```

GROUPING command. Recall that an Mplus default is to fix the first indicator listed to load on a factor as the marker indicator; thus the M1 indicator has been automatically set to be the marker indicator. In multiple-groups analysis, Mplus holds some measurement parameters to equality across groups by default—specifically, the factor loadings as well as intercepts if indicator means are included in the model. Because of the overly stringent nature of this restriction, Mplus does not hold indicator residual variances and covariances to equality by default. Moreover, all structural parameters (factor variances, covariances, latent means) are freely estimated in all groups by default. Thus, because the current model is testing for equal form, the Mplus default for holding factor loadings to equality must be overridden. This is accomplished by the MODEL MALE: command, in which all parameters listed after this keyword are freely estimated in the men's solution (however, note that the indicator list omits M1 because it has previously been fixed to 1.0 to serve as the marker indicator). Note that while additional programming is necessary for freely estimating the factor loadings in the men's solution (i.e., DEPRESS BY M2–M9), the line for correlated residuals between M1 and M2 (i.e., M1 with M2) is only included for clarity (it is redundant with the Mplus default of freely estimating residual variances and covariances in all groups).

As shown in Table 7.11, this solution provides an acceptable fit to the data. This solution serves as the baseline model for subsequent tests of measurement invariance and population heterogeneity. The parameter estimates for each group are presented in Table 7.13. Inspection of Table 7.11 shows that the df and model χ^2 of the equal form solution equal the sum of the dfs and model χ^2s of the CFAs run separately for men and women; for example, $\chi^2 = 98.91 = 45.96 + 53.95$. Although multiple-groups solutions can be evaluated when the size of the groups vary, interpretation of the analysis may be more complex if the group sizes differ markedly. This is because many aspects of the CFA are influenced by (sensitive to) sample size. For instance, recall that model χ^2 is calculated as either $F_{ML}(N-1)$ or $F_{ML}(N)$, depending on the software program. Consider the scenario where the fit function value is the same in two groups (i.e., $F_{ML1} = F_{ML2}$), but the sizes of the groups differ considerably (e.g., $n_1 = 1,000$, $n_2 = 500$). Thus, although the discrepancies between the observed and predicted covariance matrices are the same in both groups, the model χ^2s of the groups will differ greatly, and Group 1 will contribute considerably more to the equal form χ^2 than will Group 2. Specifically, in this contrived example, Group 1 will contribute two times as much to the overall χ^2 as Group 2 will. All other aspects of the CFA model that are based on χ^2 (e.g., overall fit statistics such as the CFI; modification indices) or are influenced by sample size (e.g., standard errors, power to detect parameter estimates as significantly different from zero, standardized residuals) will also be differentially influenced by the unbalanced group sizes. Thus, although it is permissible to conduct multiple-groups CFA with unequal sample sizes, it is preferable for the size of the groups to be as balanced as possible. In instances where the group ns differ considerably, the researcher must be mindful of this issue when interpreting the results.

The next analysis evaluates whether the factor loadings (unstandardized) of the MDD indicators are equivalent in men and women. The test of equal factor loadings is a critical test in multiple-groups CFA. In tandem with other aspects of measurement invariance evaluation (e.g., equal form), this test determines whether the measures have the same meaning and structure for different groups of respondents. Moreover, this test establishes the suitability of other group comparisons that may be of substantive interest (e.g., group equality of factor variances, factor means, or regressive paths among latent variables). In the current data set, the equal factor loading models had an overall good fit to the data, and does not significantly degrade fit relative to the equal form solution, $\chi^2_{diff}(8) = 3.93$, ns (critical value of $\chi^2 = 15.51$, $df = 8$, $\alpha = .05$). The difference in degrees

TABLE 7.13. Parameter Estimates (Mplus 7.11) from the Equal Form Measurement Model of MDD in Men and Women

MODEL RESULTS

	Estimate	S.E.	Est./S.E.	Two-Tailed P-Value
Group FEMALE				
DEPRESS BY				
M1	1.000	0.000	999.000	999.000
M2	1.107	0.086	12.907	0.000
M3	0.729	0.101	7.221	0.000
M4	0.912	0.108	8.406	0.000
M5	0.812	0.104	7.845	0.000
M6	0.924	0.100	9.240	0.000
M7	0.611	0.098	6.220	0.000
M8	0.979	0.107	9.131	0.000
M9	0.484	0.085	5.707	0.000
M1 WITH				
M2	0.394	0.147	2.688	0.007
Variances				
DEPRESS	1.563	0.224	6.991	0.000
Residual Variances				
M1	1.376	0.155	8.856	0.000
M2	2.133	0.223	9.579	0.000
M3	3.551	0.277	12.837	0.000
M4	3.583	0.290	12.351	0.000
M5	3.501	0.278	12.609	0.000
M6	2.676	0.226	11.822	0.000
M7	3.658	0.279	13.113	0.000
M8	3.137	0.264	11.904	0.000
M9	2.831	0.214	13.223	0.000
Group MALE				
DEPRESS BY				
M1	1.000	0.000	999.000	999.000
M2	1.236	0.098	12.580	0.000
M3	0.786	0.133	5.911	0.000
M4	1.166	0.152	7.656	0.000
M5	0.959	0.139	6.915	0.000
M6	1.132	0.145	7.790	0.000
M7	0.766	0.143	5.361	0.000
M8	1.019	0.144	7.075	0.000
M9	0.632	0.113	5.617	0.000
M1 WITH				
M2	0.920	0.160	5.743	0.000

(continued)

TABLE 7.13. *(continued)*

	Estimate	S.E.	Est./S.E.	Two-Tailed P-Value
Variances				
DEPRESS	1.048	0.183	5.719	0.000
Residual Variances				
M1	1.499	0.152	9.888	0.000
M2	2.459	0.244	10.084	0.000
M3	3.727	0.290	12.830	0.000
M4	3.547	0.304	11.671	0.000
M5	3.467	0.282	12.304	0.000
M6	3.111	0.270	11.516	0.000
M7	4.599	0.353	13.030	0.000
M8	3.626	0.297	12.192	0.000
M9	2.770	0.214	12.943	0.000

STANDARDIZED MODEL RESULTS

STDYX Standardization

	Estimate	S.E.	Est./S.E.	Two-Tailed P-Value
Group FEMALE				
DEPRESS BY				
M1	0.729	0.037	19.867	0.000
M2	0.688	0.039	17.441	0.000
M3	0.435	0.049	8.909	0.000
M4	0.516	0.045	11.378	0.000
M5	0.477	0.047	10.128	0.000
M6	0.577	0.042	13.611	0.000
M7	0.371	0.051	7.224	0.000
M8	0.569	0.043	13.287	0.000
M9	0.339	0.052	6.461	0.000
M1 WITH				
M2	0.230	0.071	3.229	0.001
Variances				
DEPRESS	1.000	0.000	999.000	999.000
Residual Variances				
M1	0.468	0.054	8.740	0.000
M2	0.527	0.054	9.707	0.000
M3	0.810	0.043	19.045	0.000
M4	0.734	0.047	15.693	0.000
M5	0.772	0.045	17.191	0.000
M6	0.667	0.049	13.644	0.000
M7	0.862	0.038	22.649	0.000
M8	0.677	0.049	13.905	0.000
M9	0.885	0.036	24.923	0.000

(continued)

TABLE 7.13. (continued)

Group MALE

DEPRESS   BY

	Estimate	S.E.	Est./S.E.	P-Value
M1	0.642	0.044	14.521	0.000
M2	0.628	0.045	13.965	0.000
M3	0.385	0.053	7.212	0.000
M4	0.535	0.047	11.286	0.000
M5	0.466	0.050	9.274	0.000
M6	0.549	0.047	11.722	0.000
M7	0.344	0.055	6.280	0.000
M8	0.480	0.050	9.663	0.000
M9	0.362	0.054	6.701	0.000

M1   WITH

M2	0.479	0.051	9.436	0.000

Variances

DEPRESS	1.000	0.000	999.000	999.000

Residual Variances

M1	0.588	0.057	10.378	0.000
M2	0.605	0.057	10.716	0.000
M3	0.852	0.041	20.762	0.000
M4	0.713	0.051	14.053	0.000
M5	0.783	0.047	16.687	0.000
M6	0.699	0.051	13.580	0.000
M7	0.882	0.038	23.471	0.000
M8	0.769	0.048	16.104	0.000
M9	0.869	0.039	22.155	0.000

R-SQUARE

Group FEMALE

Observed Variable	Estimate	S.E.	Est./S.E.	Two-Tailed P-Value
M1	0.532	0.054	9.934	0.000
M2	0.473	0.054	8.721	0.000
M3	0.190	0.043	4.455	0.000
M4	0.266	0.047	5.689	0.000
M5	0.228	0.045	5.064	0.000
M6	0.333	0.049	6.805	0.000
M7	0.138	0.038	3.612	0.000
M8	0.323	0.049	6.643	0.000
M9	0.115	0.036	3.230	0.001

(continued)

TABLE 7.13. *(continued)*

Group MALE

Observed Variable	Estimate	S.E.	Est./S.E.	Two-Tailed P-Value
M1	0.412	0.057	7.261	0.000
M2	0.395	0.057	6.982	0.000
M3	0.148	0.041	3.606	0.000
M4	0.287	0.051	5.643	0.000
M5	0.217	0.047	4.637	0.000
M6	0.301	0.051	5.861	0.000
M7	0.118	0.038	3.140	0.002
M8	0.231	0.048	4.831	0.000
M9	0.131	0.039	3.350	0.001

Note. N = 750.

of freedom (*df* = 8) corresponds to the eight factor loadings (M2–M9) that have been freely estimated in both groups in the previous analysis. Because M1 is fixed to 1.0 in both groups to serve as the marker indicator, this measure is not involved in the equality constraints. As an aside, note in Table 7.11 that the equal factor loading solution produces a slight improvement in the parsimony goodness-of-fit indices, compared to the equal form solution (e.g., RMSEA = 0.044 vs. 0.049 in the equal factor loading and equal form solution, respectively). This is due to the gain in degrees of freedom (60 vs. 52), coupled with the trivial change in model χ^2 (102.84 vs. 98.91) associated with reproducing the observed covariance matrices with fewer freely estimated parameters (i.e., increased model parsimony via constraint of previously free parameters to equality).

Because the constraint of equal factor loadings does not significantly degrade the fit of the solution, it can be concluded that the indicators evidence comparable relationships to the latent construct of MDD in men and women. Figure 7.4 graphically illustrates various forms of measurement (non)invariance with respect to factor loadings and indicator intercepts. Although the equality of intercepts has yet to be evaluated, the result of invariant factor loadings is depicted by Figures 7.4A and 7.4B, which show parallel regression slopes for Groups 1 and 2; in other words, a unit change in the underlying dimension (ξ, or MDD) is associated with statistically equivalent change in the observed measure (X2, or the indicator of loss of interest in usual activities) in both groups (men and women). However, because the intercepts have not been evaluated, it cannot be concluded that men and women will evidence equivalent observed *scores* on an indicator at a given level of the latent variable (as is shown in Figure 7.4A where both the loading and intercept of an indicator, X2, is equivalent between groups).

The previous analyses have been based on covariance structures. To examine the between-group equality of indicator intercepts, the means of the indicators must be inputted to the analysis, along with the indicators' variances and covariances. As in the invariance evaluation of longitudinal measures, the analysis of mean structures poses additional identification issues. In the case of this two-group analysis, there are 18 indi-

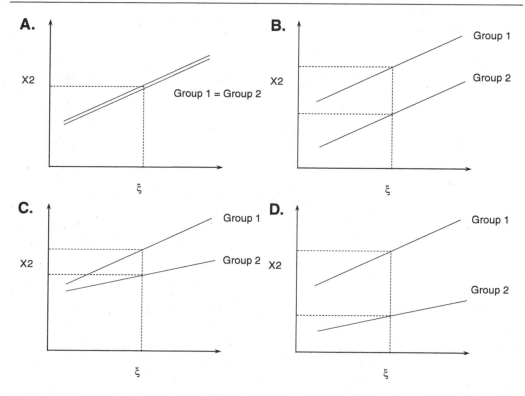

FIGURE 7.4. Graphical depictions of various forms of measurement invariance and noninvariance. (A) Equal loadings and intercepts; (B) equal loadings, unequal intercepts; (C) unequal loadings, equal intercepts; (D) unequal loadings and intercepts.

cator means (9 for men, 9 for women) but potentially 20 freed parameters of the mean structure solution (18 intercepts, 2 latent means). Moreover, latent variables must be assigned an origin in addition to a metric. Thus, as with the longitudinal invariance example, the mean structure component of the multiple-groups solution is underidentified in the absence of additional restrictions. In addition to holding the indicator intercepts to equality across groups in the measurement invariance solution, identification of the mean structure can be accomplished by any one of the three methods discussed in the single-group, longitudinal analysis: (1) fixing the latent mean to zero in one group; (2) fixing the intercept of the marker indicator of each latent variable to zero and freely estimating the latent means; or (3) using the effects coding approach. Following the default in the current version of the Mplus program, this example uses the first method of identifying the origin of the latent variable. In this approach, the group whose latent mean(s) have been fixed to zero becomes the reference group. From a statistical standpoint, selection of the reference group is arbitrary, although this choice may be guided by substantive/interpretative considerations (e.g., the reference group may be participants who have not received an intervention/experimental manipulation). The latent means in the remaining groups are freely estimated, but these parameter estimates represent deviations from the reference group's latent mean. For example, if Group 2's latent mean = 1.4, this indicates that, on average, this group scores 1.4 units higher than the reference group on the latent dimension, based on the metric of the marker indicator. The rationale of this approach to model specification is that because the measurement intercepts are constrained to equality across groups (to test for intercept invariance), the factors have an arbitrary origin (mean). Thus, latent means are not estimated in the absolute sense, but instead reflect average differences in the level of the construct across groups.

Table 7.14 presents the LISREL, Mplus, and EQS syntax for the invariant intercepts analysis (also holding the factor loadings to equality). In each program, women are used as the reference group by fixing their latent variable mean to zero, and the latent mean for men is freely estimated. In the LISREL program, note that the indicator means have been added to the input matrix (under the ME heading), and that tau-X (TX) and kappa (KA) matrices have been included on the Model line (MO) for the estimation of indicator intercepts and the factor mean, respectively. For females, the LISREL programming looks the same as a typical CFA, except for the TX and KA commands. On the Model line, the command TX=FR informs LISREL that the vectors of intercepts are to be freely estimated. The KA=FI command fixes the latent mean of MDD to zero in this group. The remaining salient aspects of the LISREL programming are found on the Model line for males. The factor loadings and indicator intercepts are constrained to equality between groups by the commands LX=IN and TX=IN, respectively (IN = constrain parameters to be invariant with those of the preceding group). Because indicator error variances and factor variances are not held to equality in this analysis, the TD=PS and PH=PS commands are used to inform LISREL that these parameters may be freely estimated by using the same pattern matrix and starting values used in the prior group. Finally, the KA=FR command allows the latent mean of MDD to be freely estimated in males. As noted earlier, this estimate will reflect a deviation (difference) of males' latent mean relative to the latent mean of females.

Most of the aforementioned model specifications are handled by default in Mplus (Table 7.14). The syntax now removes the MODEL=NOMEANSTRUCTURE command (cf. Table 7.12) to reinstate the Mplus default of estimating a mean (and covariance) structure solution. Accordingly, the indicator means must be included in the input matrix, or they can be calculated in Mplus by inputting a raw data file as in the current example. The only syntax provided for men is M1 with M2, to freely estimate the correlated error of these indicators (which in fact is redundant with another Mplus default of freely estimating residual variances and covariances in all groups). By default, the factor loadings and indicator intercepts are held to equality in Mplus.

In EQS, a key programming change is that the /EQUATIONS section now includes the intercept parameters, which are freely estimated in both groups (e.g., V1 = *V999 + F1 + E1, where *V999 = freely estimated indicator intercept; see Table 7.12). In females, the mean of the MDD factor (F1) is fixed to zero by the equation F1 = 0.0V999 + D1. "D1" reflects a residual variance (or *disturbance*, D) because in fact, the analysis of mean structures requires the regression of the factors and indicators onto a constant (denoted

TABLE 7.14. Computer Syntax (LISREL, Mplus, EQS) for Equal Factor Loadings and Equal Indicator Intercepts Multiple-Groups Model of MDD

LISREL 9.10

```
TITLE LISREL PROGRAM FOR EQUAL LOADINGS AND INTERCEPTS OF MDD (FEMALES)
DA NG=2 NI=9 NO=375 MA=CM                                      ! NOTE: NG = 2
LA
M1 M2 M3 M4 M5 M6 M7 M8 M9
KM
<Insert correlation matrix for Females from Figure 7.3>
ME
4.184  3.725  1.952  3.589  2.256  3.955  3.869  3.595  1.205
SD
1.717  2.015  2.096  2.212  2.132  2.005  2.062  2.156  1.791
MO NX=9 NK=1 PH=SY,FR LX=FU,FR TD=SY,FR TX=FR KA=FI
LK
DEPRESS
PA LX
0
1
1
1
1
1
1
1
1
VA 1.0 LX(1,1)                          ! SET THE METRIC OF THE LATENT VARIABLE
PA TD
1
1 1
0 0 1
0 0 0 1
0 0 0 0 1
0 0 0 0 0 1
0 0 0 0 0 0 1
0 0 0 0 0 0 0 1
0 0 0 0 0 0 0 0 1
OU ME=ML RS MI SC AD=OFF IT=100 ND=4
EQUAL LOADINGS AND INTERCEPTS OF MDD (MALES)          ! SEPARATE TITLE FOR MALES
DA NI=9 NO=375 MA=CM
LA
M1 M2 M3 M4 M5 M6 M7 M8 M9
KM
<Insert correlation matrix for Males from Figure 7.3>
ME
4.171  3.685  1.739  3.357  2.235  3.661  3.421  3.517  1.259
SD
1.598  2.018  2.094  2.232  2.108  2.113  2.286  2.174  1.788
MO NX=9 NK=1 PH=PS LX=IN TD=PS TX=IN KA=FR      ! PROGRAMMING SHORTCUTS
OU ME=ML RS MI SC AD=OFF IT=100 ND=4
```

Mplus 7.11

```
TITLE:      MPLUS PROGRAM FOR EQUAL LOADINGS AND INTERCEPTS OF MDD
DATA:       FILE IS MDDALL.DAT;
```

(continued)

TABLE 7.14. *(continued)*

```
VARIABLE:     NAMES ARE SEX M1-M9;
              USEVAR ARE M1-M9;
              GROUPING IS SEX (0=FEMALE 1=MALE); ! SPECIFY GROUPING FACTOR
ANALYSIS:     ESTIMATOR=ML;
MODEL:        DEPRESS BY M1-M9;
              M1 WITH M2;
MODEL MALE: M1 WITH M2;    ! BY DEFAULT, LOADINGS & INTERCEPTS HELD EQUAL
OUTPUT:       SAMPSTAT MODINDICES(10.00) STAND RESIDUAL;
```

EQS 6.2

```
/TITLE
 EQS SYNTAX FOR GENDER INVARIANCE OF MAJOR DEPRESSION (FEMALES)
/SPECIFICATIONS
 CASES=375; VAR=9; ME=ML; MA=COR; ANALYSIS=MOM; GROUPS=2;
/MATRIX
<Insert correlation matrix for Females from Figure 7.3>
/MEANS
4.184  3.725  1.952  3.589  2.256  3.955  3.869  3.595  1.205
/STANDARD DEVIATIONS
1.717  2.015  2.096  2.212  2.132  2.005  2.062  2.156  1.791
/LABELS
 V1=depmood; V2=anhedon; V3=weight; V4=sleep; V5=motor; V6=fatigue;
 V7=guilt; V8=concent; V9=suicide;
 F1 = DEPRESS;
/EQUATIONS
 V1 = *V999 +  F1 + E1;
 V2 = *V999 + *F1 + E2;
 V3 = *V999 + *F1 + E3;
 V4 = *V999 + *F1 + E4;
 V5 = *V999 + *F1 + E5;
 V6 = *V999 + *F1 + E6;
 V7 = *V999 + *F1 + E7;
 V8 = *V999 + *F1 + E8;
 V9 = *V999 + *F1 + E9;
 F1 = 0.0V999 + D1;             ! FEMALES ARE REFERENCE GROUP
/VARIANCES
 E1 TO E9= *;
 D1 = *;
/COVARIANCES
 E1,E2 = *;
/END
/TITLE
 EQS SYNTAX FOR GENDER INVARIANCE OF MAJOR DEPRESSION (MALES)
/SPECIFICATIONS
 CASES=375; VAR=9; ME=ML; MA=COR; ANALYSIS=MOM;
/MATRIX
<Insert correlation matrix for Males from Figure 7.3>
/MEANS
4.171  3.685  1.739  3.357  2.235  3.661  3.421  3.517  1.259
/STANDARD DEVIATIONS
1.598  2.018  2.094  2.232  2.108  2.113  2.286  2.174  1.788
/LABELS
 V1=depmood; V2=anhedon; V3=weight; V4=sleep; V5=motor; V6=fatigue;
 V7=guilt; V8=concent; V9=suicide;
 F1 = DEPRESS;
```

(continued)

TABLE 7.14. (continued)

```
/EQUATIONS
 V1 = *V999 +  F1 + E1;
 V2 = *V999 + *F1 + E2;
 V3 = *V999 + *F1 + E3;
 V4 = *V999 + *F1 + E4;
 V5 = *V999 + *F1 + E5;
 V6 = *V999 + *F1 + E6;
 V7 = *V999 + *F1 + E7;
 V8 = *V999 + *F1 + E8;
 V9 = *V999 + *F1 + E9;
 F1 = *V999 + D1;              ! MALE LATENT MEAN FREELY ESTIMATED
/VARIANCES
 E1 TO E9= *;
 D1 = *;
/COVARIANCES
 E1,E2 = *;
/CONSTRAINTS
 (1,V2,F1)=(2,V2,F1);         ! FACTOR LOADING EQUALITY
 (1,V3,F1)=(2,V3,F1);
 (1,V4,F1)=(2,V4,F1);
 (1,V5,F1)=(2,V5,F1);
 (1,V6,F1)=(2,V6,F1);
 (1,V7,F1)=(2,V7,F1);
 (1,V8,F1)=(2,V8,F1);
 (1,V9,F1)=(2,V9,F1);
 (1,V1,V999)=(2,V1,V999);     ! INDICATOR INTERCEPT EQUALITY
 (1,V2,V999)=(2,V2,V999);
 (1,V3,V999)=(2,V3,V999);
 (1,V4,V999)=(2,V4,V999);
 (1,V5,V999)=(2,V5,V999);
 (1,V6,V999)=(2,V6,V999);
 (1,V7,V999)=(2,V7,V999);
 (1,V8,V999)=(2,V8,V999);
 (1,V9,V999)=(2,V9,V999);
/PRINT
 fit=all;
/LMTEST
/END
```

in EQS as V999; Byrne, 2006). In the /VARIANCES section of the program, this disturbance variance is freely estimated in both groups (D1 = *), along with the indicator error variances (E1 TO E9 = *). In males, the only programming change is that the latent mean of MDD is freely estimated, F1 = *V999 + D1. The factor loadings and indicator intercepts are constrained to equality in the /CONSTRAINTS section of the syntax. For example, the command (1,V1,V999) = (2,V1,V999) informs EQS to hold the intercept of the V1 indicator (depressed mood) in the first group (women) equal to the intercept of the V1 indicator in the second group (men).

The equal measurement intercepts model is found to be good-fitting, and does not result in a significant degradation of fit relative to the equal factor loadings solution, $\chi^2_{diff}(8) = 12.47$, ns (see Table 7.11). The gain of 8 degrees of freedom (to a total of df =

68) is due to the additional 18 new elements of the input matrices (i.e., the 9 indicator means for men and women) minus the 10 mean structure parameters (9 intercepts held to equality, 1 freely estimated latent mean; for identification purposes, women are used as the reference group by fixing their latent mean to zero). Because the factor loadings and indicator intercepts are invariant in men and women, comparison of the groups on the latent mean of MDD is interpretable. In men, the unstandardized parameter estimate for the latent mean is −0.13 (not shown in the tables), indicating that, on average, men score .13 units below women on the dimension of MDD—a difference that is not statistically significant (z = 1.38). This lack of difference is upheld in a subsequent, more precise analysis of population heterogeneity (the final analysis in Table 7.11) that constrains the latent means to equality; that is, χ^2_{diff} = 1.91, which is roughly the same as z = 1.38^2 (cf. Wald test, Chapter 4).

As mentioned earlier, Figure 7.4 graphically displays different combinations of factor loading and intercept (non)invariance. Figure 7.4A is graphically consistent with the results of the applied example presented thus far. In this graph, both the loading and the intercept of the indicator are equivalent between groups. Interpreted in isolation from factor loading invariance, the finding of intercept invariance suggests that both groups are expected to have the same observed value of the indicator (X2) when the latent variable (ξ) is zero. However, the result of factor loading invariance *and* intercept invariance can be interpreted as suggesting that, *for any given factor value*, the observed values of the indicator are expected to be statistically equivalent between groups (see dotted line in Figure 7.4A). This concept also upholds the statement made earlier in this chapter that group comparisons of the means of factors should only be conducted in the context of factor loading *and* indicator intercept invariance.

The remaining three graphs in Figure 7.4 show that if either the loading or the intercept is noninvariant, the observed values of the indicator will differ between groups at a given level of the latent variable (see dotted lines).[3] For example, Figure 7.4B illustrates the situation of an equal factor loading but an unequal intercept. In this graph, although the indicator evidences the same relationship (regression slope) in both groups (i.e., a unit change in the factor is associated with the same amount of change in the indicator in both groups), the groups differ in the location parameter (origin) of the indicator, meaning that all predicted observed scores will differ at various levels of the latent variable. Group 1's predicted scores on the indicator will be higher than Group 2's across all levels of the "true score," suggesting that the indicator is biased. This is an example of *differential item functioning*—a term that is used to describe situations where an item yields a different mean response for the members of different groups with the same value of the underlying attribute (McDonald, 1999). An illustration of differential item functioning is presented in the MIMIC section of this chapter.

The final potential aspect of measurement invariance that can be tested is the equality of the indicator error variances. As seen in Table 7.11, nested χ^2 evaluation indicates that the residual variances are equivalent in men and women patients, $\chi^2_{\text{diff}}(9)$ = 9.71, *ns* (critical value of χ^2 = 16.92, *df* = 9, α = .05). The gain of 9 degrees of freedom corresponds to the 9 residual variances held to equality.

Although this example has used a real data set, this outcome is rare in applied research data. In fact, most methodologists regard equality of error variances and covariances to be an overly restrictive test that is not important to the endeavor of measurement invariance evaluation (e.g., Bentler, 1995; Byrne, 2014; T. D. Little, 2013). Indeed, although conducted in this example for illustration purposes, the evaluation of equal error variances is really not necessary in most applied research scenarios. Recall from Chapter 2 that the error variance of an indicator has two components that cannot be distinguished in the estimation process: (1) variance specific to the indicator; and (2) random error. For instance, there is no reason to expect, from either a conceptual or practical standpoint, that the random error of each indicator will be the same across groups (or across time in longitudinal models). Moreover, the illustration in Figure 7.4A shows that prediction of a group equivalent observed score (X2) by the latent variable model (i.e., factor loadings, indicator intercepts) does not rely on the condition of equal indicator error variances. Because the equality of error variances is not a substantively important condition of measurement invariance, and because it is not a prerequisite of subsequent invariance testing (i.e., tests of population heterogeneity), these constraints can be omitted from most applied research measurement models. In fact, some researchers (e.g., T. D. Little, 2013) argue that enforcing the equality of error variances can be problematic to invariance evaluation because the ill fit introduced by these constraints may bias the remaining parameter estimates (e.g., factor loadings, factor variances).

The remaining analyses pertain to group comparisons on the structural parameters of the CFA model (i.e., tests of population heterogeneity). As noted earlier, the viability of these comparisons rests on the evaluation of measurement invariance. In other words, it is not useful to compare groups on aspects of the factors (factor variances, factor covariances, latent means) without first ascertaining that the factors measure the same constructs in the same fashion in each group. Specifically, group comparisons on factor variances are meaningful only if the factor loadings have been invariant. Comparisons of the factor covariances (in CFA models with >1 factor) are meaningful if both the factor loadings and factor variances are invariant. Finally, evaluation of group equality of latent means rests on the condition of invariant factor loadings and indicator intercepts. The conceptual logic of these statements should be apparent upon review of the equations related to these parameters (e.g., $COV = r_{1,2}SD_1SD_2$; $SD^2 = \sigma^2$), presented in this and earlier chapters (e.g., Chapter 3).

Evaluation of the equality of a factor variance examines whether the amount of within-group variability (dispersion) of the construct differs across groups. Although crucial to all aspects of invariance evaluation, the test for equal factor variances best exemplifies why comparisons made by multiple-groups CFA rely on the unstandardized solution. The test of invariant factor variances will be meaningless if the metric of the factor is defined by fixing its variance to 1.0. Because the factor variance (ϕ) is strongly determined by the amount of variance in the marker indicator when this approach to latent variable scaling is used (i.e., ϕ = marker indicator variance multiplied by the marker indicator's squared completely standardized factor loading), the test of group equality of a factor variance can be regarded as an evaluation of whether the groups draw from similar ranges of the underlying construct to respond to the indicators of that construct. The question addressed by the test of factor variance equality often does not have clear substantive implications in applied research, although such evaluation is needed to establish the suitability of the potentially more interesting test of the invariance of factor covariances (i.e., are the latent variables more strongly related to each other in one group than in another?).

In the example of the measurement model of MDD, the factor variances are equal in men and women, $\chi^2_{diff}(1) = 1.13$, ns. Because the restriction of equal indicator error variances has been relaxed in this model, the χ^2 change value (and associated df) reflects the difference between the current solution and the equal indicator intercepts solution (see Table 7.11). This example entails a one-factor measurement model, and thus the invariance evaluation of factor covariances is not relevant.

Table 7.15 presents the LISREL, Mplus, and EQS syntax for the fully invariant measurement model of MDD (excluding indicator error variances, which were freely estimated in both groups). In LISREL, all of the equality constraints are made by commands on the Model line for males (i.e., equal factor loadings: LX=IN; equal indicator intercepts: TX=IN; equal factor variances: PH=IN; equal latent means: KA=IN).

In Mplus, the equality constraint for the factor variance is placed on the solution by using the number in parentheses that follows the name of the factor. This overrides the Mplus default of freely estimating the variances in both groups. As noted earlier, the factor loadings and indicator intercepts are held to equality in Mplus by default, and thus no additional programming is required unless the user wishes to override these defaults. Finally, the latent mean of males is held equal to females by the command [DEPRESS@0]; that is, the deviation between males' and females' latent mean is constrained not to differ significantly from zero. In EQS, all of the equality constraints are made in the /CONSTRAINTS section of the syntax. In the sections of the program corresponding to males' and females' measurement model, all parameters other than the marker indicator are freely estimated (i.e., the syntax is identical for men and women). In the /CONSTRAINTS section that follows, all of these parameters are then held to equality except for the factor loading of the marker indicator, and the error covariance of V1 and V2.

The equality constraint on the factor means examines whether groups differ in their levels of the underlying construct. As shown in Table 7.11, this constraint does not significantly degrade the fit of the model, indicating that men and women outpatients did not differ in their average levels of the underlying dimension of MDD, $\chi^2_{diff}(1) = 1.91$, ns. As noted earlier in this section, this constraint is essentially redundant with the significance test of the freely estimated latent mean of male outpatients ($z = 1.38$, ns). This is because only two groups are used in the analysis. Hence the omnibus test, which is the equality constraint on all factor means (in the current case, $\kappa = 2$), is approximately the same as the significance test of this single parameter; in both, $df = 1$.

In instances involving more than two groups, a significant omnibus test (i.e., a significant increase in χ^2 when the means of a given latent variable are held to equality across groups) is typically followed by post hoc evaluation to determine the nature of

TABLE 7.15. Computer Syntax (LISREL, Mplus, EQS) for Fully Invariant Multiple-Groups Model of MDD

LISREL 9.10

```
TITLE LISREL PROGRAM FOR INVARIANCE OF MAJOR DEPRESSION CRITERIA (FEMALES)
DA NG=2 NI=9 NO=375 MA=CM
LA
M1 M2 M3 M4 M5 M6 M7 M8 M9
KM
<Insert correlation matrix for Females from Figure 7.3>
ME
4.184   3.725   1.952   3.589   2.256   3.955   3.869   3.595   1.205
SD
1.717   2.015   2.096   2.212   2.132   2.005   2.062   2.156   1.791
MO NX=9 NK=1 PH=SY,FR LX=FU,FR TD=SY,FR TX=FR KA=FI
LK
DEPRESS
PA LX
0
1
1
1
1
1
1
1
1
VA 1.0 LX(1,1)                          ! SET THE METRIC OF THE LATENT VARIABLE
PA TD
1
1 1
0 0 1
0 0 0 1
0 0 0 0 1
0 0 0 0 0 1
0 0 0 0 0 0 1
0 0 0 0 0 0 0 1
0 0 0 0 0 0 0 0 1
OU ME=ML RS MI SC AD=OFF IT=100 ND=4
INVARIANCE OF MAJOR DEPRESSION CRITERIA (MALES)
DA NI=9 NO=375 MA=CM
LA
M1 M2 M3 M4 M5 M6 M7 M8 M9
KM
<Insert correlation matrix for Males from Figure 7.3>
ME
4.171   3.685   1.739   3.357   2.235   3.661   3.421   3.517   1.259
SD
1.598   2.018   2.094   2.232   2.108   2.113   2.286   2.174   1.788
MO NX=9 NK=1 PH=IN LX=IN TD=SY,FR TX=IN KA=IN      ! IN = INVARIANT
PA TD
1
1 1
0 0 1
0 0 0 1
0 0 0 0 1
0 0 0 0 0 1
```

(continued)

TABLE 7.15. *(continued)*

```
0 0 0 0 0 0 1
0 0 0 0 0 0 0 1
0 0 0 0 0 0 0 0 1
OU ME=ML RS MI SC AD=OFF IT=100 ND=4
```

Mplus 7.11

```
TITLE:        MPLUS PROGRAM FOR EQUAL LOADINGS AND INTERCEPTS OF MDD
DATA:         FILE IS MDDALL.DAT;
VARIABLE:     NAMES ARE SEX M1-M9;
              USEVAR ARE M1-M9;
              GROUPING IS SEX (0=FEMALE 1=MALE);
ANALYSIS:     ESTIMATOR=ML;
MODEL:        DEPRESS BY M1-M9;
              M1 WITH M2;
              DEPRESS (1);      ! EQUAL FACTOR VARIANCE
MODEL MALE:   M1 WITH M2;       ! BY DEFAULT, LOADINGS & INTERCEPTS HELD EQUAL
              [DEPRESS@0];      ! CONSTRAINT ON LATENT MEAN
OUTPUT:       SAMPSTAT MODINDICES(10.00) STAND RESIDUAL;
```

EQS 6.2

```
/TITLE
 EQS SYNTAX FOR GENDER INVARIANCE OF MAJOR DEPRESSION (FEMALES)
/SPECIFICATIONS
 CASES=375; VAR=9; ME=ML; MA=COR; ANALYSIS=MOM; GROUPS=2;
/MATRIX
<Insert correlation matrix for Females from Figure 7.3>
/MEANS
4.184  3.725  1.952  3.589  2.256  3.955  3.869  3.595  1.205
/STANDARD DEVIATIONS
1.717  2.015  2.096  2.212  2.132  2.005  2.062  2.156  1.791
/LABELS
 V1=depmood; V2=anhedon; V3=weight; V4=sleep; V5=motor; V6=fatigue;
 V7=guilt; V8=concent; V9=suicide;
 F1 = DEPRESS;
/EQUATIONS
 V1 = *V999 +  F1 + E1;
 V2 = *V999 + *F1 + E2;
 V3 = *V999 + *F1 + E3;
 V4 = *V999 + *F1 + E4;
 V5 = *V999 + *F1 + E5;
 V6 = *V999 + *F1 + E6;
 V7 = *V999 + *F1 + E7;
 V8 = *V999 + *F1 + E8;
 V9 = *V999 + *F1 + E9;
 F1 = *V999 + D1;
/VARIANCES
 E1 TO E9= *;
 D1 = *;
/COVARIANCES
 E1,E2 = *;
/END
/TITLE
 EQS SYNTAX FOR GENDER INVARIANCE OF MAJOR DEPRESSION (MALES)
```

(continued)

TABLE 7.15. *(continued)*

```
/SPECIFICATIONS
 CASES=375; VAR=9; ME=ML; MA=COR; ANALYSIS=MOM;
/MATRIX
<Insert correlation matrix for Males from Figure 7.3>
/MEANS
4.171  3.685  1.739  3.357  2.235 - 3.661  3.421  3.517  1.259
/STANDARD DEVIATIONS
1.598  2.018  2.094  2.232  2.108  2.113  2.286  2.174  1.788
/LABELS
 V1=depmood; V2=anhedon; V3=weight; V4=sleep; V5=motor; V6=fatigue;
 V7=guilt; V8=concent; V9=suicide;
 F1 = DEPRESS;
/EQUATIONS
 V1 = *V999 +  F1 + E1;
 V2 = *V999 + *F1 + E2;
 V3 = *V999 + *F1 + E3;
 V4 = *V999 + *F1 + E4;
 V5 = *V999 + *F1 + E5;
 V6 = *V999 + *F1 + E6;
 V7 = *V999 + *F1 + E7;
 V8 = *V999 + *F1 + E8;
 V9 = *V999 + *F1 + E9;
 F1 = *V999 + D1;
/VARIANCES
 E1 TO E9= *;
 D1 = *;
/COVARIANCES
 E1,E2 = *;
/CONSTRAINTS
 (1,V2,F1)=(2,V2,F1);            ! FACTOR LOADING EQUALITY
 (1,V3,F1)=(2,V3,F1);
 (1,V4,F1)=(2,V4,F1);
 (1,V5,F1)=(2,V5,F1);
 (1,V6,F1)=(2,V6,F1);
 (1,V7,F1)=(2,V7,F1);
 (1,V8,F1)=(2,V8,F1);
 (1,V9,F1)=(2,V9,F1);
 (1,V1,V999)=(2,V1,V999);        ! INDICATOR INTERCEPT EQUALITY
 (1,V2,V999)=(2,V2,V999);
 (1,V3,V999)=(2,V3,V999);
 (1,V4,V999)=(2,V4,V999);
 (1,V5,V999)=(2,V5,V999);
 (1,V6,V999)=(2,V6,V999);
 (1,V7,V999)=(2,V7,V999);
 (1,V8,V999)=(2,V8,V999);
 (1,V9,V999)=(2,V9,V999);
 (1,D1,D1) = (2,D1,D1);          ! EQUAL FACTOR VARIANCE
 (1,F1,V999) = 0;
 (2,F1,V999) = 0;                ! EQUAL LATENT MEANS
/PRINT
 fit=all;
/LMTEST
/END
```

the overall effect, along the lines of simple effects testing in ANOVA. If the analysis uses three groups, three possible follow-up analyses, in which the factor means of two groups are constrained to equality at a time, can be conducted to identify which groups differ in their latent means (perhaps with a control for experiment-wise error in the multiple comparisons, such as the modified Bonferroni procedure; Jaccard & Wan, 1996). Although the similarities to ANOVA and the independent t test are apparent, it should be reemphasized that the CFA-based approach to group mean comparison has several advantages over these more traditional methods. A key strength of the CFA approach is that it establishes whether the group comparisons are appropriate. Although traditional analyses simply assume that this is the case, the multiple-groups CFA may reveal considerable measurement noninvariance (e.g., unequal form, a preponderance of noninvariant factor loadings and indicator intercepts) that contraindicates groups comparison on the factor mean. (As will be discussed shortly, in some instances it may be possible to proceed with such comparisons in the context of partial measurement invariance.) Moreover, group comparisons have more precision and statistical power in CFA because the structural parameters (factor means, variances, covariances) have been adjusted for measurement error. Traditional tests such as ANOVA assume perfect reliability.

Two decades ago, it was rare to see a CFA of mean structures in the applied research literature, primarily because the syntax specification of such models was very complex in early versions of latent variable software packages. Such analyses have now become much more common in the literature. As shown in the prior example, the analysis of mean structures is straightforward in the latest releases of software programs such as LISREL, Mplus, and EQS. Given the advantages of such analysis over traditional statistics (e.g., ANOVA), and its capability to evaluate other important substantive questions (e.g., psychometric issues such as differential item functioning), investigators are encouraged to incorporate mean structures in their applied CFA-based research.

In fact, the Mplus program has made it particularly easy to conduct measurement invariance evaluation, beginning with Version 7.1. As shown in Table 7.16, the tests of equal form, equal factor loadings, and equal intercepts can be conducted in a single run by using the command MODEL = CONFIGURAL METRIC SCALAR. As a result, χ^2 difference testing for the three models is carried out automatically (see Table 7.16, and compare to results in Table 7.11), and the remaining goodness-of-fit statistics and unstandardized parameter estimates for the three solutions are provided for each group. However, standardized estimates and modification indices are currently not available when this shortcut option is employed.

Selected Issues in Single- and Multiple-Groups CFA Invariance Evaluation

In this section, four miscellaneous issues in CFA invariance evaluation are discussed. Although each of these issues can affect the conduct and interpretation of CFA invariance evaluation considerably, the extant SEM literature has provided minimal guidance on how these issues should be managed in applied research. Nevertheless, it is impor-

TABLE 7.16. Mplus 7.11 Syntax and Selected Results for Analyzing Equal Form, Equal Loadings, and Equal Intercepts in a Single Program Run

```
TITLE:        MPLUS PROGRAM FOR EQUAL FORM LOADINGS AND INTERCEPTS
DATA:         FILE IS MDDALL.DAT;
VARIABLE:     NAMES ARE SEX M1-M9;
              USEVAR ARE M1-M9;
              GROUPING IS SEX (0=FEMALE 1=MALE);
ANALYSIS:     ESTIMATOR=ML;
              MODEL=CONFIG METRIC SCALAR;   ! MPLUS SHORT-CUTS
MODEL:        DEPRESS BY M1-M9;
              M1 WITH M2;
OUTPUT:       SAMPSTAT RESIDUAL;


MODEL FIT INFORMATION

Invariance Testing
```

Model	Number of Parameters	Chi-square	Degrees of Freedom	P-value
Configural	56	98.911	52	0.0001
Metric	48	102.839	60	0.0005
Scalar	40	115.309	68	0.0003

Models Compared	Chi-square	Degrees of Freedom	P-value
Metric against Configural	3.929	8	0.8635
Scalar against Configural	16.398	16	0.4255
Scalar against Metric	12.470	8	0.1314

tant that the researcher be aware of these issues, along with some tentative remedial strategies offered to date.

Partial Measurement Invariance

In their review of the applied CFA literature over 25 years ago, Byrne et al. (1989) observed a widespread belief among researchers that when evidence of noninvariant measurement parameters (e.g., unequal factor loadings) is encountered, further testing of measurement invariance and population heterogeneity is not possible. As noted earlier, an omnibus test of invariance is conducted by placing equality constraints on a family of unstandardized parameters of the CFA model (e.g., factor loadings) and determining whether these restrictions produce a significant increase in model χ^2. If a significant increase in χ^2 is observed, then the null hypothesis is rejected (e.g., the factor loadings are not equal across groups). Fit diagnostics (e.g., modification indices, expected parameter change values) can assist the researcher in identifying which parameters are noninvariant across groups, across time, and so forth.[4] Indeed, a significant omnibus χ^2 should not be interpreted as indicating that all parameters are noninvariant; for instance, this result may be obtained when there is a single noninvariant parameter within a complex measurement model.

Byrne et al. (1989) reminded researchers that in many instances, invariance evaluation can proceed in context of *partial measurement invariance*—that is, in CFA models where some but not all of the measurement parameters are equivalent. For ease of illustration, consider a unidimensional model entailing five indicators (X1–X5, where X1 is specified as the marker indicator), which is tested for invariance between two groups. Given evidence of equal form, the researcher evaluates the between-groups equality of factor loadings. The more restricted solution produces a significant increase in χ^2, suggesting that at least one of the factor loadings is noninvariant. Fit diagnostics suggest that the factor loading of X5 is noninvariant (e.g., associated with high modification index; the modification indices for X2, X3, and X4 are below 4.0). This is verified in a respecified multiple-groups solution where all factor loadings other than the loading of X5 (which is freely estimated in both groups) and X1 (which has previously been fixed as the marker indicator) are constrained to equality between groups; that is, this respecified model does not produce a significant increase in χ^2 relative to the equal form solution. From a statistical perspective, the invariance evaluation may proceed to examine the equality of other measurement (e.g., indicator intercepts) and structural parameters (e.g., latent means) in context of this partial invariance. The researcher will freely estimate the factor loading of X5 in both groups in subsequent analyses. Indeed, Byrne et al. (1989) note that such analyses may proceed as long as there exists at least one noninvariant parameter other than the marker indicator; for instance, in the current example, invariance evaluation can continue if, say, only the factor loading of X2 is invariant in the two groups. The same logic will apply to other measurement parameters. For example, recall that comparison of latent means is meaningful only if factor loadings and indicator intercepts are invariant. In the present illustration, a between-groups comparison of the latent mean can be conducted, provided that there exist at least partial factor loading and partial intercept invariance (e.g., at least one indicator other than the marker indicator has an invariant factor loading and intercept).

The main advantage of this strategy is that it allows the invariance evaluation to proceed after some noninvariant parameters are encountered. This is very helpful in cases where the evaluation of structural parameters is of greatest substantive interest. For instance, consider the situation where the researcher wishes to evaluate a complex longitudinal structural equation model involving several latent variables. Of most interest are the structural parameters of this model (i.e., the paths reflecting regressions within and among latent variables over time). Before these paths can be evaluated, the researcher must establish longitudinal measurement invariance (e.g., to rule out the spurious influence of temporal change in measurement). Without knowledge of the partial measurement invariance strategy, the researcher may abandon the evaluation of structural parameters if measurement noninvariance is detected (e.g., the factor loadings of a few indicators differ over time). However, if a preponderance of invariant measurement parameters is assumed, the substantively more important analyses (e.g., cross-lagged effects among latent variables) can be conducted in context of partial measurement invariance.

The primary disadvantages of partial invariance analysis are its exploratory nature and its risk of capitalization on chance. These issues may be of greater concern when the purpose of the analysis is psychometric in nature (e.g., evaluation of the measurement invariance of a psychological questionnaire in demographic subgroups). Like other scenarios, psychometrically based applications of invariance evaluation are often conducted in the absence of explicit hypothesis testing. In fact, although doing so is counter to the traditions of the scientific method, researchers typically conduct invariance evaluation with the hope of retaining the null hypothesis—that is, H_0: All parameters are the same across groups (thereby supporting the notion that the testing instrument is unbiased, has equivalent measurement properties, etc.). Given the number of parameters that are constrained in invariance evaluations (especially when the measurement model contains a large number of indicators), it is possible that some parameters will differ by chance. In addition, the large sample sizes used in CFA have considerable power to detect small differences as statistically significant, especially because invariance testing relies heavily on χ^2 (see a later section of this chapter). Moreover, the guidelines and procedures for relaxing invariance constraints have not been fully developed or studied by SEM methodologists. Vandenberg and Lance (2000) recommend that the partial invariance strategy should not be employed when a large number of indicators are found to be noninvariant. Indeed, although it is statistically possible to proceed when a single indicator other than the marker indicator is invariant, this outcome should prompt the researcher to question the suitability of the measurement model for further invariance testing. Although partial invariance evaluation is a post hoc procedure, this limitation can be allayed to some degree if the researcher can provide a substantively compelling account for the source of noninvariance (e.g., see the example of a MIMIC model later in this chapter). Finally, Byrne et al. (1989) underscore the importance of cross-validation with an independent sample. For example, if the overall sample is large enough, it can be randomly divided into two subsamples. In this strategy, the first sample is used to develop a good-fitting solution where some parameters are freely estimated (e.g., across groups) when evidence of noninvariance arises (ideally, fostered by substantive arguments for why these parameters differ across groups). The final model is then fitted in the second sample to determine its replicability with independent data. Although cross-validation is a compelling method to address the limitations of partial invariance evaluation (and post hoc model revision in general), Byrne et al. (1989) highlighted some of its practical concerns—such as the feasibility of obtaining more than one sufficiently large sample, and the fact that cross-validation is likely to be unsuccessful when multiple parameters are relaxed in the first sample.

Selection of Marker Indicators

Earlier it has been suggested that selection of a latent variable's marker indicator should not be taken lightly (see Chapter 4). The selection of marker indicators can also greatly influence measurement and structural invariance evaluation. Difficulties will arise in multiple-groups CFA when the researcher inadvertently selects a marker indicator that is noninvariant across groups. First, the researcher may not detect this noninvariance because the unstandardized factor loadings of the marker indicator are fixed to 1.0 in all groups. Second, subsequent tests of partial invariance may be poor-fitting because the unstandardized measurement parameters of the remaining indicators (e.g., factor loadings, intercepts) are influenced by the noninvariant marker indicator. The differences found in the remaining indicators may not reflect "true" differences among groups, but rather may be artifacts of scaling the metric of the latent variable with an indicator that has a different relationship to the factor in two or more groups. Although somewhat cumbersome, one approach to exploring this possibility is to rerun the multiple-groups CFA with different marker indicators. Cheung and Rensvold (1999) have also introduced procedures for addressing this issue. In addition, this issue can be addressed by using one of the other two methods of defining the metric of the latent variable (i.e., fixing the factor variance to 1.0, effects coding).

Reliance on χ^2

As the previous data-based examples illustrate, invariance evaluation relies strongly on the χ^2 statistic. For example, the omnibus test of equality of indicator intercepts across groups is conducted by determining whether the constrained solution (in which the intercepts are held to be equal across groups) produces a significant increase in χ^2 relative to a less constrained model (e.g., an equal factor loadings solution). When a significant degradation in model fit is encountered, procedures to identify noninvariant parameters rely in part on χ^2-based statistics (i.e., modification indices). As discussed in prior chapters, both model χ^2 and modification indices are sensitive to sample size. Researchers have noted that a double standard exists in the SEM literature (e.g., Cheung & Rensvold, 2000; Vandenberg & Lance, 2000). That is, given the limitations of χ^2, investigators are encouraged to use a variety of fit indices to evaluate the overall fit of a CFA solution (e.g., RMSEA, TLI, CFI, SRMR). However, in invariance evaluation, the χ^2 statistic is relied on exclusively to detect differences in more versus less constrained solutions. The reason why χ^2 is the only fit statistic used for this purpose is that its distributional properties are known, and thus critical values can be determined at various degrees of freedom. This cannot be done for other fit indices; for example, a more constrained solution may produce an increase in the SRMR, but there is no way of determining at what magnitude this increase is statistically meaningful.

Researchers have begun to recognize and address this issue. Two papers (Cheung & Rensvold, 2000; Meade, Johnson, & Braddy, 2008) report the results of large Monte Carlo simulation studies that were conducted to determine whether critical values of other goodness-of-fit statistics (e.g., the CFI) could be identified to reflect the presence–absence of measurement invariance in multiple-groups solutions; for instance, what point reduction in the CFI reliably rejects the null hypothesis that the measurement parameters are the same across groups? Although critical values were proposed in both studies (e.g., CFI reductions of no more than .01 and .002 units in Cheung & Rensvold, 2000, and Meade et al., 2008, respectively), the validity of these proposals awaits further research (cf. Fan & Sivo, 2009).

Another problem that the CFA researcher may encounter on occasion is that the omnibus test of invariance is statistically significant (i.e., the constraints result in a significant increase in model χ^2), but fit diagnostics reveal no salient strains with regard to any specific parameter (e.g., modification indices for all constrained parameters are close to or below 4.0; for an example of this outcome in the applied literature, see Campbell-Sills et al., 2004). Again, such a result may be indicative of χ^2's oversensitivity to sample size, and the problem of relying exclusively on χ^2 in invariance evaluation. This outcome would suggest that the differences suggested by the statistically significant increase in model χ^2 are trivial and thus have no substantive importance. To verify this conclusion, Byrne et al. (1989) have suggested that a *sensitivity analysis* can be helpful. Sensitivity analysis is a post hoc method of determining whether the addition of minor parameters (e.g., relaxation of the constraints on parameters that have previously been held to equality) results in a clear change in the major parameters of the model (e.g., the factor loadings). For example, if the major parameters do not change when minor parameters are included in the model, this can be considered evidence for the robustness of the initial solution. The less parsimonious model may now be statistically equivalent to the baseline solution, but these additional minor parameters are not needed because they do not appreciably affect the major parameters of the solution. On the other hand, if the additional parameters significantly alter the major parameters, this will suggest that exclusion of these post hoc parameters (e.g., factor loadings or intercepts that have been freely estimated in some or all groups) will lead to biased estimates of the major parameters. However, sensitivity analysis is limited somewhat by its subjective nature (i.e., the definition of what constitutes "salient" change in the model's major parameters when post hoc parameters are included or excluded from the solution). To increase the objectivity of this approach, Byrne et al. (1989) have suggested correlating the major parameters (e.g., factor loadings) in the initial model with those of the best-fitting post hoc solution. Although coefficients close to 1.00 may provide clear support for the more constrained model and argue against the importance of the additional freed parameters, this conclusion becomes more vague when the correlation is well below 1.00. Moreover, a high correlation may still mask the fact that one or two major parameters differ considerably in context of a multiple-indicator model where all other parameters are roughly the same. The caveats of partial invariance evaluation should also be considered (i.e., exploratory nature, risk of chance effects, importance of cross-validation).

Nonlinear Factor–Indicator Relationships

Finally, another complication that may arise in measurement invariance evaluation is the situation where the true factor–indicator relationship is nonlinear (e.g., quadratic) when the solution is in fact invariant across groups. If a linear model is pursued, the factor loadings and indicator intercepts will increasingly differ across groups as the factor mean difference increases. Diagnostic procedures to identify nonlinear factor–indicator and factor–factor relationships have been developed (Bauer, 2005; Pek, Sterba, Kok, & Bauer, 2009), as well as methods for fitting nonlinear factor models (Wall & Amemiya, 2007).

MIMIC Modeling (CFA with Covariates)

A less commonly used method of examining invariance in multiple groups entails regressing the latent variables and indicators onto covariates that represent group membership (e.g., Sex: 0 = female, 1 = male). This approach has been referred to as *CFA with covariates* or *MIMIC modeling*. (As noted earlier, MIMIC stands for "multiple indicators, multiple causes"; see Jöreskog & Goldberger, 1975; Muthén, 1989.) Unlike in multiple-groups CFA, a single input matrix is used in the MIMIC analysis. The input matrix contains the variances and covariances of both the latent variable indicators and the covariates that denote group membership. Indicator means are not included as input, as might be the case for multiple-groups CFA. The two basic steps of MIMIC modeling are these: (1) Establish a viable measurement model using the full sample (i.e., collapsing across groups); and (2) add one or more covariates to the model to examine their direct effects on the factors and selected indicators. A significant direct effect of the covariate on the factor represents population heterogeneity; that is, the factor means are different at different levels of the covariate (akin to the test of equal latent means in multiple-groups CFA). A significant direct effect of the covariate on an indicator of a factor represents measurement noninvariance; that is, when the factor is held constant, the means of the indicator are different at different levels of the covariate, thereby pointing to differential item (indicator) functioning (analogous to the test of equal indicator intercepts in multiple-groups CFA). Although both aspects of invariance evaluation in MIMIC modeling correspond to the mean structure component of multiple-groups CFA, indicator intercepts and factor means are not estimated in the MIMIC analysis, and indicator means are not included in the input matrix. Rather, as in the analysis of covariance structure, these group mean differences are deviations conveyed by the parameter estimates of the direct effects (e.g., Sex → Latent Variable), where the means of the indicators and factors are assumed to be zero. For instance, if the unstandardized direct effect of Sex → Latent Variable is 0.75, then the latent means of males and females differ by 0.75 units, as per the metric used to scale and identify the latent variable.

Unlike multiple-groups CFA, MIMIC modeling can only test the invariance of indicator intercepts and factor means. Thus it assumes that all other measurement and structural parameters (i.e., factor loadings, error variances–covariances, factor variances–covariances) are the same across all levels of the covariates (groups). A primary advantage of MIMIC modeling is that it usually has smaller sample size requirements than multiple-groups CFA. Whereas multiple-groups CFA entails the simultaneous analysis of two or more measurement models, MIMIC involves a single measurement model and input matrix. For example, in the case where a researcher wishes to conduct an invariance analysis with three groups but has a total sample size of 150, the size of each group may not be sufficient for each within-group CFA in the multiple-groups approach; that is, the analysis would require three separate CFAs, each with an n = 50. However, the N = 150 may be suitable for a single CFA in which aspects of the measurement model are regressed onto covariates, although sample size considerations must also be brought to bear with regard to the statistical power of the direct effects of the covariates (see Chapter 10).[5]

Another potential strength of MIMIC modeling over multiple-groups CFA arises when there are many groups involved in the comparison. Multiple-groups CFA with three or more groups can be very cumbersome (depending also on the complexity of the measurement model) because of the number of parameters that must be estimated and held equal across all groups. In addition, post hoc testing is more complex when an omnibus test of the null hypothesis of a given aspect of invariance is rejected. In contrast, MIMIC modeling is more parsimonious because measurement model parameters are not estimated in each group.

In most applications of MIMIC, the covariate is a nominal variable that represents levels of known groups (e.g., Sex: 0 = female, 1 = male). When the levels of groups (k) are three or more, group membership can be reflected as dummy codes, according to the procedures found in multiple regression textbooks (e.g., Cohen et al., 2003). If k = 3, two (k – 1) binary codes are created that identify two of the three levels of the nominal variable, and the remaining level is treated as the reference group that does not receive its own code. Although categorical covariates are typically used in MIMIC models, a normal theory estimator such as ML may still be used if the indicators of the latent variables satisfy asymptotic theory assumptions (cf. Chapter 9). This is because the categorical variables are used as predictors instead of outcomes (cf. multiple regression vs. logistic regression). However, the covariates used in MIMIC models can also be dimensional (e.g., age). The ability to accommodate continuous predictors can be viewed as another potential advantage of MIMIC over multiple-groups CFA because the latter necessitates imposing categorical cutoffs on a dimensional variable to form "groups" for the analysis (cf. MacCallum, Zhang, & Preacher, 2002).

Ordinarily, the covariate is assumed to be free of measurement error (i.e., its error variance is fixed to zero). This assumption is reasonable when the covariate represents known groups (e.g., male vs. female). However, the desired amount of measurement error can be modeled in a dimensional covariate by using the procedures described in Chapter 4; that is, the unstandardized error of the covariate can be fixed to some nonzero value on the basis of the sample variance estimate and known reliability information (see Eqs. 4.14 and 4.15, Chapter 4).

Unlike other examples of CFA presented in this book thus far, the MIMIC approach reflects a latent Y (endogenous) variable specification (cf. Figure 3.4, Chapter 3). This is because the latent variables (and their indicators) are specified as outcomes predicted by the covariates (see Figure 7.5). This distinction has minimal impact on syntax programming in programs such as Mplus and EQS, which accommodate the latent Y specification "behind the scenes," but it does require several alterations if the analysis is conducted with the LISREL matrix programming language.

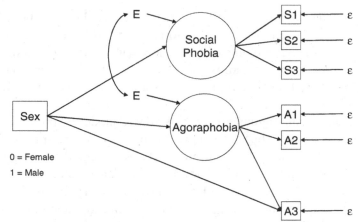

Sample Correlations and Standard Deviations (SDs); N = 730 (365 males, 365 females)

	S1	S2	S3	A1	A2	A3	Sex
S1	1.000						
S2	0.705	1.000					
S3	0.724	0.646	1.000				
A1	0.213	0.195	0.190	1.000			
A2	0.149	0.142	0.128	0.521	1.000		
A3	0.155	0.162	0.135	0.557	0.479	1.000	
Sex	-0.019	-0.024	-0.029	-0.110	-0.074	-0.291	1.000
SD:	2.260	2.730	2.110	2.320	2.610	2.440	0.500

FIGURE 7.5. MIMIC model of Social Phobia and Agoraphobia. S1, giving a speech; S2, meeting strangers; S3, talking to people; A1, going long distances from home; A2, entering a crowded mall; A3, walking alone in isolated areas. (All questionnaire items rated on 0–8 scales, where 0 = no fear and 8 = extreme fear.)

The MIMIC methodology is illustrated with the model presented in Figure 7.5. In this example, the researcher wishes to examine selected aspects of population heterogeneity (factor means) and measurement invariance (indicator intercepts) associated with an established questionnaire measure of two diagnostic constructs: Social Phobia (fear of social situations due to possibility of negative evaluation by others) and Agoraphobia (fear of public situations due to possibility of experiencing unexpected panic attacks). This example is loosely based on a psychometric study conducted by Brown et al. (2005). Indicators of the Social Phobia and Agoraphobia constructs are provided in Figure 7.5 (items self-rated on 0–8 scales; higher scores = higher fear). In particular, the researcher expects to find sex differences with respect to the latent construct of Agoraphobia (population heterogeneity), based on the prior clinical literature that women are more inclined than men to respond to unexpected panic attacks with situational fear and avoidance. Moreover, the researcher is concerned that an indicator of Agoraphobia (A3) functions differently for men and women. Specifically, it is anticipated that, regardless of the level of the underlying factor of Agoraphobia, women will score higher on the A3 indicator than men (A3: "walking alone in isolated areas") because of sex differences in the range of activities relating to personal safety; that is, regardless of the presence–absence of the symptoms of agoraphobia, women are less likely than men to walk alone in isolated areas. To address this question, data have been collected from 730 outpatients (365 men, 365 women) presenting for assessment and treatment of various anxiety disorders.

The path diagram and input matrix of this MIMIC model are presented in Figure 7.5. A path from the covariate to the Social Phobia factor has also been specified to test for sex differences on this construct. The covariate (Sex) is represented by a single dummy code (0 = female, 1 = male), and its standard deviation and correlations with the indicators of Agoraphobia and Social Phobia are included in the input matrix. As seen in Figure 7.5, the variances of Social Phobia and Agoraphobia are not estimated in the MIMIC model. Instead, because the model attempts to explain variance in these factors by the Sex covariate, these parameters are residual variances (E; often referred to as *disturbances*). Also note that the residual variances of Social Phobia and Agoraphobia are specified to be correlated. This specification is justified by the argument that the constructs of Social Phobia and Agoraphobia are not completely orthogonal, and that this overlap cannot be fully accounted for by the Sex covariate; for example, Sex is not a "third variable" responsible for the overlap between the dimensions of Social Phobia and Agoraphobia.

The first step is to ensure that the two-factor model of Social Phobia and Agoraphobia is reasonable and good-fitting in the full sample (N = 730). In this step, the Sex covariate is not included in the CFA, and the variances (ϕ_{11}, ϕ_{22}) and covariance (ϕ_{21}) of the two factors are freely estimated (i.e., a typical CFA model is specified). This model provides a good fit to the data: $\chi^2(8)$ = 3.06, p = .93, SRMR = .012, RMSEA = 0.00 (90% CI = 0.00–0.13, CFit = 1.00), TLI = 1.005, CFI = 1.00. There are no salient areas of strain in the solution (e.g., no modification indices > 4.0), and all parameter estimates are reasonable and statistically significant (e.g., range of completely standardized factor loadings = .67–.89; correlation between Social Phobia and Agoraphobia = .28).

Next, the Sex covariate is added to the model (see Figure 7.5). Table 7.17 provides Mplus, LISREL, and EQS syntax for this model specification. As can be seen in Table 7.16, the programming is straightforward in Mplus. Regressing the latent variables of Social Phobia and Agoraphobia, as well as the A3 indicator, onto the Sex covariate is accomplished with the ON keyword (e.g., SOCIAL ON SEX). The correlated residual of Social Phobia and Agoraphobia is specified by the SOCIAL WITH AGORAPH syntax. In EQS, the MIMIC portion of the model is represented in the last three lines of the /EQUA-TIONS command. The line V6 = *F2 + *V7 + E6 represents the fundamental equation that variance of A3 (V6) is to be accounted for by the Agoraphobia factor (F2), the Sex covariate (V7), and residual variance (E6). The variance of the factors (F1, F2) is reproduced by variance explained by the Sex covariate and residual variance (e.g., F1 = *V7 + D1). The /VARIANCE commands indicate that all residual variances (indicators: E1–E6; latent variables: D1, D2), and X-variable (V7) variances should be freely estimated. The /COVARIANCE command, D1,D2 = *, indicates that the correlated disturbance (residual) between Social Phobia and Agoraphobia should also be freely estimated.

LISREL matrix programming is more complicated (see Table 7.17). Although other programs (e.g., Mplus) rely on the same programming logic exemplified by the following LISREL programming, their software contains convenience features that allow the user to set up the analysis in a less complex manner. This is also true for the SIMPLIS sublanguage of LISREL. First, note that latent Y programming is used (lambda-Y and theta-epsilon for factor loadings and indicator errors, respectively; psi for factor variances and covariances) because the CFA portion of the MIMIC model is endogenous (cf. Chapter 3). Although Sex is really an X variable, it is treated as a Y variable so that the overall model can be specified in LISREL. An "all Y" programming strategy is often used to accomplish other special analyses in the LISREL framework (e.g., see scale reliability estimation in Chapter 8). Another programming trick is that A3 is regarded as another factor (see LE line, and pattern matrix for LY); thus the model is programmed to have four factors (Social Phobia, Agoraphobia, A3, Sex). This programming (in tandem with PA BE matrix programming, discussed below) actually produces a solution equivalent to a typical CFA solution (see "Equivalent CFA Solutions" section in Chapter 5). That is, although it will not be found in the estimates for lambda-Y, the solution will generate a factor loading for A3 onto Agoraphobia that is identical to what would be obtained in an ordinary CFA specification. For the pattern matrix programming of lambda-Y (PA LX), the model specification for Social Phobia is typical (S1 is used as the marker indicator, and the factor loadings of S2 and S3 are freely estimated). For Agoraphobia, A1 is used as the marker variable, and the factor loading of A2 is freely estimated. In the remaining part of PA LY (and its associated VA commands), the metric of the A3 indicator is passed onto the A3 "pseudofactor"; the metric of the Sex covariate (0/1) is passed on to the Sex "factor." In the theta-epsilon pattern matrix (PA TE), the error variances of the first five indicators (S1, S2, S3, A1, A2) are freely estimated. The measurement errors of the A3 and Sex indicators are fixed to zero. Although the Sex variable will be assumed to have no measurement error in the analysis (i.e., it corresponds to known groups), the error variance of the A3 indicator will be estimated in another portion of the solution.

Because an "all Y" specification is employed, variances and covariances (not including the error variances of indicators S1 through A2) must be estimated by the psi (PS) matrix (Ψ). The first two diagonal elements of PS (i.e., ψ_{11}, ψ_{22}) freely estimate the residual variances of Social Phobia and Agoraphobia. The first off-diagonal element of PS (ψ_{21}) estimates the correlated disturbance of these factors. The third (ψ_{33}) diagonal element of PS estimates the residual variance of the A3 pseudofactor; this parameter estimate will in fact represent the measurement error variance of the A3 indicator. The fourth (ψ_{44}) diagonal element of PS freely estimates the variance of Sex.

The MIMIC portion of the LISREL specification resides in the pattern matrix programming of Beta (BE). The Beta (**B**) matrix has not been employed in prior examples in this book because it represents a structural component of a structural equation model; in this context, *structural* refers to directional relationships among latent variables. Specifically, the Beta matrix focuses on directional relationships among endogenous (Y) variables. Alternatively, the Gamma matrix (Γ) focuses on the directional relationships between exogenous and endogenous variables. Because four latent Y variables have been

TABLE 7.17. Computer Syntax (Mplus, LISREL, EQS) for MIMIC Model of Social Phobia and Agoraphobia (Regressing the Latent Variables and the A3 Indicator on the Sex Covariate)

Mplus 7.11

```
TITLE:        CFA MIMIC MODEL
DATA:         FILE IS FIG7.5.DAT;
              TYPE IS STD CORR;
              NOBS ARE 730;
VARIABLE:     NAMES ARE S1 S2 S3 A1 A2 A3 SEX;
ANALYSIS:     ESTIMATOR=ML;
MODEL:        SOCIAL BY S1-S3;
              AGORAPH BY A1-A3;
              SOCIAL ON SEX; AGORAPH ON SEX; A3 ON SEX; SOCIAL WITH AGORAPH;
OUTPUT:       SAMPSTAT MODINDICES(4.00) STAND RESIDUAL;
```

LISREL 9.10

```
TITLE LISREL PROGRAM FOR MIMIC MODEL
DA NI=7 NO=730 MA=CM
LA
S1 S2 S3 A1 A2 A3 SEX                            ! SEX: 0 = FEMALE, 1 = MALE
KM
1.000
 0.705   1.000
 0.724   0.646   1.000
 0.213   0.195   0.190   1.000
 0.149   0.142   0.128   0.521   1.000
 0.155   0.162   0.135   0.557   0.479   1.000
-0.019  -0.024  -0.029  -0.110  -0.074  -0.291   1.000
SD
2.26 2.73 2.11 2.32 2.61 2.44 0.50
MO NY=7 NE=4 LY=FU,FR TE=SY,FR PS=SY,FR BE=FU,FR
LE
SOCIAL AGORAPH A3 SEX
PA LY
0 0 0 0
1 0 0 0
1 0 0 0
0 0 0 0
0 1 0 0
0 0 0 0                                          ! A3 IS A PSEUDOFACTOR
0 0 0 0
VA 1.0 LY(1,1) LY(4,2) LY(6,3) LY(7,4)           ! SET METRIC OF VARIABLES
PA TE
1
0 1
0 0 1
0 0 0 1
0 0 0 0 1
0 0 0 0 0 0
0 0 0 0 0 0 0           ! FIX ERROR OF SEX COVARIATE TO ZERO
PA PS
1
1 1                     ! CORRELATED DISTURBANCE BETWEEN SOCIAL AND AGORAPH
0 0 1                   ! ERROR VARIANCE OF A3
0 0 0 1                 ! VARIANCE OF SEX
```

(continued)

TABLE 7.17. *(continued)*

```
PA BE
0  0  0  1                 ! PATH FROM SEX TO SOCIAL LATENT VARIABLE
0  0  0  1                 ! PATH FROM SEX TO AGORAPH LATENT VARIABLE
0  1  0  1                 ! PATH (FACTOR LOADING) FROM AGORAPH TO A3 INDICATOR
0  0  0  0                 ! AND PATH FROM SEX TO A3 INDICATOR
OU ME=ML RS MI SC AD=OFF IT=100 ND=4
```

EQS 6.2

```
/TITLE
 EQS SYNTAX FOR MIMIC MODEL
/SPECIFICATIONS
 CASES=730; VAR=7; ME=ML; MA=COR;
/MATRIX
1.000
 0.705    1.000
 0.724    0.646    1.000
 0.213    0.195    0.190    1.000
 0.149    0.142    0.128    0.521    1.000
 0.155    0.162    0.135    0.557    0.479    1.000
-0.019   -0.024   -0.029   -0.110   -0.074   -0.291    1.000
/STANDARD DEVIATIONS
2.26 2.73 2.11 2.32 2.61 2.44 0.50
/LABELS
 V1=s1; V2=s2; V3=s3; V4=a1; V5=a2; V6=a3; V7=sex;
 F1 = SOCIAL;
 F2 = AGORAPH;
/EQUATIONS
 V1 =   F1 + E1;
 V2 = *F1 + E2;
 V3 = *F1 + E3;
 V4 =   F2 + E4;
 V5 = *F2 + E5;
 V6 = *F2 + *V7 + E6;
 F1 = *V7 + D1;
 F2 = *V7 + D2;
/VARIANCES
 E1 TO E6 = *;
 D1,D2 = *;
 V7 = *;
/COVARIANCES
 D1,D2 = *;
/PRINT
 fit=all;
/LMTEST
/END
```

specified, **B** will be a 4×4 full matrix; the row and column order of variables is SOCIAL AGORAPH A3 SEX, as specified by earlier programming. Thus the β_{14} and β_{24} parameters in PA BE inform LISREL to freely estimate the regressive path of Sex to Social Phobia and Sex to Agoraphobia, respectively (population heterogeneity). The β_{34} element corresponds to the regressive path of Sex to A3 (measurement invariance). Finally, the β_{32} element informs LISREL to freely estimate a path from Agoraphobia to A3. Although this estimate is obtained in a structural matrix (**B**), it should be interpreted as the factor loading of the A3 indicator on the Agoraphobia factor; in fact, this estimate will equal the LY estimate if the analysis is not conducted as a MIMIC model.

The MIMIC model provides a good fit to the data: $\chi^2(11) = 3.80$, $p = .98$, SRMR = .011, RMSEA = 0.00 (90% CI = 0.00–0.00, CFit = 1.00), TLI = 1.008, CFI = 1.00. Inclusion of the Sex covariate does not alter the factor structure or produce salient areas of strain in the solution (e.g., all modification indices < 4.0). Table 7.18 provides selected results of this solution. Of particular interest are the regressive paths linking Sex to the latent variables and the A3 indicator. As predicted, the path of Sex $\rightarrow$ Agoraphobia is statistically significant ($z = 2.97$, $p < .01$). Given how the Sex covariate is coded (0 = females, 1 = males), and the negative sign of this parameter estimate (e.g., unstandardized estimate = -0.475), it can be concluded that males have a lower mean than females on the Agoraphobia factor; more specifically, the mean of females is 0.475 units higher than the mean of males. The completely standardized estimate of this parameter (StdYX = -0.13, not shown in Table 7.18) is not readily interpretable because of the binary predictor (i.e., the Sex covariate). In other words, it is not meaningful to discuss this relationship in terms of a standardized score change in Sex when the level of this variable is either male or female. But the partially standardized estimate (Std = -0.261) can convey useful information about this effect. In this partially standardized solution, only the latent variable is standardized. Thus the estimate of -0.261 can be interpreted as indicating that a unit increase in Sex is associated with a 0.261 standardized score decrease in the latent variable of Agoraphobia; or, more directly, women score 0.261 standardized scores higher than men on the latent dimension of Agoraphobia. This value can be interpreted akin to Cohen's d (Cohen, 1988, 1992). Following Cohen's guidelines ($d = .20$, .50, and .80 for small, medium, and large effects, respectively; cf. Cohen, 1992), the sex difference for Agoraphobia is a small effect. The results in Table 7.18 also reveal that men and women do not differ with respect to Social Phobia ($z = 0.69$, ns).

Consistent with the researcher's predictions, the results of the MIMIC model show that the A3 indicator is not invariant for males and females (akin to intercept noninvariance in multiple-groups CFA). This is reflected by the significant direct effect of Sex on the A3 indicator ($z = 6.65$, $p < .001$) that is not mediated by Agoraphobia. In other words, when the latent variable of Agoraphobia is held constant, there is a significant direct effect of Sex on the A3 indicator. Thus, at any given value of the factor, women score significantly higher on the A3 indicator than men (by .985 units, or nearly a full point on the 0–8 scale). This is evidence of *differential item functioning*; that is, the item behaves differently as an indicator of Agoraphobia in men and women. For the substantive rea-

**TABLE 7.18. Selected Mplus 7.11 Results of MIMIC Model
of Social Phobia and Agoraphobia**

MODEL RESULTS

		Estimate	S.E.	Est./S.E.	Two-Tailed P-Value
SOCIAL	BY				
S1		1.000	0.000	999.000	999.000
S2		1.079	0.045	23.967	0.000
S3		0.855	0.035	24.534	0.000
AGORAPH	BY				
A1		1.000	0.000	999.000	999.000
A2		0.956	0.066	14.388	0.000
A3		0.917	0.063	14.495	0.000
SOCIAL	ON				
SEX		-0.109	0.158	-0.690	0.491
AGORAPH	ON				
SEX		-0.475	0.160	-2.973	0.003
A3	ON				
SEX		-0.985	0.148	-6.653	0.000
SOCIAL	WITH				
AGORAPH		0.999	0.171	5.857	0.000
Residual Variances					
S1		1.072	0.126	8.533	0.000
S2		2.750	0.195	14.087	0.000
S3		1.501	0.114	13.169	0.000
A1		2.062	0.217	9.498	0.000
A2		3.777	0.264	14.301	0.000
A3		2.705	0.214	12.642	0.000
SOCIAL		4.026	0.284	14.175	0.000
AGORAPH		3.257	0.317	10.269	0.000

STANDARDIZED MODEL RESULTS

STD Standardization

		Estimate	S.E.	Est./S.E.	Two-Tailed P-Value
SOCIAL	BY				
S1		2.007	0.071	28.353	0.000
S2		2.166	0.089	24.358	0.000
S3		1.716	0.068	25.159	0.000

(continued)

TABLE 7.18. (continued)

AGORAPH BY				
A1	1.820	0.088	20.647	0.000
A2	1.739	0.099	17.589	0.000
A3	1.669	0.090	18.511	0.000
SOCIAL ON				
SEX	-0.054	0.079	-0.690	0.490
AGORAPH ON				
SEX	-0.261	0.087	-3.010	0.003
A3 ON				
SEX	-0.985	0.148	-6.653	0.000
SOCIAL WITH				
AGORAPH	0.276	0.042	6.541	0.000
Residual Variances				
S1	1.072	0.126	8.533	0.000
S2	2.750	0.195	14.087	0.000
S3	1.501	0.114	13.169	0.000
A1	2.062	0.217	9.498	0.000
A2	3.777	0.264	14.301	0.000
A3	2.705	0.214	12.642	0.000
SOCIAL	0.999	0.002	466.638	0.000
AGORAPH	0.983	0.011	86.729	0.000

sons noted earlier (sex differences in the range of activities relating to personal safety), the A3 item is biased against females. Even when their level of the underlying dimension of Agoraphobia is the same as in men, women will have higher scores on the A3 indicator (cf. Figure 7.4B) because women's responses to this item ("walking alone in isolated areas") are affected by other influences that are less relevant to men (i.e., perceptions of personal safety, in addition to the underlying construct of Agoraphobia).

Although the current example involves a specific hypothesis about the gender noninvariance of the social phobia–agoraphobia questionnaire (re: item A3), measurement invariance is frequently evaluated in an exploratory fashion. In context of a MIMIC model, this can be done by fixing all direct effects between the covariate and the indicators to zero, and then inspecting modification indices (and associated expected parameter change values) to determine whether salient direct effects may be present in the data. The researcher should not pursue the alternative of freely estimating all of these direct effects because the model will be underidentified. This can be accomplished by modifying the Mplus syntax in Table 7.17 with the following MODEL commands (the boldfaced syntax fixes the direct effects of the Sex covariate on the six indicators to zero):

```
MODEL: SOCIAL BY S1-S3;
       AGORAPH BY A1-A3;
       SOCIAL ON SEX; AGORAPH ON SEX; S1-A3 ON SEX@0; SOCIAL WITH AGORAPH;
```

This analysis produces the following modification indices in regard to covariate–indicator direct effects:

		M.I.	E.P.C.	Std E.P.C.	StdYX E.P.C.
ON Statements					
A1	ON SEX	13.302	0.559	0.559	0.120
A2	ON SEX	10.102	0.549	0.549	0.105
A3	ON SEX	42.748	-1.046	-1.046	-0.214

As in the planned analysis, this result clearly suggests the salience of the direct effect of the Sex covariate on the A3 indicator (e.g., modification index = 42.78). Because a substantive argument can be made in regard to the source of this noninvariance, this parameter can be freed in a subsequent analysis (i.e., the model depicted in the path diagram in Figure 7.5). Although the initial analysis suggests that direct effects may exist between Sex and the A1 and A2 indicators (e.g., modification indices = 13.30 and 10.10, respectively), these relationships are less interpretable and are not salient (i.e., modification indices drop below 4.0) after the direct effect of Sex $\rightarrow$ A3 is added to the model (cf. Jöreskog, 1993, and guidelines for model revision in Chapter 5).

SUMMARY

This chapter has introduced the reader to the methods of evaluating the equivalence of CFA parameters within and across groups. These procedures provide a sophisticated approach to examining the measurement invariance and population heterogeneity of CFA models. Invariance evaluation may be the primary objective of an investigation (e.g., determining the generalizability of a test instrument) or may establish the suitability of subsequent analyses (e.g., verifying longitudinal measurement invariance to justify group comparisons on structural parameters such as the equivalence of regressive paths among latent variables). The chapter has also discussed the methodology for the analysis of mean structures. Incorporating means into CFA allows for other important analyses such as the between-group equivalence of indicator intercepts (cf. differential item functioning) and between-group comparisons on latent means. Although the latter is analogous to ANOVA, CFA provides a much stronger approach because the group comparisons are conducted in context of a measurement model (e.g., adjustment for measurement error and an error theory).

In Chapter 8, three new specific applications of CFA are discussed: higher-order factor models, CFA evaluation of scale reliability, and models with formative indicators. Although the specific applications and issues covered in the second half of this book are presented in separate chapters (i.e., Chapters 6–11), it is important to underscore the fact that these procedures can be integrated into the same analysis. For instance, the multiple-groups approach presented in this chapter can be employed to examine the between-groups equivalence of a higher-order CFA, a CFA evaluation of scale reliability, or a model containing indicators that cause the latent construct (formative indicators). Missing and non-normal data are common to all types of CFA and must be dealt with appropriately (Chapter 9). These topics are covered in the next two chapters.

NOTES

1. In addition, the term *strictly parallel* has been used for indicators that in addition to possessing invariant factor loadings and error variances, have equivalent intercepts.

2. Although somewhat small samples are used in this example for ease of illustration, simulation research (e.g., Meade & Bauer, 2007) has indicated that sample sizes in multiple-groups CFA should be larger to ensure sufficient power to detect group differences in measurement parameters (e.g., differential magnitude of factor loadings). The issues of statistical power and sample size requirements are discussed in detail in Chapter 10.

3. In Figure 7.4C, the predicted value of X2 will be the same for Groups 1 and 2 at $\xi = 0$, because the intercepts are invariant. However, predicted values of X2 will be expected to differ between groups at all nonzero values of ξ.

4. Interestingly, recent simulation research (e.g., Lee, Little, & Preacher, 2011) has shown that fixing factor variances to 1.0 outperforms other methods of scaling latent variables (e.g., marker indicators) in the endeavor of identifying noninvariant indicators in a poor-fitting invariance solution (e.g., better statistical power and control for Type I error). Thus, when a noninvariant solution has been encountered in the course of using a different latent variable scaling method, T. D. Little (2013) has recommended that the researcher respecify the model by using the fixed factor variance approach to diagnose the source(s) of noninvariance.

5. The level of statistical power is the same for between-groups comparisons of parameters tested by both the MIMIC and multiple-groups approaches (e.g., group differences in factor means). The difference lies in the fact that MIMIC assumes that other measurement and structural parameters (e.g., factor loadings) are equivalent across groups, and thus the measurement model is not first tested separately in each group (where sample size may not be sufficient to ensure adequate power and precision of parameter estimates within each group).

Reproduction of the Observed Variance–Covariance Matrix with Tau-Equivalent Indicators of Auditory Memory

Running the model specified by the syntax listed in Table 7.2 produces the following model-estimated variance–covariance matrix (obtained from LISREL output):

	X1	X2	X3	X4	X5	X6
X1	6.8011					
X2	4.4302	6.8467				
X3	4.4302	4.4302	6.9595			
X4	1.4125	1.4125	1.4125	3.7636		
X5	1.4445	1.4445	1.4445	3.1681	4.1209	
X6	1.4449	1.4449	1.4449	3.1689	3.2407	4.2025

And it produces the following residual (unstandardized) matrix:

	X1	X2	X3	X4	X5	X6
X1	0.0110					
X2	0.1588	0.2289				
X3	-0.1715	-0.0004	-0.2514			
X4	-0.0454	0.1356	-0.0659	0.0000		
X5	0.1291	-0.0136	-0.2615	0.0022	0.0000	
X6	0.1068	0.1201	-0.1281	-0.0032	0.0011	0.0000

A few observations are made, using the X1 indicator as an example (sample SD = 2.61, factor loading = 2.1048, error variance = 2.3708):

1. Unlike other CFA examples discussed in previous chapters of this book, the model-estimated variances of the indicators loading on Auditory Memory (whose factor loadings are held to equality) may not be reproduced perfectly. For instance, the sample variance of the X1 indicator is 6.8121 (SD^2 = 2.61^2 = 6.8121), but the model-estimated variance of this measure is 6.8011 (resulting in a residual of 0.011). Thus these model-implied variances, as well as the model-implied covariances, will count toward any discrepancies between the observed and predicted matrices (cf. Appendix 3.3 in Chapter 3 on the calculation of F_{ML}). Note that the model perfectly reproduces the variances of the Visual Memory indicators because the factor loadings of these measures have not been constrained to equality in this solution.

2. In the present case, where the metric of the latent variable has been defined by fixing its variance to 1.0, the model estimate of the variances of the indicators whose factor loadings have been constrained to equality is the sum of the indicator's squared unstandardized factor loading and model estimate of the indicator's error variance. For instance, the model estimate of the variance of X1 = $\lambda_{11}^2 + \delta_{11}$ = 2.1048^2 + 2.3708 = 6.801 (same as X1 variance estimate in the predicted matrix presented above). The model estimate of the covariances of the indicators loading on Auditory Memory is simply the unstandardized factor loading squared (i.e., 2.1048^2 = 4.4302). Thus note that the estimated covariances of X1–X2, X1–X3, and X2–X3 are the same (4.4302) because the factor loadings of these indicators have been constrained to be the same (see also the predicted covariances of these indicators with indicators that are specified to load on Visual Memory).

3. Inspection of the selected Mplus output presented in the text of this chapter indicates that while the unstandardized loadings for X1, X2, and X3 are the same (all λs = 2.10; because this portion of the unstandardized solution has been constrained to equality), the completely standardized estimates of these loadings differ (λs range from .798 to .807). This is because the error variances of these indicators are permitted to vary (unlike in the model that tests for parallel indicators). Recall from Chapter 4 that unstandardized factor loadings can be readily converted to completely standardized coefficients by using a formula from multiple regression, $b^* = (bSD_x)/(SD_y)$, where b is the unstandardized coefficient (in this case, the unstandardized factor loading); SD_x is the SD of the predictor (in this case, the SD of the factor); and SD_y is the SD of the outcome variable (in this case, the predicted SD of the indicator). Using the X1 indicator and model estimates from the model of tau equivalence of Auditory Memory, b = 2.1048, SD_x = 1.00 (because the variance of Auditory Memory was fixed to 1.0), and SD_y = 2.6079 (the square root of the model estimated variance of X1, 6.8011). Thus the completely standardized factor loading of X1 is calculated as 2.1048 (1.0) / 2.6079 = .807 (the same as the completely standardized factor loading for X1 presented in this chapter). The completely standardized loadings for the remaining two indicators of Auditory Memory differ because the error variances (and hence model-implied variances) are allowed to differ; for example, completely standardized loading for X3 = 2.1048 (1.0) / 2.6381 = .798.

4. In models that test for parallel indicators, all completely standardized estimates for indicators loading on a given factor will be the same (i.e., factor loadings, errors, communalities). This is because the error variances (and hence model-implied variances of the indicators) are the same. Thus all values of the equation $b^* = (bSD_x)/(SD_y)$ are identical (unlike in the tau-equivalent model, where SD_y may differ for each indicator).

8

Other Types of CFA Models
Higher-Order Factor Analysis, Scale Reliability Evaluation, and Formative Indicators

---

CFA provides a unifying analytic framework for addressing a wide range of questions commonly asked by social and behavioral scientists. This notion is further evidenced in the present chapter on other types of CFA models. Three different applications of CFA are presented: higher-order factor analysis and bifactor models, scale reliability estimation, and constructs defined by formative indicators. Higher-order factor analysis is a theory-driven procedure in which the researcher imposes a more parsimonious structure to account for the interrelationships among factors established by CFA. Bifactor analysis is another type of higher-order factor analysis, where the higher-order factor has a direct effect on the observed measures. CFA can also be employed to evaluate the reliability of a testing instrument in a manner that overcomes limitations of traditional methods (e.g., Cronbach's alpha). The concluding section describes models in which it is more reasonable to believe that a set of indicators is the cause of a construct than vice versa. The presentation of these models underscores several concepts introduced in earlier chapters (e.g., model identification, equivalent models).

HIGHER-ORDER FACTOR ANALYSIS

Empirically, it may be of interest to examine the *higher-order structure* of a CFA measurement model. Hierarchical factor analysis is often used for theory testing. For example, this analytic procedure is popular in intelligence research, where it is believed that more specialized facets of ability (e.g., verbal comprehension, perceptual organization, memory) are influenced by a broader dimension of general intelligence (*g*). Another major reason why higher-order factor analysis is used in applied research is to rescue a construct. It is often the case that for a construct initially predicted to be unidimensional, the research evidence reveals that multiple factors are required to explain the covariation among the set of indicators of the construct. In these instances, a single higher-order factor can often be specified to account for the covariation among the multiple factors; that is, the construct consists of a single broader dimension, and several subdimensions (for an applied example in the clinical psychology literature, see factor analytic research on the construct of anxiety sensitivity; e.g., Taylor, 1999; Zinbarg, Barlow, & Brown, 1997).

All CFA examples presented in this book thus far have been *first-order* measurement models. These analyses have entailed specification of the number of factors, the pattern of indicator–factor relationships (i.e., factor loadings), and a measurement error theory (random or correlated indicator error variances). In multiple-factor CFA models, the factors have been specified to be intercorrelated (oblique). In other words, the factors are presumed to be interrelated, but the nature of these relationships is unanalyzed; that is, the researcher makes no substantive claims about the directions or patterns of factor interrelationships. In higher-order factor analysis (in both EFA and CFA), the focus is on the intercorrelations among the factors. In essence, these factor correlations represent the input matrix for the higher-order factor analysis. A goal of higher-order factor analysis is to provide a more parsimonious account for the correlations among lower-order factors. Higher-order factors account for the correlations among lower-order factors, and the number of higher-order factors and higher-order factor loadings is less than the number of factor correlations. Accordingly, the rules of identification used in first-order CFA apply to the higher-order component of a hierarchical solution. For instance, the number of higher-order factors that can be specified is dictated by the number of lower-order factors (discussed below). Unlike first-order CFA, higher-order CFA tests a theory-based account for the patterns of relationships among the first-order factors. These specifications assert that higher-order factors have direct effects on lower-order factors; these direct effects (and the correlations among higher-order factors) are responsible for the covariation of the lower-order factors.

Second-Order Factor Analysis

An example of a higher-order CFA model is presented in Figure 8.1. In this example, the researcher is evaluating the latent structure of a questionnaire measure of coping styles (the illustration is loosely based on the stress and coping literature—Folkman & Lazarus, 1980; Tobin, Holroyd, Reynolds, & Wigal, 1989). The questionnaire was administered to 275 college undergraduates. The latent structure of this questionnaire is predicted to be characterized by four first-order factors, which represent four distinctive ways of coping with stressful events: Problem Solving, Cognitive Restructuring, Express Emotions, and Social Support. The four factors are presumed to be intercorrelated. If no relationships were observed among the first-order factors, there would be no justification to pursue higher-order factor analysis. Two higher-order factors are predicted to account for the six correlations among the first-order factors: Problem-Focused Coping and Emotion-Focused Coping. The conceptual basis for this specification is that problem solving and cognitive restructuring are believed to represent styles of coping aimed directly at resolving the stressor (problem-focused). Conversely, expressing emotions

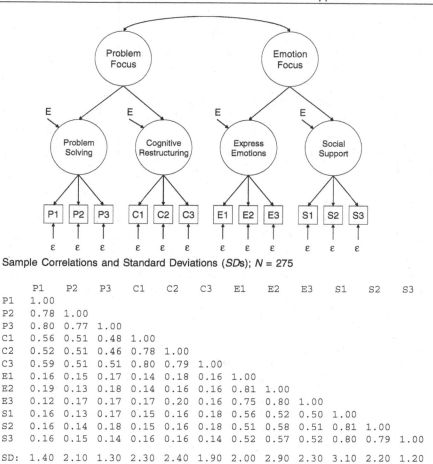

Sample Correlations and Standard Deviations (*SD*s); *N* = 275

	P1	P2	P3	C1	C2	C3	E1	E2	E3	S1	S2	S3
P1	1.00											
P2	0.78	1.00										
P3	0.80	0.77	1.00									
C1	0.56	0.51	0.48	1.00								
C2	0.52	0.51	0.46	0.78	1.00							
C3	0.59	0.51	0.51	0.80	0.79	1.00						
E1	0.16	0.15	0.17	0.14	0.18	0.16	1.00					
E2	0.19	0.13	0.18	0.14	0.16	0.16	0.81	1.00				
E3	0.12	0.17	0.17	0.17	0.20	0.16	0.75	0.80	1.00			
S1	0.16	0.13	0.17	0.15	0.16	0.18	0.56	0.52	0.50	1.00		
S2	0.16	0.14	0.18	0.15	0.16	0.18	0.51	0.58	0.51	0.81	1.00	
S3	0.16	0.15	0.14	0.16	0.16	0.14	0.52	0.57	0.52	0.80	0.79	1.00
SD:	1.40	2.10	1.30	2.30	2.40	1.90	2.00	2.90	2.30	3.10	2.20	1.20

FIGURE 8.1. Higher-order factor model of a questionnaire measure of coping styles. P1–P3, Problem Solving items; C1–C3, Cognitive Restructuring items; E1–E3, Express Emotion items; S1–S3, Social Support items.

and seeking social support are coping strategies not directed toward the stressor, but instead aimed at managing emotional reactions to the stressor (emotion-focused). The empirical feasibility of the higher-order model should be evidenced by the patterning of correlations among factors in the first-order model; for example, the latent variables of Problem Solving and Cognitive Restructuring should be more strongly correlated with Express Emotions and Social Support, and vice versa.

On a terminological note, Problem-Focused Coping and Emotion-Focused Coping can be more specifically referred to as *second-order factors* (i.e., a second level of factors that accounts for the correlations among first-order factors), as opposed to the more general term, *higher-order factors*. In context of substantive theory and statistical identification (i.e., a sufficient number of correlated, second-order factors), higher-order factor analysis can proceed to the third order and beyond, although such analyses are rare in the applied literature. It is also important to note that higher-order factors are not defined by observed measures; this aspect requires some minor alterations to typical CFA syntax programming.

The general sequence of CFA-based higher-order factor analysis is as follows: (1) Develop a well-behaved (e.g., good-fitting, conceptually valid) first-order CFA solution; (2) examine the magnitude and pattern of correlations among factors in the first-order solution; and (3) fit the second-order factor model, as justified on conceptual and empirical grounds. Thus, for the data set presented in Figure 8.1, the first step is to fit a four-factor CFA model, allowing the correlations among the factors to be freely estimated. The four-factor solution provides a good fit to the data: χ^2 (48) = 82.970, $p < .01$, SRMR = .017, RMSEA = 0.049 (90% CI = 0.03–0.07, CFit = .43), TLI = 0.99, CFI = .99 (based on LISREL 9.10 output). The completely standardized parameter estimates of this solution are presented in Table 8.1. As seen in this table, all 12 items are reasonable indicators of their respective factors (range of factor loadings = .86–.93).

The results from the phi matrix in Table 8.1 provide the correlations among the factors. Consistent with prediction, all factors are significantly interrelated (range of zs = 2.93–9.57, not shown in Table 8.1). Importantly, the pattern of correlations speaks to the viability of the posited second-order model. The model depicted in Figure 8.1 asserts that Problem Solving and Cognitive Restructuring, and Express Emotions and Social Support, are more specific dimensions of Problem-Focused Coping and Emotion-Focused Coping, respectively. If this is true in the data, the magnitude of factor correlations should show a clear pattern; that is, Problem Solving and Cognitive Restructuring will be more strongly correlated with each other than with Express Emotions and Social Support, and Express Emotions and Social Support will be more highly interrelated than either factor is with Problem Solving and Cognitive Restructuring. As shown in Table 8.1, the factor correlations follow this pattern. For example, the magnitude of the correlation between Problem Solving and Cognitive Restructuring is considerably higher (ϕ_{21} = .66) than the correlations between Problem Solving or Cognitive Restructuring and Express Emotions or Social Support (e.g., ϕ_{31} = .20).

Other patterns of factor correlations will contradict the posited higher-order model. For instance, if all factor correlations are roughly the same (e.g., ϕs ≈ .60), this will favor a single second-order factor. Although a single second-order factor will adequately account for these factor correlations, it should not be specified in the absence of conceptual justification. As with CFA in general, analysis of a higher-order solution should be fully confirmatory. This is particularly important, considering the fact that higher-order factors are specified without indicators. In addition to specifying the structure of higher-order factors, the researcher must name these factors in a manner that can be defended by theory.

Moreover, the higher-order portion of the solution must be statistically identified. The rules of identification used for a typical CFA model readily generalize to higher-order solutions (see Chapter 3, Figures 3.6 and 3.7). Like lower-order factors, higher-order factors must have a metric. Because interest is primarily in the completely standardized solution, the metric of higher-order factors is usually defined by standardizing

TABLE 8.1. Selected LISREL Results of the Four-Factor (First-Order) Measurement Model of Coping

```
Completely Standardized Solution

      LAMBDA-X

              PROBSLV     COGRES    EXPREMOT     SOCSPT
              --------   --------   --------    --------
        P1     0.9108      - -        - -         - -
        P2     0.8643      - -        - -         - -
        P3     0.8787      - -        - -         - -
        C1      - -       0.8876      - -         - -
        C2      - -       0.8739      - -         - -
        C3      - -       0.9045      - -         - -
        E1      - -        - -       0.8733       - -
        E2      - -        - -       0.9285       - -
        E3      - -        - -       0.8600       - -
        S1      - -        - -        - -        0.9008
        S2      - -        - -        - -        0.8965
        S3      - -        - -        - -        0.8863

      PHI

              PROBSLV     COGRES    EXPREMOT     SOCSPT
              --------   --------   --------    --------
   PROBSLV    1.0000
    COGRES    0.6624     1.0000
  EXPREMOT    0.2041     0.2005     1.0000
    SOCSPT    0.1959     0.2018     0.6699      1.0000

      THETA-DELTA

                 P1         P2         P3         C1         C2         C3
              --------   --------   --------   --------   --------   --------
               0.1705     0.2529     0.2279     0.2122     0.2362     0.1819

      THETA-DELTA

                 E1         E2         E3         S1         S2         S3
              --------   --------   --------   --------   --------   --------
               0.2374     0.1379     0.2604     0.1885     0.1963     0.2145
```

the entire higher-order portion of the solution. It is also possible to generate an unstandardized higher-order solution by specifying lower-order factors as "marker indicators" for the higher-order factors, although this method is less often used (both approaches are illustrated below). This approach will be favored in some situations, such as when the researcher is interested in evaluating the measurement invariance of the higher-order solution in multiple groups (e.g., Chen, Sousa, & West, 2005). Both strategies produce an identical completely standardized solution and fit to the data. In addition, the number of freely estimated parameters in the higher-order portion of the model must not exceed the total number of factor variances and covariances in the first-order solution. For this aspect of identification, it can be helpful to view the lower-order factors as "indicators" of the higher-order factors. For example, a single higher-order factor cannot be specified to account for the factor correlation from a first-order CFA model with two factors because it will be underidentified (cf. Figure 3.6A, Chapter 3), unless other (potentially unreasonable) constraints are placed on the solution (e.g., constraining the higher-order factor loadings to equality will result in a just-identified solution). If the first-order model has three factors, a solution that specifies a single higher-order factor will be just-identified (cf. Figure 3.6B, Chapter 3); that is, the higher-order solution will produce the same goodness of fit as the first-order model, in which the three factors are allowed to covary freely. Despite its just-identified nature, it may nonetheless be substantively meaningful to evaluate such a solution in order to examine the magnitude (and statistical significance) of the higher-order factor loadings and relationships of the higher-order factors to the observed measures (discussed in more detail below; e.g., Schmid–Leiman transformation).

The higher-order portion of the model presented in Figure 8.1 is overidentified by a single degree of freedom. Specifically, in the first-order model there are four factor variances and six factor covariances. In contrast, the higher-order portion of the solution contains nine freely estimated parameters: four higher-order factor loadings, four residual variances (also called *disturbances*, depicted as "E"s in Figure 8.1), and the correlation between Problem-Focused Coping and Emotion-Focused Coping. The variances of Problem-Focused Coping and Emotion-Focused Coping are not included in this tally because they will be fixed to 1.0 to define the metric of these higher-order factors. Accordingly, the higher-order solution has one degree of freedom (10 − 9; cf. Figure 3.7B, Chapter 3).

However, the researcher must be mindful of the possibility of empirical underidentification in the higher-order portion of a solution. For instance, empirical underidentification will occur in the Figure 8.1 model if the correlations of the Problem Solving and Cognitive Restructuring factors with the Express Emotions and Social Support factors are (close to) zero (cf. Figure 3.7D, Chapter 3). In this case, the correlation between the higher-order factors of Problem-Focused Coping and Emotion-Focused Coping will be zero, and there will be infinite pairs of higher-order factor loadings that will reproduce the correlations between Problem Solving and Cognitive Restructuring, and between Express Emotions and Social Support, if these parameters are freely estimated (see Chapter 3, "CFA Model Identification" section).

Table 8.2 presents computer syntax from various software packages to program the higher-order model in Figure 8.1. As in MIMIC models (Chapter 7), the factors that were developed in the initial CFA (Problem Solving, Cognitive Restructuring, etc.) are now latent Y variables (endogenous) because they are being predicted by the higher-order factors. This is reflected in several aspects of the LISREL programming: The 12 items from the coping questionnaire are treated as Y indicators (NY = 12); 4 eta (endogenous) variables are specified (NE = 4); and the lambda-Y (LY) and theta-epsilon (TE) matrices are used to program the first-order measurement model. Because these factors are now

LISREL 9.10 (higher-order factors standardized)

```
TITLE LISREL PROGRAM FOR HIGHER-ORDER FACTOR MODEL OF COPING
DA NI=12 NO=275 MA=CM
LA
P1 P2 P3 C1 C2 C3 E1 E2 E3 S1 S2 S3
KM
<Insert correlation matrix from Figure 8.1 here>
SD
1.40  2.10  1.30  2.30  2.40  1.90  2.00  2.90  2.30  3.10  2.20  1.20
MO NY=12 NK=2 NE=4 LY=FU,FR TE=DI GA=FU,FR PS=DI PH=ST
LK
PROBFOC EMOTFOC
LE
PROBSLV COGRES EXPREMOT SOCSPT
PA LY
0 0 0 0
1 0 0 0
1 0 0 0
0 0 0 0
0 1 0 0
0 1 0 0
0 0 0 0
0 0 1 0
0 0 1 0
0 0 0 0
0 0 0 1
0 0 0 1
VA 1.0 LY(1,1) LY(4,2) LY(7,3) LY(10,4)
PA GA
1 0
1 0
0 1
0 1
OU ME=ML RS MI SC AD=OFF IT=100 ND=4
```

LISREL 9.10 (higher-order factors unstandardized)

Model line

```
MO NY=12 NK=2 NE=4 LY=FU,FR TE=DI GA=FU,FR PS=DI PH=SY,FR
```

Gamma matrix

```
PA GA
0 0
1 0
0 0
0 1
VA 1.0 GA(1,1) GA(3,2)
```

Mplus 7.11 (higher-order factors unstandardized)

```
TITLE:      HIGHER ORDER CFA MODEL OF COPING
DATA:       FILE IS FIG8.1.DAT;
            TYPE IS STD CORR;
            NOBS ARE 275;
VARIABLE:   NAMES ARE P1-P3 C1-C3 E1-E3 S1-S3;
ANALYSIS:   ESTIMATOR=ML;
```

(continued)

TABLE 8.2. *(continued)*

```
MODEL:          PROBSLV BY P1-P3;
                COGRES BY C1-C3;
                EXPREMOT BY E1-E3;
                SOCSPT BY S1-S3;
                PROBFOC BY PROBSLV COGRES;
                EMOTFOC BY EXPREMOT SOCSPT;
OUTPUT:         SAMPSTAT MODINDICES(10.00) STAND RESIDUAL;
```

EQS 6.2 (higher-order factors standardized)

```
/TITLE
 HIGHER-ORDER MODEL OF COPING: Chapter 8
/SPECIFICATIONS
 CASES=275; VARIABLES=12; METHODS=ML; MATRIX=COR; ANALYSIS=COV;
/LABELS
 v1=item1; v2= item2; v3= item3; v4= item4; v5= item5; v6= item6;
 v7= item7; v8= item8; v9= item9; v10= item10; v11= item11; v12= item12;
 f1 = PROBSLV; f2 = COGRES; f3 = EXPREMOT; f4 = SOCSPT; f5 = PROBFOC;
 f6 = EMOTFOC;
/EQUATIONS
 V1  =     F1+E1;
 V2  =    *F1+E2;
 V3  =    *F1+E3;
 V4  =     F2+E4;
 V5  =    *F2+E5;
 V6  =    *F2+E6;
 V7  =     F3+E7;
 V8  =    *F3+E8;
 V9  =    *F3+E9;
 V10 =     F4+E10;
 V11 =    *F4+E11;
 V12 =    *F4+E12;
 F1  =    *F5+D1;
 F2  =    *F5+D2;
 F3  =    *F6+D3;
 F4  =    *F6+D4;
/VARIANCES
 F5 = 1;
 F6 = 1;
 D1 TO D4 = *;
 E1 TO E12 = *;
/COVARIANCES
 F5, F6 = *;
/MATRIX
<Insert correlation matrix from Figure 8.1 here>
/STANDARD DEVIATIONS
1.40  2.10  1.30  2.30  2.40  1.90  2.00  2.90  2.30  3.10  2.20  1.20
/PRINT
 fit=all;
/LMTEST
/WTEST
/END
```

SAS 9.4 PROC CALIS (higher-order factors standardized)

```
Title "Higher-Order CFA Model of Coping";
Data COPE (type=CORR);
 input _TYPE_ $ _NAME_ $ V1-V12;
```

(continued)

TABLE 8.2. *(continued)*

```
 label V1  = 'item1'
       V2  = 'item2'
       V3  = 'item3'
       V4  = 'item4'
       V5  = 'item5'
       V6  = 'item6'
       V7  = 'item7'
       V8  = 'item8'
       V9  = 'item9'
       V10 = 'item10'
       V11 = 'item11'
       V12 = 'item12';
cards;
mean  .     0     0     0     0     0     0     0     0     0     0     0     0
 std  .  1.40  2.10  1.30  2.30  2.40  1.90  2.00  2.90  2.30  3.10  2.20  1.20
   N  .   275   275   275   275   275   275   275   275   275   275   275   275
corr V1  1.00    .     .     .     .     .     .     .     .     .     .     .
corr V2  0.78  1.00    .     .     .     .     .     .     .     .     .     .
corr V3  0.80  0.77  1.00    .     .     .     .     .     .     .     .     .
corr V4  0.56  0.51  0.48  1.00    .     .     .     .     .     .     .     .
corr V5  0.52  0.51  0.46  0.78  1.00    .     .     .     .     .     .     .
corr V6  0.59  0.51  0.51  0.80  0.79  1.00    .     .     .     .     .     .
corr V7  0.16  0.15  0.17  0.14  0.18  0.16  1.00    .     .     .     .     .
corr V8  0.19  0.13  0.18  0.14  0.16  0.16  0.81  1.00    .     .     .     .
corr V9  0.12  0.17  0.17  0.17  0.20  0.16  0.75  0.80  1.00    .     .     .
corr V10 0.16  0.13  0.17  0.15  0.16  0.18  0.56  0.52  0.50  1.00    .     .
corr V11 0.16  0.14  0.18  0.15  0.16  0.18  0.51  0.58  0.51  0.81  1.00    .
corr V12 0.16  0.15  0.14  0.16  0.16  0.14  0.52  0.57  0.52  0.80  0.79  1.00
;
run;


proc calis data=COPE cov method=ml pall pcoves;
var = V1-V12;
lineqs
 V1  = 1.0   f1 + e1,
 V2  = lam2  f1 + e2,
 V3  = lam3  f1 + e3,
 V4  = 1.0   f2 + e4,
 V5  = lam5  f2 + e5,
 V6  = lam6  f2 + e6,
 V7  = 1.0   f3 + e7,
 V8  = lam8  f3 + e8,
 V9  = lam9  f3 + e9,
 V10 = 1.0   f4 + e10,
 V11 = lam11 f4 + e11,
 V12 = lam12 f4 + e12,
 f1  = ga1 f5 + D1,
 f2  = ga2 f5 + D2,
 f3  = ga3 f6 + D3,
 f4  = ga4 f6 + D4;
std
  f5 = 1.0,
  f6 = 1.0,
  d1-d4 = ps1-ps4,
  e1-e12 = te1-te12;
cov
  f5-f6 = ph1;
run;
```

endogenous variables, their residual variances (disturbances) are estimated by using the psi (PS) matrix (as opposed to estimation of factor variances, using the phi matrix, PH). On the model line, the command PS=DI accomplishes this by informing LISREL that psi is a diagonal (DI) matrix; that is, LISREL should freely estimate residual variances and fix all off-diagonal relationships (residual covariances) to zero.

In LISREL matrix programming, the two second-order factors are ksi (exogenous) variables (NK = 2) that have no observed measures. However, these factors must be provided with a metric. As noted earlier, because the tradition of factor analysis is to focus on the completely standardized solution (higher-order factor analysis was first developed in EFA), the most common approach to setting the metric of higher-order factors is to fix their variances to 1.0. In the LISREL syntax presented in Table 8.2, this is done by using the PH=ST command on the model line (i.e., to standardize the phi matrix). The gamma (GA) matrix is needed to specify the directional relationships between the second-order and first-order factors.

Table 8.2 also provides LISREL syntax for occasions when the user wishes to generate an unstandardized solution for the higher-order component of the model. In this case, the phi matrix should not be standardized (i.e., PH=SY,FR replaces PH=ST). Rather, the scale of the higher-order factors is set by passing the metric of a lower-order factor up to them. In this example, Problem Solving and Express Emotion are selected to serve as "marker indicators" for Problem-Focused Coping and Emotion-Focused Coping, respectively. The same logic for specifying marker indicators in first-order CFA is used for higher-order factors. However, instead of the lambda-X or lambda-Y matrices, the gamma matrix is used for this purpose. Table 8.2 shows that in the pattern matrix for gamma, the paths of Problem-Focus → Problem Solving and Emotion-Focused Coping → Express Emotions are fixed to zero; this is then overridden by the value (VA) statement, which fixes these unstandardized paths to 1.0.

If the reader has read the preceding chapters of this book, the programming logic for the Mplus, EQS, and SAS/CALIS syntax should be self-explanatory. As with LISREL, note that the EQS and SAS/CALIS programming differentiate disturbance variance (D1–D4 in all examples) and measurement error variance. In the EQS and CALIS examples, the higher-order solution is standardized (e.g., F5 = 1.0, F6 = 1.0). The Mplus example will also produce an unstandardized solution, although this syntax could be readily modified to generate a completely standardized solution only (i.e., to fix the variances of the higher-order factors to 1.0). As noted earlier, regardless of whether an unstandardized solution is generated, the specifications will produce identical results in terms of overall fit, modification indices, completely standardized parameter estimates, and so forth.

The fit of the higher-order solution is as follows: $\chi^2(49) = 83.004$, $p < .01$, SRMR = .017, RMSEA = 0.050 (90% CI = 0.03–0.07, CFit = .47), TLI = 0.99, CFI = .99 (LISREL 9.10 output). A higher-order solution cannot improve goodness of fit relative to the first-order solution, where the factors are freely intercorrelated. If the higher-order aspect of the model is overidentified, it is attempting to reproduce these factor correlations with a smaller number of freely estimated parameters. When the higher-order model is overidentified, the nested χ^2 test can be used to determine whether the specification produces a significant degradation in fit relative to the first-order solution. In this example, the higher-order solution is found to be equally good-fitting, $\chi^2_{diff}(1) = 0.034$, ns (i.e., $83.004 - 82.970$, $df = 49 - 48$). The gain of one degree of freedom in this particular higher-order model is because the model is attempting to account for the six correlations among the lower-order factors with one less freely estimated parameter (i.e., 4 higher-order factor loadings, 1 correlation between higher-order factors = 5).

In addition to goodness of fit, the acceptability of the higher-order model must be evaluated with regard to the magnitude of the higher-order parameters (i.e., size of higher-order factor loadings and higher-order factor correlations). The completely standardized estimates from this solution are presented in Table 8.3. Each of the first-order factors loads strongly onto the second-order factors (range of loadings = .81–.83). The estimates provided in the psi matrix indicate the proportion of variance in the lower-order factors that is not explained by the second-order factors (i.e., completely standardized disturbances). With these estimates, it can be seen that the higher-order factors account for 66–68% of the variance in the first-order factors; for example, Express Emotion, $1 - .3185 = .6815$. The correlation between the higher-order factors is estimated to be .301 (seen in "Correlation Matrix of Eta and Ksi" section of output).

Because the higher-order solution does not result in a significant decrease in model fit, it can be concluded that the model provides a good account for the correlations among the first-order factors. This can be demonstrated by using the tracing rules presented in earlier chapters (e.g., Chapter 3). For example, in the first-order CFA model, the correlation between Problem Solving and Cognitive Restructuring is .66 (see Table 8.1). Multiplying the higher-order factor loadings of Problem-Focused Coping → Problem Solving and Problem-Focused Coping → Cognitive Restructuring perfectly reproduces this correlation; that is, $.812(.816) = .66$. Similarly, in the initial CFA solution, the correlation between Problem Solving and Express Emotions is estimated to be .196. This relationship is accounted for multiplying the following three parameters: Problem-Focused Coping → Problem Solving (.812), Emotion-Focused Coping → Express Emotion (.823), correlation between Problem-Focused Coping and Emotion-Focused Coping (.301); that is, $.812(.823)(.301) = .20$.

Schmid–Leiman Transformation

The researcher can use the estimates provided by the completely standardized solution to estimate the relationship of the observed measures to the higher-order factors. This is calculated by multiplying the appropriate higher-order and first-order factor loadings. For instance, the completely standardized effect of Problem-Focused Coping on indicator P1 is .74; that is, $.812(.911)$. This can be interpreted as the indirect effect of the higher-order factor on the indicator; for example, the effect of Problem-Focused Coping on P1 is mediated by Problem Solving. However, because the higher-order factors have no direct effects on the indicators, these values can also be regarded as total effects; that is, the completely standardized relationship between Problem-Focused Coping and P1 is

TABLE 8.3. Selected Output (LISREL) of Higher-Order CFA Model of Coping

Completely Standardized Solution

LAMBDA-Y

	PROBSLV	COGRES	EXPREMOT	SOCSPT
P1	0.9108	- -	- -	- -
P2	0.8644	- -	- -	- -
P3	0.8787	- -	- -	- -
C1	- -	0.8876	- -	- -
C2	- -	0.8740	- -	- -
C3	- -	0.9044	- -	- -
E1	- -	- -	0.8733	- -
E2	- -	- -	0.9284	- -
E3	- -	- -	0.8601	- -
S1	- -	- -	- -	0.9008
S2	- -	- -	- -	0.8965
S3	- -	- -	- -	0.8863

GAMMA

	PROBFOC	EMOTFOC
PROBSLV	0.8118	- -
COGRES	0.8159	- -
EXPREMOT	- -	0.8255
SOCSPT	- -	0.8116

Correlation Matrix of ETA and KSI

	PROBSLV	COGRES	EXPREMOT	SOCSPT	PROBFOC	EMOTFOC
PROBSLV	1.0000					
COGRES	0.6624	1.0000				
EXPREMOT	0.2018	0.2028	1.0000			
SOCSPT	0.1984	0.1994	0.6699	1.0000		
PROBFOC	0.8118	0.8159	0.2485	0.2443	1.0000	
EMOTFOC	0.2444	0.2457	0.8255	0.8116	0.3011	1.0000

PSI

PROBSLV	COGRES	EXPREMOT	SOCSPT
0.3410	0.3343	0.3185	0.3414

THETA-EPS

P1	P2	P3	C1	C2	C3
0.1705	0.2529	0.2279	0.2122	0.2361	0.1820

THETA-EPS

E1	E2	E3	S1	S2	S3
0.2374	0.1380	0.2602	0.1886	0.1963	0.2145

.74. Squaring this result provides the proportion of the variance in P1 explained by the second-order factor, Problem-Focused Coping (i.e., $.74^2 = .55$).

With a few additional simple calculations, the hierarchical solution can be transformed to further elucidate the relationships of the first- and second-order factors with the observed measures. This procedure was introduced by Schmid and Leiman (1957). Although it was developed for use within the EFA framework, the procedure readily generalizes to CFA. The transformation treats first-order factors as residualized factors; that is, second-order factors explain as much variance as possible, and the variance in the indicators that cannot be accounted for by second-order factors is ascribed to the first-order factors. As Loehlin (2004) notes, whereas the explanatory power of first-order factors is lessened by the transformation, the substantive interpretation of first-order factors may be fostered. This is because the residualized factor loadings of the first-order factors represent the unique contribution of the first-order factors to the prediction of the indicators. In addition, as a result of the transformation, the first-order factors are not correlated with second-order factors.

To illustrate this procedure, a Schmid–Leiman transformation has been conducted on the higher-order solution of coping. The results of this analysis are presented in Table 8.4 (a spreadsheet prepared in Microsoft Excel). The two columns following the item names contain the first-order factor loadings and second-order factor loadings obtained from the higher-order CFA (see Table 8.3). In the fourth column, each item's loading on the second-order factor is calculated by the product of its loading on the first-order factor (column 2) and the first-order factor's loading on the second-order factor (column 3); for example, the loading of indicator P1 on Problem-Focused Coping = .9108(.8118) = .7394 (the same calculation is presented in the first paragraph of this section). The residualized first-order loading (column 6) is computed by multiplying the first-order loading (column 2) by the square root of the uniqueness of the higher-order factor. Recall that in a completely standardized solution, a uniqueness represents the proportion of variance in an outcome (in this case, a first-order factor) that is not explained by a factor (in this case, the second-order factor). The terms *disturbance* and *residual variance* are synonymous in this context (cf. "E," Figure 8.1). For example, the uniqueness of Problem-Focused Coping is estimated to be $1 - .8118^2 = .341$ (cf. psi matrix, Table 8.3). The square roots of the uniquenesses are presented in the fifth column of Table 8.4. Thus the residualized first-order loading of P1 is computed: .9108(.5839) = .532.

The second row of calculations indicates how much variance in each indicator is explained by the second-order factors ("Higher-Order Rsq" column) and first-order factors ("Residualized Loading Rsq" column). These results are obtained by squaring the indicator's transformed loadings on the first- and second-order factors. For example, for indicator P1, 55% of its variance is explained by the second-order factor, Problem-Focused Coping ($.739^2 = .55$), and 28% of its variance is explained by the first-order factor, Problem Solving ($.531^2 = .28$). From this transformation, it can be concluded that whereas slightly over half (55%) of the variance in the P1 indicator is accounted for by a broader trait reflecting the tendency to engage in coping efforts aimed directly at resolving the stressor (Problem-Focused Coping), slightly over a quarter of its variance

TABLE 8.4. Spreadsheet of Schmid–Leiman Transformation of the Higher-Order Model of Coping

Item	Factor Loading	Higher Order Loading	Item: High-Order Factor	SQRT(Uniq)	Residualized Primary Loading
P1	0.9108	0.8118	0.73938744	0.583935579	0.531848525
P2	0.8644	0.8118	0.70171992	0.583935579	0.504753914
P3	0.8787	0.8118	0.71332866	0.583935579	0.513104193
C1	0.8876	0.8159	0.72419284	0.578193039	0.513204141
C2	0.874	0.8159	0.7130966	0.578193039	0.505340716
C3	0.9044	0.8159	0.73789996	0.578193039	0.522917784
E1	0.8733	0.8255	0.72090915	0.564402117	0.492892369
E2	0.9284	0.8255	0.7663942	0.564402117	0.523990926
E3	0.8601	0.8255	0.71001255	0.564402117	0.485442261
S1	0.9008	0.8116	0.73108928	0.584213523	0.526259541
S2	0.8965	0.8116	0.7275994	0.584213523	0.523747423
S3	0.8863	0.8116	0.71932108	0.584213523	0.517788445

Item	Loading Rsq	High-Order Rsq	Residual Loading Rsq	Sum
P1	0.82955664	0.546693786	0.282862854	0.82955664
P2	0.74718736	0.492410846	0.254776514	0.74718736
P3	0.77211369	0.508837777	0.263275913	0.77211369
C1	0.78783376	0.52445527	0.26337849	0.78783376
C2	0.763876	0.508506761	0.255369239	0.763876
C3	0.81793936	0.544496351	0.273443009	0.81793936
E1	0.76265289	0.519710003	0.242942887	0.76265289
E2	0.86192656	0.58736007	0.27456649	0.86192656
E3	0.73977201	0.504117821	0.235654189	0.73977201
S1	0.81144064	0.534491535	0.276949105	0.81144064
S2	0.80371225	0.529400887	0.274311363	0.80371225
S3	0.78552769	0.517422816	0.268104874	0.78552769

(28%) is explained by the influence of a more specific problem-focused coping strategy (Problem Solving). The remaining 17% of P1's variance (i.e., $1 - .55 - .28 = .17$) is not accounted for by either the first- or second-order factors (i.e., unique variance or measurement error; cf. theta-epsilon matrix in Table 8.3).

It is important to note that this transformation does not alter the explanatory power of the original CFA solution; for example, the proportion of unexplained variance in P1 is .17 before and after the transformation. This is also demonstrated in Table 8.4, where it is shown that the sum of variance explained by the transformed first- and second-order factor loadings (e.g., P1: $.547 + .283 = .83$; see "Sum" column) is equal to the communalities of the first-order factor loadings (communality = squared factor loading) obtained in the initial hierarchical solution (e.g., P1 communality = $.9108^2 = .83$; see "Loading Rsq" column). Rather, the Schmid–Leiman procedure is simply a method of calculating the contribution of lower- and higher-order factors to the prediction of observed measures (with explanatory preference afforded to the higher-order factors). Applied examples of this procedure can be found in Campbell-Sills et al. (2004) and Brown et al. (2004).

Bifactor Models

Another, less commonly used approach to higher-order factor analysis is the bifactor model (Harman, 1976; Holzinger & Swineford, 1937). A path diagram for a bifactor model specification is provided in Figure 8.2. In a bifactor model, there exists a general factor that accounts for significance covariance in all the observed measures. In addition, there is more than one domain-specific factor that accounts for unique variance in the indicators of a specific domain over and beyond the general factor. In the model illustrated in Figure 8.2, there is a general Verbal Intelligence factor that underlies each of the specific verbal intelligence tests. Thus, unlike second-order models, the bifactor model specifies direct effects of the higher-order dimension on the indicators.[1] In addition, there are two domain-specific factors (Verbal Comprehension, Working Memory) that each account for unique variance within the specific subdomain. Consistent with the typical parameterization of the bifactor model, the general and domain-specific factors are specified to be uncorrelated; that is, the contribution of the domain-specific factors to explaining variability in the indicators is independent of the variance accounted for by the general factor. Generally speaking, bifactor models are most appropriate for constructs or measures that are posited to be principally unidimensional, but there also exist smaller latent dimensions (subdomains) that have substantive value and must be specified to achieve a good-fitting solution (i.e., specification of a general factor alone would not fit the data well).[2]

Chen, West, and Sousa (2006) have outlined several potential advantages of the bifactor model. Because the second-order model is nested within the bifactor model (Yung, Thissen, & McLeod, 1999), the bifactor model can be used as a baseline model to which the second-order model can be compared (via χ^2 difference evaluation). Moreover, the bifactor model can be used to evaluate the importance of domain-specific factors. For example, it is possible that a domain-specific factor will not be relevant to the pre-

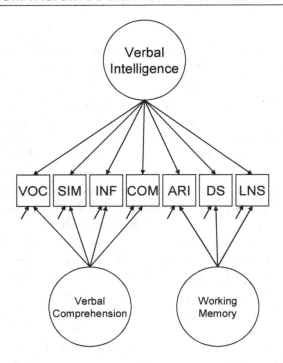

FIGURE 8.2. Example of a bifactor hierarchical model of verbal intelligence. VOC, Vocabulary; SIM, Similarities; INF, Information; COM, Comprehension; ARI, Arithmetic; DS, Digit Span; LNS, Letter–Number Sequencing.

diction of the observed measures when the general factor is included in the model; that is, once the general factor is partialed out, the domain-specific factor does not account for unique variance in the indicators. If this is the case, estimation problems may be encountered because either the factor loadings of the irrelevant domain-specific factor will be small (e.g., close to zero) or the variance of the domain-specific factor will not significantly differ from zero. In addition, the bifactor model can be used in instances where the researcher is interested in examining whether the domain-specific factors predict external variables (i.e., outcome variables not part of the measurement model) when holding the general factor constant. This cannot be done in the typical second-order model because the domain-specific factors are represented by residual variances (disturbances) of the first-order factors. That is, because the second-order factors have direct effects on the first-order factors, the "variance" of the first-order factors does not reflect the total variance of these dimensions, but rather variability that is unexplained by the second-order factors. For similar reasons, measurement invariance evaluation cannot be conducted on the first-order factors in second-order models, whereas all aspects of the bifactor model can be tested for equivalence across multiple groups (including structural parameters such as differences in the latent means).

The bifactor model is now illustrated with an actual data set in which 700 outpatients were administered a 15-item questionnaire that measures a person's judgments about his or her ability to control emotional states and stressful or threatening situations (the example is loosely based on the measure reported in Brown et al., 2004). In prior research, CFA has shown that the questionnaire is composed of three correlated factors: Stress Control (perceptions of one's ability to cope with everyday stressors; defined by items 1–4), Threat Control (beliefs about whether threatening events are outside one's control; items 5–10), and Emotional Control (self-perceptions of one's ability to control adverse emotional reactions to negative events; items 11–15). Moreover, a method effect (correlated error) has been reliably shown to exist between items 8 and 9. In the current analysis, it is posited that in addition to these three subdomains, there exists a broader dimension of Perceived Control on which all 15 items load.

Before the bifactor model is run, a first-order, three-factor model is fitted to the data to ensure an acceptable solution in the current data set. This model fits the data well: $\chi^2(86) = 200.237$, $p < .001$, SRMR = .039, RMSEA = 0.044 (90% CI = 0.04–0.05, CFit = .91), TLI = 0.94, CFI = .95 (Mplus 7.11 results). All factor loadings are large and statistically significant (range of completely standardized loadings = .43–.71, $ps < .001$); the three factors are moderately intercorrelated (range of rs = .50–.63).

Next, a bifactor model is fitted to the data, in which all 15 items load onto a general factor in addition to the three more specific dimensions of perceived control (i.e., Stress Control, Threat Control, Emotion Control). The covariances of all factors are fixed to zero. Mplus syntax for this model and selected results are provided in Table 8.5. This model fits the data well: $\chi^2(74) = 146.03$, $p < .001$, SRMR = .031, RMSEA = 0.037 (90% CI = 0.03–0.05, CFit = .99), TLI = 0.96, CFI = .97. However, although the parameter estimates for the general factor and for two of the three subdomains are reasonable, the factor loadings and factor variance for the Stress Control factor are small and statistically nonsignificant (see Table 8.5). This exemplifies an observation made by Chen et al. (2006) that bifactor analysis will occasionally indicate the psychometric irrelevance of a domain-specific factor when the general factor is included in the model; that is, after adjustment for the general factor of Perceived Control, the subdomain of Stress Control does not account for additional, unique variance in the indicators. Thus the bifactor model is respecified so that the Stress Control factor is removed and its items (X1–X4) are specified to load only on the general factor; specification is unchanged for the Threat Control and Emotion Control factors. The Mplus, LISREL, and EQS syntax for this model is presented in Table 8.6. The respecified bifactor model fits the data well: $\chi^2(78) = 167.63$, $p < .001$, SRMR = .034, RMSEA = 0.041 (90% CI = 0.03–0.05, CFit = .97), TLI = 0.95, CFI = .96; all parameter estimates are statistically significant ($ps < .001$). The completely standardized solution of this model is presented in Figure 8.3 (path diagram created by using the Mplus diagrammer introduced in Version 7.0). Collectively, these findings provide initial support for the existence of a broader dimension of Perceived Control, in addition to two subdomains of this construct (Emotion Control, Threat Control). In an applied research setting, further work would be needed to examine the validity of these dimensions (e.g., do the subdomains contribute in the expected manner to the prediction of salient clinical outcomes in addition to the general factor?) and the measurement properties (e.g., replication and invariance across samples and subgroups).

TABLE 8.5. Mplus 7.11 Syntax and Selected Output for Bifactor Model of Perceived Control (One General Factor, Three Domain-Specific Factors)

```
TITLE:   PERCEIVED CONTROL QN -- ONE GENERAL FACTOR, THREE SUBDOMAIN FACTORS
DATA:    FILE IS TAB8.5.DAT;
VARIABLE:
   NAMES ARE SUBJ X1-X15;
   USEV ARE X1-X15;
ANALYSIS: ESTIMATOR IS ML;
MODEL:
   GFACT BY X1-X15;
   STRESS BY X1-X4;
   THREAT BY X5-X10;
   EMOTION BY X11-X15;
   X8 WITH X9;
   GFACT WITH STRESS@0; GFACT WITH THREAT@0; GFACT WITH EMOTION@0;
   STRESS WITH THREAT@0; EMOTION WITH THREAT@0; STRESS WITH EMOTION@0;
   OUTPUT:  STANDARDIZED MODINDICES(10.00) SAMPSTAT FSDETERMINACY;
```

MODEL RESULTS

		Estimate	S.E.	Est./S.E.	Two-Tailed P-Value
GFACT	BY				
X1		1.000	0.000	999.000	999.000
X2		0.607	0.079	7.725	0.000
X3		0.892	0.078	11.439	0.000
X4		0.932	0.071	13.195	0.000
X5		0.480	0.068	7.036	0.000
X6		0.717	0.080	8.928	0.000
X7		0.671	0.080	8.434	0.000
X8		0.493	0.061	8.154	0.000
X9		0.558	0.069	8.115	0.000
X10		0.363	0.072	5.052	0.000
X11		0.448	0.068	6.633	0.000
X12		0.542	0.071	7.583	0.000
X13		0.597	0.072	8.261	0.000
X14		0.572	0.070	8.214	0.000
X15		0.518	0.064	8.093	0.000
STRESS	BY				
X1		1.000	0.000	999.000	999.000
X2		3.325	2.228	1.493	0.136
X3		-0.216	0.819	-0.264	0.792
X4		2.424	1.507	1.609	0.108
THREAT	BY				
X5		1.000	0.000	999.000	999.000
X6		0.867	0.111	7.800	0.000
X7		0.775	0.115	6.770	0.000
X8		0.608	0.086	7.062	0.000
X9		0.530	0.091	5.822	0.000
X10		0.690	0.105	6.547	0.000

(continued)

TABLE 8.5. *(continued)*

EMOTION BY				
X11	1.000	0.000	999.000	999.000
X12	0.619	0.097	6.376	0.000
X13	0.608	0.097	6.240	0.000
X14	0.637	0.090	7.089	0.000
X15	0.442	0.077	5.777	0.000
GFACT WITH				
STRESS	0.000	0.000	999.000	999.000
THREAT	0.000	0.000	999.000	999.000
EMOTION	0.000	0.000	999.000	999.000
STRESS WITH				
THREAT	0.000	0.000	999.000	999.000
EMOTION	0.000	0.000	999.000	999.000
EMOTION WITH				
THREAT	0.000	0.000	999.000	999.000
X8 WITH				
X9	0.404	0.066	6.147	0.000
Variances				
GFACT	1.130	0.145	7.791	0.000
STRESS	0.049	0.060	0.812	0.417
THREAT	0.748	0.125	5.979	0.000
EMOTION	0.854	0.138	6.174	0.000

Bifactor models have become increasingly common in the applied research literature in recent years. The reader is referred to the following papers for recent applications of these models: Brouwer, Meijer, Weekers, and Baneke (2008); Lahey et al. (2012); Meyer and Brown (2013); Osman et al. (2010); and Patrick, Hicks, Nichol, and Krueger (2007).

SCALE RELIABILITY ESTIMATION

Point Estimation of Scale Reliability

Multiple-item measures (e.g., questionnaires) are commonly used in the social and behavioral sciences. A key aspect of the psychometric development of such measures is the evaluation of *reliability*. Reliability refers to the precision or consistency of measurement (i.e., the overall proportion of true-score variance to total observed variance of the measure). Early in the process of scale development, researchers will often submit the measure's items to factor analysis (EFA, CFA). The resulting latent structure is then used to guide the scoring of the measure for further psychometric analyses (e.g., convergent and discriminant validity evaluation) and, ultimately, applied use; for instance, if factor

TABLE 8.6. Syntax for Respecified Bifactor Model of Perceived Control (One General Factor, Two Domain-Specific Factors)

Mplus 7.11

```
TITLE:  PERCEIVED CONTROL QN -- ONE GENERAL FACTOR, TWO SUBDOMAIN FACTORS
DATA:    FILE IS TAB8.5.DAT;
VARIABLE:
    NAMES ARE SUBJ X1-X15;
    USEV ARE X1-X15;
ANALYSIS: ESTIMATOR IS ML;
MODEL:
    GFACT BY X1-X15;
    ! STRESS BY X1-X4;  ! STRESS FACTOR HAS BEEN COMMENTED OUT OF SYNTAX
    THREAT BY X5-X10;
    EMOTION BY X11-X15;
    X8 WITH X9;
    GFACT WITH THREAT@0; GFACT WITH EMOTION@0; ! GFACT WITH STRESS@0;
    EMOTION WITH THREAT@0; ! STRESS WITH THREAT@0; STRESS WITH EMOTION@0;
OUTPUT:   STANDARDIZED MODINDICES(10.00) SAMPSTAT FSDETERMINACY;
```

LISREL 9.10

```
TITLE BIFACTOR MODEL OF PERCEIVED CONTROL QN
DA NI=16 NO=700 MA=CM
LA
SUBJ X1 X2 X3 X4 X5 X6 X7 X8 X9 X10 X11 X12 X13 X14 X15
RA FI = TAB8.5.DAT
SE
X1 X2 X3 X4 X5 X6 X7 X8 X9 X10 X11 X12 X13 X14 X15 /
MO NY=15 NE=3 LY=FU,FR TE=SY,FR PS=SY,FR
LE
GENF THREAT EMOTION
PA LY
0 0 0
1 0 0
1 0 0
1 0 0
1 0 0
1 1 0
1 1 0
1 1 0
1 1 0
1 1 0
1 0 0
1 0 1
1 0 1
1 0 1
1 0 1
VA 1.0 LY(1,1) LY(5,2) LY(11,3)
PA TE
1
0 1
0 0 1
0 0 0 1
0 0 0 0 1
0 0 0 0 0 1
```

(continued)

TABLE 8.6. *(continued)*

```
0 0 0 0 0 0 1
0 0 0 0 0 0 0 1
0 0 0 0 0 0 0 1 1
0 0 0 0 0 0 0 0 0 1
0 0 0 0 0 0 0 0 0 0 1
0 0 0 0 0 0 0 0 0 0 0 1
0 0 0 0 0 0 0 0 0 0 0 0 1
0 0 0 0 0 0 0 0 0 0 0 0 0 1
0 0 0 0 0 0 0 0 0 0 0 0 0 0 01
PA PS
1
0 1
0 0 1
OU ME=ML SC AD=OFF ND=4
```

EQS 6.2

```
/TITLE
 bifactor model of perceived control
/SPECIFICATIONS
 CASES=700; VARIABLES=16; METHODS=ML; MATRIX=RAW; ANALYSIS=COV;
 DATA = 'TAB8.5.DAT';
/LABELS
 v1=subj; v2=x1; v3=x2; v4=x3; v5=x4; v6=x5; v7=x6; v8=x7; v9=x8;
 v10=x9; v11=x10; v12=x11; v13=x12; v14=x13; v15=x14; v16=x15;
 f1 = genf; f2 = threat; f3 = emotion;
/EQUATIONS
 V2  =    F1+E1;
 V3  =   *F1+E2;
 V4  =   *F1+E3;
 V5  =   *F1+E4;
 V6  =   *F1+F2+E5;
 V7  =   *F1+*F2+E6;
 V8  =   *F1+*F2+E7;
 V9  =   *F1+*F2+E8;
 V10 =   *F1+*F2+E9;
 V11 =   *F1+*F2+E10;
 V12 =   *F1+F3+E11;
 V13 =   *F1+*F3+E12;
 V14 =   *F1+*F3+E13;
 V15 =   *F1+*F3+E14;
 V16 =   *F1+*F3+E15;
/VARIANCES
 F1 TO F3 = *;
 E1 TO E15 = *;
/COVARIANCES
 F1 TO F3 = 0;
 E8 TO E9 = *;
/PRINT
 fit=all;
/LMTEST
/WTEST
/END
```

307

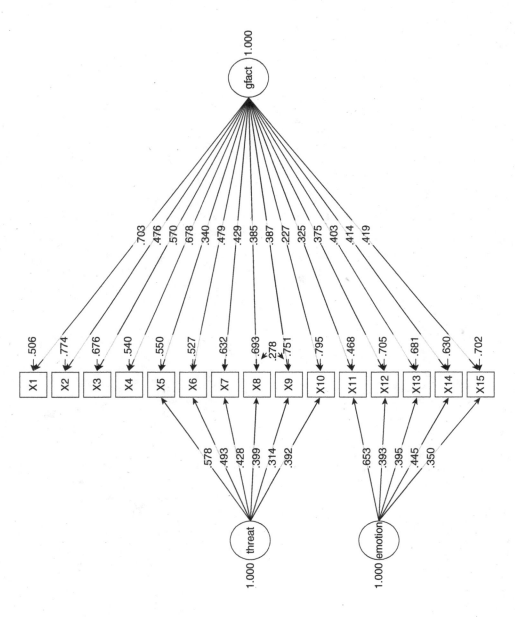

FIGURE 8.3. Final bifactor model of perceived control (completely standardized estimates). gfact, general factor of Perceived Control.

analyses reveal two factors, the measure is deemed to have two subscales. Most often, such scoring entails unrefined composites, where items found to load on a given factor are simply summed or averaged (cf. Grice, 2001). The reliability of these composite scores is usually estimated by Cronbach's (1951) coefficient alpha (α). Despite its widespread popularity, researchers have long known that α is a misestimator of scale reliability. If the scale contains no correlated measurement errors, α will underestimate scale reliability unless the condition of tau equivalence holds (Lord & Novick, 1968; Zimmerman, 1972; see Chapter 7 for definition and evaluation of tau equivalence). However, the condition of tau equivalence is frequently not realized in actual data sets, in part because the units of measurement are often arbitrary (Raykov, 2001a, 2004). Furthermore, if the measure contains correlated measurement errors, α can either underestimate or overestimate scale reliability, depending on the underlying measurement parameters (Raykov, 2001b; Zimmerman, 1972). Thus this and other research (e.g., Green & Hershberger, 2000; Komaroff, 1997; Raykov, 1997) has shown that Cronbach's coefficient α does not provide a dependable estimate of scale reliability of multiple-item measures.

Raykov (2001a, 2004) has developed a CFA-based method of estimating scale reliability that reconciles the problems with Cronbach's coefficient α. As shown in the next example, an appealing feature of this approach is that it provides an estimate of scale reliability directly in context of the CFA measurement model. CFA estimates of factor loadings, error variances, and error covariances are used to calculate the scale's true-score and error variance. As noted earlier, scale reliability represents the proportion of true-score variance to total observed variance in the measure. This is expressed by the following equation (Lord & Novick, 1968):

$$\rho_Y = Var(T) / Var(Y) \tag{8.1}$$

where ρ_Y is the scale reliability coefficient; $Var(T)$ is the true score variance of the measure; and $Var(Y)$ is the total variance of the measure. $Var(Y)$ is the sum of the true-score variance and error variance of the measure. In the case of a congeneric measurement model (Jöreskog, 1971a) without correlated measurement errors, this equation can be reexpressed by using CFA measurement parameters as follows:

$$\rho = (\Sigma\lambda_i)^2 / [(\Sigma\lambda_i)^2 + \Sigma\theta_{ii}] \tag{8.2}$$

where $(\Sigma\lambda_i)^2$ is the squared sum of unstandardized factor loadings, and $\Sigma\theta_{ii}$ is the sum of unstandardized measurement error variances. In instances where the measurement model contains correlated measurements errors, scale reliability would be calculated as follows:

$$\rho = (\Sigma\lambda_i)^2 / [(\Sigma\lambda_i)^2 + \Sigma\theta_{ii} + 2\Sigma\theta_{ij}] \tag{8.3}$$

where $2\Sigma\theta_{ij}$ is the sum of nonzero error covariances multiplied by 2. In latent software programs that accommodate nonlinear constraints (e.g., LISREL, Mplus), Raykov's

(2001a) procedure entails specifying three dummy latent variables whose variances are constrained to equal the numerator (true-score variance), denominator (total variance), and corresponding ratio of true-score variance to total score variance (ρ), as per the classic formula for scale reliability estimation (Lord & Novick, 1968). Beginning in Version 6.0 of EQS, estimates of scale reliability (and Cronbach's α) can be requested by using the /RELIABILITY command; for example,

> /RELIABILITY
> SCALE = V1,V2,V3,V4;

In Version 6.2 of EQS, these estimates are provided by default.

This procedure is now illustrated with the measurement model and data provided in Figure 8.4. In this example, the researcher has developed a multiple-item questionnaire designed to assess subjective distress following exposure to traumatic stressors such as natural disasters, sexual assault, and motor vehicle accidents (the example is based loosely on a measure developed by Horowitz, Wilner, & Alvarez, 1979). The questionnaire is purported to assess two related dimensions of psychological reactions to extreme stress: (1) Intrusions, or intrusive thoughts, images, dreams, or flashbacks of

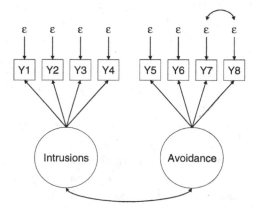

Sample Correlations and Standard Deviations (*SD*s); *N* = 500

	Y1	Y2	Y3	Y4	Y5	Y6	Y7	Y8
Y1	1.000							
Y2	0.594	1.000						
Y3	0.607	0.613	1.000					
Y4	0.736	0.765	0.717	1.000				
Y5	0.378	0.321	0.360	0.414	1.000			
Y6	0.314	0.301	0.345	0.363	0.732	1.000		
Y7	0.310	0.262	0.323	0.337	0.665	0.583	1.000	
Y8	0.317	0.235	0.276	0.302	0.632	0.557	0.796	1.000
SD:	1.150	1.200	1.570	2.820	1.310	1.240	1.330	1.290

FIGURE 8.4. Latent structure of a questionnaire measure of distress following exposure to traumatic stress.

the traumatic event; and (2) Avoidance, or overt and covert avoidance of the traumatic event (e.g., avoidance of situations, thoughts, feelings, reminders of the trauma). A correlated error is predicted to exist between items Y7 and Y8 due to a method effect (i.e., reverse wording). The measure was administered to 500 participants.

A CFA reveals that the two-factor model provides a good fit to the data: $\chi^2(18)$ = 20.52, p = .304, SRMR = .021, RMSEA = 0.017 (90% CI = 0.00–0.04, CFit = .98), TLI = 1.00, CFI = 1.00 (LISREL 9.10 output). The unstandardized parameter estimates of the factor loadings, error variances, and error covariance from this solution are provided in Table 8.7 (standard errors and test ratios have been deleted). These estimates are used shortly to illustrate the calculation of point estimates of scale reliability.

Table 8.8 presents the LISREL and Mplus syntax that can be used to estimate the scale reliability of the dimensions of Intrusions and Avoidance. In LISREL, this matrix

TABLE 8.7. Selected Unstandardized Parameter Estimates of Measurement Model of the Reactions to Traumatic Stress Questionnaire

```
LISREL Estimates (Maximum Likelihood)

        LAMBDA-Y

              INTRUS      AVOID
            --------    --------
     Y1       0.8912      - -
     Y2       0.9595      - -
     Y3       1.1968      - -
     Y4       2.6757      - -
     Y5        - -       1.1942
     Y6        - -       0.9954
     Y7        - -       0.9701
     Y8        - -       0.8938


        THETA-EPS

                 Y1          Y2          Y3          Y4          Y5          Y6
            --------    --------    --------    --------    --------    --------
     Y1       0.5284
     Y2        - -       0.5194
     Y3        - -        - -       1.0325
     Y4        - -        - -        - -       0.7931
     Y5        - -        - -        - -        - -       0.2901
     Y6        - -        - -        - -        - -        - -       0.5468
     Y7        - -        - -        - -        - -        - -        - -
     Y8        - -        - -        - -        - -        - -        - -

        THETA-EPS

                 Y7          Y8
            --------    --------
     Y7       0.8277
     Y8       0.4986      0.8652
```

TABLE 8.8. LISREL and Mplus Syntax for Estimating Scale Reliability of the Reactions to Traumatic Stress Questionnaire

LISREL 9.10

```
TITLE CFA SCALE RELIABILITY
DA NI=8 NO=500 MA=CM
LA
Y1 Y2 Y3 Y4 Y5 Y6 Y7 Y8
KM
1.000
0.594  1.000
0.607  0.613  1.000
0.736  0.765  0.717  1.000
0.378  0.321  0.360  0.414  1.000
0.314  0.301  0.345  0.363  0.732  1.000
0.310  0.262  0.323  0.337  0.665  0.583  1.000
0.317  0.235  0.276  0.302  0.632  0.557  0.796  1.000
SD
1.15 1.20 1.57 2.82 1.31 1.24 1.33 1.29
MO NY=8 NE=8 LY=FU,FR TE=SY,FR PS=SY,FR
LE
INTRUS INTNUM INTDEN INTREL AVOID AVNUM AVDEN AVOIDREL
PA LY
1 0 0 0 0 0 0 0
1 0 0 0 0 0 0 0
1 0 0 0 0 0 0 0
1 0 0 0 0 0 0 0
0 0 0 0 1 0 0 0
0 0 0 0 1 0 0 0
0 0 0 0 1 0 0 0
0 0 0 0 1 0 0 0
PA TE
1
0 1
0 0 1
0 0 0 1
0 0 0 0 1
0 0 0 0 0 1
0 0 0 0 0 0 1
0 0 0 0 0 0 1 1
PA PS
0
0 1
0 0 1
0 0 0 1
1 0 0 0 0
0 0 0 0 0 1
0 0 0 0 0 0 1
0 0 0 0 0 0 0 1
VA 1.0 PS(1,1) PS(5,5)
CO PS(2,2) = LY(1,1)**2+LY(2,1)**2+LY(3,1)**2+LY(4,1)**2+ C
2*LY(1,1)*LY(2,1)+2*LY(1,1)*LY(3,1)+2*LY(1,1)*LY(4,1)+2*LY(2,1)*LY(3,1)+ C
2*LY(2,1)*LY(4,1)+2*LY(3,1)*LY(4,1)
CO PS(3,3) = LY(1,1)**2+LY(2,1)**2+LY(3,1)**2+LY(4,1)**2+ C
2*LY(1,1)*LY(2,1)+2*LY(1,1)*LY(3,1)+2*LY(1,1)*LY(4,1)+2*LY(2,1)*LY(3,1)+ C
2*LY(2,1)*LY(4,1)+2*LY(3,1)*LY(4,1)+TE(1,1)+TE(2,2)+TE(3,3)+TE(4,4)
CO PS(4,4) = PS(2,2)*PS(3,3)**-1
```

(continued)

TABLE 8.8. *(continued)*

```
CO PS(6,6) = LY(5,5)**2+LY(6,5)**2+LY(7,5)**2+LY(8,5)**2+ C
2*LY(5,5)*LY(6,5)+2*LY(5,5)*LY(7,5)+2*LY(5,5)*LY(8,5)+2*LY(6,5)*LY(7,5)+ C
2*LY(6,5)*LY(8,5)+2*LY(7,5)*LY(8,5)
CO PS(7,7) = LY(5,5)**2+LY(6,5)**2+LY(7,5)**2+LY(8,5)**2+ C
2*LY(5,5)*LY(6,5)+2*LY(5,5)*LY(7,5)+2*LY(5,5)*LY(8,5)+2*LY(6,5)*LY(7,5)+ C
2*LY(6,5)*LY(8,5)+2*LY(7,5)*LY(8,5)+ TE(5,5)+TE(6,6)+TE(7,7)+TE(8,8)+ C
2*TE(8,7)
CO PS(8,8) = PS(6,6)*PS(7,7)**-1
OU ME=ML SC AD=OFF ND=4
```

Mplus 7.11

```
TITLE: MPLUS RELIABILITY ESTIMATION FOR CONTINUOUS INDICATORS
DATA: FILE IS FIG8.4.DAT;
      TYPE IS STD CORR;
      NOBS ARE 500;
VARIABLE: NAMES = Y1-Y8;
MODEL: INTRUS BY Y1* (B_1)
       Y2-Y4 (B_2-B_4);
       AVOID BY Y5* (B_5)
       Y6-Y8 (B_6-B_8);
       Y1-Y8 (THETA_1-THETA_8);
       Y7 WITH Y8 (THETA87);          ! ITEM ERROR COVRIANCE
       INTRUS@1; AVOID@1;
MODEL CONSTRAINT:
       NEW(RELINTR RELAVD);
       RELINTR =(B_1+B_2+B_3+B_4)**2 /
          ((B_1+B_2+B_3+B_4)**2+THETA_1+THETA_2+THETA_3+THETA_4);
       RELAVD = (B_5+B_6+B_7+B_8)**2 /
          ((B_5+B_6+B_7+B_8)**2+THETA_5+THETA_6+THETA_7+THETA_8+ 2*THETA87);
OUTPUT: CINTERVAL;
```

programming requires a latent Y variable specification, due to the use of nonlinear constraints (dummy latent variables). Note that while there are only two substantive factors (Intrusions, Avoidance), eight latent variables are specified (see LE and MO lines, Table 8.6). The remaining six latent variables represent the dummy latent variables needed to convey the true-score variance, total variance, and ratio of true-score to total variance for the two latent dimensions of substantive interest. Accordingly, lambda-Y is an 8 (indicator) × 8 (latent variable) matrix, but only eight elements of this matrix are freely estimated in order to relate the observed measures (Y1–Y8) to the substantive latent factors (see PA LY). The theta-epsilon (TE) matrix is specified in the typical fashion (indicator measurement errors are freely estimated; note error covariance of Y7 and Y8). Because eight latent variables have been specified, psi (PS) is 8 × 8 symmetric matrix. Elements PS(1,1) and PS(5,5) are fixed to 1.0 to set the metric of the factors, Intrusions and Avoidance. The off-diagonal element, PS(5,1) is freely estimated to obtain the covariance of Intrusions and Avoidance. The values of the remaining diagonal elements of the psi matrix are determined by the constraints (CO) that follow (see Table 8.8).

The first constraint holds the value of PS(2,2) equal to the true-score variance of Intrusions. Although the mathematical operations to the right of the equals sign do not readily convey this fact, the constrained value of PS(2,2) is simply the squared sum of the unstandardized factor loadings of indicators of Intrusions—that is, $(\Sigma\lambda_i)^2$. Because of syntax programming restrictions in LISREL (i.e., parentheses are not permitted in CO commands except in matrix elements), the squared sum of factor loadings must be calculated by summing the squared factor loadings [e.g., LY(1,1)**2] with the sum of the product of each factor loading pair multiplied by 2 [e.g., 2*LY(1,1)*LY(2,1)]. For instance, if $x_1 = 4$, $x_2 = 3$,

$$(\Sigma x_i)^2 = 4^2 + 3^2 + 2(4 * 3) = 49$$

The C at the end of this and other constraint commands (preceded by a space) is used to inform LISREL that the statement continues on the next line.

The next constraint, CO PS(3,3), represents the denominator of the scale reliability formula—in this case, the total variance of Intrusions. This is calculated by adding the sum of indicator error variances (unstandardized theta-epsilon estimates for Y1, Y2, Y3, and Y4) to the true score variance of Intrusions. The constraint on PS(4,4) provides the proportion of true score variance to the total variance of Intrusion (i.e., the point estimate of the scale reliability of Intrusions). Division is not permitted in LISREL (or in matrix algebra), so this estimate is calculated by using a negative exponent; that is, CO PS(4,4) = PS(2,2) * PS(3,3) ** –1.

The calculation of the true-score variance, total variance, and proportion of true-score to total variance (scale reliability) is performed in the same fashion for the dimension of Avoidance; see constraints for PS(6,6), PS(7,7), and PS(8,8), respectively (Table 8.6). The only difference is that the error covariance of indicators Y7 and Y8 must be included in the total variance of Avoidance. This is reflected by the addition of 2*TE(8,7) at the end of CO PS(7,7) command.

The Mplus syntax for estimating the scale reliability of Intrusions and Avoidance is also presented in Table 8.8. As in the use of effects coding to latent variable scaling described in Chapter 7, Mplus programming requires the MODEL CONSTRAINT command to execute the equations for scale reliability estimation. The various parameter estimates from the measurement model that are used in the scale reliability estimation are named in parentheses within the MODEL command; for example, B_1 is the label for the factor loading of the Y1 indicator. The NEW option is used to assign labels (and starting values, if needed) to parameters that are not in the analysis model—in this case, RELINTR and RELAVD for the scale reliability estimates of Intrusions and Avoidance, respectively. Otherwise, the two equations specified under the MODEL CONSTRAINT command follow Eqs. 8.2 and 8.3 for Intrusions and Avoidance, respectively. The CINTERVAL option is listed on the OUTPUT command in case the researcher wishes to obtain the confidence intervals for the scale reliability point estimates (discussed in the next section of this chapter).

Selected results of this analysis are presented in Table 8.9. Although not shown in the table, this specification provides the same fit to the data as the CFA without non-linear constraints; for example, $\chi^2(18) = 20.52$. Estimates provided in the psi matrix

TABLE 8.9. Selected Results (LISREL 9.10) for Scale Reliability Estimation of the Reactions to Traumatic Stress Questionnaire

LISREL Estimates (Maximum Likelihood)

PSI

	INTRUS	INTNUM	INTDEN	INTREL	AVOID	AVNUM
	--------	--------	--------	--------	--------	--------
INTRUS	1.0000					
INTNUM	- -	32.7543				
		(2.2588)				
		14.5009				
INTDEN	- -	- -	35.6277			
			(2.2555)			
			15.7962			
INTREL	- -	- -	- -	0.9193		
AVOID	0.4857	- -	- -	- -	1.0000	
	(0.0389)					
	12.5013					
AVNUM	- -	- -	- -	- -	- -	16.4307
						(1.2664)
						12.9740
AVDEN	- -	- -	- -	- -	- -	- -
AVOIDREL	- -	- -	- -	- -	- -	- -

PSI

	AVDEN	AVOIDREL
	--------	--------
AVDEN	19.9577	
	(1.2631)	
	15.8008	
AVOIDREL	- -	0.8233

THETA-EPS

	Y1	Y2	Y3	Y4	Y5	Y6
	--------	--------	--------	--------	--------	--------
Y1	0.5284					
	(0.0385)					
	13.7236					
Y2	- -	0.5194				
		(0.0392)				
		13.2658				
Y3	- -	- -	1.0325			
			(0.0742)			
			13.9118			
Y4	- -	- -	- -	0.7931		
				(0.1549)		
				5.1216		
Y5	- -	- -	- -	- -	0.2901	
					(0.0511)	
					5.6810	
Y6	- -	- -	- -	- -	- -	0.5468
						(0.0485)
						11.2837

(continued)

TABLE 8.9. *(continued)*

Y7	- -	- -	- -	- -	- -	- -
Y8	- -	- -	- -	- -	- -	- -

THETA-EPS

	Y7	Y8
Y7	0.8277	
	(0.0629)	
	13.1672	
Y8	.0.4986	0.8652
	(0.0536)	(0.0634)
	9.3074	13.6503

include the true-score variance, total variance, and proportion of true to total variance (scale reliability) of Intrusions and Avoidance. For example, the true score and total variance of Intrusions is 32.75 and 35.63, respectively. The point estimate of the scale reliability (ρ) of Intrusions is .919 (i.e., 32.75 / 35.63); 92% of the total variance of Intrusions is true-score variance. The scale reliability (ρ) estimate for Avoidance is .823 (i.e., 16.43 / 19.96).

Because the computation of ρ and the current CFA model are not complex, the scale reliability of Intrusions and Avoidance could be readily calculated by hand. With the estimates provided in Table 8.5, the true score variance of Intrusions would be

$$(0.8912 + 0.9595 + 1.1968 + 2.6757)^2 = (5.74)^2 = 32.75 \qquad (8.4)$$

(cf. INTNUM, Table 8.9). The total variance of Intrusions would be computed as follows:

$$32.75 + 0.5284 + 0.5194 + 1.0325 + 0.7931 = 35.63 \qquad (8.5)$$

(cf. INTDEN, Table 8.9).

In addition to estimating scale reliability within complex CFA models (e.g., multiple latent factors, indicators, and error covariances), the aforementioned LISREL parameterization can be adapted for various extensions of this methodology (e.g., testing for differences in scale reliability across groups; Raykov, 2002a, 2012).

In applied research, the reliability of the dimensions ("subscales") of a questionnaire is usually estimated by Cronbach's α. In the current data set, Cronbach's α for the Intrusions indicators is .840. For Avoidance, Cronbach's α is .845. Although the population values of the measurement parameters in this example are not known, it can be presumed that Cronbach's α underestimates the reliability of Intrusions (cf. ρ = .919) because the condition of tau equivalence is not present (see the *SD* and factor loading of Y4). On the other hand, Cronbach's α overestimates the reliability of Avoidance (cf. ρ = .823) due to the presence of an error covariance involving items Y7 and Y8.

Standard Error and Interval Estimation of Scale Reliability

The preceding section has illustrated how one may obtain a point estimate of scale reliability within the CFA framework. Although representing the "optimal" estimate, this point estimate should not be regarded as the "true" (population) value of a scale's reliability coefficient. Indeed, a point estimate contains no information about how closely (or distantly) it approximates the population value. Thus, as with other parameter estimates (e.g., means, factor loadings), it is helpful to calculate a confidence interval (in equations, CI) for a point estimate of scale reliability. Interval estimation provides the range of plausible values that the population value is within. The degree of confidence that the population value is contained within the confidence interval is stated in terms of a probability or percentage; the confidence interval most often used in the social and behavioral sciences is the 95% confidence interval.

Confidence intervals are typically calculated by adding and subtracting from the point estimate the product of an appropriate z value and the standard error (in equations, SE) of the estimate. For instance, in Chapter 4, it has been shown that the 95% confidence interval of a factor loading can be calculated as follows:

$$95\% \; CI(\lambda) = \lambda \pm 1.96 SE_\lambda \tag{8.6}$$

If $\lambda = 1.05$, and $SE_\lambda = .05$, the 95% confidence interval is 0.952–1.148 (the value of 1.96 is taken from the z distribution; i.e., 95% of all scores in a normal distribution occur between −1.96 and +1.96 standard deviations from the mean). Thus, with a high degree of confidence (95%), we can assert that the true population value of λ lies between 0.952 and 1.148.

These calculations are straightforward because latent variable software programs provide standard errors in addition to the point estimates of the various parameters. To compute the confidence interval of a point estimate of scale reliability, its standard error must be obtained. Raykov (2002b) has developed a method for the estimation of standard error and confidence interval of scale reliability within the CFA framework. This method furnishes an approximate standard error of ρ in this form:

$$SE(\rho) = SQRT[D_1{}^2 Var(u) + D_2{}^2 Var(v) + 2D_1 D_2 Cov(u,v)] \tag{8.7}$$

where u is the sum of estimated unstandardized factor loadings; v is the sum of estimated error variances; $Cov(u,v)$ is the covariance of u and v (if any); and D_1 and D_2 are the partial derivatives of the scale reliability coefficient (ρ) with respect to u and v. D_1 and D_2 are obtained by the following formulas:

$$D_1 = 2uv \, / \, (u^2 + v)^2 \tag{8.8}$$

$$D_2 = u^2 \, / \, (u^2 + v)^2 \tag{8.9}$$

Once the standard error is obtained, the 95% confidence interval of ρ is easily computed by using this formula:

$$95\% \text{ CI } (\rho) = \rho \pm 1.96SE(\rho) \qquad (8.10)$$

Other confidence intervals can be estimated by substituting the appropriate z value; for example, 90% $\text{CI}(\rho) = \rho \pm 1.645SE(\rho)$.

As noted in the preceding section of this chapter, confidence intervals are readily obtained in Mplus by requesting the CINTERVAL option on the OUTPUT command (Table 8.8). In other programs, the Raykov (2002b) procedure for interval estimation requires two analytic steps. In LISREL, nonlinear constraints are specified to obtain the estimates of u and v, and their variances and covariance. LISREL syntax for this step is shown in Table 8.10, again using the two-factor measurement model of the reactions to traumatic stress questionnaire. Using a latent-Y specification, six latent variables are specified: Intrusions and Avoidance, and four dummy latent variables of their corresponding us and vs. The programming of the LY, TE, and PS matrices follows along the same lines as the approach used to estimate scale reliability (cf. Table 8.8). In this case, certain diagonal elements of the psi (PS) matrix are constrained to equal the values of u and v for Intrusions and Avoidance. For example, PS(2,2) equals the sum of the unstandardized factor loadings of the Intrusion indicators (u, labeled UINT in the syntax). PS(3,3) equals the sum of the error variances of these indicators (v, labeled VINT). The remaining two constraints produce u and v for Avoidance (UAVD and VAVD, respectively); note that the error covariance of Y7 and Y8 is included in the calculation of v. On the output (OU) line, the keyword ALL is included (Table 8.10). When ALL is requested, the LISREL output includes the variance–covariance matrix of the parameter estimates, which are also needed in the calculation of $SE(\rho)$.

Table 8.10 presents selected output of this analysis. Although not shown in Table 8.10, this model specification provides the same fit to the data as the initial CFA without nonlinear constraints; for example, $\chi^2(18) = 20.52$. In the psi matrix, u and v values for Intrusions and Avoidance are on the diagonal (SEs and test statistics have been deleted from the output); for instance, u for Intrusions = 5.723 (the same value that was hand-calculated in the preceding section of this chapter, with the unstandardized factor loadings in Table 8.7). Below this output is the portion of the covariance matrix of parameter estimates needed for computation of $SE(\rho)$. The first two diagonal elements are the variances of u and v (i.e., 0.0389 and 0.0223, respectively) for Intrusions; see $Var(u)$ and $Var(v)$ in the formula for $SE(\rho)$. The remaining two diagonal elements are the variances of u and v (0.0244 and 0.0425, respectively) for Avoidance. The immediate off-diagonal elements of these values are the covariances of u and v; see $Cov(u,v)$ in the $SE(\rho)$ formula. The covariances are −0.0016 and −0.0031 for Intrusions and Avoidance, respectively.

In the second step, these values are used in a set of simple calculations that, for convenience, can be performed in a major software program such as SPSS or SAS. Table 8.11 provides SPSS syntax for calculating D_1, D_2, scale reliability (SR), the standard error (SE), and the 95% confidence interval (CI95LO, CI95HI) by using the values of u,

TABLE 8.10. LISREL Syntax and Selected Results for Computing Standard Errors and Confidence Intervals for Point Estimates of Scale Reliabilities of the Reactions to Traumatic Stress Questionnaire (Step 1)

```
TITLE CFA SCALE RELIABILITY: STD ERRORS AND CONFIDENCE INTERVALS (STEP ONE)
DA NI=8 NO=500 MA=CM
LA
Y1 Y2 Y3 Y4 Y5 Y6 Y7 Y8
KM
1.000
0.594  1.000
0.607  0.613  1.000
0.736  0.765  0.717  1.000
0.378  0.321  0.360  0.414  1.000
0.314  0.301  0.345  0.363  0.732  1.000
0.310  0.262  0.323  0.337  0.665  0.583  1.000
0.317  0.235  0.276  0.302  0.632  0.557  0.796  1.000
SD
1.15 1.20 1.57 2.82 1.31 1.24 1.33 1.29
MO NY=8 NE=6 LY=FU,FR TE=SY,FR PS=SY,FR
LE
INTRUS UINT VINT AVOID UAVD VAVD
PA LY
1 0 0 0 0 0
1 0 0 0 0 0
1 0 0 0 0 0
1 0 0 0 0 0
0 0 0 1 0 0
0 0 0 1 0 0
0 0 0 1 0 0
0 0 0 1 0 0
PA TE
1
0 1
0 0 1
0 0 0 1
0 0 0 0 1
0 0 0 0 0 1
0 0 0 0 0 0 1
0 0 0 0 0 0 1 1
PA PS
0
0 1
0 0 1
1 0 0 0
0 0 0 0 1
0 0 0 0 0 1
VA 1.0 PS(1,1) PS(4,4)
CO PS(2,2) = LY(1,1)+LY(2,1)+LY(3,1)+LY(4,1)
CO PS(3,3) = TE(1,1)+TE(2,2)+TE(3,3)+TE(4,4)
CO PS(5,5) = LY(5,4)+LY(6,4)+LY(7,4)+LY(8,4)
CO PS(6,6) = TE(5,5)+TE(6,6)+TE(7,7)+TE(8,8)+2*TE(8,7)
OU ME=ML ALL ND=4
```

(continued)

TABLE 8.10. *(continued)*

Selected results

LISREL Estimates (Maximum Likelihood)

PSI

	INTRUS	UINT	VINT	AVOID	UAVD	VAVD
INTRUS	1.0000					
UINT	- -	5.7231				
VINT	- -	- -	2.8734			
AVOID	0.4857	- -	- -	1.0000		
UAVD	- -	- -	- -	- -	4.0535	
VAVD	- -	- -	- -	- -	- -	3.5270

Covariance Matrix of Parameter Estimates

	PS 2,2	PS 3,3	PS 5,5	PS 6,6
PS 2,2	0.0389			
PS 3,3	-0.0016	0.0223		
PS 5,5	0.0055	0.0000	0.0244	
PS 6,6	0.0000	0.0000	-0.0031	0.0425

v, *Var(u)*, *Var(v)*, and *Cov(u,v)* obtained in the prior LISREL analysis (and the formulas presented earlier in this section). As shown in this table, the SPSS calculations reproduce the point estimates of scale reliability obtained previously in LISREL; ρs = .919 and .823 for Intrusions and Avoidance, respectively. For Intrusions, the standard error of ρ is 0.0066. With this value, the 95% confidence interval for scale reliability point estimate of .919 is found to be .9065–.9322; that is, 95% CI = .919 ± 1.96(0.0066). Thus, while the population reliability of Intrusions is unknown, with a considerable degree of confidence we can assert that it lies between the values of .9065 and .9322, with .919 being the optimal estimate. It is noteworthy that the Cronbach's α estimate of reliability (.84) falls well below the lower range of this confidence interval. For Avoidance, the standard error of ρ is 0.0147, and thus the 95% confidence interval of this point estimate (.8233) is .7944–.8521. In this case, the Cronbach's α estimate of .845 falls within the upper range of this confidence interval.

These CFA-based procedures (Raykov, 2001a, 2002b) represent a very useful and more dependable method of calculating point and interval estimates of scale reliability. Raykov has developed a number of important extensions to this work, including the estimation of scale reliability in context of noncongeneric scales (Raykov & Shrout, 2002), weighted scales (Raykov, 2001a), measures with binary outcomes (Raykov, Dimitrov, & Asparouhov, 2010), and scales with higher-order factor structure (Raykov & Marcoulides, 2012; Raykov & Shrout, 2002). In addition, these procedures can be adapted to test for significant change in composite reliability as the result of scale refinement

TABLE 8.11. SPSS Syntax for Computing Standard Errors and Confidence Intervals for Point Estimates of Scale Reliabilities of the Reactions to Traumatic Stress Questionnaire (Step 2)

```
TITLE SPSS SYNTAX FOR SE AND CI OF SR POINT ESTIMATES: INTRUSION SUBSCALE.

COMPUTE U = 5.7231.
COMPUTE V = 2.8734.
COMPUTE D1 = (2*u*v) / (U**2 + V)**2.
COMPUTE D2 = -U**2 / (U**2 + V)**2.
COMPUTE VARU = 0.0389.
COMPUTE VARV = 0.0223.
COMPUTE COVUV = -0.0016.
COMPUTE SR = U**2 / (U**2 + V).
COMPUTE SE = SQRT(D1**2*VARU+D2**2*VARV+2*D1*D2*COVUV).
COMPUTE CI95LO = SR-1.96*SE.
COMPUTE CI95HI = SR+1.96*SE.

LIST SR SE CI95LO CI95HI.


SR        SE        CI95LO      CI95HI

0.9193    0.0066    0.9065      0.9322


TITLE SPSS SYNTAX FOR SE AND CI OF SR POINT ESTIMATES: AVOIDANCE SUBSCALE.

COMPUTE U = 4.0535.
COMPUTE V = 3.527.
COMPUTE D1 = (2*u*v) / (U**2 + V)**2.
COMPUTE D2 = -U**2 / (U**2 + V)**2.
COMPUTE VARU = 0.0244.
COMPUTE VARV = 0.0425.
COMPUTE COVUV = -0.0031.
COMPUTE SR = U**2 / (U**2 + V).
COMPUTE SE = SQRT(D1**2*VARU+D2**2*VARV+2*D1*D2*COVUV).
COMPUTE CI95LO = SR-1.96*SE.
COMPUTE CI95HI = SR+1.96*SE.

LIST SR SE CI95LO CI95HI.


SR        SE        CI95LO      CI95HI

0.8233    0.0147    0.7944      0.8521
```

(e.g., addition or deletion of items; Raykov & Grayson, 2003), and to test whether scale reliabilities significantly differ across groups (Raykov, 2002a). Uses of some of these extensions in applied research can be found in Brown et al. (2004) and Campbell-Sills et al. (2004).

MODELS WITH FORMATIVE INDICATORS

All CFA examples discussed thus far in this book have entailed models with *reflective indicators* (also referred to as *effects indicators*). In a model with reflective indicators, the paths relating the indicators to the factor (i.e., factor loadings) emanate from the latent variable to the indicator (e.g., Figure 8.4). In accord with classical test theory (Lord & Novick, 1968), the direction of causality is from the latent construct to the observed measure. For indicators specified to load on a given factor, the latent variable should explain most of the covariation among the observed measures; for example, in Figure 8.4, the only reason that indicators Y1–Y4 are intercorrelated is they share a common cause—the underlying construct of Intrusions. Related to this is the fact that measurement error in reflective models is taken into account at the indicator level (i.e., the εs in Figure 8.4). Moreover, given that the direction of causality is from the factor to the observed measures, a change in the latent variable will result in a change in all constituent indicators. In such models, the indicators of a given latent factor are considered to be interchangeable. For instance, eliminating the Y4 indicator from the Figure 8.4 model would not alter the meaning of the Intrusions factor.

In many scenarios, it may be more natural to define a construct as being influenced by a set of indicators. In *formative indicator* models (also referred to as *composite cause* or *cause indicator* models), the direction of causality is from the observed measures to the construct (Bollen, 1989; Bollen & Lennox, 1991; Edwards & Bagozzi, 2000; MacCallum & Browne, 1993). An often-used example is that of socioeconomic status (SES), whose causal indicators might be income, education level, and occupational status. In this case, it seems more reasonable to assert that these three variables are the cause of one's SES than the converse, which would be the claim that income, education level, and occupational status are interrelated because they share the common underlying cause of SES. The term *composite cause* is sometimes used in reference to such models because the latent variable is posited to be a construct that is a weighted, linear combination of its observed measures, plus error. Although formative indicator models are considered interpretationally problematic by some methodologists (e.g., Howell, Breivik, & Wilcox, 2007; cf. Bagozzi, 2007; Bollen, 2007) and rarely considered in some substantive domains (e.g., mental health research), these models are very common in such fields as marketing, economics, and consumer research (e.g., Diamantopoulos & Winklhofer, 2001; Jarvis, MacKenzie, & Podsakoff, 2003). Path diagrams of models containing formative indicators are presented in Figure 8.5. In these examples, Chronic Life Stress is the composite latent variable influenced by a variety of stressful events (e.g., financial difficulties, interpersonal conflict). In practice, causal measures are usually single indicators, although higher-order factor analysis can be extended to specify first-order factors as causes of second-order constructs (cf. Jarvis et al., 2003).

In addition to the direction of construct–indicator causality, the assumptions of formative indicator models differ from those of reflective indicator models in other important ways. First, in a formative model, although the cause indicators are specified to be intercorrelated (see Figure 8.5), the correlations among the indicators are not relevant to the goodness of fit and conceptual viability of the model (except for the issue of multicollinearity, which would indicate undue redundancy in the indicator set used to form the composite latent variable). In other words, the formative indicators may influence the composite construct independently of one another. For example, SES will increase if income increases, even when education and occupation status remain the same. Similarly, financial difficulties should increase chronic life stress when job stress and interpersonal conflicts are static (Figure 8.5). Thus causal indicators of a concept can be positively or negatively correlated, or may have no association whatsoever. Second, unlike in reflective models, eliminating a formative indicator from the measurement model is apt to change the meaning of the composite construct (i.e., because the construct is a weighted, linear combination of all its observed measures). Formative indicators are not interchangeable. If the income indicator is omitted from the SES composite, part of the construct of SES is not represented. Elimination of the financial difficulties indicator removes an important variable that contributes to chronic life stress. Finally, although identification issues must be considered (as discussed below), measurement error in the formative indicator model is taken into account at the latent construct level (i.e., the disturbance of the composite variable, E; see Figure 8.5), not at the indicator level (i.e., εs in Figure 8.4).

Identification problems are issues in models with formative indicators (Bollen & Davis, 2009; MacCallum & Browne, 1993; Treiblmaier, Bentler, & Mair, 2011). The core aspects of model identification discussed in Chapter 3 apply to formative indicator models; for example, the degrees of freedom of the model must be at least zero, and the latent variable must be provided a scale. The metric of the latent composite variable can be defined either by fixing a formative indicator path to one (e.g., see Figure 8.5A) or by fixing the factor variance to unity. Although necessary, these conditions are not sufficient to identify a model containing formative indicators. Indeed, as in some CFAs of multitrait–multimethod matrices (e.g., correlated methods models, Chapter 6), identification problems are common in some types of models with formative indicators.

Figure 8.5 presents several examples of identified and unidentified models that contain formative indicators. Measurement models that consist solely of formative indicators are not identified (Figure 8.5A). MacCallum and Browne (1993) note that many identification problems of formative indicator constructs stem from indeterminacies associated with the construct-level error term (i.e., the "E" in Figure 8.5A). Depending on the nature of the model, these indeterminacies can be resolved in various ways. As shown in Figure 8.5B, the identification problem can be addressed by adding two reflective indicators to the formative construct. This strategy should not be employed simply to identify the model. Indeed, this approach may not be appropriate if it cannot be justi-

**Model A: Single Construct Defined Solely
by Formative Indicators (Unidentified)**

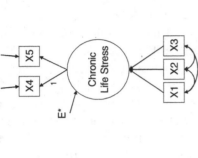

**Model B: Single Construct Defined
by Formative and Reflective Indicators (Identified)**

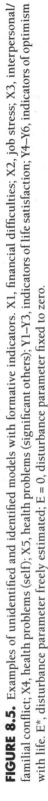

**Model C: Formative Indicator Construct
Predicting a Reflective Indicator Construct (Unidentified)**

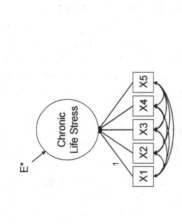

**Model D: Formative Indicator Construct Predicting a Reflective Indicator Construct
(Identified by Fixing Disturbance of Composite Latent Variable to Zero)**

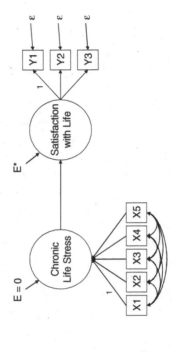

FIGURE 8.5. Examples of unidentified and identified models with formative indicators. X1, financial difficulties; X2, job stress; X3, interpersonal/familial conflict; X4, health problems (self); X5, health problems (significant others); Y1–Y3, indicators of life satisfaction; Y4–Y6, indicators of optimism with life. E*, disturbance parameter freely estimated; E = 0, disturbance parameter fixed to zero.

324

**Model E: Formative Indicator Construct
Predicting a Reflective Indicator Construct (Identified by Including
Reflective Indicators on the Latent Composite Variable)**

**Model F: Formative Indicator Construct Predicting More than One
Reflective Indicator Construct (Identified)**

**Model G: Formative Indicator Construct Predicting
More Than One Reflective Indicator Construct (Unidentified)**

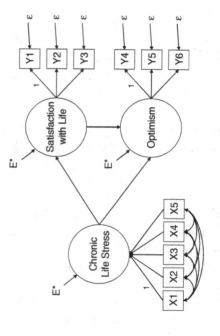

FIGURE 8.5. *(continued)*

325

fied on substantive grounds. In the current example (Figure 8.5B), note that the X4 and X5 indicators (which really should be labeled as Y indicators in Figure 8.5B because they are predicted by the formative construct) are measures of health problems in the respondent and his or her significant others. A compelling conceptual argument for treating X4 and X5 as reflective indicators could perhaps be made in this instance; that is, although health problems are often construed as stressors, they can be caused by exposure to chronic life stress (e.g., job and family stress; cf. Figure 8.5). If the identification issue can be remedied in this manner, the formative construct is identified on its own and can reside anywhere in the measurement or structural model (e.g., as a latent X or latent Y variable, as a factor in a CFA model). This strategy carries the advantage of modeling flexibility (e.g., Figure 8.5E). For instance, if the formative construct is included in a broader CFA model, its discriminant validity and measurement parameters can be fully evaluated (Jarvis et al., 2003).

Models are also not identified if the formative construct predicts a single reflective construct (Figure 8.5C). One way the indeterminacy of the construct error term can be addressed is by fixing the disturbance of the composite latent variable to zero (Figure 8.5D).[3] However, this approach is often unreasonable because it assumes that the formative indicators represent the composite latent construct perfectly (cf. principal components analysis, Chapter 2; canonical correlation, Thompson, 1984). It is noted in the preceding paragraph that the error term of the formative construct can be freely estimated if the latent composite contains some reflective indicators. In fact, unlike the Figure 8.5B model, only one reflective indicator is needed to identify the model depicted in Figure 8.5E. But if either the strategy shown in Figure 8.5D or the one shown in 8.5E is employed, the issue of equivalent models must be considered (discussed below).

The residual variances of composite latent variables may be identified if the formative construct emits paths to two or more latent constructs defined by reflective indicators. An identified model is shown in Figure 8.5F, where the formative construct of Chronic Life Stress predicts the reflective constructs of Satisfaction with Life and Optimism. In order for the identification problem to be resolved, there must be no relationship specified between the two latent Y reflective constructs. Figure 8.5G depicts an unidentified model where both Chronic Life Stress and Satisfaction with Life have direct effects on Optimism. The same problem will arise if a correlated disturbance is specified between Satisfaction with Life and Optimism. Thus models along the lines of Figure 8.5F, while more apt to be empirically identified, may also be substantively unreasonable. For instance, the model specification in Figure 8.5F indicates that the relationship between Satisfaction with Life and Optimism is spurious (i.e., fully accounted for by Chronic Life Stress)—an assertion that may be at odds with theory (e.g., a conceptual model that posits Satisfaction with Life to be a predictor of Optimism). The identification issue can also be addressed by including reflective indicators on the composite latent variable, if justified.

In addition to identification, the issue of equivalent models (see Chapter 5) must be considered in solutions involving formative indicators. Models with formative indicators may be statistically equivalent to MIMIC models (Chapter 7), where reflective latent constructs are predicted by single indicators, but no direct effects of the covariates on the reflective indicators are specified (cf. Figure 7.5, Chapter 7). Figure 8.6 provides two MIMIC models equivalent to formative construct models presented in Figure 8.5. Figure 8.6A depicts a model where the Satisfaction with Life latent variable is regressed onto the five single indicators of different forms of life stress. Although a different set of regressive parameter estimates will be obtained (i.e., five direct effects between the single indicators and Satisfaction with Life), the Figure 8.6A model will provide exactly the same predicted covariance matrix and goodness-of-fit statistics as the model shown in Figure 8.5D. Similarly, the model shown in Figure 8.5E can be alternatively construed as a model with two reflective constructs (Health Problems, Satisfaction with Life; see Figure 8.6B), in which one of the factors (Health Problems) is predicted by three background variables (X1, X2, X3). These single indicators may have indirect effects on Satisfaction with Life, as fully mediated by Health Problems. Although quite different conceptually, the models depicted in Figure 8.5E and Figure 8.6B will produce the same fitted covariance matrix. As noted in Chapter 5, the procedures of CFA/SEM cannot inform the researcher on the comparative validity of these solutions.

MIMIC Model A (Equivalent to Model D in Figure 8.5)

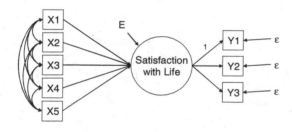

MIMIC Model B (Equivalent to Model E in Figure 8.5)

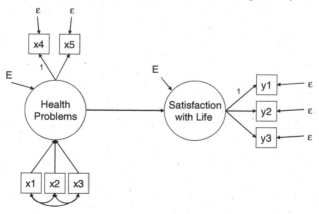

FIGURE 8.6. Equivalence of formative indicator models to MIMIC models. X1, financial difficulties; X2, job stress; X3, interpersonal/familial conflict; X4, health problems (self); X5, health problems (significant others); Y1–Y3, indicators of life satisfaction.

To foster the reader's understanding of how models with formative constructs are specified, a data-based example is provided in Figure 8.7. This model is identical to the model presented in Figure 8.5F, except that the formative construct of Chronic Life Stress is caused by three indicators instead of five. Mplus syntax is provided in Table 8.12. The latent endogenous variables of Satisfaction with Life and Optimism are programmed in the usual way (e.g., the Mplus default is to automatically set the first indicator listed—in this case, Y1 and Y4—as the marker indicator). In order for Mplus to recognize chronic stress (Stress) as a latent variable, the programming indicates that Stress is "measured by" (i.e., the BY keyword) Life Satisfaction (Satis) and Optimism (Optim). Although the BY keyword is typically used in reference to factor loadings, the actual parameters involved are paths (beta, β) of the regressions of Satisfaction with Life and Optimism on the composite latent variable, Chronic Life Stress. In addition, a * is used to override the Mplus default of fixing the unstandardized parameter of Stress $\rightarrow$ Satis to 1.0; instead, this regressive path is freely estimated (*). The statement STRESS ON X1@1 X2 X3 regresses the formative construct of Chronic Life Stress onto its three causal indicators. The parameter relating X1 to Chronic Life Stress is fixed to 1.0 to

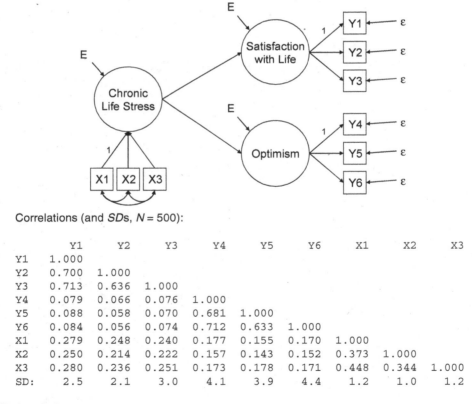

Correlations (and *SDs*, *N* = 500):

	Y1	Y2	Y3	Y4	Y5	Y6	X1	X2	X3
Y1	1.000								
Y2	0.700	1.000							
Y3	0.713	0.636	1.000						
Y4	0.079	0.066	0.076	1.000					
Y5	0.088	0.058	0.070	0.681	1.000				
Y6	0.084	0.056	0.074	0.712	0.633	1.000			
X1	0.279	0.248	0.240	0.177	0.155	0.170	1.000		
X2	0.250	0.214	0.222	0.157	0.143	0.152	0.373	1.000	
X3	0.280	0.236	0.251	0.173	0.178	0.171	0.448	0.344	1.000
SD:	2.5	2.1	3.0	4.1	3.9	4.4	1.2	1.0	1.2

FIGURE 8.7. Path diagram and input data for example of a model containing a formative construct. X1, financial difficulties; X2, job stress; X3, interpersonal/familial conflict; Y1–Y3, indicators of life satisfaction; Y4–Y6, indicators of optimism.

TABLE 8.12. Mplus 7.11 Syntax and Selected Results for Model with a Formative Construct

Syntax file

```
TITLE:        MPLUS PROGRAM FOR MODEL WITH A FORMATIVE CONSTRUCT
DATA:         FILE IS STRESS.DAT;
              TYPE IS STD CORR;
              NOBS ARE 500;
VARIABLE:     NAMES ARE Y1 Y2 Y3 Y4 Y5 Y6 X1 X2 X3;
ANALYSIS:     ESTIMATOR=ML;
MODEL:        SATIS BY Y1 Y2 Y3;
              OPTIM BY Y4 Y5 Y6;
              STRESS BY SATIS* OPTIM;
              STRESS ON X1@1 X2 X3;
              X1 WITH X2 X3; X2 WITH X3;
OUTPUT:       SAMPSTAT MODINDICES(4.0) STAND FSDETERMINACY RESIDUAL;
```

Selected results

MODEL RESULTS

		Estimate	S.E.	Est./S.E.	Two-Tailed P-Value
SATIS	BY				
Y1		1.000	0.000	999.000	999.000
Y2		0.746	0.038	19.571	0.000
Y3		1.086	0.054	19.931	0.000
OPTIM	BY				
Y4		1.000	0.000	999.000	999.000
Y5		0.848	0.045	18.733	0.000
Y6		1.000	0.051	19.441	0.000
STRESS	BY				
SATIS		0.317	0.083	3.805	0.000
OPTIM		0.338	0.101	3.357	0.001
STRESS	ON				
X1		1.000	0.000	999.000	999.000
X2		1.054	0.445	2.369	0.018
X3		1.073	0.435	2.469	0.014
X1	WITH				
X2		0.447	0.057	7.815	0.000
X3		0.644	0.070	9.142	0.000
X2	WITH				
X3		0.412	0.057	7.274	0.000
Variances					
X1		1.437	0.091	15.811	0.000
X2		0.998	0.063	15.811	0.000
X3		1.437	0.091	15.811	0.000

(continued)

TABLE 8.12. *(continued)*

Residual Variances

Y1	1.321	0.186	7.082	0.000
Y2	1.661	0.142	11.735	0.000
Y3	3.181	0.284	11.214	0.000
Y4	3.964	0.528	7.509	0.000
Y5	5.971	0.506	11.795	0.000
Y6	6.510	0.622	10.459	0.000
SATIS	4.121	0.498	8.279	0.000
OPTIM	11.905	1.130	10.534	0.000
STRESS	0.488	3.435	0.142	0.887

STANDARDIZED MODEL RESULTS

STDYX Standardization

		Estimate	S.E.	Est./S.E.	Two-Tailed P-Value
SATIS	BY				
Y1		0.888	0.018	49.927	0.000
Y2		0.789	0.022	36.328	0.000
Y3		0.804	0.021	38.198	0.000
OPTIM	BY				
Y4		0.874	0.019	45.891	0.000
Y5		0.779	0.023	34.397	0.000
Y6		0.814	0.021	38.511	0.000
STRESS	BY				
SATIS		0.402	0.096	4.200	0.000
OPTIM		0.266	0.073	3.656	0.000
STRESS	ON				
X1		0.425	0.136	3.117	0.002
X2		0.374	0.126	2.955	0.003
X3		0.456	0.139	3.278	0.001
X1	WITH				
X2		0.373	0.038	9.689	0.000
X3		0.448	0.036	12.533	0.000
X2	WITH				
X3		0.344	0.039	8.724	0.000

R-SQUARE

Observed Variable	Estimate	S.E.	Est./S.E.	Two-Tailed P-Value
Y1	0.788	0.032	24.964	0.000
Y2	0.622	0.034	18.164	0.000

(continued)

TABLE 8.12. *(continued)*

	Estimate	S.E.	Est./S.E.	P-Value
Y3	0.646	0.034	19.099	0.000
Y4	0.764	0.033	22.945	0.000
Y5	0.607	0.035	17.198	0.000
Y6	0.663	0.034	19.255	0.000
Latent Variable	Estimate	S.E.	Est./S.E.	Two-Tailed P-Value
SATIS	0.162	0.077	2.100	0.036
OPTIM	0.071	0.039	1.828	0.068
STRESS	0.939	0.405	2.317	0.020

define the metric of the composite latent variable. All three disturbances (E) are freely estimated by Mplus default. It is also an Mplus default to freely intercorrelate exogenous variables—in this case, the relationships among X1, X2, and X3. But to make the details of formative construct specification more explicit, these interrelationships are written out in the Mplus syntax example.

The Figure 8.7 model fits the data well; for example, $\chi^2(22)$ = 2.17, p = 1.00. The parameter estimates provided in Table 8.12 should be self-explanatory if the reader has become comfortable with interpreting prior examples in this book (e.g., MIMIC models). Although the measurement model holds up well, and the composite latent variable is significantly related to Satisfaction with Life and Optimism (i.e., zs = 3.805 and 3.357, respectively), the substantive significance of the results may be questioned on the basis of the magnitude of these structural relations (e.g., Chronic Life Stress accounts for 7.1% of the variance in Optimism). Because the error term of the formative construct is freely estimated, the paths linking Chronic Life Stress to Satisfaction with Life and Optimism are adjusted for measurement error. The paths linking the X indicators to Chronic Life Stress are interpreted as regression coefficients; for example, when X2 and X3 are held constant, a 1-standardized-unit increase in X1 is associated with a .425-standardized-unit increase in Chronic Life Stress. Collectively, the three formative indicators in this example have considerable strength in predicting Chronic Life Stress (R^2 = .939).

SUMMARY

This chapter has dealt with three specialized applications of CFA: higher-order factor analysis (second-order factor analysis, bifactor analysis), scale reliability evaluation, and models with formative indicators. Each analysis provides a compelling method to address substantively important research questions in a manner unparalleled by alternative approaches (e.g., CFA-based scale reliability vs. Cronbach's alpha; formative latent constructs vs. coarse composites). As with the other types of models discussed throughout this book (e.g., MTMM models, Chapter 6), higher-order factor models and formative indicator models can be embedded in broader SEM analyses (e.g., Figure 8.7), as determined by the research context.

Now that the major types of CFA models have been presented, this book turns to considering analytic issues relevant to any form of CFA. Chapter 9 addresses the complications of missing, non-normal, and categorical data. Chapter 10 describes methods for determining the appropriate sample size of a CFA study.

NOTES

1. However, as noted earlier, a Schmid–Leiman transformation can be applied to a second-order factor solution to elucidate the strength of the effects of the first- and second-order factors on the observed measures. In fact, the bifactor model and second-order model are statistically equivalent when the Schmid–Leiman transformation is applied to the latter (Yung, Thissen, & McLeod, 1999).

2. Beginning with Version 7, Mplus can also conduct bifactor modeling within the context of EFA and ESEM (see Chapter 5 for a discussion of ESEM), in addition to CFA.

3. A related strategy is to impose equality constraints on the formative construct and reflective construct error terms. Often this approach is not conceptually well defended.

9

Data Issues in CFA
Missing, Non-Normal, and Categorical Data

This chapter presents several data issues frequently encountered in CFA with applied data sets and ways to deal with them (e.g., missing data, non-normality, parceling). In addition, procedures for conducting CFA in single and multiple groups with categorical indicators are presented. In context of this discussion, the parallels between CFA and item response theory (IRT) are illustrated. The topics discussed in this chapter are equally germane to broader applications of SEM.

CFA WITH MISSING DATA

Rarely will an applied researcher have the luxury of conducting a statistical analysis (CFA or otherwise) with a data set in which all variables are present for all cases. Indeed, missing data are the norm in research data sets. However, applied research reports often fail to mention how missing data were handled. In addition, the missing data strategies often used by researchers (e.g., listwise or pairwise deletion) are inappropriate and have deleterious effects, such as loss of statistical power, and bias in parameter estimates, standard errors, and test statistics (Allison, 2002, 2003; Enders, 2010; Little & Rubin, 2002; Schafer & Graham, 2002).

Mechanisms of Missing Data

Missing data may occur for a variety of reasons. For example, data may be missing by design (*planned missingness*), as would occur in cohort-sequential research designs (a longitudinal design frequently used in developmental research), matrix sampling (used in educational testing where all students are not administered identical item sets), or other designs where participants in the sample are randomly selected to complete various subsets of the full assessment battery (possibly due to practical considerations such as financial or time constraints).

In addition, missing data may be unplanned and may occur for a variety of reasons. There are three different mechanisms of unplanned missingness, described below. In real data sets, it is likely that all three mechanisms are responsible for missing data to varying degrees, although it is impossible to determine the extent to which each mechanism is at play. For instance, data may be missing by chance for reasons other than the research design (e.g., accidental omission of a questionnaire in some assessment packets). In such conditions, data are assumed to be *missing completely at random* (MCAR). The assumption of MCAR holds if the probability of missing data on Y is unrelated to Y or to the values of any other variable in the data set. Using a two-variable (X = gender, Y = life satisfaction) illustration similar to one presented by Allison (2003), we can see that the MCAR assumption will hold if the probability of missing data on life satisfaction (Y) is not related to either life satisfaction or gender (X). The condition of MCAR can be readily tested in the research data; for example, do males and females differ in the rate of missing data on life satisfaction? If the assumption holds for all variables in the data set, the cases in the sample with complete data can be considered as a random subsample of the total sample (as would occur if missing data were planned as part of the research design).

The probability of missing data may be related to other variables in the data set. The assumption of *missing at random* (MAR) is met when the probability that data are missing on Y may depend on the value of X, but is not related to the value of Y when X is held constant. Although a weaker assumption, MAR is more likely than MCAR to hold in applied data sets when missing data are not planned in the research design. Again using the two-variable example, we can see that MAR will hold if males and females differ in their rates of missing data on life satisfaction, but within both levels of gender (i.e., for males and females separately), the likelihood of missing data on life satisfaction does not depend on the level of life satisfaction. It is impossible to test the condition of MAR in research data because the values of missing data are unknown. As Allison (2003) notes, "in essence, MAR allows missingness to depend on things that are observed, but not on things that are not observed" (p. 545). This will be evident in a research study that uses a sampling procedure of planned missingness by selective design (e.g., participants are selected for additional testing on the basis of initial testing scores). MAR is closely related to (and sometimes used interchangeably with) the missing data mechanism referred to as *ignorable* (for further discussion, see Allison, 2002, 2003; Enders, 2010).

Missing data are said to be *nonignorable* (also referred to as *missing not at random*) if the missingness is related to values that would have been observed (i.e., the condition of MAR does not hold). Missing data are nonignorable if cases with missing data on a given variable would have higher or lower values on that variable than cases with data present, after all other variables in the data set are controlled for. For instance, in the example of a longitudinal research design (e.g., a treatment outcome study where participants are assessed at pretreatment, posttreatment, and various follow-up points), the data will be MAR if dropouts are found to be related to variables collected at testing occasions prior to the dropout (Schafer & Graham, 2002). The data will be nonignorable if dropouts are related to unseen responses on variables that have been collected after participants drop out (e.g., missing posttreatment data from participants who have failed to respond to treatment; treatment nonresponse is unrelated to variables obtained at pretreatment).

Conventional Approaches to Missing Data

Over the past several years, sophisticated methods for handling missing data have been developed when MCAR or MAR is true (e.g., direct maximum likelihood (ML), multiple imputation; Allison, 1987; Enders, 2010; Little & Rubin, 2002; Schafer, 1997; Schafer & Graham, 2002). Before these state-of-the-art methods are discussed, more common approaches to missing data management are reviewed.

One such approach is *listwise deletion*, in which cases with missing data on any variable used in the analysis are removed from the sample. Listwise deletion has several advantages, including its ease of implementation, the fact that it can be used for any form of statistical analysis, and the fact that the same sample is used for all analyses. In fact, when data are MCAR, listwise deletion will produce *consistent* (i.e., unbiased) parameter estimates, standard errors, and test statistics (e.g., model χ^2). As noted earlier, when MCAR holds, the listwise complete sample can be regarded as a random subsample of the original sample. However, estimates produced using listwise deletion are usually not *efficient*. This is because listwise deletion often results in loss of a considerable proportion of the original sample. Thus standard errors will frequently be larger when listwise deletion is used, compared to alternative methods (e.g., multiple imputation) that use all of the available data. The inflation in standard errors will thus decrease statistical power and lower the precision of the parameter estimates (wider confidence intervals). Indeed, Schafer and Graham (2002) note that listwise deletion can be a very effective (and straightforward) strategy for missing data only when the MCAR assumption holds and only when a very small part of the sample is removed. When data are MAR, listwise deletion may produce results that are neither consistent or efficient. To return to the two-variable example, if males are more likely to have missing data on the life satisfaction variable than females, and if males report less life satisfaction than females, listwise deletion will result in a positively biased estimate of average life satisfaction (although Allison, 2002, discusses scenarios where listwise deletion may outperform other missing data methods when the assumptions of MAR are violated).

Another common missing data strategy is *pairwise deletion*. A variety of statistical analyses (e.g., EFA, CFA, multiple regression, ANOVA) can be performed by using means, variances, and covariances as input data (i.e., they do not require raw data as input). When pairwise deletion is used, the input vectors (means) and matrices (variances–covariances) are estimated, and all cases that have data present for each variable (means, variances) or each pair of variables (covariances) are used. For example, in a sample with 300 total participants who were administered variables X1, X2, and X3, the frequencies with which data are present for the three pairs of indicators are as follows: X1,X2 = 226; X1,X3 = 56; and X2,X3 = 55. In pairwise deletion, correlations (or covariances) will be estimated for the input matrix by using three different Ns (226, 56, 55). This approach can be considered superior to listwise deletion because more data are preserved. When the data are MCAR, pairwise deletion typically produces consistent parameter estimates (fairly unbiased in large samples). Although one might expect that pairwise deletion is more efficient than listwise deletion (because fewer data are lost), this is not always the case (see Allison, 2002).

Pairwise deletion has a few other more serious problems. For one, pairwise deletion produces biased standard errors. A related issue is the decision of what sample size to specify when the input matrix is created under pairwise deletion. In the example provided above, N might be specified as 300 (the total sample), 226 (the highest number of cases that have data present on two indicators), or 55 (the smallest number of cases that have data present on two indicators). No single number is satisfactory: The largest N is too liberal (negatively biased standard errors), but the smallest N is too conservative (inflated standard errors). Second, if the data are MAR, the parameter estimates (as well as the standard errors) are often severely biased. Finally, covariance matrices prepared by using pairwise deletion may not be positive definite, and thus the statistical model cannot be estimated (Wothke, 1993). Nonpositive definite matrices have been discussed in Chapters 4 and 5. For instance, an input matrix will be nonpositive definite when one or more of its elements have out-of-bounds values. The range of possible values that a correlation (or covariance) may possess is dependent on all other relationships in the input matrix. To illustrate, for three variables x, y, and z, the correlation between x and y ($r_{x,y}$) must be within a certain range, as determined by the following equation:

$$r_{x,z}r_{y,z} \pm \text{SQRT}[(1 - r_{x,z}^2)(1 - r_{y,z}^2)] \tag{9.1}$$

For example, if $r_{x,z} = .70$ and $r_{y,z} = .80$, then the value of $r_{x,y}$ must be within the range of .13–.99 (i.e., .56 ± .43). If $r_{x,y} < .13$, then the input matrix will not be positive definite. Thus out-of-range values may occur when correlations (covariances) are estimated via pairwise deletion.

A less frequently used conventional method for managing missing data is *simple (or single) imputation* (mean and regression imputation). For example, regression imputation (referred to by Allison, 2002, as "conditional mean imputation") entails regressing the variable with missing data on other variables in the data set for cases with complete data. The resulting regression equation is used to predict scores for cases with missing data. However, this approach frequently produces underestimates of variances, overestimates of correlations, and underestimates of standard errors. As discussed later in this chapter, the method of *multiple imputation* reconciles these problems in part by introducing random variation into the imputations, and by repeating the imputation process several times.

Recommended Strategies for Missing Data

More sophisticated methods for handling missing data have been developed in recent years. Direct ML and multiple imputation are the most widely preferred methods for handling missing data in SEM and other data analytic contexts (Allison, 2003; Schafer & Graham, 2002). Both approaches use all the available data; that is, N = the total sample size, including cases with missing data. When either MCAR or MAR is true (and the data have a multivariate normal distribution), ML and multiple imputation produce parameter estimates, standard errors, and test statistics (e.g., model χ^2) that are consistent and efficient. These two approaches are now discussed.

The ML estimator can be readily used for the estimation of CFA and SEM approaches with missing data. When ML is used in context of missing data, it is often referred to as *full information ML*, or *FIML*. However, this term is not descriptive because ML is a full information estimator regardless of whether the data set is (in)complete. Thus some methodologists (e.g., Allison, 2003) prefer the term *direct ML* or *raw ML*, because ML estimation with missing data requires that raw data be input to the analysis rather than a variance–covariance matrix (and means).

Before direct ML became widely available, some researchers relied on the *expectation–maximization (EM) algorithm* (Little & Rubin, 2002) for handling missing data in CFA/SEM (under the assumptions of MAR and multivariate normality). The EM algorithm is a computational device for obtaining ML estimates of the means and the covariance matrix (for details on its computation, see Allison, 2002). These estimates (e.g., variance–covariance matrix) are then used as the input matrix in the CFA/SEM. An advantage of the EM algorithm is that it can be easily implemented in a variety of popular statistical software packages. However, a key limitation of using the EM algorithm to calculate input matrices for CFA/SEM is that the resulting standard errors of the parameter estimates are not consistent. Thus confidence intervals and significance tests may be compromised. As with pairwise deletion, this is due in part to the problem of specifying the proper sample size (Allison, 2003).

Direct ML

Direct ML is free of these problems (Allison, 1987; Arbuckle, 1996; Muthén, Kaplan, & Hollis, 1987). Methodologists generally regard direct ML as the best method for handling missing data in most CFA and SEM applications (e.g., Allison, 2003; Duncan, Duncan, & Li, 1998). Direct ML is available in most current latent variable software packages (e.g., Amos, Mplus, LISREL, Mx) and is very straightforward to use. In fact, in many programs (e.g., Mplus 7.11, LISREL 9.10), direct ML is the default estimation option when missing data are encountered in the raw data input file. The only requirement, in addition to inputting a raw data file, is to define missing values in the program syntax; for example, including MI = 9 on the Data (DA) line (in tandem with programming LISREL to read a raw data file; e.g., RA = FILE.DAT) will inform LISREL to treat the value of 9 in the raw data file as missing data.

Table 9.1 presents Mplus 7.11 syntax for estimating a longitudinal measurement model of the construct of negative affectivity, measured by three questionnaires (represented by the variable names PANAS, BIS, and NFFI) on two testing occasions several months apart ($N = 826$). As would be expected in most applied time series research, some participants were lost to the follow-up assessment (attrition); in addition, complete questionnaire data were not available for some participants within a given testing occasion (nonresponse).

TABLE 9.1. Computer Syntax (Mplus 7.11) and Selected Output for Estimation of CFA Model with Missing Data, Using Direct ML and Auxiliary Variables

```
TITLE:   EXAMPLE OF DIRECT ML MISSING DATA ESTIMATION IN MPLUS;
DATA:
  FILE IS DEPSTR.DAT;
  FORMAT IS F6,F3,F2,F2,3F3,6F5.1/F6,24F3;
VARIABLE:
  NAMES ARE SUBJ AGE SEX RACE MDDSUMA MDDSUMB MDDSUMC
  CSTRA CSTRB CSTRC ESTRA ESTRB ESTRC
  AD2 DASDEP DASDEPB DASDEPC BDIR BDIRB BDIRC PANASN PANASNB PANASNC
  PANASP PANASPB PANASPC BIS BISB BISC BAS BASB BASC NFFIN NFFINB NFFIC
  NFFIE NFFIEB NFFIEC;
  MISSING = BLANK;
  USEV = PANASN PANASNB BIS BISB NFFIN NFFINB;
  AUXILIARY = (M) MDDSUMA DASDEP;
ANALYSIS: ESTIMATOR=ML;
MODEL:   NEGAFF1 BY PANASN BIS NFFIN;
         NEGAFF2 BY PANASNB BISB NFFINB;
         PANASN WITH PANASNB; BIS WITH BISB; NFFIN WITH NFFINB;
OUTPUT:  SAMPSTAT STANDARDIZED PATTERNS;
```

SUMMARY OF DATA

      Number of missing data patterns   14

SUMMARY OF MISSING DATA PATTERNS

    MISSING DATA PATTERNS (x = not missing)

	1	2	3	4	5	6	7	8	9	10	11	12	13	14
PANASN	x	x	x	x	x	x	x	x	x					
PANASNB	x	x	x	x	x	x				x	x			
BIS	x	x	x	x			x	x		x		x		
BISB	x	x	x		x		x			x	x			
NFFIN	x	x		x	x	x	x	x	x		x	x		
NFFINB	x		x	x	x	x				x	x			

    MISSING DATA PATTERN FREQUENCIES

Pattern	Frequency	Pattern	Frequency	Pattern	Frequency
1	505	6	1	11	11
2	2	7	1	12	3
3	1	8	265	13	1
4	8	9	4	14	12
5	9	10	3		

COVARIANCE COVERAGE OF DATA

Minimum covariance coverage value   0.100

(continued)

TABLE 9.1. *(continued)*

PROPORTION OF DATA PRESENT

Covariance Coverage

	PANASN	PANASNB	BIS	BISB	NFFIN
PANASN	0.964				
PANASNB	0.637	0.654			
BIS	0.947	0.628	0.954		
BISB	0.627	0.643	0.620	0.644	
NFFIN	0.962	0.639	0.953	0.630	0.971
NFFINB	0.636	0.651	0.627	0.642	0.638

Covariance Coverage

	NFFINB
NFFINB	0.653

Only minor revisions are needed to the Mplus syntax for CFA models with missing data. First, in this example, the statement MISSING = BLANK informs Mplus that for all variables, missing values are represented by blanks in the raw data file. The use of blanks to declare missing values can only be used when the raw data file is in a fixed format (see the FORMAT IS line in Table 9.1). Missing values can be defined in other manners (e.g., use of out-of-range data values such as 999) that work equally well regardless of the structure of the raw data file (e.g., fixed or free format). Note that the syntax in Table 9.1 does not include a statement to request direct ML estimation. As noted above, direct ML is the default estimator in recent releases of Mplus (in earlier versions of Mplus, the TYPE = MISSING option on the ANALYSIS command had to be requested to obtain direct ML). On the OUTPUT line, the PATTERNS option is requested to obtain a summary of missing data patterns. Otherwise, the programming and output of CFA results are the same as for a CFA without missing data.

However, direct ML under the assumption of MAR is often aided by the inclusion of other variables that are not part of the analytic model, but are believed to influence the missing data. Recall that MAR is based on the assumption that the process of missingness can be potentially known and thus predicted by other variables. Thus the use of correlates of missing data can make direct ML estimation more plausible (i.e., can produce parameter estimates closer to the values that would be obtained if no data were lost) because more information about the missing data process is provided to the analysis. The success of this approach depends on whether the missing data correlates were assessed and included in the data set, and on the strength of the relationships these correlates have with the variables in the analytic model with missing data (Enders, 2010; T. D. Little, 2013).

In Mplus, correlates of missingness can be specified by using the AUXILIARY option of the VARIABLE command. In this example, two measures of depression that were collected at the first testing occasion are used as auxiliary variables; that is, AUXILIARY = (M) MDDSUMA DASDEP (Table 9.1). The AUXILIARY option has seven different uses in Mplus 7.11; setting M in parentheses after the equal sign informs Mplus that the auxiliary variables will be used as missing data correlates.

Table 9.1 presents output of the analysis produced by the PATTERNS option. It is important to examine the missing data patterns to determine how well the model can be estimated. The results show that there are 14 missing data patterns (x = data are present; a blank = data are missing). In the first pattern, data are present for all questionnaires at both testing occasions (n = 505). When data are missing, the second pattern is most common (n = 265); that is, all Time 2 measures are missing, suggesting that the participant was lost to follow-up (attrition). Note that 12 cases have missing data on all six variables (Pattern 14), but are nonetheless included in the analysis because missingness is predicted by the auxiliary variables. These cases will be omitted from the analysis if the AUXILIARY option is not used (the researcher may still wish to exclude these cases from the analysis). The next section of the output presents the covariance coverage in the data. Covariance coverage indicates the proportion of data available for each indicator and pair of indicators. For example, the PANASN indicator is available for 96.4% of the sample. Nearly 64% (.637) of the sample have data present for this indicator at both testing occasions (PANASN, PANASNB). Muthén and Muthén (1998–2012) have noted that coverage of around 50% usually causes no problems with direct ML estimation in Mplus. By default, Mplus stops the analysis if coverage is less than 10% for any covariance, although this default can be overridden by an option on the ANALYSIS command (e.g., COVERAGE = .05).

Direct ML assumes that the data are MCAR or MAR and multivariate normal. However, when data are non-normal, direct ML can be implemented to provide standard errors and test statistics that are robust to non-normality by using the MLR estimator (one of the robust ML estimators; Yuan & Bentler, 2000). This feature has been added to recent releases of many programs such as Mplus and EQS.

Multiple Imputation

In most cases, direct ML represents the best and easiest way to manage missing data in CFA and SEM. However, the procedure of *multiple imputation* (Rubin, 1987) is a valuable alternative to direct ML. Multiple imputation possesses statistical properties that closely approximate ML (Allison, 2002, 2003; Enders, 2010; Schafer, 1997). As Allison (2003) notes, multiple imputation is a useful approach to missing data when the researcher does not have access to a program capable of direct ML or wishes to estimate a CFA/SEM approach with a fitting function other than ML or MLR. In addition, T. D. Little (2013) has noted potential advantages of multiple imputation over direct ML in instances where auxiliary variables are involved (as in direct ML, auxiliary variables can and, wherever possible, should be used in multiple imputation to assist the missing data estimation process). Unlike direct ML, multiple imputation is a data-based process that occurs as a separate step before estimation of the CFA model. The advantages of conducting multiple imputation in a separate step include the abilities to incorporate a multitude of auxiliary variables (in direct ML, inclusion of multiple missing data correlates increases estimation demands and the chances of model nonconvergence), and to estimate more complex, conceptually based predictive models of missingness (e.g., to include a missing data correlate that is nonlinearly related to indicators in the analytic model).

As noted earlier, simple imputation procedures such as mean or regression imputation are problematic because they produce underestimates of variances and overestimates of correlations among the variables with imputed data. For instance, if regression imputation is used to supply values on variable Y from data that are available on variable X (i.e., $\hat{Y} = a + bX$), the correlation between X and Y will be overestimated (i.e., for cases with imputed X values, X is perfectly correlated with Y). Multiple imputation reconciles this problem by introducing random variation into the process. In other words, missing values for each case are imputed on the basis of observed values (as in regression imputation), but random noise is incorporated to preserve the proper degree of variability in the imputed data. In the bivariate example, inclusion of this random variation is reflected by the equation

$$\hat{Y} = a + bX + S_{x,y}E \tag{9.2}$$

where $S_{x,y}$ is the estimated standard deviation of the regression's error term (root mean squared error), and E is a random draw (with replacement) from a standard normal distribution. However, if this procedure is conducted a single time, the resulting standard errors will be too small because the standard error calculation will not account adequately for the additional variability in the imputed data. This problem is solved by repeating the imputation process multiple times. Because random variation is included in the process, the resulting data sets will vary slightly. The variability across imputed data sets is used to adjust the standard errors upwardly.[1]

Thus, there are three basic steps of multiple imputation. The first step is to impute multiple data sets. When the amount of missing data is minimal, generating five imputed data sets ($M = 5$) can be sufficient (Allison, 2003), and five has been the default in some software packages (e.g., SAS PROC MI; Yuan, 2000). However, given the ease of conducting multiple imputation in current software (e.g., Mplus), it is recommended that at least 20 imputations be computed to improve the estimates of standard errors and stability of parameter estimates. The number of imputations should be increased much further when the amount of missing data is high and strong correlates of missingness (auxiliary variables) are not available (Enders, 2010). In the second step, the M data sets are analyzed via standard analytic procedures (CFA, ANOVA, etc.). In the third step, the results from the M analyses are combined into a single set of parameter estimates, standard errors, and test statistics. Parameter estimates are combined by simply averaging the estimates across the M analyses. Standard errors are combined using the average of the standard errors over the set of analyses and the between-analysis parameter estimate variation (Schafer, 1997); for the exact formula of this calculation, see either Allison (2003) or Rubin (1987).

Several programs for multiple imputation are now available. The first widely used program for multiple imputation, NORM (a free, downloadable program for Windows), was developed by Schafer (1997). Using the algorithms contained in NORM, SAS has introduced a relatively new procedure (PROC MI; Yuan, 2000) for multiple imputation. Multiple imputation procedures are also available in recent releases of LISREL and Mplus.

To illustrate the three steps of multiple imputation, the first example estimates a CFA model by using PROC MI in SAS. As seen in Table 9.2, the measurement model in this example involves a single latent variable defined by four indicators (S1–S4); in addition, an error covariance is specified for the S2 and S4 indicators. The syntax listed in Step 1 generates multiple imputations with SAS PROC MI. The PROC MI syntax is very straightforward. The names of the input and output data files are specified, along with the list of variables to be used in the imputation process. A couple of PROC MI options are also illustrated in the syntax. The NIMPU option specifies the number of imputations to be conducted (in this example, 20 imputations are requested). The SEED option specifies a positive integer that is used to start the pseudorandom number generator (the SAS default is a value generated from reading the time of day from the computer's clock). In this example, a seed value is specified so that the results can be duplicated in separate computer runs. This option touches on a minor drawback of multiple imputation: Because random variation is introduced, multiple imputation will yield slightly different results each time it is conducted.[2] Various other options are available in PROC MI, including options to make the imputed values consistent with the observed variable values (i.e., set minimum and maximum values, round imputed values to desired units), and to specify the method of imputation (the SAS default is Markov chain Monte Carlo or MCMC, which in essence is a method of improving the estimates of imputed data in context of complex missing data patterns by using the imputed data to produce optimal estimates of the regression coefficients; see Chapter 11 for more information about MCMC). See Yuan (2000) and Allison (2003) for a discussion of other options and complications with multiple imputation.

At a minimum, all variables that are to be used in the substantive analysis (e.g., CFA) should be included in the multiple imputation process. Multiple imputation can be assisted by including correlates of missingness (auxiliary variables), although these are not shown in the Table 9.2 example. In SAS PROC MI, these auxiliary variables are simply added to the variable list.

In the second step of Table 9.2, SAS PROC CALIS is used to conduct a CFA for each of the 20 completed data sets produced by PROC MI. PROC MI writes the 20 data sets to a single file, adding the variable _IMPUTATION_ (with range of values = 1–20) to distinguish each data set. As can be seen in Table 9.2, the CALIS syntax looks typical except for the final two lines. The by _imputation_ line requests a separate CFA for each value of the variable _IMPUTATION_ (i.e., 20 separate CFAs). The final line (ods = output delivery system) informs SAS to write the parameter estimates (Estmates) and covariance matrix (covmat) to SAS data sets (named a and b) for further analysis.

The third step of Table 9.2 demonstrates the use of SAS PROC MIANALYZE, an SAS procedure that combines the results of the multiple analyses and generates valid statistical inferences about each parameter. The keyword PARMS is used to indicate a SAS data set that contains parameter estimates from imputed data sets; the keyword COVB

TABLE 9.2. Computer Syntax (SAS) and Selected Output for Estimation of CFA Model with Missing Data, Using Multiple Imputation

Step 1: Creation of data sets with imputed data

```
proc mi data=sedata nimpu=20 seed=44176 out=seimp;
  var s1 s2 s3 s4;
run;
```

Step 2: CFA of data sets created by multiple imputation

```
libname sys 'c:\missing\';
data esteem;
 set sys.seimp;

proc calis data=esteem method=ml cov privec pcoves;
var = s1-s4;
lineqs
  s1 = 1.0 f1 + e1,
  s2 = lam2 f1 + e2,
  s3 = lam3 f1 + e3,
  s4 = lam4 f1 + e4;
std
  f1 = ph1,
  e1-e4 = th1 th2 th3 th4;
cov
  e2 e4 = th5;
by _imputation_;
ods output Estimates=a covmat=b;
run;
```

Step 3: Combining parameter estimates and standard errors

```
proc mianalyze parms=a covb=b;
var lam2 lam3 lam4 ph1 th1 th2 th3 th4 th5;
run;
```

Selected output

Multiple Imputation Parameter Estimates

Parameter	Estimate	Std Error	95% Confidence Limits		DF
lam2	1.375094	0.062479	1.252596	1.497593	3486.7
lam3	1.354973	0.057990	1.241247	1.468698	2076
lam4	1.265969	0.061006	1.145971	1.385966	339.46
ph1	0.834902	0.078748	0.680498	0.989305	3095.1
th1	0.690190	0.044470	0.602668	0.777712	292.53
th2	0.288787	0.044836	0.200662	0.376912	429.23
th3	0.451633	0.041280	0.370563	0.532703	602.7
th4	0.298328	0.050084	0.195041	0.401615	24.359
th5	0.254208	0.043327	0.164350	0.344070	21.985

Multiple Imputation Parameter Estimates

Parameter	Minimum	Maximum	t for H0: Parameter=Theta0	Pr > |t|
lam2	1.364608	1.387795	22.01	<.0001
lam3	1.345343	1.371108	23.37	<.0001

(continued)

TABLE 9.2. *(continued)*

lam4	1.246292	1.288201	20.75	<.0001
ph1	0.814763	0.852187	10.60	<.0001
th1	0.679365	0.712501	15.52	<.0001
th2	0.266607	0.298306	6.44	<.0001
th3	0.435737	0.463256	10.94	<.0001
th4	0.256843	0.330129	5.96	<.0001
th5	0.233794	0.298554	5.87	<.0001

names a SAS data set that contains covariance matrices of the parameter estimates from imputed data sets (if the COVB option is used, so must PARMS, and vice versa). The variable (VAR) statement is required to list the names of the parameters to be combined (in this example, the freely estimated factor loadings, factor variance, measurement error variances, and error covariance).

In addition, Table 9.2 presents selected output generated by PROC MIANALYZE. Included in this output are the averaged unstandardized parameter estimates, their standard errors, their 95% confidence intervals, range of values across imputations, and test statistics (see the t for H0: column; i.e., parameter estimate divided by its standard error) and associated p values. These results can be interpreted in the same fashion as a standard CFA of a single, complete data set (e.g., unstandardized factor loading of S2 = 1.375, SE = 0.0625, z = 22.01, p < .001). Through simple manipulations (e.g., see Chapter 4), a completely standardized solution can be readily calculated.

In the current version of Mplus (Version 7.11), the three steps of multiple imputation can be carried out in a single syntax program. Table 9.3 presents Mplus syntax for estimating the longitudinal measurement model that was initially estimated by direct ML in Table 9.1. Multiple imputation in Mplus is carried out by using Bayesian estimation (Bayesian estimation is discussed in Chapter 11). However, when the ESTIMATOR=ML option is requested on the ANALYSIS command, the CFA models will be estimated by ML (see Table 9.3). The DATA IMPUTATION command is used to request imputed data sets via multiple imputation. The IMPUTE option indicates the analysis variables for which missing values will be imputed, and the NDATASETS option is used to specify the number of imputed data sets that will be created. The AUXILIARY options can be used to include missing data correlates to foster the recovery of missing values, although these options are not used in this illustration. The format of the Mplus output is identical to the output of model that is estimated without missing data (not shown in Table 9.3). However, the ML parameter estimates are averaged over the number of data sets that have been imputed, and the standard errors of the estimates are computed by using the average of the standard errors across imputations and the between-analysis parameter estimate variation (cf. Rubin, 1987; Schafer, 1997). Moreover, a χ^2 test and other fit statistics for the overall model are provided (Enders, 2010).

In summary, direct ML and multiple imputation are strong methodologies for handling missing data when the data are either MCAR or MAR. If missing data are nonignor-

TABLE 9.3. Computer Syntax (Mplus 7.11) for Estimation of CFA Model with Missing Data, Using Multiple Imputation

```
TITLE:    EXAMPLE OF MULTIPLE IMPUTATION MISSING DATA ESTIMATION IN MPLUS;
DATA:
  FILE IS C:\SPSSWIN\BU2002\FU2002\NxS\DEPSTR.DAT;
  FORMAT IS F6,F3,F2,F2,3F3,6F5.1/F6,24F3;
VARIABLE:
  NAMES ARE ADIS AGE SEX RACE MDDSUMA MDDSUMB MDDSUMC
  CSTRA CSTRB CSTRC ESTRA ESTRB ESTRC
  AD2 DASDEP DASDEPB DASDEPC BDIR BDIRB BDIRC PANASN PANASNB PANASNC
  PANASP PANASPB PANASPC BIS BISB BISC BAS BASB BASC NFFIN NFFINB NFFIC
  NFFIE NFFIEB NFFIEC;
  MISSING = BLANK;
  USEV = PANASN PANASNB BIS BISB NFFIN NFFINB;
 ! AUXILIARY = (M) MDDSUMA DASDEP;
DATA IMPUTATION:
  IMPUTE = PANASN PANASNB BIS BISB NFFIN NFFINB;
  NDATASETS = 20;
ANALYSIS: ESTIMATOR=ML;
MODEL:    NEGAFF1 BY PANASN BIS NFFIN;
          NEGAFF2 BY PANASNB BISB NFFINB;
          PANASN WITH PANASNB; BIS WITH BISB; NFFIN WITH NFFINB;
OUTPUT: STANDARDIZED ;
```

able (i.e., the MAR assumption does not hold), these procedures will yield misleading results (unfortunately, although there are often reasons to believe that data are not MAR, there is no way to test this assumption). Methodologists have developed procedures of estimation with nonignorable missing data (e.g., pattern mixture models). However, such procedures require firm knowledge of the missing data mechanism. In addition, such models are prone to underidentification and are very difficult to estimate properly (Allison, 2002, 2003; Enders, 2010; Little & Rubin, 2002). Allison (2002, 2003) urges that methods for nonignorable missing data should be interpreted with great caution, and should be accompanied by sensitivity analyses to explore the effects of different modeling assumptions (e.g., mechanisms of missingness).

CFA WITH NON-NORMAL OR CATEGORICAL DATA

In previous examples in this book, the ML estimator has been used. The vast majority of CFA and SEM studies reported in the applied research literature use ML. However, an alternative to ML for normal, continuous data is generalized least squares (GLS). GLS is a computationally simpler fitting function, and produces approximately the same goodness of fit as ML (i.e., $F_{ML} = F_{GLS}$), especially when sample size is large. Nevertheless, ML and GLS are only appropriate for multivariate normal, interval-type data (i.e., data in which the joint distribution of the continuous variables is normal). When continuous data depart markedly from normality (i.e., marked skewness or kurtosis), or when some of the indicators are not interval-level (i.e., binary, polytomous, ordinal), alternatives to standard ML estimation should be used.[3] In this section, some of these alternatives are discussed for such situations.

Non-Normal, Continuous Data

Research has shown that ML and GLS are robust to minor departures in normality (e.g., Chou & Bentler, 1995). However, when non-normality is more pronounced, an estimator other than ML should be used to obtain reliable statistical results (i.e., accurate goodness-of-fit statistics and standard errors of parameter estimates). ML is particularly sensitive to excessive kurtosis. The consequences of using ML under conditions of severe non-normality include (1) spuriously inflated model χ^2 values (i.e., overrejection of solutions); (2) modest underestimation of fit indices such as the TLI and CFI; and (3) moderate to severe underestimation of the standard errors of the parameter estimates (inflating the risk of Type I error—i.e., concluding that a parameter is significantly different from zero when that is not the case in the population) (West, Finch, & Curran, 1995). These deleterious effects are exacerbated as sample size decreases (i.e., the risk for nonconverging or improper solutions increases). The two most commonly used estimators for non-normal continuous data are (1) robust ML (Bentler, 1995; Satorra & Bentler, 1994); and (2) weighted least squares (WLS; Browne, 1984b). For reasons discussed in the next section (e.g., requirement of extremely large samples), WLS is not recommended. In contrast, research has shown that robust ML is a very well-behaved estimator across different levels of non-normality (except in instances of severe floor or ceiling effects), model complexity, and sample size (e.g., Chou & Bentler, 1995; Curran et al., 1996).

The robust ML estimator provides ML parameter estimates with standard errors and a χ^2 test statistic that are robust to non-normality. There are various types of robust ML estimators. The two most commonly used estimators are MLM and MLR. MLM, the first robust ML estimator to be developed, provides ML parameter estimates with standard errors and a mean-adjusted χ^2 statistic that are robust to non-normality. The χ^2 statistic produced by MLM is often referred to as the Satorra–Bentler scaled χ^2 (SB χ^2; Satorra & Bentler, 1994). MLR has properties similar to those of MLM, but within the Mplus program it has broader applications, such as the estimation of models with missing data or models where the data have violated the independence-of-observations assumption (this topic is discussed further in Chapter 11). The χ^2 statistic produced by MLR is equivalent to the Yuan–Bentler T2* test statistic (Yuan & Bentler, 2000). Robust ML estimators are available in most latent variable software packages. To use these estimators in Mplus or EQS, raw data must be inputted. In LISREL releases prior to Version 9.10, the preprocessor companion, PRELIS, had to be used to generate a covariance matrix and asymptotic covariance matrix for use as input in the subsequent CFA. The asymptotic covariance matrix is used to compute a weight matrix (W, to adjust fit statistics and standard errors for non-normality) in subsequent LISREL analyses that rely on a non-normal theory estimator such as MLM (Jöreskog & Sörbom, 1996b). Beginning with Version 9.10, LISREL automatically performs robust estimation of standard errors and χ^2 statistics, if a raw data file is used as input and this request is made by using the new RO command (not to be confused with the RO *option*, which invokes a ridge constant for the analysis of nonpositive definite matrices). In addition, as the next example illustrates, robust ML estimation in LISREL 9.10 can still be carried out with the assistance of the PRELIS program, which also provides useful descriptive statistics for the sample data.

Table 9.4 presents Mplus, EQS, and PRELIS/LISREL syntax for conducting a simple CFA model (one factor, five indicators, $N = 870$) by using the MLM estimator. The table shows how raw data can be read into Mplus and EQS (see DATA and /SPECIFICATIONS lines, respectively). The MLM is requested in Mplus and EQS by the ESTIMATOR IS MLM and METHODS=ML, ROBUST syntax, respectively. Otherwise, the syntax is like a typical CFA analysis. In this first example, a congeneric model is specified (one factor, no correlated measurement errors). A later example entails a respecified CFA model of this data set (i.e., correlated error of X1 and X3) to illustrate nested χ^2 evaluation when the MLM or MLR estimator is employed.

The LISREL preprocessing program, PRELIS, serves a variety of functions. These include tests of normality, multiple imputation, bootstrapping, Monte Carlo studies, and generating various types of matrices (e.g., covariance, tetrachoric correlations, polychoric correlations) from raw text-file data or data imported from other software programs such as SPSS. In this example, PRELIS is used to test for normality and create matrices to be used as input for LISREL (e.g., covariance and asymptotic covariance matrices). The line CO ALL is included to declare all 5 indicators as continuous. PRELIS regards any variable with fewer than 16 distinct values as ordinal by default. Thus this command is necessary in instances where indicators with a sample range of 15 or fewer are desired to be treated as interval-level. Syntax on the PRELIS output line (OU) saves the covariance and asymptotic covariance matrices to external files (in addition, a file containing indicator means is saved). The LISREL programming is the same as for a typical CFA, except that the covariance matrix (CM) and asymptotic covariance matrix (AC) are used as input for the analysis. Note that ML is still requested as the estimator on the output (OU) line (i.e., ME=ML). However, because an asymptotic covariance matrix has been inputted, LISREL will provide the SB χ^2 (along with several other χ^2 statistics) and standard errors that are robust to non-normality.

Table 9.5 provides selected results from EQS and PRELIS with regard to the univariate normality of the five indicators (normality tests are automatically produced by EQS when MLM is requested). As shown in this table, EQS and PRELIS produce very similar results. Some of the indicators evidence considerable non-normality (e.g., kurtosis of X5 = 9.4), and thus the assumption of multivariate normality does not hold (although univariate normality does not ensure multivariate normality, univariate non-normality does ensure multivariate non-normality).

When MLM is requested, EQS will output both the ML χ^2 and SB χ^2, as well as both the ML standard errors and robust standard errors of the unstandardized parameter estimates (although LISREL will also provide the ML χ^2, Mplus and LISREL only provide robust standard errors when MLM is requested). In Table 9.6, χ^2 values and the unstandardized factor loadings (and their standard errors) are presented from EQS 6.2 output (the model has been respecified to include the correlated error between X1 and X3). These results demonstrate the typical consequences of using ML in context of non-

TABLE 9.4. Mplus, EQS, and LISREL/PRELIS Syntax for Conducting CFA with Non-Normal, Continuous Data, Using Robust ML (MLM)

Mplus 7.11

```
TITLE:   CFA WITH NON-NORMAL, CONTINUOUS DATA (ROBUST ML)
DATA:
    FILE IS NONML.DAT;
VARIABLE:
    NAMES ARE x1 x2 x3 x4 x5;
ANALYSIS: ESTIMATOR IS MLM;
MODEL:
    F1 BY x1 x2 x3 x4 x5;
                        !without x1, x3 correlated residual (ADDED IN 2ND RUN);
OUTPUT:   STANDARDIZED MODINDICES(10.00) SAMPSTAT;
```

EQS 6.2

```
/TITLE
 one-factor measurement model: non-normal, continuous data
/SPECIFICATIONS
 CASES=870; VARIABLES=5; METHODS=ML, ROBUST; MATRIX=RAW; ANALYSIS=COV;
 DATA = 'NONML.DAT'; FO = '(5F2.0)';
/LABELS
 v1=item1; v2= item2; v3= item3; v4= item4; v5= item5;
 f1 = factor1;
/EQUATIONS
 V1 =  F1+E1;
 V2 = *F1+E2;
 V3 = *F1+E3;
 V4 = *F1+E4;
 V5 = *F1+E5;
/VARIANCES
 F1 = *;
 E1 TO E5= *;
 !/COVARIANCES          ! ADDED IN SECOND RUN
 ! E1, E3 = *;
/PRINT
 fit=all;
/LMTEST
/END
```

LISREL/PRELIS 9.10

```
PRELIS PROGRAM: CFA WITH NON-NORMAL CONTINUOUS DATA
DA NI=5 NO=870
RA FI=NONML.DAT
CO ALL
LA
X1 X2 X3 X4 X5
OU MA=CM ME=NNML.ME CM=NNML.CM AC=NNML.ACC


TITLE LISREL CFA WITH NON-NORMAL, CONTINUOUS DATA
DA NI=5 NO=870 MA=CM
CM=NNML.CM
AC=NNML.ACC
ME=NNML.ME    ! NOT NEEDED UNLESS CONDUCTING MEANSTRUCTURE ANALYSIS
LA
X1 X2 X3 X4 X5
```

(continued)

TABLE 9.4. (continued)

```
MO NX=5 NK=1 PH=SY,FR LX=FU,FR TD=SY,FR
LK
FACTOR1
PA LX
0
1
1
1
1
VA 1.0 LX(1,1)
PA TD
1
0 1
0 0 1          ! X1, X3 CORRELATED ERROR ADDED ON SECOND RUN, TD(3,1)
0 0 0 1
0 0 0 0 1
OU ME=ML RS MI SC AD=OFF IT=100 ND=4
```

normal data. Specifically, the ML χ^2 (25.88) is considerably larger than the SB χ^2 (10.14), reflecting the tendency for ML to produce inflated χ^2 values when data are non-normal. In addition, the standard errors of the ML estimates are noticeably smaller than those based on MLM (e.g., .035 vs. .062 for the V2 indicator), illustrating the propensity for ML to underestimate standard errors in this context. The underestimation of standard errors results in inflated test statistics (e.g., zs for V2 = 22.12 and 11.34 for ML and MLM, respectively), thus increasing the risk of Type I error. Note that the parameter estimates are not affected by the type of estimator used (e.g., λ_{21} = .703 in both ML and MLM). Although not shown in Table 9.6, the values of other commonly used fit indices (e.g., TFI, CFI, RMSEA, SRMR) also differ to varying degrees (e.g., RMSEA = 0.079 and 0.042 in ML and MLM, respectively).

Chi-square difference testing can be conducted with the χ^2 statistics produced by MLM and MLR. However, unlike ML-based analysis, this testing cannot be conducted by simply calculating the difference in χ^2 values produced by the nested and comparison models. This is because a difference between two χ^2 values for nested models estimated by MLM or MLR is not distributed as χ^2. Thus a *scaled difference in χ^2s* (SDCS) test should be used (Satorra & Bentler, 2001, 2010). The SDCS test statistic, T_S, is computed as

$$T_S = (T_0 - T_1) / c_d \qquad (9.3)$$

where T_0 is the regular ML χ^2 for the nested model; T_1 is the regular ML χ^2 for the comparison (less restricted) model; and c_d is the difference test scaling correction. c_d is defined as

$$c_d = [(d_0 * c_0) - (d_1 * c_1)] / (d_0 - d_1) \qquad (9.4)$$

where d_0 is the degrees of freedom of the nested model; d_1 is the degrees of freedom of the comparison model; c_0 is the scaling correction factor for the nested model; and c_1

TABLE 9.5. EQS and PRELIS Selected Results of Tests of Normality

EQS

UNIVARIATE STATISTICS

VARIABLE	ITEM1	ITEM2	ITEM3	ITEM4	ITEM5
MEAN	1.4701	0.8230	1.2655	1.0264	0.6069
SKEWNESS (G1)	1.5090	2.4025	1.8010	2.1612	3.1093
KURTOSIS (G2)	1.2624	5.6901	2.3555	3.9936	9.4023
STANDARD DEV.	2.1728	1.6015	2.0700	1.9280	1.5192

MULTIVARIATE KURTOSIS

MARDIA'S COEFFICIENT (G2,P) =   62.5153
NORMALIZED ESTIMATE =   110.1963

PRELIS

Univariate Summary Statistics for Continuous Variables

Variable	Mean	St. Dev.	T-Value	Skewness	Kurtosis	Minimum	Freq.	Maximum	Freq.
X1	1.470	2.173	19.957	1.512	1.277	0.000	472	8.000	24
X2	0.823	1.601	15.158	2.407	5.730	0.000	591	8.000	8
X3	1.266	2.070	18.032	1.804	2.376	0.000	509	8.000	24
X4	1.026	1.928	15.703	2.165	4.024	0.000	575	8.000	22
X5	0.607	1.519	11.783	3.115	9.464	0.000	669	8.000	7

Test of Univariate Normality for Continuous Variables

	Skewness		Kurtosis		Skewness and Kurtosis	
Variable	Z-Score	P-Value	Z-Score	P-Value	Chi-Square	P-Value
X1	13.682	0.000	5.020	0.000	212.393	0.000
X2	17.956	0.000	10.666	0.000	436.185	0.000
X3	15.269	0.000	7.222	0.000	285.303	0.000
X4	16.957	0.000	9.275	0.000	373.562	0.000
X5	20.434	0.000	12.571	0.000	575.587	0.000

TABLE 9.6. Selected EQS 6.2 Results of CFA with Robust ML Estimation

```
GOODNESS OF FIT SUMMARY FOR METHOD = ML

INDEPENDENCE MODEL CHI-SQUARE = 2523.941 ON 10 DEGREES OF FREEDOM

CHI-SQUARE =        25.883 BASED ON       4 DEGREES OF FREEDOM
PROBABILITY VALUE FOR THE CHI-SQUARE STATISTIC IS        0.00003

GOODNESS OF FIT SUMMARY FOR METHOD = ROBUST

ROBUST INDEPENDENCE MODEL CHI-SQUARE = 612.165 ON 10 DEGREES OF FREEDOM

SATORRA-BENTLER SCALED CHI-SQUARE = 10.1364 ON 4 DEGREES OF FREEDOM
PROBABILITY VALUE FOR THE CHI-SQUARE STATISTIC IS        0.03819


MEASUREMENT EQUATIONS WITH STANDARD ERRORS AND TEST STATISTICS
(ROBUST STATISTICS IN PARENTHESES)

ITEM1   =V1  =    1.000 F1    + 1.000 E1


ITEM2   =V2  =     .703*F1    + 1.000 E2
                   .035
                  20.122
                 (   .062)
                 ( 11.340)

ITEM3   =V3  =    1.068*F1    + 1.000 E3
                   .034
                  31.711
                 (   .044)
                 ( 24.119)

ITEM4   =V4  =     .918*F1    + 1.000 E4
                   .042
                  21.762
                 (   .063)
                 ( 14.614)

ITEM5   =V5  =     .748*F1    + 1.000 E5
                   .033
                  22.402
                 (   .055)
                 ( 13.588)
```

Note. Unstandardized parameter estimates are derived from a revised solution that adds the parameter of an error covariance between items 1 and 3 (see Table 9.4).

is the scaling correction factor for the comparison model. In Mplus, scaling correction factors are automatically provided when the MLM or MLR estimator is used. However, Mplus and EQS users can readily hand-compute the scaling correction factors by dividing the regular ML χ^2 by the SB χ^2:

$$c_0 = T_0 / T_0{}^* \tag{9.5}$$

where $T_0{}^*$ is the SB χ^2 value (or the χ^2 value produced by MLR if this estimator is used). However, a slightly different approach must be used in LISREL. Based on the way LISREL defines SB χ^2, the correct scaling correction value is hand-computed in this program by dividing the model's normal theory WLS χ^2 by its SB χ^2 (Bryant & Satorra, 2012).

The process of computing the SDCS test statistic (T_S) is illustrated with the previous one-factor CFA model (MLM estimation in Mplus 7.11).[4] After the fitting of the initial CFA model, the results indicate that the fit of the solution can be improved by specifying a correlated error between indicators X1 and X3. Because this modification can be defended on substantive grounds, a revised solution is pursued. Table 9.7 breaks

TABLE 9.7. Computing the Scaled Difference in χ^2s (SDCS) Test for Nested Models Estimated with Robust ML (MLM)

Step 1: Obtain T_0 and T_1 values and scaling correction factors for nested and comparison models.

Nested model (one-factor, no correlated measurement errors)

	T_0	$T_0{}^*$	d_0	c_0
ML	87.578		5	
MLM		34.406	5	2.545[*]

$^*c_0 = T_0 / T_0{}^* = 87.578 / 34.406 = 2.545$

Comparison model (one-factor, one correlated measurement error)

	T_1	$T_1{}^*$	d_1	c_1
ML	25.913		4	
MLM		10.356	4	2.502[*]

$^*c_1 = T_1 / T_1{}^* = 25.913 / 10.356 = 2.502$

Step 2: Compute the difference test scaling correction (c_d).

$c_d = [(d_0 * c_0) - (d_1 * c_1)] / (d_0 - d_1)$

$c_d = [(5 * 2.545) - (4 * 2.502)] / (5 - 4) = 2.717$

Step 3: Compute the Satorra–Bentler SDCS test.

$T_S = (T_0 - T_1) / c_d$

$T_S = (87.578 - 25.913) / 2.717 = 22.70$

$T_S = 22.70$, $df = 1$, $p < .001$

Note. T_0, ML χ^2 for nested model; $T_0{}^*$, Satorra–Bentler scaled χ^2 for nested model; T_1, ML χ^2 for comparison model; $T_1{}^*$, Satorra–Bentler scaled χ^2 for comparison model; d_0, degrees of freedom of nested model; d_1, degrees of freedom of comparison model; c_0, scaling correction factor for nested model; and c_1, scaling correction factor for comparison model. Values derived from Mplus 7.11 output.

down the process of computing T_S into three steps. First, the χ^2s must be obtained from ML and MLM (or MLR) for the nested and comparison models (the initial CFA model is nested within the CFA model with a correlated error with a single degree of freedom; $d_0 - d_1 = 5 - 4 = 1$). These four χ^2 values are then used to calculate the scaling correction factors for the nested and comparison models; for example, $c_0 = T_0 / T_0{}^* = 87.578 / 34.406 = 2.545$ (see Table 9.7). In the second step, the difference test scaling correction (c_d) is computed by using the scaling correction factors and degrees of freedom from the nested and comparison models. As shown in Table 9.7, c_d in this example equals 2.717. In the third and final step, T_S is obtained by dividing the difference between the ML χ^2 values of the nested and comparison models by c_d; that is, $T_S = (87.578 - 25.913) / 2.717 = 22.70$. T_S is interpreted in the same fashion as the regular χ^2 difference test. Because the T_S value is statistically significant ($df = 1$, $p < .001$), it can be concluded that the revised one-factor model provides a significantly better fit to the data than the original one-factor solution. In this example, the traditional χ^2 difference test yields the same conclusion (i.e., $T_0 - T_1 = 87.578 - 25.913 = 61.665$, $p < .001$), although the differential magnitude of the χ^2 differences is appreciable (22.70 vs. 61.665). However, there are many situations where using the standard χ^2 difference test to compare nested models estimated by MLM or MLR will yield misleading results (i.e., will provide false evidence of the statistical equivalence or difference between the fit of the nested and comparison model solutions). Thus the SDCS test should always be employed when a researcher is comparing nested solutions estimated by MLM or MLR.

Due to its asymptotic nature, the SDCS procedure will occasionally produce a negative test statistic (because the scaling correction factor is negative). This is most likely to occur in small samples or in situations where the nested model is highly misspecified. Thus, as a follow-up to the original SDCS test, Satorra and Bentler (2010) have developed an improved scaling correction procedure that avoids negative χ^2 difference results. This alternative procedure should be used when negative test statistics are encountered (or when the scaling correction factor is very small). Applied illustrations of the revised SDCS test statistic can be found in Bryant and Satorra (2012) and Asparouhov and Muthén (2013).

Categorical Data

When at least one factor indicator is categorical (i.e., dichotomous, polytomous, ordinal), ordinary ML should not be used to estimate CFA models. The potential consequences of treating categorical variables as continuous variables in CFA are manifold: (1) They produce attenuated estimates of the relationships (correlations) among indicators, especially when there are floor or ceiling effects; (2) they lead to "pseudofactors" that are artifacts of item difficulty or extremeness; and (3) they produce incorrect test statistics and standard errors. ML can also produce incorrect parameter estimates, such as in cases where marked floor or ceiling effects exist in purportedly interval-level measurement scales (i.e., because the assumption of linear relationships does not hold). Thus it is important that an estimator other than traditional ML be used with categorical outcomes or severely non-normal data.

There are several estimators that can be used with categorical indicators: for example, WLS, robust WLS (WLSMV), and unweighted least squares (ULS). Historically, WLS (Browne, 1984b) has been the most frequently used estimator for categorical outcomes (despite the fact that it was originally intended for use with non-normal, continuous data). WLS is available in the major latent variable software programs (in EQS, WLS is referred to as arbitrary generalized least squares, or AGLS). WLS is closely related to the GLS estimator. Like ML, GLS minimizes the discrepancy between the observed (S) and predicted (Σ) covariance matrices. However, GLS uses a weight matrix (W) for the residuals. In GLS, W is typically the inverse of S. WLS uses a different W—specifically, one that is based on estimates of the variances and covariances of each element of S, and fourth-order moments based on multivariate kurtosis (Jöreskog & Sörbom, 1996b; Kaplan, 2000). Thus, unlike GLS, the WLS fit function is weighted by variances–covariances and kurtosis to adjust for violations in multivariate normality; that is, if there is no kurtosis, WLS and GLS will produce the same minimum fit function value, $F_{WLS} = F_{GLS}$.

Because W in WLS is based on the variances and covariances of each element of S (in other words, the "covariances of the covariances"), it can be extremely large, especially when there are many indicators in the model. Consider a three-factor CFA model in which each factor is defined by 6 indicators ($p = 18$). Thus there are 171 elements of S; that is, $a = 18(19) / 2$ (see Eq. 3.14, Chapter 3). In this example, W is of the order $a \times a$ (171×171) and has 14,706 distinct elements; $a(a + 1) / 2 = 171(172) / 2 = 14,706$. The need to store and invert such large matrices in the iterative model estimation process may pose serious difficulties in practical applications (e.g., it is quite demanding of computer resources). Moreover, WLS requires very large samples for accurately estimating the matrix of fourth-order moments (Jöreskog & Sörbom, 1996b). In addition, WLS requires a sample size exceeding $a + p$ (number of elements of S plus number of indicators) to ensure that W is nonsingular (EQS will not perform WLS if $N < a$). Unless the sample size is quite large, very skewed items can make W not invertible; W will frequently be nonpositive definite in small to moderate samples with variables that evidence floor or ceiling effects. Consequently, WLS behaves very poorly in small or moderately sized samples. Moreover, in Monte Carlo studies that have evaluated the performance of various estimators of non-normally distributed continuous data, results have suggested that WLS does not perform as well as MLR or MLM (e.g., Chou & Bentler, 1995). Evidence also suggests that the performance of the WLS estimator with categorical outcomes is not favorable (e.g., oversensitivity of χ^2 and considerable negative bias in standard errors as model complexity increases; Muthén & Kaplan, 1992). Thus, as with non-normal continuous data, WLS is not a good estimator choice with categorical outcomes, especially in small to moderate samples (Flora & Curran, 2004).

At this writing, the Mplus program appears to provide the best options for CFA modeling with categorical data. This is due in part to the availability of the WLSMV estimator and the ability to estimate models with ML via numerical integration (ML estimation is discussed later in this chapter). The WLSMV estimator provides WLS parameter estimates by using a *diagonal* weight matrix (W) and robust standard errors, and a mean- and variance-adjusted χ^2 test statistic (Muthén & Muthén, 1998–2012). Unlike WLS, WLSMV does not require W to be positive definite because W is not inverted as part of the estimation procedure. In WLSMV, the number of elements in the diagonal W equals the number of sample correlations in S, but this matrix is not inverted during estimation. Nevertheless, WLSMV estimation is fostered when N is larger than the number of rows in W. In the computation of the χ^2 test statistic and standard errors, the full W is used but not inverted. Thus, although CFA models using categorical indicators necessitate larger samples than comparably sized models using continuous indicators, the sample size requirements of WLSMV are far less restrictive than those of WLS. For example, Muthén has conducted unpublished simulation studies and has found that Ns of 150 to 200 may be sufficient for medium-sized models (e.g., 10–15 indicators). Flora and Curran (2004) have confirmed these results by showing that WLSMV produces accurate test statistics, parameter estimates, and standard errors of CFA models under a variety of conditions (e.g., sample sizes ranging from 100 to 1,000; varying degrees of non-normality and model complexity). In addition, preliminary simulation research has shown that WLSMV performs well with samples as low as 200 for variables with floor or ceiling effects (although, as with continuous indicators, very skewed categorical indicators call for larger Ns). More studies are needed to more fully establish the performance of WLSMV under various sample sizes and other conditions (e.g., skewness, model complexity, size of indicator relationships, behavior of goodness-of-fit statistics). Because requisite sample size is closely tied to the specific model and data of a given study, general rules of thumb are of limited utility. Instead, researchers are encouraged to use the Monte Carlo routines available in the recent versions of most latent variable programs to determine the necessary N for their particular CFA investigation (see Chapter 10).

The framework and procedures of CFA differ considerably from those of normal-theory CFA when categorical indicators are used. For instance, S is a correlation matrix rather than a covariance matrix (e.g., a tetrachoric correlation matrix is used for binary indicators; a polychoric correlation matrix is used for polytomous indicators). Within the Mplus framework, various response models for categorical indicators are placed into a unifying framework by the use of latent continuous response variables, y^* (e.g., Muthén & Asparouhov, 2002). In the latent response variable framework, y^* reflects the amount of an underlying, continuous, and normally distributed characteristic (e.g., intelligence, attitude, personality trait) that is required to respond in a certain category of an observed categorical variable. For example, in the case of an IQ test, y^* expresses the level of underlying intelligence needed to provide the correct response on a binary test item. This framework assumes that the latent variable can be measured in a more refined fashion (e.g., individual differences in intelligence can be more precisely measured by items that render more units of discrimination than dichotomous or polytomous indicators). Correlations of the underlying y^* variables are used for S rather than the correlations of observed variables. The underlying y^* variables are related to observed categorical variables by means of threshold parameters (τ).[5] In the case of a binary indicator ($y = 0$ or 1), the threshold is the point on y^* where $y = 1$ if the threshold is exceeded (and where $y = 0$ if the threshold is not exceeded). Polytomous items have more than one threshold parameter. Specifically, the number of thresholds is equal to the number of categories minus 1; for example, an ordinal item with three categories ($y = 0$, 1, or 2) has two thresholds (i.e., point on y^* where $y = 1$, point on y^* where $y = 2$). Thresholds are part of the mean structure of a CFA model, and can be used in multiple-groups comparisons (Chapter 7; cf. intercept invariance) or in instances where the researcher wishes to convert the parameters of the CFA model into IRT parameters (described later in this chapter).

Moreover, because a correlation matrix for the y^*s is used as S, the observed variances of the indicators are not analyzed. There are two ways that y^*s can be scaled. In the first and more common method, the variance of y^* is fixed to 1.0 for all items. Although the choice of scale standardization to 1.0 is arbitrary, this approach is in accord with the notion that the input correlation matrix (S) is a covariance matrix of y^*s with unit variance. As a result, the residual variances of categorical indicators are not identified and are not part of the CFA model (unlike CFA with continuous indicators). Thus the measurement errors (θ) of the CFA model with categorical indicators are not free parameters, but instead reflect the remainder of 1 minus the product of the squared factor loading and factor variance; that is,

$$\theta = 1 - \lambda^2 \phi \qquad (9.6)$$

(or simply 1 minus the squared completely standardized factor loading). In the Mplus framework, this approach is referred to as *delta parameterization* (Muthén & Asparouhov, 2002).

The second approach to y^* scaling is *theta parameterization*. In this method, the indicator residual variances are part of the CFA model but are fixed to unity. Consequently, the variances of y^* are computed as the sum of the residual variance plus the variance due to the latent variable; that is,

$$VAR(y^*) = \lambda^2 \phi + \theta \qquad (9.7)$$

(where $\theta = 1$ for all indicators). Although used less often than delta parameterization, theta parameterization is more closely aligned to standard two-parameter IRT modeling (Kamata & Bauer, 2008; IRT is discussed in the next section of this chapter). Moreover, this method is useful when the structure of the residual variances may be an important aspect of the measurement model (e.g., to obtain fit diagnostic information regarding the possible presence of method effects). The delta and theta approaches are equivalent parameterizations of the CFA model and thus provide identical goodness-of-fit statistics and nested model test results.

Estimation of a CFA model with categorical indicators is now illustrated with an applied example (using delta parameterization). In this example, the researcher wishes to verify a unifactorial model of Alcoholism in a sample of 750 outpatients. Indicators of Alcoholism are binary items reflecting the presence–absence of six diagnostic criteria for Alcoholism (0 = criterion not met, 1 = criterion met). Mplus 7.11 syntax for this one-factor CFA is presented in Table 9.8. As with CFA with non-normal, continuous indicators, a raw data file must be used as input. The third line of the VARIABLE command informs Mplus that all six indicators are categorical. Mplus determines the number of categories for each indicator (in this case, all indicators are dichotomous) and then calculates the appropriate type of correlation matrix (in this case, tetrachoric). On the ANALYSIS command, the estimator is specified as WLSMV (although this is the Mplus default with categorical outcomes). Delta parameterization is also the Mplus default, and so this option is not explicitly requested on the ANALYSIS command (the default can be overridden by the option PARAMETERIZATION = THETA). In the current version of Mplus, a mean structure analysis is conducted by default. Consequently, the output will include estimates of the indicator thresholds. The MODEL and OUTPUT commands are written in the same fashion as CFA models with continuous indicators; for example, as before, the unstandardized loading of the first indicator, Y1, is fixed to 1.0 by Mplus default to define the metric of the factor. From Version 3.1 on, Mplus provides modification indices (and nested χ^2 evaluation; see below) when the WLSMV estimator is employed. WLSMV-based modification indices and associated values (e.g., expected parameter change) are interpreted in the same manner as CFA with continuous outcomes (see Chapter 5). The final command in Table 9.8 (SAVEDATA) is not necessary for the current CFA, but is to be used in a forthcoming illustration of nested model evaluation in context of the WLSMV estimator.

Selected Mplus output of the one-factor CFA results appear in Table 9.9. The first section of the output provides the proportions of the sample that endorsed each diagnostic criterion of alcoholism (e.g., 77.6% of the sample met the criterion assessed by the Y1 indicator). As noted earlier, the estimation process benefits from an absence of highly skewed indicators (e.g., to ensure that univariate and bivariate distributions contain several observations per cell). In the next section, the sample tetrachoric correlations reflect the zero-order relationships among the six y^* variables. Because these coefficients are based on the latent response variable underlying the binary indicators, they differ in value from phi correlations, which are based on observed measures. The results indicate that the one-factor model fits the data well: $\chi^2(9) = 9.53$, $p = .39$, RMSEA = 0.009, TLI = 0.999, CFI = .999. Note that Mplus does not provide an SRMR fit statistic. Presumably, this is due to evidence from simulation research (Yu, 2002) that the SRMR does not perform well with binary indicators.[6]

TABLE 9.8. Mplus 7.11 Syntax for Conducting CFA with Categorical Indicators, Using WLSMV Estimation (Delta Parameterization)

```
TITLE:      UNIFACTORIAL MODEL OF ALCOHOLISM
DATA:       FILE IS BINARY.DAT;
            FORMAT IS 6F1;
VARIABLE:   NAMES ARE Y1-Y6;
            USEV = Y1-Y6;
            CATEGORICAL ARE Y1-Y6;
ANALYSIS:   ESTIMATOR=WLSMV;          ! Mplus default with categorical indicators
MODEL:      ETOH BY Y1-Y6;
OUTPUT:     SAMPSTAT MODINDICES(10.00) STAND RESIDUAL TECH2;
SAVEDATA:   DIFFTEST = DERIV.DAT;  ! this command is to be used for nested chi2
PLOT:       TYPE = PLOT3;
```

TABLE 9.9. Selected Mplus Output of One-Factor CFA of Alcoholism, Using Binary Indicators

```
UNIVARIATE PROPORTIONS AND COUNTS FOR CATEGORICAL VARIABLES

   Y1
     Category 1    0.224      168.000
     Category 2    0.776      582.000
   Y2
     Category 1    0.345      259.000
     Category 2    0.655      491.000
   Y3
     Category 1    0.107       80.000
     Category 2    0.893      670.000
   Y4
     Category 1    0.213      160.000
     Category 2    0.787      590.000
   Y5
     Category 1    0.351      263.000
     Category 2    0.649      487.000
   Y6
     Category 1    0.207      155.000
     Category 2    0.793      595.000

SAMPLE STATISTICS

   ESTIMATED SAMPLE STATISTICS

         SAMPLE THRESHOLDS
         Y1$1          Y2$1          Y3$1          Y4$1          Y5$1
         _____      _____      _____      _____      _____
   1      -0.759        -0.398        -1.244        -0.795        -0.384

         SAMPLE THRESHOLDS
         Y6$1
         _____
   1      -0.818


         SAMPLE TETRACHORIC CORRELATIONS
         Y1            Y2            Y3            Y4            Y5
         _____      _____      _____      _____      _____
   Y1
   Y2    0.494
   Y3    0.336         0.336
   Y4    0.610         0.513         0.479
   Y5    0.632         0.515         0.362         0.572
   Y6    0.437         0.317         0.277         0.521         0.468


MODEL FIT INFORMATION

Number of Free Parameters                    12

Chi-Square Test of Model Fit

         Value                            9.532*
         Degrees of Freedom                  9
         P-Value                          0.3897
```

(continued)

TABLE 9.9. *(continued)*

* The chi-square value for MLM, MLMV, MLR, ULSMV, WLSM and WLSMV cannot be used for chi-square difference testing in the regular way. MLM, MLR and WLSM chi-square difference testing is described on the Mplus website. MLMV, WLSMV, and ULSMV difference testing is done using the DIFFTEST option.

RMSEA (Root Mean Square Error Of Approximation)

```
        Estimate                       0.009
        90 Percent C.I.                0.000  0.043
        Probability RMSEA <= .05       0.984
```

CFI/TLI

```
        CFI                            0.999
        TLI                            0.999
```

Chi-Square Test of Model Fit for the Baseline Model

```
        Value                         923.336
        Degrees of Freedom                 15
        P-Value                        0.0000
```

WRMR (Weighted Root Mean Square Residual)

```
        Value                          0.519
```

MODEL RESULTS

	Estimate	S.E.	Est./S.E.	Two-Tailed P-Value
ETOH BY				
Y1	1.000	0.000	999.000	999.000
Y2	0.822	0.072	11.407	0.000
Y3	0.653	0.092	7.106	0.000
Y4	1.031	0.075	13.721	0.000
Y5	1.002	0.072	13.880	0.000
Y6	0.759	0.076	10.024	0.000
Thresholds				
Y1$1	-0.759	0.051	-14.910	0.000
Y2$1	-0.398	0.047	-8.448	0.000
Y3$1	-1.244	0.061	-20.305	0.000
Y4$1	-0.795	0.051	-15.457	0.000
Y5$1	-0.384	0.047	-8.158	0.000
Y6$1	-0.818	0.052	-15.796	0.000
Variances				
ETOH	0.601	0.063	9.609	0.000

IRT PARAMETERIZATION IN TWO-PARAMETER PROBIT METRIC
WHERE THE PROBIT IS DISCRIMINATION*(THETA - DIFFICULTY)

Item Discriminations

ETOH BY				
Y1	1.227	0.160	7.668	0.000
Y2	0.827	0.096	8.605	0.000

(continued)

TABLE 9.9. *(continued)*

Y3	0.587	0.102	5.736	0.000
Y4	1.330	0.180	7.381	0.000
Y5	1.232	0.145	8.482	0.000
Y6	0.728	0.097	7.514	0.000

Item Difficulties

Y1$1	-0.979	0.085	-11.449	0.000
Y2$1	-0.624	0.087	-7.211	0.000
Y3$1	-2.459	0.352	-6.995	0.000
Y4$1	-0.994	0.083	-12.053	0.000
Y5$1	-0.494	0.065	-7.624	0.000
Y6$1	-1.390	0.153	-9.082	0.000

Variances

ETOH	1.000	0.000	0.000	1.000

STANDARDIZED MODEL RESULTS

STDYX Standardization

	Estimate	S.E.	Est./S.E.	Two-Tailed P-Value
ETOH BY				
Y1	0.775	0.040	19.218	0.000
Y2	0.637	0.044	14.492	0.000
Y3	0.506	0.066	7.711	0.000
Y4	0.799	0.039	20.444	0.000
Y5	0.776	0.036	21.358	0.000
Y6	0.589	0.051	11.500	0.000

Thresholds

Y1$1	-0.759	0.051	-14.910	0.000
Y2$1	-0.398	0.047	-8.448	0.000
Y3$1	-1.244	0.061	-20.305	0.000
Y4$1	-0.795	0.051	-15.457	0.000
Y5$1	-0.384	0.047	-8.158	0.000
Y6$1	-0.818	0.052	-15.796	0.000

Variances

ETOH	1.000	0.000	999.000	999.000

R-SQUARE

Observed Variable	Estimate	S.E.	Est./S.E.	Two-Tailed P-Value	Residual Variance
Y1	0.601	0.063	9.609	0.000	0.399
Y2	0.406	0.056	7.246	0.000	0.594
Y3	0.256	0.066	3.856	0.000	0.744
Y4	0.639	0.063	10.222	0.000	0.361
Y5	0.603	0.056	10.679	0.000	0.397
Y6	0.347	0.060	5.750	0.000	0.653

The unstandardized factor loadings are probit coefficients (because a linear equation can be written for the normal, continuous y^*s), which can be converted to probabilities if desired (see below). In delta parameterization, the y^* variances are standardized to 1.0, and thus parameter estimates should be interpreted accordingly. Squaring the completely standardized factor loadings yields the proportion of variance in y^* that is explained by the factor (e.g., Y1 = $.775^2$ = .601), not the proportion of variance explained in the observed measure (e.g., Y1), as in the interpretation of CFA with continuous indicators. Similarly, the residual variances convey the proportion of y^* variance that is not accounted for by the factor; e.g., for Y1: $1 - .775^2$ = .399 (see R-SQUARE section of output for indicator residual variances). The threshold estimates are considered in the next section of this chapter.

As with model χ^2 test statistics estimated with robust ML (e.g., MLM, MLR), the difference in χ^2 values for nested models estimated with WLSMV is not distributed as χ^2 (although beginning with Mplus Version 6.0, the model df under WLSMV are computed in the same fashion as traditional ML). To obtain the correct χ^2 difference test with the WLSMV estimator, a two-step procedure is required. In the first step, the less restricted model is estimated and the DIFFTEST option of the SAVEDATA command is used to save the derivatives needed for the nested χ^2 test. In the second, the more constrained model is fitted to the data, and the χ^2 difference test is calculated by using the derivatives from both analyses.

To illustrate this procedure, a second model is estimated in which the five previously freely estimated factor loadings are constrained to equality (tau equivalence; see Chapter 7). The syntax for this model is provided in Table 9.10. In addition to this constraint, the Mplus syntax is modified so that the DIFFTEST option of the ANALYSIS command is used to give the program the name of the data file that contains the derivatives from the less restricted model. As shown in Table 9.10, Mplus provides the model χ^2 of the new solution as well as the χ^2 difference test for the two solutions. These results are interpreted in the same fashion as χ^2 difference testing with the ML estimator; that is, $\chi^2_{\text{diff}}(4)$ = 27.44, $p < .001$ indicates that the restriction of equal factor loadings significantly degrades the fit of the model.

Comparison with IRT Models

It is well known that factor analysis with binary outcomes is equivalent to a two-parameter normal ogive IRT model (e.g., Ferrando & Lorenza-Sevo, 2005; Glöckner-Rist & Hoijtink, 2003; Kamata & Bauer, 2008; MacIntosh & Hashim, 2003; Moustaki, Jöreskog, & Mavridis, 2004; Muthén, Kao, & Burstein, 1991; Reise, Widaman, & Pugh, 1993; Takane & de Leeuw, 1987). Although a detailed review of IRT is beyond the scope of this book, a few fundamental aspects are noted to demonstrate the comparability of IRT and CFA.

IRT, which has also been referred to as *latent trait theory*, relates characteristics of items (item parameters) and characteristics of individuals (latent traits) to the probability of endorsing a particular response category (Bock, 1997; Lord, 1980). Whereas CFA

TABLE 9.10. Nested Model Comparison with WLSMV: One-Factor CFA of Alcoholism with Binary Indicators (Factor Loadings Constrained to Equality)

Syntax file

```
TITLE:       UNIFACTORIAL MODEL OF ALCOHOLISM (EQUAL FACTOR LOADINGS)
DATA:        FILE IS BINARY.DAT;
             FORMAT IS F4,F2,5F1;
VARIABLE:    NAMES ARE ID Y1-Y6;
             USEV = Y1-Y6;
             CATEGORICAL ARE Y1-Y6;
ANALYSIS:    ESTIMATOR=WLSMV;
             DIFFTEST = DERIV.DAT;              ! line added for chisq diff test
MODEL:       ETOH BY Y1@1 Y2-Y6 (1);           ! equal Y2-Y6 lambdas
OUTPUT:      SAMPSTAT MODINDICES(10.00) STAND RESIDUAL TECH2;
```

Selected output

```
MODEL FIT INFORMATION

Chi-Square Test of Model Fit

            Value                          47.819*
            Degrees of Freedom                 13
            P-Value                        0.0000

Chi-Square Test for Difference Testing

            Value                          27.442
            Degrees of Freedom                  4
            P-Value                        0.0000
```

*   The chi-square value for MLM, MLMV, MLR, ULSMV, WLSM and WLSMV cannot be
     used for chi-square difference testing in the regular way. MLM, MLR and
     WLSM chi-square difference testing is described on the Mplus website. MLMV,
     WLSMV, and ULSMV difference testing is done using the DIFFTEST option.

aims to explain the correlations among test items (or the y^*s underlying the test items), IRT models account for participants' item responses. In other words, an IRT model specifies how both the level of the latent trait and the item properties are related to a person's item responses (which can be measured by either binary or polytomous items). The probability of answering correctly or endorsing a particular response category is graphically depicted by an *item response function* (IRF, also referred to as an *item characteristic curve*, or ICC). IRFs reflect the nonlinear (logit) regression of a response probability on the latent trait. Most often, applied IRT models entail a single latent trait, although multidimensional IRT models can be analyzed (Bock, Gibbons, & Muraki, 1988; Embretson & Reise, 2000). Many objectives of IRT are similar to those of CFA. For instance, IRT can be used to explore the latent dimensionality of categorical outcomes, to evaluate the psychometric properties of a test, and to conduct differential item functioning analysis (DIF; see also Chapter 7). The results of an IRT model can be used to assign sample participants a latent trait level estimate (akin to a factor score) on a standard z-score metric (test scoring). IRT is frequently used in the domains of computerized and educational testing (e.g., the SAT [formerly the Scholastic Aptitude Test], the Graduate Record Examinations) for item parameter estimation, test calibration, and test equating and scoring (e.g., identification of items that yield the highest measurement precision/ information about the examinee in a given trait domain, development of parallel tests or shorter test forms, equating scores across different subsets of items).

In addition to the latent trait level (denoted θ in the IRT literature), either one, two, or three item parameters can be estimated in an IRT model. The choice of IRT model should be based on substantive and empirical considerations (e.g., model fit, although IRT currently provides limited information in regard to goodness of model fit). Examples with binary items are discussed below, although these models can be readily extended to polytomous outcomes. The simplest model is the *one-parameter logistic model* (1PL), also known as the *Rasch model* (Rasch, 1960). In the 1PL model, the probability of responding positively on an item (e.g., correct response on an ability test, meeting diagnostic criterion in the alcoholism example) is predicted by the latent trait (θ) and a single item parameter, *item difficulty* (denoted as either b or β in the IRT literature). This is represented by the logistic function

$$P(y_{is} = 1 \mid \theta_s, b_i) = \exp(\theta_s - b_i) / [1 + \exp(\theta_s - b_i)] \qquad (9.8)$$

In other words, the probability (P) that y equals 1 for a specific item (i) and participant (s), given ($\mid$) the participant's trait level (θ_s) and the item's difficulty (b_i), is the exponent (exp) of the difference of θ_s and b_i, divided by 1 plus the exponent (exp) of the difference of θ_s and b_i.

An item difficulty conveys the level of the latent trait (θ) where there is a 50% chance of a positive response on the item; for example, if $b = .75$, there is a .50 probability that a person with a trait level of .75 will respond positively to the item. The item difficulty parameter (b) represents the location of the curve along the horizontal axis of the IRF; for instance, relatively "easier" items have lower b values and are represented by curves closer to the horizontal axis. Accordingly, b is inversely related to a proportion-correct score (p, or proportion of items endorsed in the presence–absence of a symptom, trait, attitude, etc.). Item difficulties have been alternatively referred to in the IRT literature as *item threshold* or *item location* parameters. In fact, item difficulty parameters are analogous to item thresholds (τ) in CFA with categorical outcomes (cf. Muthén et al., 1991).

In a *two-parameter logistic model* (2PL), an *item discrimination* parameter is included (denoted as either a or α in the IRT literature). Thus the probability of a positive response is predicted by the logistic function

$$P(y_{is} = 1 \mid \theta_s, b_i, a_i) = \frac{\exp[a_i(\theta_s - b_i)]}{1 + \exp[a_i(\theta_s - b_i)]} \qquad (9.9)$$

Item discrimination parameters are analogous to factor loadings in CFA and EFA because they represent the relationship between the latent trait and the item responses.

Thus discrimination parameters influence the steepness of the *slope* of the IRF curves. For instance, items with relatively high *a* parameter values are more strongly related to the latent variable (θ) and have steeper IRF curves. As can be seen in Eq. 9.9, item discrimination (*a*) is a multiplier of the difference between trait level (θ) and item difficulty (*b*). This reflects the fact that the impact of the difference θ and *b* on the probability of a positive response (*P*) depends on the discriminating power of the item (Embretson & Reise, 2000).

A *three-parameter logistic model* (3PL), which includes a "guessing" parameter (denoted as either *c* or γ), can also be estimated in IRT. This additional parameter (also referred to as an *asymptote*) is used to represent IRF curves that do not fall to zero on the vertical axis (i.e., > .00 probability of a positive response for persons with very low θ levels). In other words, if an item can be correctly answered by guessing (as in true–false or multiple-choice items on an aptitude test), the probability of a positive response is greater than zero even for persons with low levels of the latent trait characteristic. Accordingly, 3PL models are most germane to IRT analyses of aptitude or ability tests (e.g., intelligence and educational testing).

Unlike 1PL and 2PL models, the 3PL IRT model currently does not have a CFA counterpart (i.e., there is no analogous CFA parameter for the IRT "guessing" parameter). As will be seen shortly, the one-factor CFA model of alcoholism (Tables 9.8 and 9.9) is analogous to a 2PL IRT model (cf. factor loadings and item thresholds to item discrimination and item difficulty parameters, respectively). A CFA counterpart to a 1PL IRT model can be estimated by holding the factor loadings equal across items.

To illustrate these concepts, a unidimensional 2PL IRT model is fitted to the alcoholism data by using the BILOG-MG software program (Version 3.0; Zimowski, Muraki, Mislevy, & Bock, 2003). Item parameters are estimated by a marginal ML method (MML), the most commonly used estimator in IRT analyses (Embretson & Reise, 2000). The syntax and selected output of this analysis are presented in Table 9.11. The first section of the selected output provides descriptive and classical test statistics. #TRIED indicates that there were no missing responses in the sample of 750 outpatients. #RIGHT provides the frequency with which the item (criterion) was endorsed in the sample. PCT provides the proportion of cases that responded positively to the item (cf. "Univariate Proportions and Counts" in Mplus output, Table 9.9). The last two columns present item–total correlations, both Pearson coefficients and point–biserial correlations that have been adjusted for the base rate of responses.

The second portion of the selected output presents the item calibrations (i.e., IRT parameter estimates). The item discrimination (*a*) and item difficulty (*b*) parameter estimates are provided under the "Slope" and "Threshold" columns, respectively. All asymptote parameters (*c*, "guessing") equal zero because a 2PL model is specified. As in CFA, the item loadings are interpreted as the correlation between the item and the latent trait (θ). Loadings can be calculated by the equation

$$a / SQRT(1 + a^2) \tag{9.10}$$

TABLE 9.11. BILOG-MG 3.0 Estimation of a Unidimensional Two-Parameter Logistic IRT Model of Alcoholism

Syntax file

```
IRT MODEL OF
ALCOHOLISM CRITERIA
>GLOBAL DFName = 'C:\BINARY.DAT',
        NPArm = 2;
>LENGTH NITems = (6);
>INPUT NTOtal = 6,
       NIDchar = 4;
>ITEMS ;
>TEST1 TNAme = 'ETOHDEP',
       INUmber = (1(1)6);
(4A1, 1X, 9A1)
>CALIB ACCel = 1.0000;
>SCORE ;
```

Selected output

ITEM STATISTICS FOR SUBTEST ETOHDEP

						ITEM*TEST CORRELATION	
ITEM	NAME	#TRIED	#RIGHT	PCT	LOGIT/1.7	PEARSON	BISERIAL
1	ITEM0001	750.0	582.0	77.6	-0.73	0.486	0.678
2	ITEM0002	750.0	491.0	65.5	-0.38	0.409	0.527
3	ITEM0003	750.0	670.0	89.3	-1.25	0.267	0.449
4	ITEM0004	750.0	590.0	78.7	-0.77	0.514	0.724
5	ITEM0005	750.0	487.0	64.9	-0.36	0.500	0.644
6	ITEM0006	750.0	595.0	79.3	-0.79	0.358	0.508

SUBTEST ETOHDEP;   ITEM PARAMETERS

ITEM	INTERCEPT	SLOPE	THRESHOLD	LOADING	ASYMPTOTE
ITEM0001	1.213	1.218	-0.996	0.773	0.000
	0.123*	0.155*	0.085*	0.099*	0.000*
ITEM0002	0.517	0.831	-0.622	0.639	0.000
	0.067*	0.097*	0.085*	0.075*	0.000*
ITEM0003	1.515	0.656	-2.309	0.549	0.000
	0.104*	0.099*	0.276*	0.083*	0.000*
ITEM0004	1.319	1.292	-1.021	0.791	0.000
	0.136*	0.171*	0.086*	0.105*	0.000*
ITEM0005	0.608	1.211	-0.502	0.771	0.000
	0.088*	0.159*	0.068*	0.101*	0.000*
ITEM0006	1.020	0.736	-1.386	0.593	0.000
	0.080*	0.092*	0.144*	0.074*	0.000*

* STANDARD ERROR

For example, loading for Y1 = 1.218 / SQRT(1 + 1.218^2) = .773. Item intercepts can be computed by the equation

$$-ab \qquad (9.11)$$

For example, intercept for Y1 = $-(1.218 \times -0.996)$ = 1.213. Item thresholds (b) always have the opposite sign of item intercepts. On the basis of these results, it appears that items Y1, Y4, and Y5 have the highest discrimination; that is, they are most strongly related to θ, such that the probability of a positive response changes most rapidly with a change in θ. The Y3 indicator has the lowest difficulty (b_3 = -2.309); that is, lower levels of the latent dimension of Alcoholism (θ) are required for a positive response on the Y3 criterion (e.g., when θ = -2.309, there is a 50% likelihood that this criterion will be endorsed). As noted earlier, b is related to the rate of item endorsement in the sample (e.g., p_3 = .893; Table 9.11) and the type of sample (e.g., larger b values would be likely if a community sample rather than outpatient sample was used).

With the 2PL equation presented earlier, an IRF curve can be plotted for each item. IRFs for three items from BILOG are displayed in Figure 9.1. These and other IRT curves can also be obtained by using the PLOT command in Mplus 7.11 (see Table 9.8). The form of the curves depict how change in the latent trait (θ, horizontal axis) relate to change in the probability of a positive response (P, vertical axis). The three curves have some similarities. For example, the steepest section of each is the middle of the curve, where small changes in θ are associated with the greatest increase in probability of a positive response. The end of each curve indicates that once a certain trait level is reached (i.e., $\theta \approx 1.5$), increases in the trait are not associated with appreciable change in item endorsement. Several differences are also evident. For instance, although items Y1 and Y5 have similar discrimination parameters (a = 1.218 and 1.211, respectively), Y1 is a "less difficult" item (b = -0.996 and -0.502, respectively). Item difficulty describes the extent to which items differ in probabilities across trait levels. Thus, although items Y1 and Y5 have similar a parameters, the shape of their curves differs somewhat because lower levels of θ are needed for a positive response to item Y1 (its IRF curve is a bit closer to the horizontal axis than Y5's). Item Y3 has the relatively weakest relationship with θ (a = 0.656); hence the probability of a positive response on Y3 changes most slowly with a change in θ. Item 3 also has the lowest threshold (b = -2.309).

With this background, the parallels of CFA and IRT should become clear. When the CFA estimates provided in Table 9.9 are compared with the IRT estimates provided in Table 9.11, it can be seen that the item loadings are quite similar in value (e.g., Y1 = 0.775 and 0.773 for CFA and IRT, respectively). Because factor loadings are closely linked to item discrimination parameters, the CFA factor loadings can be roughly interpreted in IRT terms (i.e., item Y4 has the highest factor loading and thus evidences the highest discrimination). However, it is possible to convert CFA parameters directly into IRT parameters (cf. Kamata & Bauer, 2008; Muthén et al., 1991; Muthén & Asparouhov, 2002). By using CFA parameterization (and symbols), an IRT discrimination parameter can be calculated as follows (with the current WLSMV estimates):

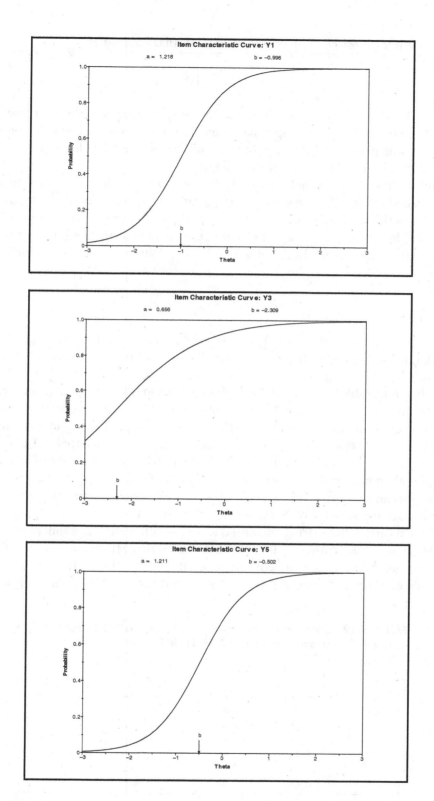

FIGURE 9.1. Item response functions for three selected alcoholism criteria (Y1, Y3, Y5).

$$a = \lambda / \text{SQRT}(\theta) \tag{9.12}$$

where λ is the factor loading, and θ is the residual variance (*not* the trait level, as in IRT parameterization). For example, based on the CFA parameter estimates presented in Table 9.9, the item discrimination of Y1 would be calculated as follows: $a_1 = 0.775 /$ SQRT(0.399) = 1.23 (cf. a_1 = 1.22 in Table 9.11).

Similarly, the item thresholds in Table 9.9 correspond to the item difficulty parameters estimated in the IRT analysis; for example, as the rank order of the item thresholds in the CFA is the same as in IRT, the results of the CFA also convey that item Y3 has the lowest difficulty. However, again by using CFA parameterization (and symbols), an IRT difficulty parameter can be directly calculated as follows (with the current WLSMV estimates):

$$b = \tau / \lambda \tag{9.13}$$

where τ is the CFA item threshold, and λ is the CFA factor loading. Thus the item difficulty parameter of item Y1 is computed as follows: $b_1 = -0.759 / .775 = -0.979$ (cf. b_1 = −0.996 in Table 9.11).

As shown in Table 9.9, these IRT discrimination and difficulty parameters are automatically provided in the latest release of Mplus. In addition, the standard errors for the IRT parameter estimates are presented in this section of the Mplus output (cf. MacIntosh & Hashim, 2003). Table 9.12 presents a summary of all CFA-converted IRT parameters along with the item parameters estimated in BILOG-MG. In this table it can be seen that although both approaches yield very similar results, the CFA-converted estimates and IRT estimates do not correspond exactly. This can be attributed mainly to the use of differing estimation methods (WLSMV in CFA, MML in IRT).

As noted earlier, IRT is frequently used to assess differential item functioning of test items. Muthén (1988; Muthén et al., 1991) has shown that MIMIC models (see Chapter 7) with categorical indicators are equivalent to DIF analysis in the IRT framework (see also Meade & Lautenschlager, 2004). MacIntosh and Hashim (2003) have provided an

TABLE 9.12. Conversion of CFA Parameters to IRT Parameters in the Two-Parameter Probit Metric: IRT Model of Alcoholism

Item	CFA to IRT		IRT	
	a	b	a	b
Y1	1.23	−0.98	1.22	−1.00
Y2	0.83	−0.62	0.83	−0.62
Y3	0.59	−2.46	0.66	−2.31
Y4	1.33	−0.99	1.29	−1.02
Y5	1.23	−0.49	1.21	−0.50
Y6	0.73	−1.39	0.74	−1.39

Note. a, item discrimination parameter; *b*, item difficulty parameter.

applied illustration of converting MIMIC model parameters (and standard errors) to IRT in context of a differential item functioning analysis. The MIMIC parameter estimates pertaining to covariate → item direct effects can be used to calculate item conditional probabilities (i.e., likelihood of item endorsement, given various levels of the covariate and factor). Moreover, Muthén (1988; Muthén et al., 1991) notes that the MIMIC framework offers several potential advantages over IRT. These include the ability to (1) use either continuous covariates (e.g., age) or categorical background variables (e.g., gender); (2) model a direct effect of the covariate on the latent variable (in addition to direct effects of the covariate on test items); (3) readily evaluate multidimensional models (i.e., measurement models with more than one factor); and (4) incorporate an error theory (e.g., measurement error covariances). Indeed, a general advantage of the covariance structure analysis approach is that the IRT model can be embedded in a larger structural equation model (e.g., Lu, Thomas, & Zumbo, 2005).

Measurement Invariance Evaluation

Multiple-groups CFA can be used to evaluate the measurement invariance of instruments composed of categorical outcomes. Like single-group CFA, multiple-groups CFA can be conducted by using either delta or theta parameterization in Mplus, although delta parameterization is the default and more commonly used approach (Muthén & Asparouhov, 2002). In the multiple-groups framework, the variances of y^* are referred to as *scale factors* (symbolized as Δ). As noted in the previous discussion of variances of y^*, scale factors contain information about the residual variance of y^*, the factor loadings, and the factor variances (cf. Eq. 9.7). In multiple-groups analysis, by Mplus default, the scale factors for each indicator are fixed to 1.0 in the first group and freely estimated in the remaining groups because the variances of the y^*s are not presumed to be equal across groups. Thus the scale factors reflect group differences in the variances of the y^*s for the observed categorical dependent variables (specifically, the inverted latent response variable standard deviation). Even though correlations among the y^*s (e.g., tetrachoric correlations with binary indicators) are the units of analysis in CFA with categorical outcomes (which, if analyzed alone, would involve assuming that the model is scale-free), across-groups variation in the factor loadings, factor variances, and residual variances is captured by the corresponding across-groups variation in the scale factors. Although the delta approach reconciles the problem of a scale-free, correlation-based analysis, a drawback is that when noninvariance of scale factors is encountered, it cannot be discerned whether this is due to group differences in the factor loadings, factor variances, or residual variances, or some combination. When the factor loading and threshold for a given indicator are freely estimated across groups (i.e., when an indicator is found to be noninvariant), the scale factor for that indicator must be fixed to 1.0 in all groups for identification purposes.

Akin to single-group analyses, the residual variances of the y^*s of the observed categorical indicators are not allowed to be parameters of the multiple-groups model when the delta approach is used. Generally, this is not seen as a disadvantage because residual variances are rarely invariant and because the test of residual variance equality is usually of no substantive importance (cf. Chapter 7). Nonetheless, the alternative theta parameterization can be used for this purpose because the residual variances are part of the multiple-groups model (but the scale factors are not). Moreover, as in the single-group scenario, theta parameterization will provide fit diagnostic information (modification indices) pertaining to the potential need for the specification of error covariances in one or more group(s). By Mplus default, the residual variances of the y^*s of the categorical indicators are fixed to 1.0 in the first group and are freely estimated in the remaining groups. Invariance of the residual variances can subsequently be tested by fixing the residuals to unity in the remaining groups (with the DIFFTEST option).

An applied example of a multiple-groups analysis using WLSMV estimation and delta parameterization is now provided. This example uses an actual data set of 1,056 participants (426 males, 630 females) who were administered a brief personality questionnaire (measuring the trait of Extraversion) consisting of five true–false items. The invariance of a one-factor measurement model (with no covariances among the five indicator residual variances) is evaluated. The analysis should be carried out in three major steps. In the first step, the CFA should be conducted separately in each group to ensure that the measurement model is acceptable for each group (e.g., good model fit, strong and statistically significant parameter estimates). Although not presented in the tables, these CFAs indicate that the one-factor measurement model fits the data well in both groups; for example, $\chi^2(5) = 2.143$ and 29.756 for males and females, respectively.

The second step entails a baseline multiple-groups model whereby the factor loadings and thresholds are freely estimated in all groups, the factor means are fixed to zero in all groups, and the scale factors are fixed to 1.0 in all groups (as noted earlier, the latter restriction must be imposed for identification purposes when the factor loadings and thresholds are freely estimated across groups). As in multiple-groups CFA with continuous indicators, this model can be referred to as the *equal form* model. Mplus 7.11 syntax for this step is presented in Table 9.13. The programming follows the same approach used for multiple-groups CFA with continuous outcomes (see Chapter 7). In this example, females are the first group. Because the factor loadings and thresholds should be freely estimated in both groups (except for the marker indicator, Y1, which is fixed to 1.0 in both groups), this portion of the model specification is repeated under the MODEL MALE: command to override the Mplus default of fixing these measurement parameters to equality. In the Mplus language, scale factors are represented by curly brackets; thus the statement {Y1-Y5@1} under the MODEL: command fixes the scale factors to 1.0 in the measurement model for females (and also for males because this statement is not repeated under MODEL MALES:). The SAVEDATA command is used to save the model derivatives for nested χ^2 evaluation.

Although not shown in Table 9.13, the equal form model fits the data well: $\chi^2(10) = 31.899$, $p < .001$, RMSEA = 0.064 (CFit $p = .152$), TLI = 0.991, CFI = .996. As in multiple-groups CFA with continuous indicators, the equal form model χ^2 and *df* are the sum of the model χ^2 and *df* from the separate CFA solutions for males and females (e.g., 2.143 + 29.756 = 31.899). The Mplus output indicates that there are no modification

TABLE 9.13. Mplus 7.11 Syntax for Testing the Full and Partial Measurement Invariance of a Personality Measure with Five Categorical Items

Step 1: Fitting the measurement model separately in each group

```
               <see Table 9.8 for single-group CFA syntax>
```

Step 2: Equal form model

```
TITLE:        PERSONALITY QN INVARIANCE - STEP 2
DATA:         FILE IS PXQN2.DAT;
              FORMAT IS F6,6F2;
VARIABLE:     NAMES ARE SUBJ Y1-Y5 SEX;
              USEV = Y1-Y5 ;
              GROUPING IS SEX (0=FEMALE 1=MALE);
              CATEGORICAL ARE Y1-Y5;
ANALYSIS:     ESTIMATOR=WLSMV;
MODEL:        EXTRAV BY Y1-Y5;
              [EXTRAV@0];
              {Y1-Y5@1};
MODEL MALE:   EXTRAV BY Y1@1 Y2-Y5;
              [Y1$1-Y5$1];
OUTPUT:       SAMPSTAT MODINDICES(4.00) STAND RESIDUAL TECH2;
SAVEDATA:     DIFFTEST = DERIVD.DAT;
```

Step 3: Measurement invariance model

```
TITLE:        PERSONALITY QN INVARIANCE - STEP 3
DATA:         FILE IS PXQN2.DAT;
              FORMAT IS F6,6F2;
VARIABLE:     NAMES ARE SUBJ Y1-Y5 SEX;
              USEV = Y1-Y5 ;
              GROUPING IS SEX (0=FEMALE 1=MALE);
              CATEGORICAL ARE Y1-Y5;
ANALYSIS:     ESTIMATOR=WLSMV;
              DIFFTEST=DERIVD.DAT;
MODEL:        EXTRAV BY Y1-Y5;
!                 [EXTRAV@0];
!                 {Y1-Y5@1};
!MODEL MALE:  EXTRAV BY Y1@1 Y2-Y5;
!                 [Y1$1-Y5$1];
OUTPUT:       SAMPSTAT MODINDICES(4.00) STAND RESIDUAL TECH2;
```

Step 4 (if needed): Partial invariance model

```
TITLE:        PERSONALITY QN - PARTIAL INVARIANCE
DATA:         FILE IS PXQN2.DAT;
              FORMAT IS F6,6F2;
VARIABLE:     NAMES ARE SUBJ Y1-Y5 SEX;
              USEV = Y1-Y5;
              GROUPING IS SEX (0=FEMALE 1=MALE);
              CATEGORICAL ARE Y1-Y5;
ANALYSIS:     ESTIMATOR=WLSMV;
              DIFFTEST=DERIVD.DAT;
```

(continued)

TABLE 9.13. *(continued)*

```
MODEL:       EXTRAV BY Y1-Y5;
  !          [EXTRAV@0];
  !          {Y1-Y5@1};
MODEL MALE:  EXTRAV BY Y2;
             [Y2$1];
             {Y2@1};
OUTPUT:      SAMPSTAT MODINDICES(4.00) STAND RESIDUAL TECH2;
```

indices above the minimum value specified in the syntax (4.00), suggesting the lack of salient focal areas of misspecification. However, no modification indices are provided for potential indicator error covariances because residual variances are not an aspect of the CFA model when delta parameterization is used. These fit diagnostic values could be obtained by switching to theta parameterization.

In the third step, the factor loadings and thresholds are constrained to equality across groups (measurement invariance); the factor means are fixed to zero in one group and freely estimated in other groups; and the scale factors are fixed to 1.0 in one group and freely estimated in the other groups.[7,8] Unlike multiple-groups CFA with continuous outcomes, the equality of factor loadings and thresholds (intercepts) cannot be tested separately by using either delta or theta parameterization (Muthén & Asparouhov, 2002). Because the difference in model χ^2s produced by the nested baseline and invariance models is not distributed as χ^2, the DIFFTEST option previously illustrated for the single-group evaluation of tau equivalence (see Tables 9.8 and 9.10) should be used to determine whether the factor loadings and thresholds are statistically equivalent across groups.

As shown in Table 9.13, the measurement invariance model is programmed primarily by using Mplus defaults. For example, because the factor loadings and indicator thresholds are held to equality across groups by default, the syntax used for MODEL MALE: in the previous model is simply commented out (using exclamation marks). The DIFFTEST option is requested on the ANALYSIS command. This output indicates that the restriction of equal loadings and thresholds resulted in a significant increase in model χ^2; that is, $\chi^2_{\text{diff}}(3) = 9.513$, $p = .023$ (not shown in Table 9.13). Inspection of modification indices suggests that the Y2 indicator is noninvariant.

As with continuous outcomes, partial measurement invariance can be pursued when full invariances does not hold. Mplus syntax for the partially measurement invariant model is presented in Table 9.13. As noted earlier, with categorical outcomes, equality constraints for thresholds and factor loadings for a given indicator must be relaxed together. Moreover, with delta parameterization, the scale factor for the noninvariant indicator must be fixed to one in all groups. This is illustrated by the three statements for the Y2 indicator under the MODEL MALE: command (Table 9.13). The results from DIFFTEST indicate that this partial measurement invariance model does not result in a significant increase in χ^2 relative to the equal form model, $\chi^2_{\text{diff}}(2) = 0.743$, $p = .690$; other than Y2, the remaining four indicators are invariant between sexes.

ML Estimation

Beginning with Mplus Version 3.0, ML can also be used as an estimator for single-group or multiple-groups CFA models with categorical outcomes when it is used in conjunction with *numerical integration*. Numerical integration is a computationally complex and time-consuming algorithm that is needed when the posterior distribution for the latent variables does not have a closed form expression. Numerical integration becomes more and more computationally demanding (and less and less feasible) as the number of factors and sample size increase (on the other hand, the computational demand of WLSMV is positively related to the number of variables in the analysis). ML analysis with numerical integration is requested by overriding the WLSMV default (e.g., for the single-group CFA syntax presented in Table 9.8, revising the ESTIMATOR option to read ESTIMATOR=MLR). By default, Mplus executes one dimension of integration with 15 integration points (these defaults can be easily overridden). Practical aspects of employing numerical integration are discussed in the Mplus software manual (Muthén & Muthén, 2008–2012); these include guidelines for altering the number of dimensions of integration and number of integration points per dimension. A key advantage of this approach is the ability to bring all the full information capabilities of ML into a categorical modeling framework (e.g., it handles missing data better than the limited information estimator WLSMV, where missingness is allowed to be a function of the observed covariates but not the observed outcomes). This method is also useful for modeling interaction effects between continuous latent variables (cf. Klein & Moosbrugger, 2000). Unlike WLSMV, which produces probit estimates only, the parameter estimates produced by ML can be either logit or probit coefficients.

Other Potential Remedies for Indicator Non-Normality

Three other remedial strategies for non-normality are briefly presented: bootstrapping, item parceling, and data transformation. The advantages and disadvantages of each method are summarized. These strategies are reviewed primarily to increase the reader's familiarity with alternative approaches that have been used for dealing with non-normal data. However, given the advent of other full information estimators (e.g., MLM, MLR, WLSMV) and limitations associated with each method, utilization of these alternative strategies is becoming less common.

Bootstrapping

Bootstrapping is a resampling procedure in which the original sample serves as the population (cf. Efron & Tibshirani, 1993; Mooney & Duval, 1993). Multiple samples (with the same *N* as the original sample) are randomly drawn from the original sample *with replacement* (i.e., a given case may be randomly selected more than once in any given bootstrapped data set); the CFA model is estimated in each data set; and the results are averaged over the data sets. The number of bootstrapped samples can be specified by the researcher, but should be sufficiently large to foster the quality of the averaged esti-mates (e.g., 500 samples is common). With the one-factor CFA model presented earlier in this chapter (Table 9.4, $N = 870$), bootstrapping might entail generating 500 random samples of $N = 870$ from the original data set; fitting the one-factor CFA model in each sample (i.e., 500 times); and then averaging the results (e.g., parameter estimates, stan-dard errors) across the 500 analyses. The procedure is most appropriate for models with non-normal, continuous indicators. Bootstrapping should not be confused with Monte Carlo simulation, although the two procedures have some similarities (see Chapter 10 for discussion of Monte Carlo simulation). In Monte Carlo simulation, multiple samples (e.g., >500) are randomly generated on the basis of population parameter values and other data aspects (e.g., sample size, amount of non-normality) that are prespecified by the researcher (i.e., unlike in bootstrapping, the researcher has population parameters on hand, but not a sample from the population). As in bootstrapping, the results of mod-els fitted in the simulated data sets are averaged to examine the behavior of the estimates (e.g., stability and precision of parameter estimates and test statistics).

Bootstrapping is based on the notion that when the distributional assumptions of normal-theory statistics are violated, an *empirical sampling distribution* can be relied upon to describe the actual distribution of the population on which the parameter esti-mates are based. Unlike theoretical sampling distributions, the bootstrapped sampling distribution is *concrete* because it is based on the multiple samples spawned from the original data set. Accordingly, the bootstrapped averaged estimates and their standard errors (and possibly fit statistics, depending on the software program used) can be com-pared against the results from the original sample to evaluate the stability of model parameters.

The bootstrapping procedure is straightforward in most latent variable software programs. Table 9.14 presents Mplus 7.11 syntax for generating bootstrapped estimates of the one-factor CFA presented earlier in Table 9.4 (one factor; $N = 870$; five non-normal, continuous indicators). In addition to freely estimating the error covariance between the X1 and X3 indicators, the Mplus syntax is the same as the program in Table 9.4 except that (1) the estimator is changed from MLM to ML; (2) bootstrapping is requested on the ANALYSIS command to compute bootstrapped standard errors (500 bootstrap sam-ples are drawn by the statement BOOTSTRAP = 500); and (3) the CINTERVAL option is listed on the OUTPUT command to obtain bootstrapped bias-corrected confidence intervals for the freely estimated parameters.

The first portion of the selected output in Table 9.14 provides the unstandardized parameter estimates and bootstrapped standard errors for the one-factor solution. The point estimates of the parameters are not affected by the estimation method (e.g., the factor loadings are the same whether robust ML estimation or ML estimation with boot-strapping is used; cf. Tables 9.6 and 9.14). The estimates under the "S.E." column of the Mplus output are the bootstrap estimates of the standard errors of the factor loadings (specifically, these values represent the standard deviations of the parameter estimates across the 500 bootstrapped samples). As in the MLM illustration (Table 9.6), the boot-strapped standard errors are considerably larger than the ML estimates; for instance, for X2, the bootstrapped standard error of 0.059 (Table 9.14) is 69% larger than the ML

TABLE 9.14. Mplus 7.11 Syntax and Selected Output for Bootstrapping a One-Factor CFA with Non-Normal, Continuous Indicators

Syntax file

```
TITLE:  CFA WITH NON-NORMAL, CONTINUOUS DATA (ROBUST ML)
DATA:
    FILE IS NONML.DAT;
VARIABLE:
    NAMES ARE x1 x2 x3 x4 x5;
ANALYSIS: ESTIMATOR IS ML;
        BOOTSTRAP = 500;
MODEL:
    F1 BY x1 x2 x3 x4 x5;
    x1 with x3;
OUTPUT:   STANDARDIZED MODINDICES(10.00) CINTERVAL SAMPSTAT;
```

Selected output

MODEL RESULTS

		Estimate	S.E.	Est./S.E.	Two-Tailed P-Value
F1	BY				
X1		1.000	0.000	999.000	999.000
X2		0.703	0.059	11.921	0.000
X3		1.068	0.049	22.024	0.000
X4		0.918	0.060	15.237	0.000
X5		0.748	0.054	13.957	0.000
X1	WITH				
X3		0.655	0.152	4.309	0.000
Intercepts					
X1		1.470	0.074	19.798	0.000
X2		0.823	0.051	16.156	0.000
X3		1.266	0.070	18.140	0.000
X4		1.026	0.065	15.739	0.000
X5		0.607	0.049	12.338	0.000
Variances					
F1		2.675	0.284	9.435	0.000
Residual Variances					
X1		2.040	0.240	8.498	0.000
X2		1.241	0.126	9.838	0.000
X3		1.227	0.179	6.863	0.000
X4		1.458	0.178	8.170	0.000
X5		0.807	0.110	7.352	0.000

CONFIDENCE INTERVALS OF MODEL RESULTS

	Lower .5%	Lower 2.5%	Lower 5%	Estimate	Upper 5%	Upper 2.5%	Upper .5%
F1 BY							
X1	1.000	1.000	1.000	1.000	1.000	1.000	1.000
X2	0.551	0.587	0.606	0.703	0.800	0.818	0.855
X3	0.943	0.973	0.988	1.068	1.148	1.163	1.193
X4	0.763	0.800	0.819	0.918	1.017	1.036	1.073
X5	0.610	0.643	0.660	0.748	0.837	0.853	0.886

(continued)

TABLE 9.14. *(continued)*

X1 WITH							
X3	0.264	0.357	0.405	0.655	0.905	0.953	1.047
Intercepts							
X1	1.279	1.325	1.348	1.470	1.592	1.616	1.661
X2	0.692	0.723	0.739	0.823	0.907	0.923	0.954
X3	1.086	1.129	1.151	1.266	1.380	1.402	1.445
X4	0.858	0.899	0.919	1.026	1.134	1.154	1.194
X5	0.480	0.510	0.526	0.607	0.688	0.703	0.734
Variances							
F1	1.945	2.119	2.209	2.675	3.142	3.231	3.406
Residual Variances							
X1	1.422	1.570	1.645	2.040	2.435	2.511	2.659
X2	0.916	0.994	1.033	1.241	1.448	1.488	1.566
X3	0.767	0.877	0.933	1.227	1.521	1.578	1.688
X4	0.998	1.108	1.164	1.458	1.751	1.808	1.917
X5	0.524	0.592	0.626	0.807	0.988	1.022	1.090

estimate of 0.035 (Table 9.6). It is noteworthy that the bootstrapped standard errors are similar in magnitude to the standard errors produced by MLM; for example, X2 = 0.059 and 0.062 for MLM and bootstrap, respectively (see Tables 9.6 and 9.14).

The primary objective of bootstrapping is often to obtain better standard errors for the purpose of significance testing, confidence intervals, and so forth. The second section of the selected Mplus output presents the bootstrapped bias-corrected confidence intervals for the freely estimated parameters (i.e., confidence intervals of the parameter estimates using standard errors that have been adjusted on the basis of bootstrapped results). These bias-corrected intervals are interpreted in the same fashion as ordinary confidence intervals; for example, it is 95% likely that the true population value of the unstandardized X2 factor loading falls between 0.587 and 0.818 (because this interval does not include zero, the parameter estimate is significantly different from zero).

Yung and Bentler (1996) warn researchers to avoid believing that bootstrapping will always yield more accurate and reliable results. The success of bootstrapping depends on a number of aspects, including (1) the original sample's representativeness of the population (if it is not representative, bootstrapping will produce misleading results); and (2) "the sampling behavior of a statistic being the same when the samples are drawn from the empirical distribution and when they are taken from the original population" (Bollen & Stine, 1993, p. 113). Bollen and Stine (1993) have also demonstrated that the bootstrap distribution will follow a noncentral χ^2 distribution rather than a central χ^2 distribution, in accord with statistical theory (due to sampling fluctuation in the original sample). These authors have introduced transformation procedures that seem to work reasonably well to minimize this problem. Moreover, bootstrapping is not a remedy for small sample size. As Yung and Bentler (1996) note, "the success of the bootstrap depends on the accuracy of the estimation of the parent distribution (and/or under a particular model) by the observed sample distribution" (p. 223). Thus a very small sample will not render an acceptable level of accuracy. Finally, Yung and Bentler (1996) assert that while bootstrapping may provide more accurate standard error estimates, this approach is not necessarily the best method. The authors acknowledge that other approaches (e.g., using a non-normal theory estimator such as MLM) may be preferred to acquire certain statistical properties such as efficiency, robustness, and so forth.

Item Parceling

Another remedial approach that has been used to address non-normality is *item parceling*. A *parcel* (also referred to as a *testlet*) is a sum or average of several items that presumably measure the same construct. For guidelines on constructing parcels, see Yuan, Bentler, and Kano (1997). The primary potential advantages of using parcels are as follows: (1) Parcels may be more apt to approximate normality than individual items (and thereby the assumptions of ML are more likely to be met); (2) reliability and relationships with other variables are improved (Kishton & Widaman, 1994); and (3) models based on parcels may be considerably less complex (i.e., smaller input matrix, fewer estimated parameters, fewer cross-loadings, and fewer method effects) than models based on individual items (it has been assumed that less model complexity will foster the stability of parameter estimates, but see below).

However, there are a variety of potential problems with using item parcels as indicators in CFA models. The most serious problem is when the underlying structure of the items in a parcel is not unidimensional. In such instances (i.e., one or more parcels is multifactorial), the use of parcels will obscure rather than clarify the latent structure of the data (Bandalos, 2008; West et al., 1995). Latent structures may also be obscured when the uniquenesses (measurement error variances) of items within a given parcel correlate with the unique or common factors of items in other parcels (Bandalos & Finney, 2001; Hall, Snell, & Foust, 1999). In many situations, the use of parcels may not be feasible, such as in instances where there are too few items to form a sufficient number of parcels. For example, in the case of binary indicators, it may take 15 or more items to create a parcel that is reliable and approximates normality. If the entire scale contains only 30 items, for example, then only a few parcels can be created. Particularly when the intent of the CFA is to validate a testing instrument (or other multicomponent measure), the use of a small handful of indicators (parcels) will result in an overly liberal and incomplete test of latent structure (West et al., 1995). In addition, simulation research has shown that the likelihood of improper or nonconvergent solutions increases as the number of parcels decreases (Nasser & Wisenbaker, 2003). Other research has shown that under many different conditions (e.g., sample size), parcels do not outperform models based on individual items (Alhija & Wisenbaker, 2006; Hau & Marsh, 2004; Marsh et al., 1998; Sass & Smith, 2006). For example, in the past parceling has been considered a good method of addressing non-normality when sample size does not permit use of a non-normal theory estimator such as WLS (Browne, 1984b). With the advent of non-normal theory estimators with less restrictive sample size requirements (e.g., WLSMV and MLR in Mplus) and hybrid measurement models such as ESEM (see Chapter 5), item-level analysis is now more feasible (Bandalos, 2008; Marsh, Lüdtke, Nagengast, Morin, & Von Davier, 2013).

Data Transformation

A final potential remedial strategy for non-normality is to transform raw scores of a variable so that they more closely approximate a normal distribution. Different transformation procedures are available for different forms of non-normality (e.g., logarithmic transformation for substantial negative skewness; for an overview, see Tabachnick & Fidell, 2013). However, this approach has a few possible drawbacks. First, transformation is not always successful at reducing the skewness or kurtosis of a variable (e.g., data with severe floor or ceiling effects). Transformed variables must be reassessed in order to verify the success of the transformation in approximating normality at the uni- and multivariate levels. Second, in addition to altering the distribution of variables, nonlinear transformation often changes the variable's relationships with other variables in the analysis. Thus the resulting fit statistics, parameter estimates, and standard errors of a CFA based on transformed indicators may differ markedly from an analysis based on the original variables (West et al., 1995). This may be problematic for at least two reasons: (1) It "strains" reality if the true population distribution is not normally distributed; and (2) it makes the interpretability of the parameter estimates more complex because the metric of the indicators has changed dramatically as the result of transformation. For example, if the variable "years of education" is log-transformed to approximate a normal distribution, all resulting parameter estimates involving this variable should be interpreted with respect to log years, not calendar years. The latter issue is less salient when the original metric of a variable is arbitrary (e.g., Likert response scales of questionnaire items). Nonetheless, the availability and ease of robust estimation procedures in current latent variable software programs (e.g., MLM or MLR, bootstrapping) essentially eliminates the need for transformation as a remedial strategy for non-normality of observed outcomes.

SUMMARY

This chapter has dealt with several situations often encountered in the analysis of applied data sets (missing data, non-normal data, categorical data). As noted at the outset of the chapter, these data situations, and the methods of addressing them, are pertinent to any form of latent variable analysis (e.g., structural regression models, latent growth models). The field of latent variable statistical analysis has evolved tremendously over the past several years. As shown in this chapter, latent variable software programs are now well equipped to accommodate these data difficulties (e.g., direct ML for missing data, estimators other than ML for non-normal or categorical indicators). The next chapter of this book addresses another issue that has received scant attention in SEM sourcebooks to date: determination of the sample size needed to obtain sufficient power and precision of estimates in CFA. As will be seen in Chapter 10, current latent variable software programs offer elegant procedures for addressing this important question (e.g., Monte Carlo evaluation).

NOTES

1. In addition, a single imputation (with added random variation) will produce parameter estimates that are not fully efficient (i.e., estimated variances of the parameter estimates will be biased toward zero) (Allison, 2002; Schafer & Graham, 2002). This problem stems from the fact that the analysis will assume that the missing data are predicted on the basis of equations containing true population values. In other words, single imputation does not incorporate the uncertainty about the predictions of the unknown missing values. Within a multiple imputation framework, uncertainty about these parameters from one imputation to the next is reflected by taking random draws from the Bayesian posterior distribution of the parameters. The form of this posterior distribution is described by Iversen (1985); procedures for random draws from this distribution (e.g., data augmentation) are discussed by Schafer (1997).

2. As Allison (2002) notes, the slight variability in multiply imputed data sets could lead to the awkward situation where different researchers obtain different results from the same data and standard analyses. In addition, such variability could promote unscrupulous uses of the procedure (e.g., a researcher might repeat the multiple imputation process until a "borderline" effect becomes statistically significant).

3. In CFA, if one or more indicators of a factor are not continuous, then an estimator other than ML must be used. However, this does not apply to exogenous predictors (covariates) of latent variables. For example, in MIMIC models (where the covariate may be categorical dummy codes reflecting levels of groups; see Chapter 7), ML is appropriate, assuming that indicators of the latent variables are normal, continuous.

4. Identical SDCS test statistic results can be obtained in Mplus by using the model log-likelihoods and their correction factors. Advantages of the log-likelihood approach include the facts that it can be used for any pair of nested models and that it does not require the computation of model goodness-of-fit statistics; see Asparouhov and Muthén (2013) for a description of this method.

5. Note that τ is also used to denote indicator intercepts in mean structure analysis of continuous indicators (see Chapter 7).

6. As an alternative to SRMR, the results of Yu (2002) suggest that a cutoff of ≤ 1.0 on a relatively new fit statistic, the weighted root mean square residual, can be used for models with binary outcomes when Ns ≥ 250.

7. If theta parameterization is used, the models in Steps 2 and 3 will be as follows. Step 2: Freely estimate factor loadings and thresholds in all groups; fix factor means to zero in all groups; and fix residual variances to 1.0 in all groups. Step 3: Hold factor loadings and thresholds to equality in all groups; fix factor means to zero in one group and freely estimate them in remaining groups; and fix residual variances to 1.0 in one group and freely estimate them in remaining groups. Invariance of the residual variances can be subsequently tested by fixing the residuals to unity in the remaining groups.

8. In Mplus 7.11, the convenience feature options of CONFIGURAL and SCALAR can be used in multiple-groups CFA with categorical outcomes (see Chapter 7 for discussion of these programming options). However, the METRIC option cannot be used with WLSMV estimation; the model with binary outcomes is not identified because the scale factors or residual variances are allowed to vary across groups.

10

Statistical Power and Sample Size

In this chapter, appropriate strategies are presented for determining the sample size required to achieve adequate statistical power and precision of parameter estimates in a CFA study. This topic is often neglected in SEM sourcebooks and widely misunderstood in the applied literature. For instance, applied researchers often cite a general guideline (e.g., a participant–indicator ratio) in support of study sample size. As discussed in this chapter, these rules of thumb are very crude and usually do not generalize to the researcher's data set and model. Thus sample size requirements should be evaluated in context of the particular data set and model at hand. As with many of the topics presented in Chapter 9, the concepts and strategies presented in this chapter are relevant to SEM approaches of any type (e.g., CFA, structural regression models).

OVERVIEW

In designing a CFA investigation, the researcher must address the critical question of how many cases (participants) should be collected to obtain an acceptable level of precision and statistical power of the model's parameter estimates, as well as reliable indices of overall model fit. The extant literature provides little guidance on this issue. Some SEM sourcebooks provide general rules of thumb based on a small set of Monte Carlo studies. Many rules of thumb have been offered, including minimum sample size (e.g., $N \geq 100–200$), minimum number of cases per each freed parameter (e.g., at least 5–10 cases per parameter), and minimum number of cases per indicator in the model (cf. Bentler & Chou, 1987; Boomsma, 1983; Ding, Velicer, & Harlow, 1995; Tanaka, 1987). Such guidelines are limited by their poor generalizability to any given research data set. That is, the models and assumptions used in Monte Carlo studies to provide sample size guidelines are often dissimilar to the types of models and data used by the applied researcher. Indeed, requisite sample size depends on a variety of aspects, such as the study design (e.g., cross-sectional vs. longitudinal); the size of the relationships among the indicators; the reliability of the indicators; the scaling (e.g., categorical, continuous) and distribution of the indicators; estimator type (e.g., ML, robust ML, WLSMV); the amount and patterns of missing data; and the size of the model (model complexity). These features will vary widely from data set to data set. Thus the goal of determining required sample size is fostered by the extent to which the actual model and data can be approximated in the power analysis.

Sample size affects the *statistical power* and *precision* of the model's parameter estimates. Statistical power is defined as one minus the probability of Type II error. Based on the work of Cohen (1988, 1992), the cutoff most frequently used to define acceptable power is .80—that is, an 80% likelihood of rejecting a false null hypothesis (thus risk of Type II error is 20%). In SEM and CFA, power pertains to both the test of the model (e.g., sensitivity of χ^2 to detect model misspecifications) and the model parameter estimates (i.e., probability of detecting a parameter estimate as significantly different from zero). The closely related concept of precision pertains to the ability of the model's parameter estimates to capture true population values. For example, in the population, if the correlation between variables X and Y is .30, do the sample data and model reasonably approximate this population value? Precision can be gauged in part by the amount of bias in the parameter estimates and their standard errors. Traditionally, model-based approaches to determining appropriate sample size have focused on the issue of statistical power. Some latent software packages also allow researchers to consider the precision of model estimates.

SATORRA–SARIS METHOD

The first, most widely known approach for conducting power analysis for multiple-indicator SEM was introduced by Satorra and Saris (1985). This approach focuses on the power of the χ^2 difference test to detect specification errors associated with a single parameter. The researcher specifies a model associated with a null hypothesis (H_0) that is evaluated in comparison to a model that represents the alternative hypothesis (H_1). In simplest terms, H_1 reflects the "true" model (i.e., it contains the true population values for all parameters), and H_0 is identical to H_1 except for the parameter(s) to be tested. Thus a population covariance matrix is generated from the parameter estimates of H_1. This covariance matrix is used as input in the test of H_0, which contains the misspecified parameter and the sample size of interest. In practice, the misspecified parameter is most often fixed to zero, and thus the analysis focuses on the power of model χ^2 to detect that a parameter is different from zero. The test of H_0 produces a nonzero model χ^2 value resulting from the misspecified parameter. This value represents a noncentrality parameter (NCP, often symbolized as λ, but not to be confused with a factor loading) of the noncentral χ^2 distribution. The noncentral χ^2 distribution reflects the χ^2 distribution when the null hypothesis is false. With the resulting NCP value, the power of the test can be determined from tabled values (e.g., Saris & Stronkhorst, 1984) or simple routines in commercial software packages such as SPSS and SAS. Various sample sizes are considered to identify the required N for power = .80 (α = .05).

To illustrate the logic and procedures of this approach, a simple example is given. The model (H_1) containing the population values of the various parameters is presented in Figure 10.1. Ideally, these values should be based on prior research (e.g., pilot data). The solution of interest is a two-factor CFA model of Self-Esteem and Depression (measured by three indicators each). The analysis will focus on the sample size needed to have adequate power (.80) to detect the factor covariance of Self-Esteem and Depression (ϕ_{21} = .35) as significantly different from zero (α = .05).

The first step is to generate a population covariance matrix from the parameter estimates of the H_1 model. In this example, standardized values are used for ease of illustration and interpretation. To accomplish this, a covariance matrix of null relationships (i.e., zero values in the off-diagonal, variances on the diagonal) is used as input, and the various model parameters (i.e., factor loadings, error variances and covariances, factor covariances) are fixed to equal the putative population values. Of course, this specification will produce a poor fit to the data because all relationships are null in the input matrix, but it is not the purpose of this step to obtain goodness-of-fit information. Rather, this analysis will produce a fitted covariance matrix that can be used as the population covariance matrix in subsequent steps of the procedure.

Table 10.1 provides LISREL syntax for conducting this initial step with the parameter estimates from Figure 10.1. Again, note that the covariance matrix (CM) used as input contains zeroes in all off-diagonal elements (i.e., no relationships among the six indicators) and the indicator variances on the diagonal. There are no freely estimated parameters in this analysis because all parameters are fixed to population values using the value (VA) command. A sample size of 500 is specified. In this step, the size of the

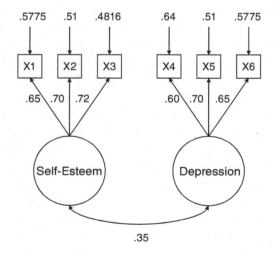

FIGURE 10.1. Hypothesized model of Self-Esteem and Depression.

TABLE 10.1. LISREL 9.10 Syntax for Satorra–Saris Method of Determining That Power to Detect Factor Covariance Is Significantly Different from Zero

Step 1: Generate population covariance matrix from H_1 model

```
TITLE SATORRA-SARIS METHOD OF POWER CALCULATION: STEP ONE
DA NI=6 NO=500 MA=CM
LA
X1 X2 X3 X4 X5 X6
CM
1.0
0.0 1.0
0.0 0.0 1.0
0.0 0.0 0.0 1.0
0.0 0.0 0.0 0.0 1.0
0.0 0.0 0.0 0.0 0.0 1.0
MO NX=6 NK=2 PH=SY,FI LX=FU,FI TD=SY,FI
LK
ESTEEM DEPRESS
VA .65 LX(1,1)
VA .70 LX(2,1)
VA .72 LX(3,1)
VA .60 LX(4,2)
VA .70 LX(5,2)
VA .65 LX(6,2)
VA .5775 TD(1,1)
VA .51   TD(2,2)
VA .4816 TD(3,3)
VA .64   TD(4,4)
VA .51   TD(5,5)
VA .5775 TD(6,6)
VA 1.0 PH(1,1) PH(2,2)
VA .35 PH(2,1)
OU ME=ML RS AD=OFF IT=100 ND=4
```

Step 2: Analyze residual covariance matrix to ensure that population values are recovered

```
TITLE SATORRA-SARIS METHOD OF POWER CALCULATION: STEP TWO
DA NI=6 NO=500 MA=CM
LA
X1 X2 X3 X4 X5 X6
CM
1.0000
0.4550  1.0000
0.4680  0.5040  1.0000
0.1365  0.1470  0.1512  1.0000
0.1592  0.1715  0.1764  0.4200  1.0000
0.1479  0.1592  0.1638  0.3900  0.4550  1.0000
MO NX=6 NK=2 PH=SY,FR LX=FU,FR TD=DI
LK
ESTEEM DEPRESS
PA LX
1 0
1 0
1 0
0 1
0 1
0 1
```

(continued)

TABLE 10.1. *(continued)*

```
PA PH
0
1 0
VA 1.0 PH(1,1) PH(2,2)
OU ME=ML RS AD=OFF IT=100 ND=4
```

Step 3: Fit H$_0$ model that contains the misspecified parameter (i.e., factor covariance fixed to zero) and target sample size (e.g., $N = 100$)

```
TITLE SATORRA-SARIS METHOD OF POWER CALCULATION: STEP THREE
DA NI=6 NO=100 MA=CM                      ! SPECIFY SAMPLE SIZE OF INTEREST
LA
X1 X2 X3 X4 X5 X6
CM
1.0000
0.4550   1.0000
0.4680   0.5040   1.0000
0.1365   0.1470   0.1512   1.0000
0.1592   0.1715   0.1764   0.4200   1.0000
0.1479   0.1592   0.1638   0.3900   0.4550   1.0000
MO NX=6 NK=2 PH=SY,FR LX=FU,FR TD=DI
LK
ESTEEM DEPRESS
PA LX
1 0
1 0
1 0
0 1
0 1
0 1
PA PH
0
0 0                                       ! FACTOR COVARIANCE FIXED TO ZERO
VA 1.0 PH(1,1) PH(2,2)
OU ME=ML RS MI SC AD=OFF IT=100 ND=4
```

Step 4: Use χ^2 from Step 3 as noncentrality parameter to estimate power at targeted sample sizes (e.g., $N = 100$)

SAS syntax

```
DATA POWER;
DF=1; CRIT = 3.841459;
LAMBDA = 6.4368;
POWER = (1-(PROBCHI(CRIT,DF,LAMBDA)));
RUN;
```

SPSS syntax

```
DATA LIST FREE /X DF NCP.
COMPUTE POWER = 1 - NCDF.CHISQ(X,DF,NCP).
BEGIN DATA.
3.841459 1 6.4368
END DATA.
LIST POWER.
```

sample is arbitrary. In addition, to reduce rounding error, the analysis should be conducted with a sufficient number of decimal points (in this case, ND = 4 on the Output line). This syntax produces the following fitted covariance matrix, which will be used as the population matrix in the next two steps. Although more time-consuming, this matrix could also be calculated by hand, using the equations presented in Chapter 3; for example, covariance of X1 and X4 = .65(.35)(.60) = .1365 (cf. Eq. 3.8).

```
     Fitted Covariance Matrix

                 X1          X2          X3          X4          X5          X6
             --------    --------    --------    --------    --------    --------
      X1       1.0000
      X2       0.4550      1.0000
      X3       0.4680      0.5040      1.0000
      X4       0.1365      0.1470      0.1512      1.0000
      X5       0.1592      0.1715      0.1764      0.4200      1.0000
      X6       0.1479      0.1592      0.1638      0.3900      0.4550      1.0000
```

The second step is simply an accuracy check. The H_1 model is freely estimated with the fitted covariance matrix from the first step as input data (see Table 10.1). If the first two steps are conducted properly, this analysis will fit the data perfectly, and the parameter estimates will be identical within rounding error to the original population values (e.g., $\lambda_{11} = .65$, $\phi_{21} = .35$).

In the third step, the prior analysis is repeated except for two key alterations. First, the parameter of interest is misspecified. In this step, the null hypothesis is specified. Specifically, we are interested in determining the power to detect that the factor correlation of Self-Esteem and Depression is significantly different from zero. Thus this parameter should be fixed to zero in this analysis (see PA PH commands in Table 10.1). Second, the sample size (N) for which power is desired must be specified. In Table 10.1, N is specified as 100. Therefore, the current analysis will estimate the power to detect a factor covariance of .35 as statistically significant when $N = 100$.

This analysis produces a model χ^2 value of 6.4368. Because all other aspects of the solution are properly specified, this nonzero χ^2 value is due to the misspecification of the factor covariance parameter. As noted earlier, this χ^2 value is the NCP (λ) of the noncentral χ^2 distribution—that is, the distribution of χ^2 when the null hypothesis is false.

In the fourth step, the χ^2 value from the preceding analysis (here, 6.4368) is used as the NCP to calculate the power to detect the model misspecification. Although tabled values are available (e.g., Saris & Stronkhorst, 1984), this step can be performed by using simple routines in commercial software packages. These routines require three pieces of information: the NCP, degrees of freedom (df), and the critical χ^2 value. Because this particular power analysis focuses on a single parameter, df is 1 (i.e., this df should not be confused with the df of the H_1 or H_0 model) and the critical χ^2 value will be 3.84 (i.e., when α is set at .05). In Table 10.1, the critical χ^2 value is written out to six decimal points to reduce rounding error. Table 10.1 provides SAS and SPSS syntax for these routines.

The analysis yields a power value of .72. Thus, when $N = 100$, there is a 72% likelihood of rejecting a false null hypothesis (i.e., detecting the factor covariance of Self-Esteem and Depression as significantly different from zero at the .05 alpha level). Because .72 is below conventional cutoffs (i.e., power = .80), the results indicate that sample size should be larger to obtain adequate statistical power. NCPs (χ^2) for other sample sizes can be calculated in a few ways (each produce identical results). The Step 3 analysis can be rerun after changing the N of 100 to a larger value. Alternatively, most programs (e.g., LISREL) provide the minimized fit function value (F_{ML}) which can be used to calculate χ^2 at various sample sizes; e.g., $\chi^2 = F_{ML}(N)$. In programs that do not provide this value, F_{ML} can be readily computed as $\chi^2 / (N)$. The NCPs for other sample sizes are then used as input in the power calculations (e.g., SAS or SPSS routines presented in Table 10.1).

In the current example, increasing sample size to 125 produces an NCP value of 8.0459. Using this NCP in the SAS or SPSS routine produces a power estimate of .81. Thus an $N = 125$ appears more suitable in terms of the power to detect the factor covariance as statistically different from zero. For the reader's information, NCPs and power estimates for a few other sample sizes are as follows:

N	NCP	Power
100	6.4368	.72
125	8.0459	.81
150	9.6551	.87
200	12.8735	.95

The Satorra–Saris (1985) method is superior to general rules of thumb (e.g., ratio of cases to freed parameters) because it is a model-based, quantitative approach to determining requisite sample size. However, this approach has several disadvantages. As noted by Kaplan (1995), the method requires that the researcher specify the alternative value(s) to be tested (e.g., $\phi_{21} = .35$). Satorra (1989) later demonstrated that modification indices can be used as approximate NCP values for each restriction (e.g., fixed parameter) in the model (recall the relationship of the modification index and χ^2 at $df = 1$; see Chapter 5). Thus alternative values for the parameters of interest do not need to be specified; instead, the modification indices are used as NCP values to determine the power to detect misspecification of the restricted parameter. In addition, the aforementioned approach focuses on a single parameter at a time. Although Saris and Satorra (1993) subsequently developed a procedure for assessing power for multiple simultaneous misspecifications (*isopower contours*), these procedures are not readily implemented in current latent variable software packages.

Some have criticized model-based approaches to power analysis because the procedures require the researcher to make exact estimates of population values for each parameter in the model. For instance, although the example above focuses on the factor covariance, all the factor loadings, factor variances, and indicator residuals have to be properly specified. This is often difficult to carry out in practice. A few misestimates of the parameter population values may undermine the power analysis. Accordingly, some methodologists (e.g., Jaccard & Wan, 1996) have suggested that OLS-based power analysis (e.g., Cohen, 1988) may be useful to determine power/requisite sample size for salient structural parameters of the model (i.e., effects involving relationships among latent variables). Bootstrapping (e.g., Bollen & Stine, 1993; see also Chapter 9) has been proposed as another approach to power analysis, although this method has limited utility in research planning (e.g., determining target N) because a sufficiently large existing data set is required to generate bootstrapped samples (for an overview of bootstrapping as a power analysis procedure, see Jaccard & Wan, 1996).

MONTE CARLO APPROACH

In addition to the aforementioned drawbacks, the Satorra–Saris and OLS methods are limited by the fact that they do not address the precision of the model parameter estimates or directly take into account other salient aspects of the research data, such as degree of non-normality, type of indicators (e.g., binary, continuous), and the amount and patterns of missing data. Recent developments in latent variable software packages permit researchers to use Monte Carlo methodology to determine the power and precision of model parameters in context of a given model, sample size, and data set. Typically, Monte Carlo studies are used in SEM research to study the behavior of statistical estimators and test statistics under various conditions manipulated by the researcher, such as sample size, degree of model misspecification, and degree of data non-normality (cf. Paxton, Curran, Bollen, Kirby, & Chen, 2001). However, Muthén and Muthén (2002) have shown how the Monte Carlo method can be conveniently used to determine sample size and power in the design of SEM studies.

As the Satorra–Saris (1985) approach does, the Monte Carlo method requires specification of an H_1 model containing the population values of all parameters (e.g., Figure 10.1). Ideally, these values should be guided by existing data (e.g., pilot studies). Numerous samples are randomly generated on the basis of the population values of the model. Appendix 10.1 provides an illustration of the data generation process used in Monte Carlo research; for further details, the reader is referred to Mooney (1997), Fan and Fan (2005), and Muthén and Muthén (1998–2012). In the absence of additional programming, the data sets generated by the Monte Carlo utility will consist of normally distributed, continuous data (with no missing data). However, programs such as Mplus and EQS offer programming features to produce the desired amount of non-normality and missing data, as well as the capability of estimating in context of noncontinuous indicators (and categorical estimators such as WLSMV; see Chapter 9).[1] The specified SEM model is estimated in each sample, and the results of these analyses (e.g., parameter values, standard errors, fit statistics) are averaged across the samples. These averages are used to determine the precision (e.g., bias and coverage) and power of the estimates (i.e., proportion of samples in which the parameter is significantly different from zero). As in the Satorra–Saris approach, various sample sizes are studied to determine the N needed to obtain parameter estimates with sufficient power and precision.

The CFA model in Figure 10.1 is used to illustrate this approach. This CFA model is first examined under the condition of normally distributed indicators without missing data (for examples of Monte Carlo evaluations with non-normal indicators, see Muthén & Muthén, 2002). As in the Satorra–Saris example, the initial analysis will focus on the suitability of a sample size of 100. Mplus syntax for this analysis is presented in Table 10.2. The MONTECARLO command describes the details of the Monte Carlo study, such as the number and names of indicators (X1–X6) and the sample size (NOBSERVA-TIONS = 100) used for data generation and analysis. The NREPS statement specifies how many samples (replications) are to be generated from the parameters of the population model. This number can be regarded as the sample size of the Monte Carlo study. Current computer technology (e.g., fast CPU speeds) allows the user to specify a large number of replications (10,000, in this example) to ensure the stability of the simulation results (e.g., average of parameter estimates across replications). The SEED statement specifies the starting point used for the random draws from the population. Muthén and Muthén (2002) recommend the use of multiple seeds to verify the consistency of results across different seed values.

The MODEL POPULATION command describes the data generation model for the Monte Carlo study. On this command, the true population parameter values used for data generation are provided. In the current example, the parameter values listed in Figure 10.1 are specified as population values (and factor variances of Self-Esteem and Depression are fixed to 1.0). The MODEL command is used to indicate the analysis model (i.e., the CFA). In some applications, the analysis model will be different from

TABLE 10.2. Mplus 7.11 Syntax for Monte Carlo Approach to Determining Power and Precision of CFA Model Parameter Estimates

```
TITLE:              CFA TWO-FACTOR, NORMAL DATA, NO MISSING
MONTECARLO:
                    NAMES ARE X1-X6;
                    NOBSERVATIONS = 100;                ! SAMPLE SIZE OF INTEREST
                    NREPS = 10000;
                    SEED = 53567;
MODEL POPULATION:
                    ESTEEM BY X1*.65 X2*.70 X3*.72;
                    DEPRESS BY X4*.60 X5*.70 X6*.65;
                    ESTEEM@1; DEPRESS@1;
                    X1*.5775; X2*.51; X3*.4816; X4*.64; X5*.51; X6*.5775;
                    ESTEEM WITH DEPRESS*.35;
MODEL:
                    ESTEEM BY X1*.65 X2*.70 X3*.72;
                    DEPRESS BY X4*.60 X5*.70 X6*.65;
                    ESTEEM@1; DEPRESS@1;
                    X1*.5775; X2*.51; X3*.4816; X4*.64; X5*.51; X6*.5775;
                    ESTEEM WITH DEPRESS*.35;
ANALYSIS:           ESTIMATOR = ML;
OUTPUT:             TECH9;
```

Note. Normal and no missing data; $N = 100$.

the data generation model; for example, the data are generated as categorical, but analyzed as continuous to study the impact on various statistical estimators. In the typical power analysis, the MODEL command will be identical to the data generation model (MODEL POPULATION). Accordingly, these commands are used to provide both population values and starting values in the estimation of the model (for computing coverage, described below). In this example, ML is specified as the statistical estimator on the ANALYSIS command. TECH9 is requested on the OUTPUT line to obtain any errors that are encountered in regard to the convergence of each Monte Carlo replication.

Selected output of the analysis is presented in Table 10.3. Muthén and Muthén (2002) have proposed the following criteria for determining sample size: (1) Bias of the parameters and their standard errors do not exceed 10% for any parameter in the

TABLE 10.3. Selected Mplus Output for Monte Carlo Approach to Determining Power and Precision of CFA Model Parameter Estimates

MODEL RESULTS

	Population	ESTIMATES Average	Std. Dev.	S. E. Average	M. S. E.	95% Cover	% Sig Coeff
ESTEEM BY							
X1	0.650	0.6439	0.1100	0.1076	0.0121	0.943	1.000
X2	0.700	0.6955	0.1127	0.1087	0.0127	0.942	1.000
X3	0.720	0.7154	0.1129	0.1093	0.0128	0.942	1.000
DEPRESS BY							
X4	0.600	0.5949	0.1189	0.1149	0.0142	0.941	0.998
X5	0.700	0.6953	0.1223	0.1197	0.0150	0.948	0.999
X6	0.650	0.6468	0.1199	0.1171	0.0144	0.943	0.999
ESTEEM WITH							
DEPRESS	0.350	0.3495	0.1334	0.1283	0.0178	0.934	0.743
Intercepts							
X1	0.000	-0.0011	0.0999	0.0992	0.0100	0.945	0.055
X2	0.000	-0.0006	0.1005	0.0992	0.0101	0.944	0.056
X3	0.000	-0.0005	0.0992	0.0991	0.0098	0.949	0.051
X4	0.000	0.0007	0.0993	0.0992	0.0099	0.946	0.054
X5	0.000	0.0008	0.0996	0.0991	0.0099	0.948	0.052
X6	0.000	-0.0008	0.1006	0.0992	0.0101	0.945	0.055
Variances							
ESTEEM	1.000	1.0000	0.0000	0.0000	0.0000	1.000	0.000
DEPRESS	1.000	1.0000	0.0000	0.0000	0.0000	1.000	0.000
Residual Variances							
X1	0.577	0.5628	0.1165	0.1139	0.0138	0.938	0.994
X2	0.510	0.4926	0.1195	0.1174	0.0146	0.950	0.969
X3	0.482	0.4625	0.1232	0.1198	0.0155	0.950	0.943
X4	0.640	0.6209	0.1302	0.1250	0.0173	0.940	0.993
X5	0.510	0.4893	0.1442	0.1384	0.0212	0.961	0.910
X6	0.577	0.5567	0.1330	0.1300	0.0181	0.948	0.966

Note. Normal and no missing data; $N = 100$.

model; (2) for parameters that are the specific focus of the power analysis (e.g., the factor covariance of Self-Esteem and Depression), bias of their standard errors does not exceed 5%; and (3) coverage is between .91 and .98. In addition to these criteria, appropriate sample size is determined when the power of salient model parameters is .80 or above (cf. Cohen, 1988). In order (by column), the Mplus summary of the analysis provides the true population value for each parameter (Population); the average of the parameter estimates across replications (Average); the standard deviation of the parameter estimates across replications (Std. Dev.); the average of the standard errors across replications (S.E. Average); the mean square error of each parameter (M.S.E., calculated as the variance of the estimate across replications plus the square of the bias); 95% coverage (95% Cover, the proportion of replications for which the 95% confidence interval contains the true population parameter value); and the proportion of replications in which the parameter is significantly different from zero at the .05 alpha level (% Sig Coeff).

The percentage of parameter bias can be calculated by subtracting the population parameter value from the average parameter value, dividing this difference by the population value, and then multiplying the result by 100. For example, if we use the results in Table 10.3, the percent bias of the factor covariance parameter is

$$\text{Bias}(\phi_{21}) = [(0.3495 - 0.35) / 0.35](100) = -.14\% \tag{10.1}$$

The bias of this estimate is less than a percentage point, which is negligible. Indeed, bias is well below 10% for each parameter in the model, which satisfies the first criterion.

Bias in the standard errors of parameters is calculated in a similar fashion. The standard deviation of the parameter estimate over replications (Std. Dev., in Mplus output) can be treated as the population standard error when the number of replications is large (as in the current example, NREPS = 10,000). The percentage of standard error bias is calculated by subtracting the average of the estimated standard errors across replications (S.E. Average) from the standard deviation of the parameter estimate (Std. Dev.), dividing this difference by the standard deviation of the parameter estimate (Std. Dev.), and multiplying the result by 100. If we again use the factor covariance (Table 10.3), the calculation is

$$SE\text{ Bias}(\phi_{21}) = [(0.1283 - 0.1334) / 0.1334](100) = -3.82\% \tag{10.2}$$

According to the guideline of 5% provided by Muthén and Muthén (2002), this result suggests that the amount of bias (imprecision) in the standard error of this parameter of interest is acceptable. Although coverage values are also satisfactory (except for the fixed estimates, all are close to the correct value of .95), evidence against the suitability of an $N = 100$ is provided by the proportion of replications for which the factor covariance parameter is significantly different from zero (% Sig Coeff = .743; Table 10.3). Specifically, the value of .743 indicates that the factor covariance of Self-Esteem and Depression is statistically significant ($\alpha = .05$) in 74.3% of the replications. Note that this result is close to the theoretical value of .72 obtained previously with the Satorra–Saris approach.

This can be regarded as the power estimate of this parameter (i.e., the probability of rejecting the null hypothesis when it is false). Alternatively, if a parameter's population value is specified to be zero, this will provide an estimate of Type I error—the likelihood of rejecting a true null hypothesis.

The analysis will proceed by examining larger sample sizes (and other seed values, to ensure stability once a suitable N has been tentatively identified). This is accomplished by changing the number of observations in the syntax provided in Table 10.2. Table 10.4 provides selected results for a Monte Carlo analysis conducted with a sample size of 125. Consistent with the results of the Satorra–Saris method, this analysis suggests that an $N = 125$ has sufficient power (.830) to reject a false null hypothesis in regard to the factor covariance.

Beginning with Version 3.0, Mplus summarizes data pertaining to the behavior of selected fit indices (e.g., χ^2, RMSEA, SRMR) for a target model, data, and sample size. In Table 10.4, the mean and standard deviation of model χ^2 across the 10,000 replications of the Monte Carlo analysis are provided. The Proportions Expected and Percentiles Expected columns convey aspects of the χ^2 distribution. For instance, with a Proportions Expected value of .05 (third row from the bottom), there is a 5% probability of obtaining a χ^2 value that exceeds the Percentiles Expected value of 15.507 at $df = 8$ (the df of the model presented in Figure 10.1 and analyzed in this Monte Carlo study). In other words, the value of 15.507 is the critical value of χ^2 at $df = 8$, $\alpha = .05$ (the same information can be obtained from tabled values of the χ^2 distribution in an appendix of a statistics textbook). However, the Proportions Observed and Percentiles Expected provide the values obtained in the Monte Carlo replications; for instance, in the preceding example, the proportion of replications for which the critical value is exceeded is .055. The Percentiles Observed value of 15.820 is the χ^2 at this percentile from the Monte Carlo analysis that has 5% of the values in replications above it. Because it is very close to the theoretical value of 15.507 (in tandem with the similarity of the observed and expected proportions values, .055 vs. .05), this supports the notion that the χ^2 distribution is well approximated in this instance ($N = 125$); that is, there is a bias of roughly 2%. Similarly, the average χ^2 across replications (8.202) is roughly the same as the model df (8); the variance of χ^2 across replications ($4.086^2 = 16.69$) is near the value of 2 df (16). This also provides evidence that the χ^2 distribution is approximated.

Mplus has extensive Monte Carlo facilities beyond those illustrated in this example. For instance, a variety of data types (e.g., categorical, non-normal continuous, clustered) and models (e.g., multiple groups, mixtures, multilevel) can be studied in this framework. In addition, the Monte Carlo routine has several other useful features, such as the option to save all or a portion of the generated data sets, and the ability to import estimates from an analysis of real data as population or coverage values in the Monte Carlo study. Another important feature is the ability to determine sample size and power in context of missing data (see also Dolan, der Sluis, & Grasman, 2005). As noted in Chapter 9, missing data must be reckoned with in most applied research (especially in longitudinal designs). In addition, missing data can be planned as part of the research design (e.g., as in the cohort sequential design to study lifespan developmental processes

TABLE 10.4. Selected Mplus 7.11 Output for Monte Carlo Approach to Determining Power and Precision of CFA Model Parameter Estimates

MODEL RESULTS

	Population	ESTIMATES Average	Std. Dev.	S. E. Average	M. S. E.	95% Cover	% Sig Coeff
ESTEEM BY							
X1	0.650	0.6449	0.0988	0.0961	0.0098	0.942	1.000
X2	0.700	0.6955	0.0986	0.0971	0.0097	0.943	1.000
X3	0.720	0.7164	0.0984	0.0975	0.0097	0.948	1.000
DEPRESS BY							
X4	0.600	0.5963	0.1050	0.1022	0.0110	0.943	1.000
X5	0.700	0.6975	0.1088	0.1061	0.0118	0.945	1.000
X6	0.650	0.6474	0.1067	0.1040	0.0114	0.943	1.000
ESTEEM WITH							
DEPRESS	0.350	0.3507	0.1196	0.1148	0.0143	0.933	0.830
Intercepts							
X1	0.000	0.0004	0.0897	0.0889	0.0081	0.946	0.054
X2	0.000	-0.0002	0.0900	0.0889	0.0081	0.946	0.054
X3	0.000	-0.0004	0.0895	0.0889	0.0080	0.948	0.052
X4	0.000	-0.0001	0.0892	0.0889	0.0080	0.949	0.051
X5	0.000	0.0003	0.0892	0.0889	0.0080	0.947	0.053
X6	0.000	0.0002	0.0895	0.0889	0.0080	0.948	0.052
Variances							
ESTEEM	1.000	1.0000	0.0000	0.0000	0.0000	1.000	0.000
DEPRESS	1.000	1.0000	0.0000	0.0000	0.0000	1.000	0.000
Residual Variances							
X1	0.577	0.5656	0.1042	0.1017	0.0110	0.942	0.998
X2	0.510	0.4980	0.1079	0.1046	0.0118	0.946	0.986
X3	0.482	0.4690	0.1089	0.1064	0.0120	0.953	0.975
X4	0.640	0.6248	0.1133	0.1106	0.0131	0.939	0.998
X5	0.510	0.4937	0.1259	0.1216	0.0161	0.956	0.947
X6	0.577	0.5615	0.1183	0.1146	0.0143	0.950	0.985

MODEL FIT INFORMATION

Number of Free Parameters   19

Chi-Square Test of Model Fit

      Degrees of freedom   8

          Mean   8.202
          Std Dev   4.086
          Number of successful computations   10000

(continued)

TABLE 10.4. (continued)

Proportions		Percentiles	
Expected	Observed	Expected	Observed
0.990	0.992	1.646	1.728
0.980	0.983	2.032	2.164
0.950	0.954	2.733	2.792
0.900	0.909	3.490	3.594
0.800	0.811	4.594	4.700
0.700	0.715	5.527	5.654
0.500	0.523	7.344	7.548
0.300	0.317	9.524	9.744
0.200	0.217	11.030	11.331
0.100	0.110	13.362	13.720
0.050	0.055	15.507	15.820
0.020	0.024	18.168	18.664
0.010	0.011	20.090	20.462

Note. Normal and no missing data; $N = 125$.

in a more compressed time frame). However, the effects of missing data are rarely considered in power analysis.

Table 10.5 presents Mplus syntax for a study with planned missingness in which half of the sample is administered all six indicators; a quarter is administered the Depression indicators only; and the remaining quarter is administered the Self-Esteem indicators only (based on the model presented in Figure 10.1). In the MONTECARLO command, the PATMISS option is used to specify the missing data patterns and the proportion missing for each indicator. In the Table 10.5 example, there are three missing data patterns (the patterns are separated by |). In the first pattern, all six indicators are present for all cases (the numbers in parentheses indicate the proportion of data missing for a given indicator; in this pattern, all proportions are 0). In the second pattern, all of the Self-Esteem indicators are missing (indicated by the value 1 in parentheses following the X1, X2, and X3 indicators—100% missing), and all of the Depression indicators are present. In the third pattern, the Depression indicators are missing, and the Self-Esteem indicators are present. The PATPROBS statement specifies the proportion of cases with each missing data pattern. In the current example, 50% of cases have data for all six indicators, 25% have data for Depression indicators only (X4–X6), and 25% have data for Self-Esteem indicators only (X1–X3). The proportions in this statement must sum to one. Direct ML estimation to accommodate missing data is executed by Mplus default.

This example is based on a cross-sectional analysis in which some data are missing by design (e.g., due to practical or financial constraints that preclude administration of a test battery to the full sample). More often, data are lost due to participant attrition, incomplete responding, noncompliance, and so on. For example, if the researcher is interested in evaluating the temporal stability of a latent construct (defined by three indicators, A, B, and C), he or she might specify the following pattern of missingness for a two-wave study (in which a quarter of the sample is expected to be lost at the second testing occasion):

TABLE 10.5. Mplus Syntax for Monte Carlo Approach to Determining Power and Precision of CFA Model Parameter Estimates

```
TITLE:               CFA TWO-FACTOR, NORMAL DATA, PLANNED MISSING
MONTECARLO:
                     NAMES ARE X1-X6;
                     NOBSERVATIONS = 200;
                     NREPS = 10000;
                     SEED = 53567;
                     PATMISS = X1 (0) X2 (0) X3 (0) X4 (0) X5 (0) X6 (0) |
                               X1 (1) X2 (1) X3 (1) X4 (0) X5 (0) X6 (0) |
                               X1 (0) X2 (0) X3 (0) X4 (1) X5 (1) X6 (1);
                     PATPROB = .5 | .25 | .25;
ANALYSIS:            ESTIMATOR = ML;
MODEL POPULATION:
                     ESTEEM BY X1*.65 X2*.70 X3*.72;
                     DEPRESS BY X4*.60 X5*.70 X6*.65;
                     ESTEEM@1; DEPRESS@1;
                     X1*.5775; X2*.51; X3*.4816; X4*.64; X5*.51; X6*.5775;
                     ESTEEM WITH DEPRESS*.35;
MODEL:
                     ESTEEM BY X1*.65 X2*.70 X3*.72;
                     DEPRESS BY X4*.60 X5*.70 X6*.65;
                     ESTEEM@1; DEPRESS@1;
                     X1*.5775; X2*.51; X3*.4816; X4*.64; X5*.51; X6*.5775;
                     ESTEEM WITH DEPRESS*.35;
OUTPUT:              TECH9;
```

Note. Normal and planned missing data; $N = 200$.

```
PATMISS = A1 (0) B1 (0) C1 (0) A2 (.25) B2 (.25) C2 (.25);
PATPROB = 1;
```

In fact, the A1, B1, and C1 indicators can be omitted from the PATMISS statement, because indicators omitted from this statement are assumed to have no missing data by Mplus default). Ideally, the specification of 25% at the Time 2 assessment will be evidence-based (e.g., pilot data, extant literature). Unlike the prior example, there is a single pattern of missing data (all cases have complete data at Time 1, 25% of the sample is lost at Time 2). Accordingly, the probability of this missing pattern is specified as 1.0 (PATPROB = 1).

SUMMARY

This chapter has focused on the important, yet often overlooked, topic of sample size determination in planning a CFA investigation. As this chapter has shown, current latent variable software programs provide elegant methods of determining the sample size required for obtaining adequate statistical power and sufficient precision of parameter estimates. Two procedures are described in detail: the Satorra–Saris method and the Monte Carlo approach. It is hoped that these illustrations have convinced readers not to rely on general rules of thumb that seem to persist in the applied research literature (e.g., minimum sample size). Rather, sample size and power determinations should be based on models and data that mirror the actual empirical context. Innovations in software packages also allow the researcher to consider the impact of other common real-world complications (e.g., non-normality, missing data) on sample size and power.

NOTE

1. However, no current program can generate data with the desired amount of multivariate non-normality. For instance, the Mplus program uses a mixture-modeling approach to generate non-normal data; that is, by varying the amount of overlap of the normal distributions of two subpopulations, and then combining the data, a researcher can obtain the desired marginal skewness and kurtosis on a trial-and-error basis. Nonetheless, the procedures in Mplus and other programs with extensive data generation facilities (e.g., EQS) do not address joint skewness and kurtosis (multivariate non-normality), distribution shape, and so forth. For more information on these issues, and their impact on power analyses based on simulations, see Yuan and Hayashi (2003).

Appendix 10.1

Monte Carlo Simulation in Greater Depth: Data Generation

As noted in this chapter, the Monte Carlo approach to power analysis involves several steps. First, a hypothesized model is established that contains the population values of the model parameters (see Figure 10.1). Population values are often derived from the research literature and pilot studies. Second, multiple random samples are generated from the population values of the hypothesized model, based on the conditions specified by the researcher (e.g., sample size, patterns/amount of missing data). The model is then analyzed in each sample, and the results are averaged across all simulated data sets. The averaged results provide information about the precision and statistical power of the model; for example, the proportion of replications in which a given parameter is statistically significant indicates the power of the estimate under the specified conditions (e.g., N). Although the rationale of this procedure is relatively straightforward, perhaps the most mysterious aspect of the Monte Carlo approach is how samples are generated from the population values of the model. There are various ways data can be simulated. To foster the reader's understanding of this process, a matrix decomposition procedure (Kaiser & Dickman, 1962) is illustrated with the Figure 10.1 model. To avoid undue complexity, only the indicators of the Self-Esteem factor (X1, X2, X3) are used, under the assumptions of multivariate normality and no missing data. A more extensive illustration of this approach under other conditions (e.g., non-normality) can be found in Fan and Fan (2005).

Step 1: Calculate correlation matrix (R) from the population values of the hypothesized model.

Because the values in Figure 10.1 are completely standardized, it is straightforward to compute correlations among the X1, X2, and X3 indicators by using basic equations presented in earlier chapters; for example, population r of X1,X2 = .65(.70) = .455 (cf. Eq. 3.7, Chapter 3). Thus the population correlation matrix for these three indicators is as follows:

```
         X1      X2      X3
X1  1.000
X2   .455   1.000
X3   .468    .504   1.000
```

396

Step 2: Obtain a factor pattern matrix (*F*) by conducting a principal components analysis (PCA) on the population correlation matrix (*R*), where the requested number of components equals the number of indicators.

Because the number of components is the same as the number of indicators, the input correlation matrix is reproduced perfectly. SAS PROC FACTOR syntax for this step is provided below:

```
Title "Principal components analysis of population correlation matrix";
Data SE (type=CORR);
input _TYPE_ $ _NAME_ $ x1-x3;
cards;
corr x1    1.000    .       .
corr x2    0.455    1.000   .
corr x3    0.468    0.504   1.000
;
run;
proc factor data=SE method=p nfactors=3;
run; \
```

The resulting pattern matrix (*F*) is this:

Factor Pattern

	Factor1	Factor2	Factor3
x1	0.79017	0.60799	0.07738
x2	0.81110	-0.36403	0.45781
x3	0.81816	-0.22629	-0.52859

Step 3: Generate *p* uncorrelated random normal variables (*M* = 0, *SD* = 1), each with *N* observations (*X*).

In the current example, three variables are generated (i.e., X1, X2, X3). To obtain a closer correspondence to the original population values, a large sample size is requested (*N* = 2,000). The line of SAS syntax below accomplishes this step.

```
X = RANNOR(J(2000,3,0));
```

Step 4: Premultiply the random data matrix (*X*) with the pattern matrix (*F*) to create the correlated data matrix (*Z*).

The fundamental equation of the Kaiser and Dickman (1992) matrix decomposition procedure is

$$Z_{(p \times N)} = F_{(p \times p)} X_{(p \times N)} \tag{10.3}$$

where *Z* is the correlated data matrix with *N* cases and *p* variables; *F* is a *p* × *p* pattern matrix from PCA (see Step 2); and *X* is an uncorrelated random normal data matrix with *p* variables, *N* cases (see Step 3).

This step imposes the population correlations (R) among the variables on the sample data (X), as if the data were sampled from a population with the intercorrelations represented by the imposed correlation matrix (reflected by F, which reproduces R exactly).

The following SAS PROC IML syntax performs Steps 3 and 4:

```
PROC IML;
F = {0.79017  0.60799  0.07738,       * 1
     0.81110 -0.36403  0.45781,
     0.81816 -0.22629 -0.52859};
X = RANNOR(J(2000,3,0));              * 2
X = X`;                               * 3
Z = F*X;                              * 4
Z = Z`;                               * 5
```

*1 = factor pattern matrix (F; Step 2); *2 = generate 3 uncorrelated random normal variables (X), each with N = 2,000 (Step 3); *3 = transpose X to a $3 \times 2,000$ matrix in preparation for multiplication with F; *4 = Eq. 10.3; *5 = transpose Z back into a $2,000 \times 3$ matrix.

If desired, the SAS syntax above can be extended to assign means and standard deviations to the sample data. The SAS syntax below executes a linear transformation on X1 and X2 (target M = 5, SD = 1); no such transformation is done on X1, so it remains standardized (M = 0, SD = 1).

```
X1 =Z[,1];
X2 =Z[,2]* 5 + 1;
X3 =Z[,3]* 5 + 1;
Z = X1||X2||X3;
CREATE EST FROM Z[COLNAME={X1 X2 X3}];     *output data file is EST
APPEND FROM Z;
```

Step 5: Obtain descriptive statistics and correlations from the simulated data.

These statistics can be obtained from the SAS syntax below:

```
proc means data=EST n mean std skewness kurtosis;
  var x1 x2 x3;
proc corr data=EST;
  var x1 x2 x3;
run;
```

The results are as follows:

```
The MEANS Procedure
```

Variable	N	Mean	Std Dev	Skewness	Kurtosis
X1	2000	0.0073703	1.0087635	-0.0441861	-0.0042880
X2	2000	0.9776742	5.0160837	-0.0363566	0.0504317
X3	2000	0.8757570	5.0265543	-0.1164463	0.0314612

The CORR Procedure

Pearson Correlation Coefficients, N = 2000
Prob > |r| under H0: Rho=0

	X1	X2	X3
X1	1.00000	0.44867 <.0001	0.48283 <.0001
X2	0.44867 <.0001	1.00000	0.49222 <.0001
X3	0.48283 <.0001	0.49222 <.0001	1.00000

The sample data obtained in this initial simulation approximate the target population characteristics in terms of their univariate distributions (e.g., X2: $M = 0.98$, $SD = 5.02$) and their intercorrelations (e.g., r of X1, X2 = .45).

In context of the Mplus illustration provided in this chapter, this process would be repeated multiple times (e.g., number of replications = 10,000; Table 10.2) for the entire set of indicators (X1–X6). The hypothesized measurement model is fitted to each generated data set, and the fit statistics (e.g., χ^2) and parameter estimates are averaged across replications (cf. Table 10.4).

11

Recent Developments Involving CFA Models

A theme echoed in the preceding chapters is that methodological developments in CFA and SEM have advanced rapidly in recent years. The aim of this book has been to provide a practical discussion and guide to CFA, including the latest methodological developments that are most germane to applied research. In this final chapter, it seems appropriate to discuss two additional CFA modeling possibilities that have emerged even more recently, in terms of their development and implementation in applied research: *Bayesian CFA* and *multilevel factor models*. An overview and applied illustration of both procedures are presented.

BAYESIAN CFA

Although Bayesian statistics have been around for centuries (Bayes's theorem dates back to the 18th century), these methods have been largely neglected by applied researchers until recently (Kaplan & Depaoli, 2013). The primary reasons for this are twofold. For one, as noted throughout this section, Bayesian statistics challenges many assumptions of traditional statistics, including ML estimation in CFA and SEM. Therefore, considerable controversy has surrounded the use of Bayesian statistics. On a terminological note, traditional statistics are typically referred to as *frequentist statistics* in the context of Bayesian analysis to denote the types of quantitative methods that are most commonplace in the applied sciences. The second reason why applied researchers have been slow to adopt Bayesian statistics stems from their complexity and the unavailability of statistical software. This drawback has been rectified in recent years by software programs such as Mplus and Amos (and the R freeware package), which now render statistical modeling from a Bayesian perspective quite feasible.

Indeed, as will be illustrated shortly, Bayesian analysis can be readily implemented in the Mplus program to estimate CFA and structural equation models (cf. Muthén & Asparouhov, 2012). Bayesian analysis has several potential advantages over traditional CFA/SEM estimation (i.e., frequentist statistical estimation such as ML). These strengths include better performance in small samples; lack of reliance on asymptotic theory (i.e., unlike ML, Bayesian analysis does not assume that the distributions of parameter estimates are normal on the basis of large-sample theory); automatic handling of missing data (multiple imputation); ability to estimate complex models that are not feasible in ML estimation (e.g., measurement models entailing categorical indicators with multiple latent variables); better factor scores; ability to provide proper solutions for models that are prone to offending estimates (i.e., Heywood cases such as negative indicator residual variances); and higher statistical power in some scenarios (e.g., in non-normal sample data compared to ML estimates, which are based on a symmetric distribution). In addition, the estimation of multilevel and mixture models can be greatly benefited by Bayesian analysis (e.g., Kaplan & Depaoli, 2012).

With regard to CFA specifically, the advantage of Bayesian analysis is the ability to replace parameters that are typically fixed to zero (e.g., cross-loadings, indicator error covariances) with values that are close to zero, but not exactly zero. As discussed earlier in this book (e.g., Chapter 5), although a measurement model may be properly specified with regard to the number of factors and the pattern of factor–indicator relationships (primary loadings), ML analysis will often result in poor fit due to the restriction of fixing cross-loadings to zero. The restriction of fixing indicator error covariances to zero may also contribute to ill fit. Moreover, these restrictions tend to result in overestimates of factor correlations (cf. Chapter 5). Although the cross-loadings may not be salient in the true model (i.e., these parameters are of trivial magnitude), poor model fit will be encountered because these parameters have been fixed to exactly zero. This will be especially the case in very large samples (e.g., χ^2 oversensitivity to slight model misspecifications) and in multidimensional measurement models that entail a large set of indicators (cf. Marsh et al., 2004, 2009). Consequently, either the measurement model may be rejected, or a series of model revisions may be pursued on the basis of fit diagnostic information (e.g., modification indices) that may exacerbate model misspecification (i.e., capitalization on chance associations in the sample data; see Chapter 5). Bayesian analysis avoids these unduly strict models by allowing the parameters that are ordinarily fixed to zero to be approximately zero. As Muthén and Asparouhov (2012) argue, the Bayesian framework offers the applied researcher a more reasonable and flexible approach to model testing that is more closely aligned to substantive theory; for example, it allows for the specification of a measurement model to avoid the unnecessarily strict hypothesis that the factor cross-loadings or the indicator error covariances are exactly zero. In addition, Bayesian analysis can be extended to longitudinal and multiple-groups measurement invariance evaluation, where it may be more reasonable to expect that group or temporal differences in the factor loadings and indicator intercepts are approximately but not exactly zero.

Bayesian Probability and Statistical Inference

Before an applied example of Bayesian CFA is presented, it will be useful to provide a brief overview of Bayesian probability and statistical inference. Although intended to foster the reader's conceptual understanding of Bayesian CFA, the forthcoming material only scratches the surface of this very large and complex topic. The reader is referred to Kaplan (2014) and Hoff (2009) for extensive, but user-friendly, discussions of Bayesian statistics.

The essence of Bayes's theorem is understanding how the probability that a theory is true is affected by new evidence. The output of Bayesian analysis is a *posterior* (more specifically, a *posterior probability* in Bayesian probability analysis, or a *posterior distribution* in Bayesian statistical inference), which conveys the probability assigned to the hypothesis after the new evidence is taken into account. With regard to probability analysis, consider the scenario where we wish to express the conditional probability of heart disease (*H*) based on whether or not an individual eats a high-fat diet (*D*). It is no accident that the variables used in this example are symbolized as *H* and *D*; this is done to bolster their associations with hypothesis and data (new evidence), respectively. Specifically, what is the probability that an individual has heart disease (*H*) after we obtain knowledge that he or she consumes a high-fat diet (*D*)? Within standard probability theory, the joint probability of *H* and *D* can be conveyed by their conditional and marginal probabilities:

$$P(H, D) = P(D|H) \, P(H) = P(H|D) \, P(D) \tag{11.1}$$

where $P(H, D)$ is the joint event *H* and *D*; $P(D|H)$ is the conditional probability of *D*, given *H*; $P(H)$ is the marginal probability of *H*; $P(H|D)$ is the conditional probability of *H*, given *D*; and $P(D)$ is the marginal probability of *D*.[1] If Eq. 11.1 is divided by $P(D)$, the result is

$$P(H|D) = [P(D|H) \, P(H)] \, / \, P(D) \tag{11.2}$$

which is Bayes's theorem. A numerical illustration of Eq. 11.2 is presented in Appendix 11.1. In this example, Bayes's theorem asserts that the conditional probability of heart disease in the presence of a high-fat diet, $P(H|D)$, is equal to the probability of a high-fat diet in the presence of heart disease, $P(D|H)$, times the probability of having heart disease, $P(H)$. As noted earlier, the denominator of Eq. 11.2 is the marginal probability of a high-fat diet, $P(D)$; this reflects the probability that an individual consumes a high-fat diet, collapsed across persons with and without heart disease. As noted by Kaplan (2014), $P(D)$ does not contain information relevant to the conditional probability $P(H|D)$ because this marginal probability is obtained across all possible outcomes of heart disease. Thus $P(D)$ can be regarded as a normalizing factor ensuring that the probability sums to one. Consequently, Bayes's theorem is often expressed as follows:

$$P(H|D) \propto P(D|H) \, P(H) \tag{11.3}$$

which states that the probability of heart disease given a high-fat diet is proportional ($\propto$) to the probability of a high-fat diet given the presence of heart disease times the marginal probability of having heart disease.

To extend the discussion to Bayesian statistical inference and modeling, H and D in Eqs. 11.2 and 11.3 are replaced by the symbols θ and y, where θ is a set of parameters representing the statistical model of interest (e.g., parameters of a measurement model), and y represents the observed data. Eq. 11.2 can thus be rewritten as follows:

$$P(\theta|y) = [P(y|\theta)\, P(\theta)] \,/\, P(y) \tag{11.4}$$

where $P(\theta|y)$ is the posterior distribution of the parameters θ, given the observed data y. Eq. 11.4 states that the posterior distribution of the parameters θ, given the observed data y, is equal to the product of the probability of observing y, given the parameters $P(y|\theta)$ and the prior distribution of parameters $P(\theta)$, divided by $P(y)$. However, because $P(y)$ does not involve model parameters (i.e., it does not contain unique information relevant to the posterior distribution), Eq. 11.4 can be reexpressed along the lines of Eq. 11.3:

$$P(\theta|y) \propto P(y|\theta)\, P(\theta) \tag{11.5}$$

which thereby indicates that the posterior distribution of the parameters θ, given the observed data y, is proportional to the product of the probability of the observed data (y), given the parameters and the prior distribution of parameters. The $P(y|\theta)$ portion of Eq. 11.5 can be reexpressed as the *likelihood*, symbolized $L(\theta|y)$, which denotes the unknown parameters θ for fixed values of y (alternatively, the likelihood can be considered as the distribution of the data, given a parameter value). A detailed discussion of the nature of likelihoods can be found in Kaplan (2014) and Kaplan and Depaoli (2013). Thus Eq. 11.5 can be rewritten as follows:

$$P(\theta|y) \propto L(\theta|y)\, P(\theta) \tag{11.6}$$

which indicates that the posterior distribution is proportional to the product of the likelihood and the prior distribution of parameters (cf. Eq. 11.3).

As stated earlier in this section, the essence of Bayes's theorem is understanding how the probability of a hypothesis or theory is changed after new evidence is obtained. Translated in context of Bayesian statistical inference, data inform us about a parameter $L(\theta|y)$ and modify the prior $P(\theta)$ into a posterior $P(\theta|y)$ that renders the Bayesian estimates. This represents one of the key distinctions between Bayesian analysis and ML estimation (frequentist analysis). In frequentist analysis, parameter values θ are unknown, but fixed. For instance, ML renders a fixed set of parameter estimates by maximizing a likelihood computed for the sample data. Frequentist analysis assumes that one true population parameter exists, and methods such as ML are used to estimate that parameter. In Bayesian analysis, parameter values θ are random and associated with a probability distribution that reflects uncertainty about the true values of θ; that is, there is not one true population parameter, and there is always uncertainty about the population parameter. Eqs. 11.5 and 11.6 convey the core of Bayes's theorem: Uncertainty about the parameters of the specified model (the hypothesis or theory), as reflected by the prior distribution $P(\theta)$, is weighted by the actual data (the new evidence), represented equivalently by either $P(y|\theta)$ or $L(\theta|y)$. The product yields updated estimates of the model parameters, as reflected by the posterior distribution $P(\theta|y)$ (i.e., the extent to which the prior distribution has been modified by the likelihood).

Priors in CFA

A fundamental element of Bayesian statistical inference (and a salient source of controversy) is the specification of the prior distributions for the model parameters, often referred to more simply as *priors*. Priors reflect previously held beliefs about what the parameter values are likely to be before new data are collected (e.g., the expected magnitude of cross-loadings in CFA). As noted by Muthén and Asparouhov (2012), priors can be specified on the basis of substantive theory, pilot data, or the results of other studies relevant to the current investigation. However, the specification of priors can be difficult in the absence of a compelling substantive or empirical background. This is a potential drawback of Bayesian analysis, and it is one reason why frequentist statisticians are reluctant to pursue this approach; for example, estimates emanating from Bayesian analysis can vary widely as a function of the priors used in the model specification.

There are two major types of priors: *informative* and *noninformative*.[2] Generally speaking, informative priors are used to build previous knowledge into the model specification. Conversely, noninformative priors are used to express the absence of previous knowledge. For both types of priors, the scale and shape of the distributions must be used to specify the location and precision of the prior distributions. A number of different distribution shapes can be used. Some of the distribution shapes most germane to CFA models are *normal*, *uniform* (flat), *inverse-Wishart*, and *inverse-gamma* (with respect to informative priors, the normal, inverse-Wishart, and inverse-gamma distributions are often used for the specification of cross-loadings, covariances, and variances, respectively). The scaling of the priors should be consistent with the metrics of the observed and latent variables in the CFA model. For measurement models, it is thus common to analyze completely standardized solutions (e.g., to fix factor variances to unity) in order to make the scaling of the priors more straightforward.

Bayesian measurement models are specified by using a combination of informative and noninformative priors. When previous knowledge is unavailable, it is important to reflect this in the model by using noninformative priors. Noninformative priors are often specified by using uniform distributions over a plausible range of values or normal distributions with large variances. As noted earlier, the variance of the prior distribution reflects the degree of (un)certainty of the parameter value (e.g., large variance = high uncertainty in the parameter value). When the variance of the prior distribution is large, the likelihood (actual data) contributes considerably more to the formation of the posterior distribution than the prior. Consequently, the Bayesian point estimate of the parameter will approximate an ML estimate. In addition, because the prior contributes less information when the variance is large, the model becomes closer to being underidentified (Muthén & Asparouhov, 2012). In CFA, noninformative priors are typically used for parameters that are not restricted in frequentist analysis (e.g., ML). For example, the primary loadings of each factor could be specified by using normally distributed, noninformative priors with large variances.

Informative priors are used for measurement model parameters that, in frequentist analysis (e.g., ML), would ordinarily be fixed to zero (e.g., factor cross-loadings or indicator error covariances in single-group CFA) or constrained to equality (e.g., factor loadings or indicator intercepts in longitudinal or multiple-groups measurement invariance evaluation). In addition, informative priors can be used to rectify offending parameter estimates, if substantively justified (e.g., specification of indicator error variances by using an inverse-gamma distribution that does not permit variance estimates below zero).

Consider informative priors with respect to cross-loadings in CFA models. The convention in frequentist CFA is to fix all cross-loadings to zero. The drawbacks of this tradition have been discussed earlier (e.g., proneness to model rejection, inflation of factor correlations). Moreover, all cross-loadings cannot be freely estimated in ML CFA because the model will then be underidentified. From a Bayesian perspective, when a cross-loading is fixed to zero, the frequentist researcher has implicitly proposed a prior distribution for that parameter with a mean and variance equal to zero. In the majority of applied research data sets, it can be argued that more plausible prior distributions for CFA cross-loadings have means equal to zero and normal distributions with small variances. Setting a small variance to the prior distribution (along with a mean of the parameter distribution equal to zero) expresses considerable certainty that the cross-loading is close to zero, but not exactly zero. On the other hand, if the variance is set to be large, the upper and lower limits of the cross-loading will deviate farther away from zero, and the prior will contribute less to the posterior. The specification of conceptually driven, small-variance priors for the cross-loadings confers more information to the analysis, which thereby avoids the problem of model underidentification associated with ML CFA (i.e., in ML CFA, the model is underidentified if all cross-loadings are freely estimated).

Figure 11.1 presents three examples of prior, likelihood, and posterior distributions. In Model A, a noninformative prior has been specified (uniform distribution), as might be used for primary factor loadings in a CFA model. Recall that the posterior distribution is produced by weighting the prior distribution by the actual data (the likelihood). In Model A, the peaks of the posterior and likelihood distributions are the same because the prior does not contain information for the formation of the posterior distribution (i.e., the posterior is determined more by the actual data). As noted earlier, in this scenario the Bayesian point estimate of the parameter (peak of the posterior distribution) will approximate an ML estimate (e.g., primary factor loadings derived from Bayesian and ML estimation will be roughly the same). Model B displays an informative prior with a normal distribution with a smaller variance (relative to Model A), along

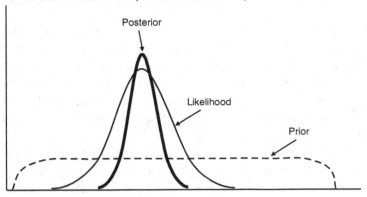

A. Noninformative Prior (Uniform Distribution)

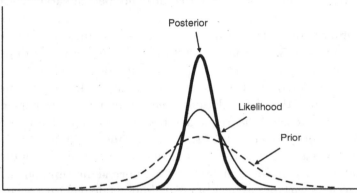

B. Informative Prior (Normal Distribution)

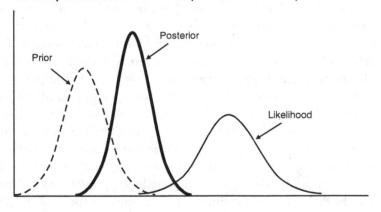

C. Misspecified Informative Prior (Normal Distribution)

FIGURE 11.1. Examples of prior, likelihood, and posterior distributions. Prior distributions have been plotted with dotted lines, and posterior distributions have been plotted with heavier lines, to foster the distinguishability of the overlapping curves.

the lines of what might be specified for cross-loadings in CFA. In Model B, the resulting posterior distribution is somewhat smaller than the data distribution (likelihood) because the prior distribution (reflecting previous knowledge) provides more information. Although informative, small-variance priors can significantly foster the statistical power of the model, a potential pitfall of Bayesian analysis is the misspecification of the prior distributions. This situation is depicted in Model C. In this model, the posterior is a compromise between the prior and the likelihood, but the peak of the posterior distribution does not correspond to the peaks of either the prior or the likelihood distributions because the prior has been misspecified. In Model C, the posterior distribution is somewhat closer to the prior distribution than the data distribution because the prior distribution has been specified to contain more information, albeit incorrect (i.e., the size of the prior distribution is relatively small). The procedures for checking the goodness of fit of the model priors are discussed in the applied example that follows.

Applied Example of Bayesian CFA

Bayesian CFA is now illustrated by using a 10-item questionnaire of political attitudes administered to a sample of 1,000 college students (simulated data). It is expected that the latent structure of the questionnaire is defined by two dimensions, Conservative Attitudes (items 1–5) and Liberal Attitudes (items 6–10), which are inversely correlated. As a comparison to Bayesian CFA, the two-factor model is first fitted to the data by using traditional CFA with ML estimation (i.e., freely estimating primary loadings, fixing cross-loadings to zero, fixing error covariances to zero). The model provides a marginal fit to the data: $\chi^2(34) = 217.90$, $p < .001$, SRMR = .059, RMSEA = 0.74 (90% CI = 0.06–0.08, CFit = .000), TLI = 0.93, CFI = .95 (Mplus 7.11 results). The syntax, parameter estimates, and modification indices for this model are presented in Table 11.1. Inspection of fit diagnostic information reveals several large modification indices indicative of potentially significant cross-loadings. Because modification indices entail freely estimating a single parameter at a time, in practice a series of model respecifications may ensue that can capitalize on chance and exacerbate the misspecification of the model.

Although the measurement model appears to be correctly specified with respect to the number of factors and the pattern of primary loadings, marginal fit appears to stem from the restriction of exact zero cross-loadings when in fact many of the test items have significant cross-loadings. Bayesian CFA may provide a more reasonable approach in this scenario by allowing for the estimation of small-magnitude (approximate zero) cross-loadings. This cannot be done in ML CFA because a model with freely estimated cross-loadings is not identified.

Mplus 7.11 syntax for the two-factor Bayesian CFA model is presented in Table 11.2. The ESTIMATOR=BAYES option is requested on the ANALYSIS command to override the Mplus default (ML). The first three statements under the MODEL command are the same that would be used for ML CFA (the factors are scaled by fixing their variances to 1.0). The last two statements of the MODEL command specify the 10 item cross-loadings (with starting values of zero) and assign labels to these parameters. The MODEL PRI-

TABLE 11.1. Mplus 7.11 Syntax and Selected Output for Two-Factor Model of Political Attitudes, Using ML Estimation

```
TITLE:   TWO-FACTOR MODEL OF POLITICAL ATTITUDES
DATA:
  FILE IS BAYESCFA.DAT;
VARIABLE:
  NAMES ARE X1-X10;
ANALYSIS: ESTIMATOR IS ML;
MODEL:
     CONSERV BY X1* X2 X3 X4 X5;
     LIBERAL BY X6* X7 X8 X9 X10;
     CONSERV-LIBERAL@1;
OUTPUT:  STANDARDIZED MODINDICES(4.00) SAMPSTAT FSDETERMINACY;
```

MODEL RESULTS

	Estimate	S.E.	Est./S.E.	Two-Tailed P-Value
CONSERV BY				
X1	0.626	0.030	20.975	0.000
X2	0.742	0.030	24.543	0.000
X3	0.782	0.029	26.865	0.000
X4	0.658	0.030	22.012	0.000
X5	0.664	0.029	22.902	0.000
LIBERAL BY				
X6	0.757	0.028	26.684	0.000
X7	0.796	0.028	28.248	0.000
X8	0.727	0.029	25.035	0.000
X9	0.579	0.035	16.528	0.000
X10	0.683	0.028	24.637	0.000
LIBERAL WITH				
CONSERV	-0.400	0.033	-11.933	0.000
Intercepts				
X1	0.056	0.031	1.812	0.070
X2	0.001	0.032	0.023	0.982
X3	0.018	0.032	0.575	0.565
X4	-0.025	0.031	-0.813	0.416
X5	0.021	0.030	0.678	0.498
X6	0.022	0.031	0.720	0.471
X7	0.014	0.031	0.458	0.647
X8	0.047	0.031	1.484	0.138
X9	0.001	0.035	0.034	0.973
X10	0.054	0.030	1.789	0.074
Variances				
CONSERV	1.000	0.000	999.000	999.000
LIBERAL	1.000	0.000	999.000	999.000

(continued)

TABLE 11.1. *(continued)*

Residual Variances				
X1	0.556	0.029	19.243	0.000
X2	0.490	0.028	17.228	0.000
X3	0.396	0.026	15.305	0.000
X4	0.539	0.029	18.829	0.000
X5	0.489	0.027	18.373	0.000
X6	0.402	0.024	16.585	0.000
X7	0.358	0.024	15.174	0.000
X8	0.459	0.026	17.682	0.000
X9	0.891	0.043	20.908	0.000
X10	0.430	0.024	18.054	0.000

MODEL MODIFICATION INDICES

Minimum M.I. value for printing the modification index   4.000

		M.I.	E.P.C.	Std E.P.C.	StdYX E.P.C.
BY Statements					
CONSERV	BY X6	49.811	0.203	0.203	0.205
CONSERV	BY X7	63.570	-0.226	-0.226	-0.227
CONSERV	BY X8	33.953	0.173	0.173	0.174
CONSERV	BY X10	27.306	-0.149	-0.149	-0.157
LIBERAL	BY X1	11.839	0.107	0.107	0.110
LIBERAL	BY X2	13.119	0.112	0.112	0.110
LIBERAL	BY X3	39.097	-0.185	-0.185	-0.184
LIBERAL	BY X4	8.546	-0.091	-0.091	-0.092
LIBERAL	BY X5	10.343	0.096	0.096	0.100

ORS command is used with ESTIMATOR=BAYES to specify the prior distributions for the parameters. Specifically, the MODEL PRIORS command overrides the Mplus default of noninformative priors for every parameter (if this default was not overridden, the Bayesian results would be very similar to those of ML). For CFA with continuous outcomes, the Mplus noninformative prior distribution defaults are normal for factor loadings, indicator intercepts, and factor means; inverse-gamma for factor variances and indicator error variances; and inverse-Wishart for factor covariances and indicator error covariances (Muthén & Muthén, 1998–2012). As shown in Table 11.2, the MODEL PRIORS command is used to specify informative priors for the 10 cross-loadings by using means of zero (0) and normal distributions (N) with small variance (.01); that is, A1–B5 ~ N(0, .01). The indicators are standardized, and the factor variances are fixed to one so that the metrics of the priors correspond to the standardized loadings (Table 11.2). Although the selection of zero for the mean of the prior distribution is straightforward (i.e., based on the belief that the cross-loadings are close to but not exactly zero), specification of the variance is not as clear. As previously discussed, the variance of the prior distribution expresses the researcher's degree of certainty about the parameter values.

TABLE 11.2. Mplus 7.11 Syntax and Selected Output for Two-Factor Model of Political Attitudes, Using Bayesian Estimation (Cross-Loading Priors with Mean = 0, Variance = 0.01, Normal Distribution)

```
TITLE:    TWO-FACTOR MODEL OF POLITICAL ATTITUDES
DATA:
  FILE IS BAYESCFA.DAT;
VARIABLE:
  NAMES ARE X1-X10;
DEFINE:   STANDARDIZE X1-X10;
ANALYSIS: ESTIMATOR=BAYES;
          PROCESSORS = 2;  ! used to increase computational speed
          FBITER = 10000;  ! over-rides Mplus convergence criterion
MODEL:
    CONSERV BY X1* X2 X3 X4 X5;
    LIBERAL BY X6* X7 X8 X9 X10;
    CONSERV-LIBERAL@1;
    ! SPECIFICATION OF CROSS-LOADINGS:
    CONSERV BY X6-X10*0 (A1-A5);
    LIBERAL BY X1-X5*0 (B1-B5);
MODEL PRIORS:
    A1-B5 ~ N(0,.01);
OUTPUT: TECH1 TECH8 STAND;
PLOT:    TYPE=PLOT2;
```

```
MODEL FIT INFORMATION

Number of Free Parameters                           41

Bayesian Posterior Predictive Checking using Chi-Square

          95% Confidence Interval for the Difference Between
          the Observed and the Replicated Chi-Square Values

                      -37.933            20.585

          Posterior Predictive P-Value         0.717

Information Criterion

          Deviance (DIC)                       24671.370
          Estimated Number of Parameters (pD)     38.793
          Bayesian (BIC)                       24876.882
```

MODEL RESULTS

	Estimate	Posterior S.D.	One-Tailed P-Value	95% C.I. Lower 2.5%	Upper 2.5%	Significance
CONSERV BY						
X1	0.680	0.037	0.000	0.609	0.754	*
X2	0.768	0.036	0.000	0.702	0.842	*
X3	0.721	0.035	0.000	0.654	0.791	*
X4	0.629	0.036	0.000	0.560	0.701	*
X5	0.721	0.037	0.000	0.653	0.796	*

(continued)

TABLE 11.2. *(continued)*

X6	0.133	0.058	0.007	0.025	0.255	*
X7	-0.158	0.052	0.002	-0.253	-0.052	*
X8	0.117	0.056	0.012	0.013	0.234	*
X9	0.029	0.045	0.261	-0.055	0.120	
X10	-0.131	0.049	0.006	-0.222	-0.031	*
LIBERAL BY						
X6	0.842	0.037	0.000	0.773	0.920	*
X7	0.734	0.034	0.000	0.669	0.802	*
X8	0.793	0.036	0.000	0.726	0.869	*
X9	0.536	0.036	0.000	0.468	0.609	*
X10	0.666	0.035	0.000	0.597	0.735	*
X1	0.070	0.050	0.074	-0.023	0.170	
X2	0.064	0.054	0.109	-0.035	0.172	
X3	-0.134	0.050	0.004	-0.230	-0.034	*
X4	-0.086	0.048	0.038	-0.177	0.008	
X5	0.059	0.051	0.123	-0.038	0.162	
LIBERAL WITH						
CONSERV	-0.362	0.088	0.000	-0.529	-0.184	*
Variances						
CONSERV	1.000	0.000	0.000	1.000	1.000	
LIBERAL	1.000	0.000	0.000	1.000	1.000	
Residual Variances						
X1	0.573	0.031	0.000	0.517	0.638	*
X2	0.448	0.027	0.000	0.397	0.503	*
X3	0.397	0.024	0.000	0.353	0.447	*
X4	0.562	0.029	0.000	0.506	0.622	*
X5	0.513	0.028	0.000	0.461	0.571	*
X6	0.361	0.025	0.000	0.315	0.412	*
X7	0.358	0.021	0.000	0.318	0.402	*
X8	0.430	0.026	0.000	0.382	0.483	*
X9	0.729	0.035	0.000	0.665	0.801	*
X10	0.484	0.026	0.000	0.436	0.537	*

In this particular example, the specification $\lambda \sim N(0, .01)$ means that 95% of the cross-loading variation is between −0.2 and 0.2. In context of a completely standardized solution, cross-loadings of 0.2 are generally considered to be small, but are still potentially statistically significant and will contribute to poor model fit if not estimated (especially in large samples). Thus this prior is very informative, but it indicates that the cross-loadings are close to but not exactly zero. The variance of the prior can be adjusted on the basis of substantive reasoning and for the purpose of sensitivity analysis (discussed later). For instance, the 95% limits of factor loadings with zero means and variances of 0.001, 0.005, 0.01, and 0.02 are ±0.06, ±0.14, ±0.20, and ±0.28, respectively (for a full table of prior variances and their corresponding ranges of parameter values, see Table 2 in Muthén & Asparouhov, 2012). If, for example, the researcher believes that the cross-loadings are higher than 0.2, then a prior variance larger than 0.01 should be specified.

Unlike ML, which focuses on the computation of point estimates of the model parameters that have sound asymptotic properties, the goal of Bayesian analysis is to estimate features of posterior distributions (which are not bound to large-sample theory). In Bayesian analysis, a numerical algorithm called *Markov chain Monte Carlo* sampling (MCMC) is used to estimate posterior distributions of the model parameters, given the data $P(\theta|y)$. MCMC is a very large and complex topic, so only a brief overview is provided here. Interested readers are referred to Gelman, Carlin, Stern, and Rubin (2004); Gilks, Richardson, and Spiegelhalter (1996); and Kaplan (2014) for more details about MCMC.

Through the use of simulation techniques entailing a very large number of iterations (random draws), MCMC approximates the joint distribution of the model parameters (posterior distribution) based on random draws of parameter values from the conditional distribution of one set of parameters, given other parameter sets. In other words, in a sequential fashion (i.e., in a Markov chain), multiple samples are drawn from the conditional distribution, and the distribution formed by those samples is summarized. An important feature of the Markov chain is that the Bayesian iterations are dependent on the immediately preceding iteration (unlike in bootstrapping; see Chapter 9). That is, the conditional probability of the parameter values, given all the past variables, depends only on the immediate prior parameter values in the iterative sequence. Over a long iterative sequence, the chain will ignore the initial starting values and iterations, and will converge on a stationary distribution $P(\theta|y)$ that does not depend on either the number of samples or the initial iterations. The term *burn-in samples* is used to refer to the set of iterations that precede the stabilization of the distribution. As discussed shortly, several techniques are available to evaluate the stability of the distribution (convergence).

There are a number of different MCMC algorithms. One of the most common MCMC algorithms is the *Gibbs sampler*, which is used as the default by Mplus for Bayesian analysis (the other Bayes estimator option in Mplus is the *Metropolis–Hastings algorithm*; see Kaplan, 2014, for description). In a cogent conceptual description of the Gibbs sampler, Kaplan (2014) offers the following two-parameter illustration:

> Consider that the goal is to obtain the joint posterior distribution of two model parameters—say θ_1 and θ_2, given some data y, written as $p(\theta_1,\theta_2|y)$. These two model parameters can, for example, be regression coefficients from a simple multiple regression model. Dropping the conditioning on y for notational simplicity, what is required is to sample from $p(\theta_1|\theta_2)$ and $p(\theta_2|\theta_1)$. In the first step, an arbitrary value for θ_2 is chosen, say θ_2^0. We next obtain a sample from $p(\theta_1|\theta_2^0)$. Denote this value as θ_1^1. With this new value, we then obtain a sample θ_2^1 from $p(\theta_2|\theta_1^1)$. The Gibbs algorithm continues to draw samples using previously obtained values until two long chains of values for both θ_1 and θ_2 are formed. After discarding the burn-in samples, the remaining samples are then considered to be drawn from the marginal distributions of $p(\theta_1)$ and $p(\theta_2)$. (Kaplan, 2014, p. 69)

The number of iterations and number of Markov chains must be selected for the Gibbs sampler (e.g., in this applied example, a fixed number of iterations—10,000—has been set by using the FBITER option; see Table 11.2). Several MCMC chains are typically used; each chain begins from different starting values, and Mplus uses different seeds when making the random draws (i.e., each chain samples from another location of the posterior distribution, based on the starting values). Independent sequences of iterations are conducted for each chain. When the chain has stabilized, the burn-in samples are discarded. Summary statistics of the posterior distribution (e.g., posterior point estimate, standard deviation, credibility interval) are calculated by using the post-burn-in iterations. When two or more chains are specified, Mplus discards the first 50% of the iterations and uses the second half of the chain for forming the posterior distribution and for evaluating convergence. The Mplus default of two chains can be overridden by the CHAINS option of the ANALYSIS command. The Mplus default for the Bayes point estimate is the median; this default can be changed by using the POINT option (e.g., POINT = MEAN).

In addition to setting informative priors, another potential drawback of Bayesian analysis involves the quantification and operationalization of MCMC convergence. This difficulty stems from the fact that the aim of MCMC is to converge on a posterior distribution rather than to a point estimate (unlike ML estimation). The use of parallel chains with different starting values provides one means of assessing convergence. By default, Mplus uses the Gelman–Rubin convergence criterion for determining convergence of the Bayesian estimation (Gelman & Rubin, 1992). This criterion assesses convergence by considering the within- and between-chain variability of the parameter estimates in terms of the *potential scale reduction* (PSR) factor. Gelman and Rubin (1992) assert that an overestimation of the target distribution variance is reflected by the between-chain variance, whereas the within-chain variance is an underestimation. These two estimates should be roughly the same at the point of convergence. Using the between-chain (B) and within-chain (W) variance estimates, the PSR is computed as follows:

$$PSR = SQRT(W + B) / W \qquad (11.7)$$

where PSR values approximately equal to 1.0 are indicative of convergence. A variance ratio close to 1.0 indicates that convergence has been achieved when the between-chain variation is small relative to the within-chain variation. Gelman et al. (2004) have recommended PSR values of 1.10 for all parameters as an operationalization of convergence. Except in very small models, the value of 1.10 is used in Mplus by default in the determination of Bayesian convergence (this default can be overridden by the BCONVERGENCE option).[3]

In addition, MCMC convergence can be assessed in a more subjective fashion by examining the *convergence plots* produced by the chains for each parameter (also referred to as *trace plots* or *history plots*), the posterior distributions of the parameters, and the autocorrelations of the chains. These can be obtained in Mplus by using the PLOT option of the OUTPUT command; TECH8 is requested to show the optimization history in the output (see Table 11.2). The most common subjective method of evaluating convergence is to examine the convergence (history) plots of the parameter for each chain. For example, convergence is gauged by whether the estimates for the parameter evidence a tight horizontal band across the history plot (i.e., across post-burn-in iterations). Autocorrelation plots may be inspected to discern whether the post-burn-in estimates are affected by starting values or by preceding sampling states in the chain; a proper solution is reflected by small autocorrelations, which indicate low dependence in the chain. Posterior probability density plots are examined to determine the extent to which the posterior distributions of the parameters approximate a normal density. Although every parameter of the model should be graphically evaluated, it is perhaps best to begin with the residual variance terms, as these parameters are the most likely to be problematic (i.e., proper convergence of the residual variances bodes well for the remaining parameters of the model).

Traditional goodness-of-fit statistics are not available in Bayesian analyses (e.g., CFI, TLI). Instead, model fit is evaluated by using *posterior predictive checking* (Gelman, Meng, Stern, & Rubin, 1996). In Mplus, posterior predictive checking builds on the standard likelihood ratio χ^2 test statistic, representing the discrepancy between the actual data and the data replicated by the model. A posterior predictive p (PPP) value is then computed on the basis of this discrepancy function (test statistic) every 10th iteration among the iterations used to describe the posterior distribution of the parameters (the same iterations are used to compute point estimates, standard errors, etc.). Specifically, the PPP represents the proportion of MCMC iterations where the test statistic based on the model-generated data exceeds the test statistic derived from the sample data. Small PPP values are indicative of poor fit (i.e., the data generated by the model across iterations do not closely approximate the observed data). The PPP value should not be interpreted in the same fashion as the p value for model χ^2 (e.g., Type I error is not 5% for the correct model). Unlike the traditional p value, PPP does not rely on asymptotic theory. In addition to the PPP value, posterior predictive checking produces a 95% confidence interval for the difference between the test statistics for the sample data and for the generated data. Muthén and Asparouhov (2012) have offered the following tentative criteria for good model fit: (1) a PPP value close to .50; and (2) a 95% confidence interval where the lower band is negative and a zero fit statistic difference (i.e., difference between the test statistics for the real and replicated data) falls close to the middle of the interval. However, further study of the PPP value is needed (e.g., how small does PPP need to be to indicate poor model fit?).

In addition to use of the PPP as an index of absolute fit, the deviation information criterion (DIC) can be used as a fit index to compare competing models (Spiegelhalter, Best, Carlin, & van der Linde, 2002). Although the Bayesian information criterion (BIC; see Chapter 5) can also be used for model comparisons, the DIC is preferred by some methodologists because it is more closely aligned to the concept of Bayesian deviance (cf. Kaplan & Depaoli, 2012). It is akin to the BIC in that, among a group of competing models, the model associated with the smallest DIC is preferred.

We now return to the applied example. The two-factor model of political attitudes is estimated by using 10,000 iterations (5,000 burn-in, 5,000 post-burn-in), as requested by the FBITER = 10,000 option (see Table 11.2). Thus, instead of relying on the Mplus default convergence criterion (BCONVERGENCE = .05, based on a PSR = 1.10), a fixed number of iterations is specified. Evidence of proper convergence is obtained by the fact that the PSRs for all parameters are close to 1.0 (although not shown in Table 11.2, the largest PSR for a parameter at the 10,000th iteration is 1.01). This conclusion is bolstered by graphical evaluations of each parameter. To illustrate, three plots are presented in Figure 11.2 for the primary factor loading of indicator X4. The first graph in this figure depicts a well-behaved convergence history for this parameter; that is, the plot is characterized by a tight, horizontal band with no sharp jumps or fluctuations. The second graph in Figure 11.2 indicates that the posterior density of the X4 primary factor loading approximates a normal distribution. The autocorrelation plot in the third graph is suggestive of a relatively small degree of dependence (formal guidelines for this determination have yet to be established).

As shown in Table 11.2, acceptable model fit is evidenced by a PPP value of .717. The 95% confidence interval for the difference between χ^2 test statistics for the real and replicated data is −37.933–20.585 (also supportive of acceptable fit). The Bayesian estimates from this solution are also presented in Table 11.2. Following procedures outlined in Muthén and Asparouhov (2012), another descriptive measure of fit can be derived by using the estimates from the Bayesian analysis as fixed parameters in an ML analysis to obtain a χ^2 test value for the Bayes solution. When an ML analysis is conducted by using estimates from a Bayesian model without cross-loadings, the resulting χ^2 value is 217.94, which is virtually the same as the value from the original ML analysis presented at the beginning of this section of the chapter (χ^2 = 217.90). Of note, the fit of the Bayesian model without cross-loadings is poor (PPP = 0.000); the lower limit of the 95% confidence band for the difference between the statistics for the real and replicated data is positive (95% CI = 153.68–208.526), which is in accord with a low PPP value and indicates that the χ^2 for the observed data is much larger than what would have been generated by the model (Kaplan & Depaoli, 2012). However, when the ML analysis is repeated by using the Bayesian estimates from Table 11.2 as fixed parameters (i.e., allowing small cross-loadings), the model χ^2 is considerably smaller (88.63), which indicates substantially better model fit. The remaining fit statistics for this model are as follows: SRMR = .055, RMSEA = 0.025 (90% CI = 0.015–0.034, CFit = 1.00), TLI = 0.99, CFI = .99.

The parameter estimates from the Bayesian CFA are presented in Table 11.2. In addition to the point estimates (medians of the posterior distributions), the standard deviations of the posterior distributions, one-tailed p values, and 95% *credibility intervals* (as explained below) are provided for each parameter estimate. If the parameter estimate is positively signed, the p value reflects the proportion of the posterior distribution that is below zero. If the parameter estimate is negative, the p value is the proportion of the posterior distribution that is above zero (Kaplan & Depaoli, 2012). All of the primary factor loadings (X1–X5 for Conservative Attitudes, X6–X10 for Liberal Attitudes) are moderate to large in magnitude (range of λs = 0.54–0.84) and statistically significant, as reflected by 95% credibility intervals that do not cover zero. In Bayesian analysis, a 95% *credibility interval* (also referred to as a *posterior probability interval*) means the probability that the true population value parameter is within the limits of the confidence

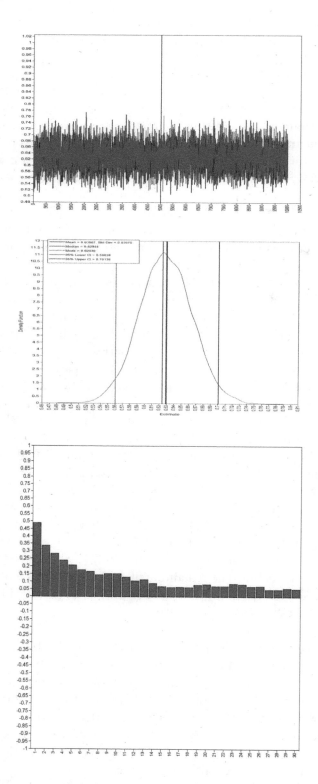

FIGURE 11.2. Convergence, posterior distribution, and autocorrelation plots for item 4 (primary factor loading).

interval is .95 (e.g., it is 95% likely that the true factor loading for X1 is between 0.609 and 0.754; Table 11.2). Although confidence intervals are often interpreted in this fashion in frequentist analysis, this interpretation, strictly speaking, is incorrect because these confidence intervals are based on one sample of data (unlike Bayesian analysis, where the credibility intervals are derived from the posterior distributions of the parameter estimates). The technically accurate interpretation of a 95% confidence interval in frequentist analysis is that over an infinity of samples, 95% of the confidence intervals generated from them contain the true population value (but in a single sample, there is no way to determine whether the point estimate and confidence interval are among those 95%). Thus the Bayesian credibility interval represents a more intuitive and user-friendly assessment of the precision of the parameter estimates.

Modification indices are not provided in Bayesian analysis. Instead, issues pertaining to the revision of fixed parameters are considered simultaneously in a single analysis by the specification of small-variance priors (Muthén & Asparouhov, 2012). In Table 11.2, it can be seen that while the cross-loadings are generally small (range of λs = −0.16–0.13), 6 of the 10 cross-loadings are statistically significant (and thus contributing to poorer fit of the ML analysis with zero cross-loadings). It is also of note that the correlation between the two latent dimensions of political attitudes is somewhat smaller in the Bayesian analysis (−.362) than in the ML analysis (−.40). This is not surprising because, in the Bayesian analysis, the non-zero cross-loadings assisted in the reproduction of the observed relationships among the indicators; in the ML analysis, these relationships are reproduced solely by the primary factor loadings and the factor correlation (cf. Chapter 5).

Because the outcomes of Bayesian analysis can be affected strongly by the choice of informative priors, it is important to conduct a sensitivity analysis to evaluate the stability of the results across a plausible range of priors. The range of selected priors and the results of the sensitivity analysis should be included in the report of research findings. In the current example, the analysis is reconducted with cross-loading prior variances of 0.001, 0.005, 0.02, and 0.03. The results are presented in Table 11.3. As noted by Muthén and Asparouhov (2012), if the variance of the prior is set to be too small, this will not

TABLE 11.3. Results of Sensitivity Analyses with Different Variances for the Informative Priors of the Cross-Loadings in Bayesian CFA

Prior variance	95% cross-loading limit	PPP	DIC	Largest cross-loading	Factor correlation
0.001	±0.06	.151	24,695	0.11	−.370
0.005	±0.14	.680	24,672	0.15	−.363
0.01	±0.20	.717	24,671	0.16	−.362
0.02	±0.28	.720	24,671	0.16	−.363
0.03	±0.38	.718	24,671	0.16	−.364

Note. PPP, posterior predictive *p*; DIC, deviance information criterion.

allow some of the cross-loadings to escape sufficiently from the zero prior mean, thereby producing a worse PPP value. This is reflected in the analysis, where the cross-loading prior variances are set to the lowest value (0.001; 95% range of loadings = ±0.06); this model results in a considerably smaller PPP value and a larger DIC value than in the other models (PPP and DIC = .151 and 24,695, respectively). Otherwise, the results of the sensitivity analysis indicate that degree of informativeness (i.e., level of certainty in the cross-loading parameters) does not have an appreciable impact on model fit or the parameter estimates. For example, at a variance of 0.01 and above, the fit indices are virtually identical (e.g., all DICs = 24,671), as are the factor correlation and highest cross-loading (-.36 and .16, respectively). It is likely that more variability in model fit and the estimates would be seen in a smaller sample. In larger samples such as in the current illustration ($N = 1,000$), the data contribute relatively more information to the formation of the posterior distributions than do the priors.

Bayesian CFA: Summary

The Bayesian approach to latent variable analysis holds considerable promise for testing more realistic models in applied data sets, as guided by substantive theory. A specific advantage of Bayesian analysis in the context of latent variable measurement models is the ability to replace parameters ordinarily fixed to zero with approximate zeroes, through the specification of informative priors. Although the example used in this chapter has focused on near-zero cross-loadings, this Bayesian method can be applied to other aspects of the CFA model such as indicator error covariances (see, e.g., Muthén & Asparouhov, 2012), as well as in evaluations of measurement invariance, where it may be more plausible to expect that between-group or temporal differences in the factor loadings and indicator intercepts are close to but not exactly zero. Despite these strengths, aspects of Bayesian analysis require further research so that they can better inform applied analysis (e.g., MacCallum, Edwards, & Li, 2012; Muthén & Asparouhov, 2012). Some areas in particular need of additional study are model fit assessment (e.g., the behavior of PPP across various types of models and data, development of a Bayesian counterpart to modification indices) and the techniques for evaluating proper model convergence.

The ability to include previous knowledge in the model specification is a hallmark feature of Bayesian analysis, but is also a potential drawback. First, informative priors may be prone to abuse by unscrupulous researchers. An example of dishonest practice would be conducting an ML analysis in order to use the resulting parameter estimates as informative priors in a Bayesian analysis of the same sample. This would build an unfounded level of certainty into the model, which would falsely augment the statistical power and precision of the resulting parameter estimates. Second, the specification of informative priors for indicator error covariances can be particularly challenging (e.g., it may be more susceptible to convergence problems and model misspecification). As discussed by Muthén and Asparouhov (2012), the specification of small residual correlations may obscure other aspects of the model, such as the need for cross-loadings (in addition to, or in place of, error covariances) or additional factors. In this respect (i.e., the ability to estimate a range of parameters that would ordinarily be fixed), it can be argued that a strong substantive foundation for model specification is even more paramount to Bayesian analysis than to traditional CFA.

Because the Bayesian method has only become accessible to latent variable researchers in the past few years, examples in the applied social sciences literature are sparse at this writing. However, a couple of recent papers have evaluated measurement models via Bayesian estimation (Fong & Ho, 2013; Golay, Reverte, Rossier, Favez, & Lecerf, 2013).

MULTILEVEL CFA

Besides Bayesian CFA, another relatively new methodology is the multilevel factor model. Although multilevel models have been around for years (cf. hierarchical linear models; Raudenbush & Bryk, 2002), only more recently has this methodology merged with CFA in a manner readily accessible to applied researchers. Multilevel models should be employed when data have been obtained by cluster or unequal probability sampling. In these instances, the data are said to have a hierarchical or nested structure (in this context, the term *hierarchical* should not be confused with higher-order factor analysis; see Chapter 8). For example, consider a study of grade school scholastic achievement where the student data have been collected statewide. The data have a multilevel hierarchical structure; that is, students are nested within classrooms, classrooms are nested within schools, and schools are nested within school districts. Other common examples of clustered data structures are family data (i.e., cases nested within families) and repeated measures data (i.e., cases nested by time). The observations may not be independent within clusters. For example, one should not employ an independent-sample *t* test to compare the means of a sample assessed at two time points because the observations are not independent. In cross-sectional data, the observations also may not be independent because of the clustered data structure. For instance, the scholastic achievement of students within a given classroom of a given school may be homogeneous because students share the same teacher and classroom dynamics, and come from similar family/socio-economic backgrounds. In data sets of this nature, multilevel modeling is employed to avoid biases in parameter estimates, standard errors, and tests of model fit. In other words, if the hierarchical structure is ignored, so is the nonindependence of observations. Consequently, for example, the standard errors of parameter estimates may be underestimated, resulting in positively biased statistical significance testing.

Moreover, a multilevel model can be estimated to learn more about within- and between-cluster relationships. Multilevel models are also referred to as *random coefficient models*. Random coefficients are parameters in a model that vary across clusters. Covariates can be included in a multilevel model to account for variability within and between clusters. In models where individuals are nested under clusters (i.e., dependent measures representing characteristics of individuals are at the lowest level of the data hierarchy), within-cluster predictors are characteristics of individuals (also referred to as Level 1 predictors). Between-cluster predictors are characteristics of the clusters (also referred to as Level 2 predictors). To continue with the example of scholastic achievement, a multilevel regression model may find that student gender is a significant predictor of reading achievement: Girls are better at reading than boys are. This is a within-level or Level 1 effect (gender is a characteristic of individuals); that is, the gender covariate accounts for variation in reading achievement among individuals. As an example of a between-level (Level 2) effect, the amount of teacher training/experience (a characteristic of classrooms, the clusters) may account for variability in reading achievement across classrooms. Moreover, the multilevel analysis may reveal that the effect of gender on reading achievement varies significantly across classrooms. Thus the gender → reading achievement effect is a *random slope* (the slope varies across classrooms). Level 2 covariates can be brought into the model to explain the variability of this coefficient across clusters (classrooms). For instance, it may be found that the level of teacher training/ experience is inversely related to this random slope (a between-level effect); that is, the effect of gender on reading achievement decreases as teacher experience increases. This *cross-level interaction* might be interpreted substantively as indicating that girls do better at reading than boys because, in the classroom, boys are less attentive, more recalcitrant, and so forth (the within-level effect). However, experienced teachers are more adept at implementing strategies to direct boys' attention to classroom learning (the cross-level interaction). If the necessary data are available, this two-level analysis can be expanded to incorporate additional levels of the data structure (e.g., classrooms within schools, schools within school districts).

Multilevel models can be employed to analyze within- and between-level EFA and CFA models. In fact, the number of factors can differ at the within and between levels. Indeed, some evidence in applied data sets suggests that fewer factors are often obtained at the between levels because less variability is usually present across clusters than among individuals. As in the single-indicator example described in the preceding paragraph, covariates at either level can be brought into the model to explain variability within and between clusters. Any parameter of the CFA solution (e.g., a factor loading, an indicator intercept) can be treated as a random coefficient, if doing so can be justified on substantive and empirical grounds. Most latent variable software programs (e.g., Mplus, EQS, LISREL) have multilevel factor modeling capabilities. Applications of this methodology are increasing in the literature, including studies by Dedrick and Greenbaum (2011); Dyer, Hanges, and Hall (2005); Kaplan and Kreisman (2000); and J. Little (2013).

To illustrate these concepts, consider an example where a five-item questionnaire assessing job satisfaction is administered to 830 employees, sampled from 85 departments of a large company (i.e., a two-level data structure where employees are nested under departments). Thus the average cluster size is 9.765 (i.e., 830 / 85). The ultimate objective is to specify a multilevel CFA model where a single factor of Job Satisfaction is specified both within and between levels to account for the variability in the five questionnaire items (measured on continuous scales). Although only multilevel CFAs are shown in this illustration, in practice this analysis should be conducted in multiple steps (cf. Hox, 2013). An initial step entails the examination of the *intraclass correlation coefficients* (ICCs) of the indicators. ICCs convey the proportion of variance in the indicators due to the clusters (e.g., variability in the questionnaire items explained by departments).[4] If the ICCs are small (e.g., <.05), then a multilevel model may not be necessary. Instead, alternative methods can be employed that do not formally model hierarchical structure, but do adjust for the dependency (however trivial) that exists in the data. For instance, Mplus has two options that can be used on the ANALYSIS command to accommodate complex data, TYPE = COMPLEX and TYPE = TWOLEVEL. The TWOLEVEL option estimates a two-level model where the within- and between-level structures are fully specified. The COMPLEX option executes a standard, nonstructured analysis in which the model goodness-of-fit statistics and standard errors of the parameter estimates are adjusted for the dependency in the data. Thus, in situations where the ICCs are small, the TYPE = COMPLEX option can be used in place of TYPE = TWOLEVEL. ICCs are considered in more detail later in this example.

To ensure a viable measurement model, the second step entails the specification of the CFA model at the within level, leaving it unstructured at the between level. If Mplus is used, the CFA model can be specified in the typical fashion by using the TYPE = COMPLEX option on the ANALYSIS command. On the assumption that an acceptable measurement model is present, the next step is to examine the between-level factor structure in a two-level model with the within-level structure fully specified.

Figure 11.3 depicts the two-level measurement model where the single factor of Job Satisfaction is specified at both the within and between levels (SATIS$_W$, SATIS$_B$). Although other approaches exist, the path diagram follows the conventions forwarded by the developers of Mplus (Muthén & Muthén, 2008–2012). Of note is the fact that the indicators (Y1–Y5) are observed measures at the within levels (employees), but are latent variables at the between levels (departments). Specifically, the filled circles in the path diagram on the Y indicators represent random intercepts that are continuous between-level latent variables. On the between levels, a single factor (SATIS$_B$) is specified to account for the variation and covariation among these random intercepts. The observed individual-level variables (Y) are reproduced by the following within- and between-level equations:

$$Y = \Lambda_W \eta_W + \varepsilon_W \qquad \text{(within)} \qquad (11.8)$$

$$\mu_B = \mu + \Lambda_B \eta_B + \varepsilon_B \qquad \text{(between)} \qquad (11.9)$$

where μ is the vector of between-level means; Λ_W is the within-level factor loading matrix; η_W is the within-level factor; Λ_B is the between-level factor loading matrix; η_B is the between-level factor; ε_W is the indicator residual variance within levels; and ε_B is the indicator residual variance between levels (in a common alternative notation system, the subscripts W and B are replaced by the subscripts i and j, respectively, to denote within- and between-level variables or coefficients; cf. Raudenbush & Bryk, 2002). The factor loading matrices (Λ_W, Λ_B) and cluster-level means (μ) are fixed effects. Importantly, μ_B represents the random intercepts of the Y variables that are the focus of the between-

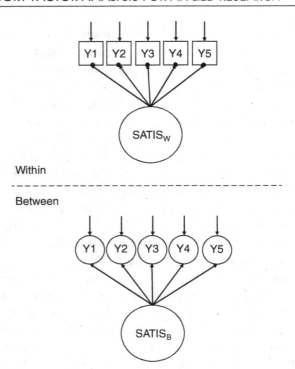

FIGURE 11.3. Two-level CFA model of Job Satisfaction. $SATIS_W$, within-level Job Satisfaction factor; $SATIS_B$, between-level Job Satisfaction factor.

level factor analytic model.[5] Eq. 11.8 models within-group variation (Level 1), whereas Eq. 11.9 models between-group variation and the between-level means (Level 2). These two equations can be combined as follows:

$$Y = \mu + \Lambda_W \eta_W + \Lambda_B \eta_B + \varepsilon_B + \varepsilon_W \tag{11.10}$$

Mplus 7.11 syntax and selected output for the two-level CFA model are presented in Table 11.4. In addition to the five items from the job satisfaction questionnaire (Y1–Y5), a variable representing cluster membership (department number, DEPT) must be input and declared as the cluster variable (CLUSTER = DEPT). On the ANALYSIS command, TYPE = TWOLEVEL is requested to estimate a two-level factor model (cf. TYPE = COMPLEX). Robust MLR is requested as the estimator, although this is the Mplus default. MLR has many advantages over its estimator predecessors, including its robustness to non-normality, ability to accommodate missing data, and computational efficiency (Muthén & Asparouhov, 2011). The within- and between-level factor models are specified under the MODEL command by using the %WITHIN% and %BETWEEN% keywords, respectively. At both levels, the unstandardized factor loading for Y1 is fixed to 1.0 (marker indicator; Mplus default), and the remaining factor loadings are freely

TABLE 11.4. Mplus 7.11 Syntax and Selected Output for Two-Level Factor Model of Job Satisfaction

```
TITLE: TWO-LEVEL CFA WITH 5 INDICATORS AND RANDOM INTERCEPT FACTOR
DATA:      FILE IS MLM2.dat;
VARIABLE: NAMES ARE Y1-Y5 DEPT;
          CLUSTER = DEPT;
ANALYSIS: TYPE = TWOLEVEL;
          ESTIMATOR = MLR;   ! Mplus default
MODEL:
          %WITHIN%
          SATISW BY Y1-Y5;
          %BETWEEN%
          SATISB BY Y1-Y5;
          Y1-Y5@0;             ! errors fixed to zero at between level
OUTPUT: SAMPSTAT STAND;
```

SUMMARY OF DATA

```
     Number of clusters                    85
     Average cluster size      9.765
```

Estimated Intraclass Correlations for the Y Variables

Variable	Intraclass Correlation	Variable	Intraclass Correlation	Variable	Intraclass Correlation
Y1	0.156	Y2	0.143	Y3	0.105
Y4	0.121	Y5	0.143		

ESTIMATED SAMPLE STATISTICS FOR WITHIN

Covariances

	Y1	Y2	Y3	Y4	Y5
Y1	2.077				
Y2	0.863	1.614			
Y3	0.881	0.697	1.641		
Y4	0.848	0.633	0.754	1.752	
Y5	0.828	0.625	0.645	0.641	1.621

ESTIMATED SAMPLE STATISTICS FOR BETWEEN

Covariances

	Y1	Y2	Y3	Y4	Y5
Y1	0.383				
Y2	0.308	0.270			
Y3	0.269	0.224	0.192		
Y4	0.301	0.240	0.211	0.241	
Y5	0.317	0.268	0.227	0.246	0.271

(continued)

TABLE 11.4. *(continued)*

MODEL RESULTS

	Estimate	S.E.	Est./S.E.	Two-Tailed P-Value
Within Level				
SATISW BY				
Y1	1.000	0.000	999.000	999.000
Y2	0.787	0.034	23.114	0.000
Y3	0.785	0.034	22.877	0.000
Y4	0.786	0.037	21.499	0.000
Y5	0.773	0.036	21.195	0.000
Variances				
SATISW	1.092	0.081	13.551	0.000
Residual Variances				
Y1	0.985	0.072	13.673	0.000
Y2	0.972	0.056	17.472	0.000
Y3	0.921	0.056	16.578	0.000
Y4	1.083	0.054	20.048	0.000
Y5	1.015	0.050	20.183	0.000
Between Level				
SATISB BY				
Y1	1.000	0.000	999.000	999.000
Y2	0.787	0.034	23.114	0.000
Y3	0.785	0.034	22.877	0.000
Y4	0.786	0.037	21.499	0.000
Y5	0.773	0.036	21.195	0.000
Intercepts				
Y1	0.012	0.084	0.144	0.885
Y2	-0.041	0.072	-0.576	0.565
Y3	-0.055	0.063	-0.870	0.384
Y4	-0.013	0.071	-0.181	0.857
Y5	0.030	0.072	0.419	0.675
Variances				
SATISB	0.380	0.093	4.085	0.000
Residual Variances				
Y1	0.000	0.000	999.000	999.000
Y2	0.000	0.000	999.000	999.000
Y3	0.000	0.000	999.000	999.000
Y4	0.000	0.000	999.000	999.000
Y5	0.000	0.000	999.000	999.000

(continued)

TABLE 11.4. *(continued)*

R-SQUARE

Within Level

Observed Variable	Estimate	S.E.	Est./S.E.	Two-Tailed P-Value
Y1	0.526	0.030	17.487	0.000
Y2	0.410	0.026	16.002	0.000
Y3	0.422	0.030	13.888	0.000
Y4	0.383	0.029	13.157	0.000
Y5	0.391	0.023	17.310	0.000

Between Level

Observed Variable	Estimate	S.E.	Est./S.E.	Two-Tailed P-Value
Y1	1.000	0.000	********	0.000
Y2	1.000	0.000	9524.510	0.000
Y3	1.000	0.000	********	0.000
Y4	1.000	0.000	9863.795	0.000
Y5	1.000	0.000	8891.305	0.000

estimated. By default, the following parameters are freely estimated: factor variances (both levels), fixed intercepts (between-level), and indicator residual variances (both levels). The latent variable means and covariances of the residuals are fixed to zero at both levels as the default. In this example, the Mplus default of freely estimating the indicator residual variances at the between levels is overridden by fixing these parameters to zero (i.e., Y1–Y5@0). Although this is done to foster the estimation and parsimony of the model, this restriction may not be realistic in many applied data sets, where between-level variation in the random intercepts exists beyond that explained by the between-level factor model (this can be formally tested with χ^2 difference evaluation).

Before we turn to the results for the two-level CFA solution, consider the selected sample statistics presented in Table 11.4. The ICCs for the five indicators are of a magnitude (range = .105–.156) that attests to the existence of Level 2 variability, warranting a between-level structural analysis (i.e., a two-level model is justified). For instance, the ICC for Y1 is .156, which indicates that 15.6% of the total variance of this observed measure is due to the clusters (i.e., the effect size of variance in Y1 explained by departments). In addition, the Mplus output parses the within- and between-level variance–covariance matrices of the five indicators (Table 11.4). From these estimates, the ICCs for each indicator can be hand-computed; for example, ICC_{Y1} = .383 / (.383 + 2.077) = .156.

The two-level measurement model fits the data well: $\chi^2(15)$ = 10.654, p = .78, RMSEA = 0.00, TLI = 1.00, CFI = 1.00 (not shown in the table). The final section of Table 11.4

presents the unstandardized parameter estimates from this solution and proportion of indicator variance explained (R-SQUARE). Whereas the R^2s = 1.00 at the between level due to fixing the Level 2 error variances to zero, the proportions of indicator variance explained by the Job Satisfaction factor at the within level ranged from .391 to .526. The lower R^2 values at Level 1 suggests that after variability in questionnaire scores due to departments is controlled, there is considerable heterogeneity in the within-level (employees) variability in Job Satisfaction. Of note is that the Job Satisfaction factor has significant variation at both the within and between levels, 1.092 and 0.380, respectively (ps < .001). The magnitude of these variances cannot be directly compared unless it can be shown that the factors possess a common metric. Because the same measurement model exists at the within and between levels (equal form), the equality of factor loadings across levels can be pursued. If the factor loadings are indeed invariant, then the metrics of the within- and between-level factors are equated, and thus the factor variances are directly comparable (Mehta & Neale, 2005).

In Table 11.5, Mplus syntax and selected results are provided for the equal factor loading two-level model. This model provides a good fit to the data: $\chi^2(19)$ = 17.549, p = .55, RMSEA = 0.00, TLI = 1.00, CFI = 1.00 (not shown). Given the use of MLR estimation, the DIFFTEST option cannot be used to compare the fit of the nested model. Instead, the χ^2 difference test is hand-computed by using the log-likelihoods, scaling correction factors, and number of freely estimated parameters from the two models (cf. Chapter 9). The results support the equivalence of factor loadings across levels, $\chi^2_{diff}(4)$ = 5.70, ns. Because the factor loadings are invariant, an ICC can be calculated for the Job Satisfaction factor by using the within- and between-level factor variance estimates; that is, $ICC_{\eta 1}$ = 0.380 / (1.092 + 0.380) = .258. Thus 25.8% of the variability in the latent construct of Job Satisfaction is due to departments (vs. individual differences in employees within departments).

In most applied research scenarios, specification of a multilevel measurement model is not the final analysis, but instead is a precursor to multilevel SEM (e.g., Hox, 2013).[6] As in the simple multilevel regression model discussed earlier, covariates of the factors can be brought into the model at either level after a good-fitting measurement model has been established. The two-level model can also be expanded to include multiple factors (at either or both levels) and cross-level interactions. In the current illustration, for example, age and a measure of leadership style can be added as predictors of the within- and between-level Job Satisfaction factors, respectively (age is a characteristic of the Level 1 units, employees; leadership style is a characteristic of the Level 2 units, departments). The Age → SATIS$_W$ relationship can be specified as a random slope to determine whether the strength of this effect varies across departments (however, a random effect of Leadership Style → SATIS$_B$ cannot be evaluated because these variables reside at Level 2, the top of the data hierarchy). If doing so is conceptually plausible and supported by the data (i.e., there is evidence of significant random effects), cross-level interactions can be pursued; for instance, does leadership style account for significant variability in the magnitude of the Age → SATIS$_W$ effect across departments?

TABLE 11.5. Mplus 7.11 Syntax and Selected Output for Two-Level Factor Model of Job Satisfaction with Invariant Factor Loadings

```
TITLE:   TWO-LEVEL CFA WITH 5 INDICATORS AND RANDOM INTERCEPT FACTOR
DATA:       FILE IS mlm2.dat;
VARIABLE: NAMES ARE Y1-Y5 DEPT;
          CLUSTER = DEPT;
ANALYSIS: TYPE = TWOLEVEL;
          ESTIMATOR = MLR;
MODEL:
      %WITHIN%
      SATISW BY Y1
      Y2 (1)
      Y3 (2)
      Y4 (3)
      Y5 (4);
      %BETWEEN%
      SATISB BY Y1
      Y2 (1)
      Y3 (2)
      Y4 (3)
      Y5 (4);
      Y1-Y5@0;
OUTPUT: SAMPSTAT STAND;
```

MODEL RESULTS

	Estimate	S.E.	Est./S.E.	Two-Tailed P-Value
Within Level				
SATISW BY				
Y1	1.000	0.000	999.000	999.000
Y2	0.787	0.034	23.114	0.000
Y3	0.785	0.034	22.877	0.000
Y4	0.786	0.037	21.499	0.000
Y5	0.773	0.036	21.195	0.000
Variances				
SATISW	1.092	0.081	13.551	0.000
Residual Variances				
Y1	0.985	0.072	13.673	0.000
Y2	0.972	0.056	17.472	0.000
Y3	0.921	0.056	16.578	0.000
Y4	1.083	0.054	20.048	0.000
Y5	1.015	0.050	20.183	0.000
Between Level				
SATISB BY				
Y1	1.000	0.000	999.000	999.000
Y2	0.787	0.034	23.114	0.000

(continued)

TABLE 11.5. (continued)

Y3	0.785	0.034	22.877	0.000
Y4	0.786	0.037	21.499	0.000
Y5	0.773	0.036	21.195	0.000
Intercepts				
Y1	0.012	0.084	0.144	0.885
Y2	-0.041	0.072	-0.576	0.565
Y3	-0.055	0.063	-0.870	0.384
Y4	-0.013	0.071	-0.181	0.857
Y5	0.030	0.072	0.419	0.675
Variances				
SATISB	0.380	0.093	4.085	0.000
Residual Variances				
Y1	0.000	0.000	999.000	999.000
Y2	0.000	0.000	999.000	999.000
Y3	0.000	0.000	999.000	999.000
Y4	0.000	0.000	999.000	999.000
Y5	0.000	0.000	999.000	999.000

SUMMARY

Bayesian estimation and multilevel factor models offer exciting new modeling possibilities to the researcher working within the realm of CFA and SEM. These developments perpetuate the hallmarks of CFA/SEM as an exceptionally versatile analytic framework for addressing a wide array of questions in the social and behavioral sciences.

NOTES

1. The left side of Eq. 11.1 can be expressed as either $P(H, D)$ or $P(D, H)$ because joint probabilities are symmetric; that is, $P(H, D) = P(D, H)$.

2. Alternative terminologies exist for informative and noninformative priors. For instance, informative priors have also been referred to as *subjective priors*. Other terms for noninformative priors include *objective, diffuse,* and *vague priors.*

3. Strictly speaking, the critical PSR value for convergence in Mplus ranges between 1.05 and 1.10, depending on the number of model parameters. Bayesian convergence (BCONVERGENCE) in Mplus is based on the following formula: $a = 1 +$ BCONVERGENCE * factor, where the factor value ranges between one and two depending on the number of parameters (e.g., with one parameter, factor = 1.0; with large number of parameters, factor = 2.0). This multiplicity factor allows for more leniency in the convergence criterion when there are many parameters in the model (Asparouhov & Muthén, 2010). The convergence criterion is checked every 100th iteration. Convergence is obtained when PSR < a for every parameter in the model. The default BCONVERGENCE value in Mplus is .05. Thus, for a large model, $a =$

$1 + (.05)(2) = 1.10$; that is, convergence will be obtained if PSRs are below 1.10. Some methodologists have argued that stricter convergence criteria should be implemented in Bayesian analyses. To request a more stringent analysis, BCONVERGENCE can be set to a smaller value; for example, BCONVERGENCE = .01 will result in the requirement that PSR values are below 1.02.

4. Unlike the Pearson correlation coefficient, the ICC should not be squared to obtain the proportion of explained variance.

5. The multilevel CFA can also be specified to allow between-level variation in the factor loadings, although this strategy is less commonly implemented because the random effects of these parameters tend to be considerably smaller than the degree of cluster-level variation in the random intercepts.

6. An exception is the specification of the multilevel measurement model to evaluate the scale reliability of a test instrument in context of a complex data structure (e.g., Geldhof, Preacher, & Zyphur, 2014).

Numerical Example of Bayesian Probability

As noted in this chapter, the essence of Bayes's theorem is understanding how the probability that a theory is true is affected by new evidence. Using Bayes's probability theorem, we wish to compute the conditional probability of heart disease (H) based on whether or not an individual eats a high-fat diet (D). Specifically, what is the probability that an individual has heart disease (H) after we obtain knowledge that he or she consumes a high-fat diet (D)? Consider the following table:

	Heart disease (10%)	No heart disease (90%)
Eats high-fat diet	70%	25%
Does not eat high-fat diet	30%	75%

Thus the rate of heart disease in this sample is 10%. In individuals with heart disease, 70% ate a high-fat diet. But what is the probability that an individual has heart disease after we obtain knowledge that he or she consumes a high-fat diet? The correct answer is not reflected by any of the cells in the table. Rather, we use Bayes's theorem to obtain the actual probability of heart disease in the presence of a high-fat diet, taking into account the fallibility (i.e., measurement error) of our predictor (e.g., 25% of individuals without heart disease consume a high-fat diet).

Bayes's theorem is as follows:

$$P(H|D) = [P(D|H)\ P(H)]\ /\ P(D)$$

which indicates that the conditional probability of heart disease in the presence of a high-fat diet, $P(H|D)$, is equal to the probability of a high-fat diet in the presence of heart disease, $P(D|H)$, times the probability of having heart disease, $P(H)$. The values for the numerator can be taken directly from the table; that is, $P(D|H) = .70$, and $P(H) = .10$. The denominator of this equation is the marginal probability of a high-fat diet, $P(D)$; this reflects the probability that an individual consumes a high-fat diet, collapsed across persons with and without heart disease. This probability can be computed as the sum of the probabilities of true positives and false positives: True positive = .70(.10) = .07; false positive = .25(.90) = .225; .07 + .225 = .295. Thus

$$P(H|D) = [.70(.10)]\ /\ .295 = .2373$$

Given the presence of a high-fat diet, the probability of heart disease is .2373.

References

[Brackets indicate the chapter(s) in which the reference is cited.]

Ahmavaara, Y. (1954). Transformation analysis of factorial data. *Annales Academiae Scientiarium Fennicae, 88*(Ser. B), 150. [3]

Akaike, H. (1987). Factor analysis and AIC. *Psychometrika, 52,* 317–322. [5]

Alhija, F. N., & Wisenbaker, J. (2006). A Monte Carlo study investigating the impact of item parceling strategies on parameter estimates and their standard errors in CFA. *Structural Equation Modeling, 13,* 204–228. [9]

Allison, P. D. (1987). Estimation of linear models with incomplete data. In C. Clogg (Ed.), *Sociological methodology 1987* (pp. 71–103). Washington, DC: American Sociological Association. [9]

Allison, P. D. (2002). *Missing data.* Thousand Oaks, CA: Sage. [9]

Allison, P. D. (2003). Missing data techniques for structural equation modeling. *Journal of Abnormal Psychology, 112,* 545–557. [9]

American Psychiatric Association. (1994). *Diagnostic and statistical manual of mental disorders* (4th ed.). Washington, DC: Author. [6]

American Psychiatric Association. (2013). *Diagnostic and statistical manual of mental disorders* (5th ed.). Arlington, VA: Author. [4, 7]

Anderson, J. C., & Gerbing, D. W. (1984). The effect of sampling error on convergence, improper solutions, and goodness-of-fit indices for maximum likelihood confirmatory factor analysis. *Psychometrika, 49,* 155–173. [5]

Arbuckle, J. L. (1996). Full information estimation in the presence of missing data. In G. A. Marcoulides & R. E. Schumacker (Eds.), *Advanced structural equation modeling: Issues and techniques* (pp. 243–277). Mahwah, NJ: Erlbaum. [9]

Arbuckle, J. L., & Wothke, W. (1999). *Amos 4.0 user's guide.* Chicago: Smallwaters. [7]

Asparouhov, T., & Muthén, B. (2009). Exploratory structural equation modeling. *Structural Equation Modeling, 16,* 397–438. [5]

Asparouhov, T., & Muthén, B. (2010). *Bayesian analysis using Mplus: Technical implementation* (Technical report). Retrieved from *www.statmodel.com/download/Bayes3.pdf* [11]

Asparouhov, T., & Muthén, B. (2013). *Conputing the strictly positive Satorra–Bentler chi-square test in Mplus* (Mplus Webnote No. 12). Retrieved from *www.statmodel.com/examples/webnotes/SB5.pdf* [9]

Bagozzi, R. P. (1993). Assessing construct validity in personality research: Applications to measures of self-esteem. *Journal of Research in Personality, 27,* 49–87. [6]

Bagozzi, R. P. (2007). On the meaning of formative measurement and how it differs from reflective measurement: Comment on Howell, Breivik, and Wilcox (2007). *Psychological Methods, 12,* 229–237. [8]

Bagozzi, R. P., & Yi, Y. (1990). Assessing method variance in multitrait–multimethod matrices: The case of self-reported affect and perceptions at work. *Journal of Applied Psychology, 75,* 547–560. [6]

Bandalos, D. L. (2008). Is parceling really necessary?: A comparison of results from item parceling and categorical variable methodology. *Structural Equation Modeling, 15,* 211–240. [9]

Bandalos, D. L., & Finney, J. S. (2001). Item parceling issues in structural equation modeling: In G. A. Marcoulides & R. E. Schumacker (Eds.), *New developments and techniques in structural equation modeling* (pp. 269–295). Mahwah, NJ: Erlbaum. [9]

Bartlett, M. S. (1937). The statistical conception of mental factors. *British Journal of Psychology, 28,* 97–104. [2]

Bauer, D. J. (2005). The role of nonlinear factor-to-indicator relationships in tests of measurement equivalence. *Psychological Methods, 10,* 305–316. [7]

Beauducel, A., & Wittman, W. W. (2005). Simulation study on fit indices in CFA based on data with slightly distorted simple structure. *Structural Equation Modeling, 12,* 41–75. [3]

Bentler, P. M. (1990). Comparative fit indices in structural models. *Psychological Bulletin, 107,* 238–246. [3]

Bentler, P. M. (1995). *EQS structural equations program manual.* Encino, CA: Multivariate Software. [3, 7, 9]

Bentler, P. M., & Chou, C. (1987). Practical issues in structural modeling. *Sociological Methods and Research, 16,* 78–117. [10]

Bock, R. D. (1997). A brief history of item response theory. *Educational Measurement: Issues and Practice, 16,* 21–33. [9]

Bock, R. D., Gibbons, R., & Muraki, E. J. (1988). Full information item factor analysis. *Applied Psychological Measurement, 12,* 261–280. [9]

Bollen, K. A. (1989). *Structural equations with latent variables.* New York: Wiley. [1, 3, 4, 8]

Bollen, K. A. (1990). Outlier screening and a distribution-free test for vanishing tetrads. *Sociological Methodology and Research, 19,* 80–92. [5]

Bollen, K. A. (2007). Interpretational confounding is due to misspecification, not to type of indicator: Comment on Howell, Breivik, and Wilcox (2007). *Psychological Methods, 12,* 219–228. [8]

Bollen, K. A., & Curran, P. J. (2006). *Latent curve models: A structural equation perspective.* Hoboken, NJ: Wiley. [7]

Bollen, K. A., & Davis, W. R. (2009). Causal indicator models: Identification, estimation, and testing. *Structural Equation Modeling, 16,* 498–522. [8]

Bollen, K., & Lennox, R. (1991). Conventional wisdom on measurement: A structural equation perspective. *Psychological Bulletin, 110,* 305–314. [8]

Bollen, K. A., & Stine, R. A. (1993). Bootstrapping goodness-of-fit measures in structural equation models. In K. A. Bollen & J. S. Long (Eds.), *Testing structural equation models* (pp. 111–135). Newbury Park, CA: Sage. [9, 10]

Bollen, K. A., & Ting, K. (1993). Confirmatory tetrad analysis. In P. M. Marsden (Ed.), *Sociological methodology* (pp. 147–175). Washington, DC: American Sociological Association. [5]

Bollen, K. A., & Ting, K. (2000). A tetrad test for causal indicators. *Psychological Methods, 5,* 3–22. [5]

Boomsma, A. (1983). *On the robustness of LISREL against small sample size and nonnormality.* Amsterdam: Sociometric Research Foundation. [10]

Brouwer, D., Meijer, R. R., Weekers, A. M., & Baneke, J. J. (2008). On the dimensionality of the Dispositional Hope Scale. *Psychological Assessment, 20,* 310–315. [8]

Brown, T. A. (2003). Confirmatory factor analysis of the Penn State Worry Questionnaire: Multiple factors or method effects? *Behaviour Research and Therapy, 41,* 1411–1426. [3, 5]

Brown, T. A., Chorpita, B. F., & Barlow, D. H. (1998). Structural relationships among dimensions of the DSM-IV anxiety and mood disorders and dimensions of negative affect, positive affect, and autonomic arousal. *Journal of Abnormal Psychology, 107*, 179–192. [2]

Brown, T. A., White, K. S., & Barlow, D. H. (2005). A psychometric reanalysis of the Albany Panic and Phobia Questionnaire. *Behaviour Research and Therapy, 43*, 337–355. [5, 7]

Brown, T. A., White, K. S., Forsyth, J. P., & Barlow, D. H. (2004). The structure of perceived emotional control: Psychometric properties of a revised Anxiety Control Questionnaire. *Behavior Therapy, 35*, 75–99. [5, 8]

Browne, M. W. (1984a). The decomposition of multitrait–multimethod matrices. *British Journal of Mathematical and Statistical Psychology, 37*, 1–21. [6]

Browne, M. W. (1984b). Asymptotic distribution free methods in the analysis of covariance structures. *British Journal of Mathematical and Statistical Psychology, 37*, 62–83. [9]

Browne, M. W., & Cudeck, R. (1993). Alternate ways of assessing model fit. In K. A. Bollen & J. S. Long (Eds.), *Testing structural equation models* (pp. 136–162). Newbury Park, CA: Sage. [3]

Bryant, F. B., & Satorra, A. (2012). Principles and practice of scaled difference chi-square testing. *Structural Equation Modeling, 19*, 372–398. [9]

Byrne, B. M. (2006). *Structural equation modeling with EQS and EQS/Windows: Basic concepts, applications, and programming* (2nd ed.). New York: Psychology Press. [7]

Byrne, B. M. (2014). *Structural equation modeling with LISREL, PRELIS, and SIMPLIS: Basic concepts, applications, and programming.* New York: Psychology Press. [4, 7]

Byrne, B., & Goffin, R. (1993). Modeling multitrait–multimethod data from additive and multiplicative covariance structures: An audit of construct validity concordance. *Multivariate Behavioral Research, 28*, 67–96. [6]

Byrne, B. M., Shavelson, R. J., & Muthén, B. (1989). Testing for the equivalence of factor covariance and mean structures: The issue of partial measurement invariance. *Psychological Bulletin, 105*, 456–466. [3, 7]

Campbell, D. T., & Fiske, D. W. (1959). Convergent and discriminant validation by the multitrait–multimethod matrix. *Psychological Bulletin, 56*, 81–105. [6]

Campbell, D. T., & O'Connell, E. J. (1967). Method factors in multitrait–multimethod matrices: Multiplicative rather than additive? *Multivariate Behavioral Research, 2*, 409–426. [6]

Campbell-Sills, L. A., Liverant, G., & Brown, T. A. (2004). Psychometric evaluation of the Behavioral Inhibition/Behavioral Activation Scales (BIS/BAS) in large clinical samples. *Psychological Assessment, 16*, 244–254. [5, 7, 8]

Castro-Schilo, L., Widaman, K. F., & Grimm, K. J. (2013). Neglect the structure of multitrait–multimethod data at your peril: Implications for associations with external variables. *Structural Equation Modeling, 20*, 181–207. [6]

Cattell, R. B. (1966). The scree test for the number of factors. *Multivariate Behavioral Research, 1*, 245–276. [2]

Chan, D. (1998). The conceptualization and analysis of change over time: An integrative approach incorporating longitudinal means and covariance structures analysis (LMACS) and multiple indicator latent growth modeling (MLGM). *Organizational Research Methods, 1*, 421–483. [7]

Chen, F., Bollen, K. A., Paxton, P., Curran, P. J., & Kirby, J. B. (2001). Improper solutions in structural equation models. *Sociological Methods and Research, 29*, 468–508. [5]

Chen, F. F., Sousa, K. H., & West, S. G. (2005). Testing measurement invariance of second-order factor models. *Structural Equation Modeling, 12*, 471–492. [8]

Chen, F. F., West, S. G., & Sousa, K. H. (2006). A comparison of bifactor and second-order models of quality of life. *Multivariate Behavioral Research, 41*, 189–225. [8]

Cheung, G. W., & Rensvold, R. B. (1999). Testing factorial invariance across groups: A reconceptualization and proposed new method. *Journal of Management, 25*, 1–27. [7]

Cheung, G. W., & Rensvold, R. B. (2000). Evaluating goodness-of-fit indices for testing measurement invariance. *Structural Equation Modeling, 9*, 233–255. [7]

Chou, C. P., & Bentler, P. M. (1995). Estimates and tests in structural equation modeling. In R. H. Hoyle (Ed.), *Structural equation modeling: Concepts, issues, and applications* (pp. 37–55). Thousand Oaks, CA: Sage. [9]

Cohen, J. (1988). *Statistical power analysis for the behavioral sciences.* Hillsdale, NJ: Erlbaum. [4, 7, 10]

Cohen, J. (1992). A power primer. *Psychological Bulletin, 112,* 155–159. [7, 10]

Cohen, J., Cohen, P., West, S. G., & Aiken, L. S. (2003). *Applied multiple regression/correlation analysis for the behavioral sciences* (3rd ed.). Mahwah, NJ: Erlbaum. [2, 5, 7]

Cole, D. A. (1987). Utility of confirmatory factor analysis in test validation research. *Journal of Consulting and Clinical Psychology, 55,* 584–594. [6]

Cole, D. A., Ciesla, J. A., & Steiger, J. H. (2007). The insidious effects of failing to include design-driven correlated residuals in latent-variable covariance structure analysis. *Psychological Methods, 12,* 381–398. [6]

Comrey, A. L., & Lee, H. B. (1992). *A first course in factor analysis.* Hillsdale, NJ: Erlbaum. [2]

Cooper, M. L. (1994). Motivations for alcohol use among adolescents: Development and validation of a four-factor model. *Psychological Assessment, 6,* 117–128. [5]

Coovert, M. D., Craiger, J. P., & Teachout, M. S. (1997). Effectiveness of the direct product versus confirmatory factor model for reflecting the structure of multimethod–multirater job performance data. *Journal of Applied Psychology, 82,* 271–280. [6]

Cox, B. J., Walker, J. R., Enns, M. W., & Karpinski, D. (2002). Self-criticism in generalized social phobia and response to cognitive-behavioral treatment. *Behavior Therapy, 33,* 479–491. [2]

Cronbach, L. J. (1951). Coefficient alpha and the internal structure of a test. *Psychometrika, 16,* 297–334. [8]

Cudeck, R. (1988). Multiplicative models and MTMM matrices. *Journal of Educational Statistics, 13,* 131–147. [6]

Curran, P. J., & Hussong, A. M. (2003). The use of latent trajectory models in psychopathology research. *Journal of Abnormal Psychology, 112,* 526–544. [7]

Curran, P. J., West, S. G., & Finch, J. F. (1996). The robustness of test statistics to nonnormality and specification error in confirmatory factor analysis. *Psychological Methods, 1,* 16–29. [3, 9]

Dedrick, R. F., & Greenbaum, P. E. (2011). Multilevel confirmatory factor analysis of a scale measuring interagency collaboration of children's mental health agencies. *Journal of Emotional and Behavioral Disorders, 19,* 27–40. [11]

Diamantopoulos, A., & Winklhofer, H. M. (2001). Index construction with formative indicators: An alternative to scale development. *Journal of Marketing Research, 38,* 269–277. [8]

Ding, L., Velicer, W. F., & Harlow, L. L. (1995). Effects of estimation methods, number of indicators per factor, and improper solutions in structural equation modeling fit indices. *Structural Equation Modeling, 2,* 119–143. [10]

Dolan, C., der Sluis, S., & Grasman, R. (2005). A note on normal theory power calculation in SEM with data missing completely at random. *Structural Equation Modeling, 12,* 245–262. [10]

Duncan, T. E., Duncan, S. C., & Li, F. (1998). A comparison of model- and multiple imputation-based approaches to longitudinal analysis with partial missingness. *Structural Equation Modeling, 5,* 1–21. [9]

Duncan, T. E., Duncan, S. C., & Strycker, L. A. (2006). *An introduction to latent variable growth curve modeling: Concepts, issues, and applications* (2nd ed.). Mahwah, NJ: Erlbaum. [7]

Dyer, N. G., Hanges, P. J., & Hall, R. J. (2005). Applying multilevel confirmatory factor analysis techniques to the study of leadership. *Leadership Quarterly, 16,* 149–167. [11]

Edwards, J. R., & Bagozzi, R. P. (2000). On the nature and direction of relationships between constructs and measures. *Psychological Methods, 5,* 155–174. [8]

Efron, B., & Tibshirani, R. J. (1993). *An introduction to the bootstrap.* New York: Chapman & Hall. [9]

Eid, M. (2000). A multitrait–multimethod model with minimal assumptions. *Psychometrika, 65,* 241–261. [6]

Eid, M., Lischetzke, T., Nussbeck, F. W., & Trierweiler, L. I. (2003). Separating trait effects from trait-specific method effects in multitrait–multimethod models: A multiple-indicator CT-C(M − 1) model. *Psychological Methods, 8,* 38–60. [6]

Eid, M., Nussbeck, F. W., Geiser, C., Cole, D. A., Gollwitzer, M., & Lischetzke, T. (2008). Structural equation modeling of multitrait–multimethod data: Different models for different types of data. *Psychological Methods, 13*, 230–253. [6]

Eliason, S. R. (1993). *Maximum likelihood estimation: Logic and practice.* Newbury Park, CA: Sage. [3]

Embretson, S. E., & Reise, S. P. (2000). *Item response theory for psychologists.* Mahwah, NJ: Erlbaum. [9]

Enders, C. K. (2010). *Applied missing data analysis.* New York: Guilford Press. [3, 9]

Fabrigar, L. R., Wegener, D. T., MacCallum, R. C., & Strahan, E. J. (1999). Evaluating the use of exploratory factor analysis in psychological research. *Psychological Methods, 4*, 272–299. [2]

Fan, X., & Fan, X. (2005). Using SAS for Monte Carlo simulation research in SEM. *Structural Equation Modeling, 12*, 299–333. [10]

Fan, X., & Sivo, S. A. (2009). Using Δgoodness-of-fit indexes in assessing mean structure invariance. *Structural Equation Modeling, 16*, 54–69. [7]

Ferrando, P. J., & Lorenza-Sevo, U. (2005). IRT-related factor analytic procedures for testing the equivalence of paper-and-pencil and Internet-administered questionnaires. *Psychological Methods, 10*, 193–205. [9]

Flamer, S. (1983). Assessment of the multitrait–multimethod matrix validity of Likert scales via confirmatory factor analysis. *Multivariate Behavioral Research, 18*, 275–308. [6]

Flora, D. B., & Curran, P. J. (2004). An empirical evaluation of alternative methods of estimation for confirmatory factor analysis with ordinal data. *Psychological Methods, 9*, 466–491. [9]

Floyd, F. J., & Widaman, K. F. (1995). Factor analysis in the development and refinement of clinical assessment instruments. *Psychological Assessment, 7*, 286–299. [2]

Folkman, S., & Lazarus, R. S. (1980). An analysis of coping in a middle-aged community sample. *Journal of Health and Social Behavior, 21*, 219–239. [8]

Fong, T. C. T., & Ho, R. T. H. (2013). Factor analyses of the Hospital Anxiety and Depression Scale: A Bayesian structural equation modeling approach. *Quality of Life Research, 22*, 2857–2863. [11]

Fresco, D. M., Heimberg, R. G., Mennin, D. S., & Turk, C. L. (2002). Confirmatory factor analysis of the Penn State Worry Questionnaire. *Behaviour Research and Therapy, 40*, 313–323. [5]

Geldhof, G. J., Preacher, K. J., & Zyphur, M. J. (2014). Reliability estimation in a multilevel confirmatory factor analysis framework. *Psychological Methods, 19*, 72–91. [11]

Gelman, A., Carlin, J. B., Stern, H. S., & Rubin, D. B. (2004). *Bayesian data analysis* (2nd ed.). Boca Raton, FL: Chapman & Hall. [11]

Gelman, A., Meng, X. L., Stern, H. S., & Rubin, D. B. (1996). Posterior predictive assessment of model fitness via realized discrepancies. *Statistica Sinica, 6*, 733–807. [11]

Gelman, A., & Rubin, D. B. (1992). Inference from iterative simulation using multiple sequences. *Statistical Science, 7*, 457–472. [11]

Gilks, W. R., Richardson, S., & Spiegelhalter, D. J. (1996). *Markov chain Monte Carlo in practice.* London: Chapman & Hall. [11]

Glöckner-Rist, A., & Hoijtink, H. (2003). The best of both worlds: Factor analysis of dichotomous data using item response theory and structural equation modeling. *Structural Equation Modeling, 10*, 544–565. [9]

Glorfeld, L. W. (1995). An improvement on Horn's parallel analysis methodology for selecting the correct number of factors to retain. *Educational and Psychological Measurement, 55*, 377–393. [2]

Golay, P., Reverte, I., Rossier, J., Favez, N., & Lecerf, T. (2013). Further insights on the French WISC-IV factor structure through Bayesian structural equation modeling. *Psychological Assessment, 25*, 496–508. [11]

Golembiewski, R. T., Billingsley, K., & Yeager, S. (1976). Measuring change and persistence in human affairs: Types of change generated by OD designs. *Journal of Applied Behavioral Science, 12*, 133–157. [7]

Gonzalez, R., & Griffin, D. (2001). Testing parameters in structural equation modeling: Every "one" matters. *Psychological Methods, 6*, 258–269. [4]

Gorsuch, R. L. (1983). *Factor analysis* (2nd ed.). Hillsdale, NJ: Erlbaum. [2]

Gorsuch, R. L. (1997). New procedures for extension analysis in exploratory factor analysis. *Educational and Psychological Measurement, 57*, 725–740. [3]

Green, D. P., Goldman, S. L., & Salovey, P. (1993). Measurement error masks bipolarity in affect ratings. *Journal of Personality and Social Psychology, 64*, 1029–1041. [6]

Green, S. B., & Hershberger, S. L. (2000). Correlated errors in true score models and their effect on coefficient alpha. *Structural Equation Modeling, 7*, 251–270. [8]

Grice, J. W. (2001). Computing and evaluating factor scores. *Psychological Methods, 6*, 430–450. [2, 8]

Hall, R. J., Snell, A. F., & Foust, M. S. (1999). Item parceling strategies in SEM: Investigating the subtle effects of unmodeled secondary constructs. *Organizational Research Methods, 2*, 233–256. [9]

Harman, H. H. (1976). *Modern factor analysis* (3rd ed.). Chicago: University of Chicago Press. [2, 8]

Hau, K. T., & Marsh, H. W. (2004). The use of item parcels in structural equation modeling: Non-normal data and small sample sizes. *British Journal of Mathematical and Statistical Psychology, 57*, 327–351. [9]

Hayton, J. C., Allen, D. G., & Scarpello, V. (2004). Factor retention decisions in exploratory factor analysis: A tutorial on parallel analysis. *Organizational Research Methods, 7*, 191–205. [2]

Hazlett-Stevens, H., Ullman, J. B., & Craske, M. G. (2004). Factor structure of the Penn State Worry Questionnaire: Examination of a method factor. *Assessment, 11*, 361–370. [5]

Hershberger, S. L. (1994). The specification of equivalent models before the collection of data. In A. von Eye & C. C. Clogg (Eds.), *Latent variables analysis* (pp. 68–105). Thousand Oaks, CA: Sage. [5]

Hoff, P. D. (2009). *A first course in Bayesian statistical methods*. New York: Springer. [11]

Holzinger, K. J., & Swineford, F. (1937). The bifactor method. *Psychometrika, 2*, 41–54. [8]

Horn, J. L. (1965). A rationale and test for the number of factors in factor analysis. *Psychometrika, 30*, 179–185. [2]

Horn, J. L., & McArdle, J. J. (1992). A practical and theoretical guide to measurement invariance in aging research. *Experimental Aging Research, 18*, 117–144. [7]

Horowitz, M. J., Wilner, N., & Alvarez, W. (1979). The Impact of Event Scale: A measure of subjective stress. *Psychosomatic Medicine, 41*, 209–218. [8]

Howell, R. D., Breivik, E., & Wilcox, J. B. (2007). Reconsidering formative measurement. *Psychological Methods, 12*, 205–218. [8]

Hox, J. J. (2013). Multilevel regression and multilevel structural equation modeling. In T. D. Little (Ed.), *The Oxford handbook of quantitative methods* (Vol. 2, pp. 281–294). New York: Oxford University Press. [11]

Hoyle, R. H., & Panter, A. T. (1995). Writing about structural equation models. In R. H. Hoyle (Ed.), *Structural equation modeling: Concepts, issues, and applications* (pp. 158–176). Thousand Oaks, CA: Sage. [4]

Hu, L., & Bentler, P. M. (1995). Evaluating model fit. In R. H. Hoyle (Ed.), *Structural equation modeling: Concepts, issues, and applications* (pp. 76–99). Thousand Oaks, CA: Sage. [3]

Hu, L., & Bentler, P. M. (1998). Fit indices in covariance structure modeling: Sensitivity to under-parameterized model misspecification. *Psychological Methods, 3*, 424–453. [3]

Hu, L., & Bentler, P. M. (1999). Cutoff criteria for fit indexes in covariance structure analysis: Conventional criteria versus new alternatives. *Structural Equation Modeling, 6*, 1–55. [3, 4]

Hu, L., Bentler, P. M., & Kano, Y. (1992). Can test statistics in covariance structure analysis be trusted? *Psychological Bulletin, 112*, 351–362. [3]

Humphreys, L. G., & Montanelli, R. G. (1975). An investigation of parallel analysis criterion for determining the number of common factors. *Multivariate Behavioral Research, 10*, 191–205. [2]

Iversen, G. (1985). *Bayesian statistical inference*. Thousand Oaks, CA: Sage. [9]

Jaccard, J., & Wan, C. K. (1996). *LISREL approaches to interaction effects in multiple regression.* Thousand Oaks, CA: Sage. [3, 4, 7, 10]

Jarvis, C. B., MacKenzie, S. B., & Podsakoff, P. M. (2003). A critical review of construct indicators and measurement model misspecification in marketing and consumer research. *Journal of Consumer Research, 30,* 199–218. [8]

Jöreskog, K. G. (1969). A general approach to confirmatory maximum likelihood factor analysis. *Psychometrika, 34,* 183–202. [2, 5]

Jöreskog, K. G. (1971a). Statistical analysis of sets of congeneric tests. *Psychometrika, 36,* 109–133. [2, 3, 8]

Jöreskog, K. G. (1971b). Simultaneous factor analysis in several populations. *Psychometrika, 36,* 409–426. [7]

Jöreskog, K. G. (1993). Testing structural equation models. In K. A. Bollen & J. S. Long (Eds.), *Testing structural equation models* (pp. 294–316). Newbury Park, CA: Sage. [4, 5, 7]

Jöreskog, K. G. (1999). *How large can a standardized coefficient be?* Chicago: Scientific Software International. Retrieved from *www.ssicentral.com/lisrel/column2.htm* [4]

Jöreskog, K. G., & Goldberger, A. S. (1975). Estimation of a model with multiple indicators and multiple causes of a single latent variable. *Journal of the American Statistical Association, 70,* 631–639. [7]

Jöreskog, K. G., & Sörbom, D. (1979). *Advances in factor analysis and structural equation models* (J. Magidson, Ed.). Cambridge, MA: Abt Books. [5]

Jöreskog, K. G., & Sörbom, D. (1996a). *LISREL 8: User's reference guide.* Chicago: Scientific Software International. [3]

Jöreskog, K. G., & Sörbom, D. (1996b). *PRELIS 2: User's reference guide.* Chicago: Scientific Software International. [4, 9]

Kaiser, H. F., & Dickman, K. (1962). Sample and population score matrices and sample correlation matrices from an arbitrary population correlation matrix. *Psychometrika, 27,* 179–182. [10]

Kamata, A., & Bauer, D. J. (2008). A note on the relation between factor analytic and item response theory models. *Structural Equation Modeling, 15,* 136–153. [9]

Kaplan, D. (1989). Model modification in covariance structure analysis: Application of the expected parameter change statistic. *Multivariate Behavioral Research, 24,* 285–305. [4]

Kaplan, D. (1990). Evaluating and modifying covariance structure models: A review and recommendation. *Multivariate Behavioral Research, 25,* 137–155. [4]

Kaplan, D. (1995). Statistical power in structural equation modeling. In R. H. Hoyle (Ed.), *Structural equation modeling: Concepts, issues, and applications* (pp. 100–117). Thousand Oaks, CA: Sage. [10]

Kaplan, D. (2000). *Structural equation modeling: Foundations and extensions.* Thousand Oaks, CA: Sage. [4, 9]

Kaplan, D. (2014). *Bayesian statistics for the social sciences.* New York: Guilford Press. [11]

Kaplan, D., & Depaoli, S. (2012). Bayesian structural equation modeling. In R. H. Hoyle (Ed.), *Handbook of structural equation modeling* (pp. 650–673). New York: Guilford Press. [11]

Kaplan, D., & Depaoli, S. (2013). Bayesian statistical methods. In T. Little (Ed.), *The Oxford handbook of quantitative methods* (Vol. 1, pp. 407–437). New York: Oxford University Press. [11]

Kaplan, D., & Kreisman, M. B. (2000). On the validation of indicators of mathematics education using TIMSS: An application of multilevel covariance structure modeling. *International Journal of Educational Policy, Research, and Practice, 1,* 217–242. [11]

Kenny, D. A. (1979). *Correlation and causality.* New York: Wiley-Interscience. [3, 6]

Kenny, D. A., & Kashy, D. A. (1992). Analysis of the multitrait–multimethod matrix by confirmatory factor analysis. *Psychological Bulletin, 112,* 165–172. [6]

Kenny, D. A., Kashy, D. A., & Bolger, N. (1998). Data analysis in social psychology. In D. T. Gilbert & S. T. Fiske (Eds.), *The handbook of social psychology* (4th ed., Vol. 2, pp. 233–265). New York: McGraw-Hill. [5]

Kishton, J. M., & Widaman, K. F. (1994). Unidimensional versus domain representative parceling of questionnaire items: An empirical example. *Educational and Psychological Measurement, 54,* 757–765. [9]

Klein, A., & Moosbrugger, H. (2000). Maximum likelihood estimation of latent interaction effects with the LMS method. *Psychometrika, 65,* 457–474. [9]

Kline, R. B. (2011). *Principles and practice of structural equation modeling* (3rd ed.). New York: Guilford Press. [4]

Kollman, D. M., Brown, T. A., & Barlow, D. H. (2009). The construct validity of acceptance: A multitrait–multimethod investigation. *Behavior Therapy, 40,* 205–218. [6]

Komaroff, E. (1997). Effect of simultaneous violations of essential tau-equivalence and correlated errors on coefficient alpha. *Applied Psychological Measurement, 21,* 337–348. [8]

Kumar, A., & Dillon, W. R. (1992). An integrative look at the use of additive and multiplicative covariance models in the analysis of MTMM data. *Journal of Marketing Research, 29,* 51–64. [6]

Lahey, B. B., Applegate, B., Hakes, J. K., Zald, D. H., Hariri, A. R., & Rathouz, P. J. (2012). Is there a general factor of prevalent psychopathology during adulthood? *Journal of Abnormal Psychology, 121,* 971–977. [8]

Lance, C. E., Noble, C. L., & Scullen, S. E. (2002). A critique of the correlated trait-correlated method and correlated uniqueness models for multitrait–multimethod data. *Psychological Methods, 7,* 228–244. [6]

Lee, J., Little, T. D., & Preacher, K. J. (2011). Methodological issues in using structural equation models for testing differential item functioning. In E. Davidov, P. Schmidt, & J. Billiet (Eds.), *Cross-cultural data analysis: Methods and applications* (pp. 55–84). New York: Routledge. [7]

Lee, S., & Hershberger, S. (1990). A simple rule for generating equivalent models in structural equation modeling. *Multivariate Behavioral Research, 25,* 313–334. [5]

Lievens, F., & Conway, J. M. (2001). Dimension and exercise variance in assessment center scores: A large-scale evaluation of multitrait–multimethod matrices. *Journal of Applied Psychology, 86,* 1202–1222. [6]

Lischetzke, T., & Eid, M. (2003). Is attention to feelings beneficial or detrimental to affective well-being?: Mood regulation as a moderator variable. *Emotion, 3,* 361–377. [6]

Little, J. (2013). Multilevel confirmatory ordinal factor analysis of the Life Skills Profile-16. *Psychological Assessment, 25,* 810–825. [11]

Little, R. J. A., & Rubin, D. B. (2002). *Statistical analysis with missing data* (2nd ed.). Hoboken, NJ: Wiley. [9]

Little, T. D. (2013). *Longitudinal structural equation modeling.* New York: Guilford Press. [7, 9]

Little, T. D., Slegers, D. W., & Card, N. A. (2006). A non-arbitrary method of identifying and scaling latent variables in SEM and MACS models. *Structural Equation Modeling, 13,* 59–72. [3, 7]

Loehlin, J. C. (2004). *Latent variable models* (4th ed.). Mahwah, NJ: Erlbaum. [3, 8]

Lord, F. M. (1980). *Applications of item response theory to practical testing problems.* Hillsdale, NJ: Erlbaum. [9]

Lord, F. M., & Novick, M. (1968). *Statistical theories of mental test scores.* Reading, MA: Addison-Wesley. [8]

Lu, I. R. R., Thomas, D. R., & Zumbo, B. D. (2005). Embedding IRT in structural equation models: A comparison with regression based on IRT scores. *Structural Equation Modeling, 12,* 263–277. [9]

MacCallum, R. C. (1986). Specification searches in covariance structure modeling. *Psychological Bulletin, 100,* 107–120. [4, 5]

MacCallum, R. C., & Austin, J. T. (2000). Applications of structural equation modeling in psychological research. *Annual Review of Psychology, 51,* 201–226. [4]

MacCallum, R. C., & Browne, M. W. (1993). The use of causal indicators in covariance structure models: Some practical issues. *Psychological Bulletin, 114,* 533–541. [8]

MacCallum, R. C., Browne, M. W., & Sugawara, H. M. (1996). Power analysis and determination of sample size for covariance structure modeling. *Psychological Methods, 1,* 130–149. [3]

MacCallum, R. C., Edwards, M. C., & Li, L. (2012). Hopes and cautions in implementing Bayesian structural equation modeling. *Psychological Methods, 17*, 340–345. [11]

MacCallum, R. C., Roznowski, M., & Necowitz, L. B. (1992). Model modifications in covariance structure analysis: The problem of capitalization on chance. *Psychological Bulletin, 111*, 490–504. [4]

MacCallum, R. C., Wegener, D. T., Uchino, B. N., & Fabrigar, L. R. (1993). The problem of equivalent models in applications of covariance structure analysis. *Psychological Bulletin, 114*, 185–199. [5]

MacCallum, R. C., Zhang, S., & Preacher, K. J. (2002). On the practice of dichotomization of quantitative variables. *Psychological Methods, 7*, 19–40. [7]

MacIntosh, R., & Hashim, S. (2003). Variance estimation for converting MIMIC model parameters to IRT parameters in DIF analysis. *Applied Psychological Measurement, 27*, 372–379. [9]

MacKinnon, D. P. (2008). *Introduction to statistical mediation analysis*. New York: Taylor & Francis. [3]

Marsh, H. W. (1989). Confirmatory factor analyses of multitrait–multimethod data: Many problems and a few solutions. *Applied Psychological Measurement, 13*, 335–361. [6]

Marsh, H. W. (1996). Positive and negative global self-esteem: A substantively meaningful distinction or artifactors? *Journal of Personality and Social Psychology, 70*, 810–819. [3, 5, 6]

Marsh, H. W., & Bailey, M. (1991). Confirmatory factor analysis of multitrait–multimethod data: A comparison of the behavior of alternative models. *Applied Psychological Measurement, 15*, 47–70. [6]

Marsh, H. W., Balla, J. R., & McDonald, R. P. (1988). Goodness-of-fit indices in confirmatory factor analysis: The effect of sample size. *Psychological Bulletin, 103*, 391–410. [3]

Marsh, H. W., & Grayson, D. (1995). Latent variable models of multitrait–multimethod data. In R. H. Hoyle (Ed.), *Structural equation modeling: Concepts, issues, and applications* (pp. 177–198). Thousand Oaks, CA: Sage. [6]

Marsh, H. W., Hau, K. T., Balla, J. R., & Grayson, D. (1998). Is more ever too much?: The number of indicators per factor in confirmatory factor analysis. *Multivariate Behavioral Research, 33*, 181–220. [3, 9]

Marsh, H. W., Hau, K. T., & Wen, Z. (2004). In search of golden rules: Comment on hypothesis testing approaches to setting cutoff values for fit indexes and dangers in overgeneralizing Hu and Bentler's (1999) findings. *Structural Equation Modeling, 11*, 320–341. [3, 11]

Marsh, H. W., & Hocevar, D. (1983). Confirmatory factor analysis of multitrait–multimethod matrices. *Journal of Educational Measurement, 20*, 231–248. [6]

Marsh, H. W., Lüdtke, O., Muthén, B., Asparouhov, T., Morin, A. J. S., Trautwein, U., & Nagengast, B. (2010). A new look at the big-five factor structure through exploratory structural equation modeling. *Psychological Assessment, 22*, 471–491. [5]

Marsh, H. W., Lüdtke, O., Nagengast, B., Morin, A. J. S., & Von Davier, M. (2013). Why item parcels are (almost) never appropriate: Two wrongs do not make a right—camouflaging misspecification with item parcels in CFA models. *Psychological Methods, 18*, 257–284. [9]

Marsh, H. W., Muthén, B., Asparouhov, A., Lüdtke, O., Robitzsch, A., Morin, A. J. S., & Trautwein, U. (2009). Exploratory structural equation modeling, integrating CFA and EFA: Application to students' evaluations of university teaching. *Structural Equation Modeling, 16*, 439–476. [5, 11]

McDonald, R. P. (1981). Constrained least squares estimators of oblique common factors. *Psychometrika, 46*, 337–341. [2]

McDonald, R. P. (1999). *Test theory: A unified treatment*. Mahwah, NJ: Erlbaum. [7]

McDonald, R. P., & Ho, M. R. (2002). Principles and practice in reporting structural equation analyses. *Psychological Methods, 7*, 64–82. [4]

Meade, A. W., & Bauer, D. J. (2007). Power and precision in confirmatory factor analytic tests of measurement invariance. *Structural Equation Modeling, 14*, 611–635. [7]

Meade, A. W., Johnson, E. C., & Braddy, P. W. (2008). Power and sensitivity of alternative fit indices in tests of measurement invariance. *Journal of Applied Psychology, 93*, 568–592. [7]

Meade, A. W., & Lautenschlager, G. J. (2004). A comparison of item response theory and confirmatory factor analytic methodologies for establishing measurement equivalence/invariance. *Organizational Research Methods, 7*, 361–388. [9]

Mehta, P. D., & Neale, M. C. (2005). People are variables too: Multilevel structural equations modeling. *Psychological Methods, 10*, 259–284. [11]

Meredith, W. (1993). Measurement invariance, factor analysis and factorial invariance. *Psychometrika, 58*, 525–543. [7]

Meyer, J. F., & Brown, T. A. (2013). Psychometric properties of the Thought–Action Fusion Scale in a large clinical sample. *Assessment, 20*, 764–775. [8]

Meyer, J. F., Frost, R. O., Brown, T. A., Steketee, G., & Tolin, D. F. (2013). A multitrait–multimethod matrix investigation of hoarding. *Journal of Obsessive–Compulsive and Related Disorders, 2*, 273–280. [6]

Meyer, T. J., Miller, M. L., Metzger, R. L., & Borkovec, T. D. (1990). Development and validation of the Penn State Worry Questionnaire. *Behaviour Research and Therapy, 28*, 487–495. [5]

Mooney, C. Z. (1997). *Monte Carlo simulation.* Thousand Oaks, CA: Sage. [10]

Mooney, C. Z., & Duval, R. D. (1993). *Bootstrapping: A nonparametric approach to statistical inference.* Thousand Oaks, CA: Sage. [9]

Moustaki, I., Jöreskog, K. G., & Mavridis, D. (2004). Factor models for ordinal variables with covariate effects on the manifest and latent variables: A comparison of LISREL and IRT approaches. *Structural Equation Modeling, 11*, 487–513. [9]

Muthén, B. (1989). Latent variable modeling in heterogeneous populations. *Psychometrika, 54*, 557–585. [7]

Muthén, B., & Asparouhov, T. (2002). *Latent variable analysis with categorical outcomes: Multiple-group and growth modeling in Mplus.* Unpublished manuscript. Retrieved from *www.statmodel.com/mplus/examples/webnote.html#web4* [9]

Muthén, B., & Asparouhov, T. (2011). Beyond multilevel regression modeling: Multilevel analysis in a general latent variable framework. In J. Hox & J. K. Roberts (Eds.), *Handbook of advanced multilevel analysis* (pp. 15–40). New York: Taylor & Francis. [11]

Muthén, B., & Asparouhov, T. (2012). Bayesian structural equation modeling: A more flexible representation of substantive theory. *Psychological Methods, 17*, 313–335. [11]

Muthén, B., & Kaplan, D. (1992). A comparison of some methodologies for the factor analysis of non-normal Likert variables: A note on the size of the model. *British Journal of Mathematical and Statistical Psychology, 45*, 19–30. [9]

Muthén, B., Kaplan, D., & Hollis, M. (1987). On structural equation modeling with data that are not missing completely at random. *Psychometrika, 52*, 431–462. [9]

Muthén, B. O. (1988). Some uses of structural equation modeling in validity studies: Extending IRT to external variables. In H. Wainer & H. Braun (Eds.), *Test validity* (pp. 213–238). Hillsdale, NJ: Erlbaum. [9]

Muthén, B. O., Kao, C., & Burstein, L. (1991). Instructional sensitivity in mathematics achievement test items: Applications of a new IRT-based detection technique. *Journal of Educational Measurement, 28*, 1–22. [9]

Muthén, L. K., & Muthén, B. O. (1998–2012). *Mplus user's guide.* Los Angeles: Authors. [5, 9, 10, 11]

Muthén, L. K., & Muthén, B. O. (2002). How to use a Monte Carlo study to decide on sample size and determine power. *Structural Equation Modeling, 4*, 599–620. [10]

Nasser, F., & Wisenbaker, J. (2003). A Monte Carlo study investigating the impact of item parceling on measures of fit in confirmatory factor analysis. *Educational and Psychological Measurement, 63*, 729–757. [9]

Neale, M. C., & Miller, M. B. (1997). The use of likelihood-based confidence intervals in genetic models. *Behavior Genetics, 27*, 113–120. [4]

O'Boyle, E. H., & Williams, L. J. (2011). Decomposing model fit: Measurement vs. theory in organizational research using latent variables. *Journal of Applied Psychology, 96*, 1–12. [3]

O'Connor, B. P. (2001). Extension: SAS, SPSS, and MATLAB programs for extension analysis. *Applied Psychological Measurement, 25*, 88. [2]

Osman, A., Gutierrez, P. M., Smith, K., Fang, Q., Lozano, G., & Devine, A. (2010). The Anxiety Sensitivity Index–3: Analyses of dimensions, reliability estimates, and correlates in non-clinical samples. *Journal of Personality Assessment, 92*, 45–52. [8]

Patrick, C. J., Hicks, B. M., Nichol, P. E., & Krueger, R. F. (2007). A bifactor approach to modeling the structure of the Psychopathy Checklist—Revised. *Journal of Personality Disorders, 21*, 118–141. [8]

Paxton, P., Curran, P. J., Bollen, K. A., Kirby, J., & Chen, F. (2001). Monte Carlo experiments: Design and implementation. *Structural Equation Modeling, 8*, 287–312. [10]

Pek, J., Sterba, S. K., Kok, B. E., & Bauer, D. J. (2009). Estimating and visualizing nonlinear relations among latent variables: A semi-parametric approach. *Multivariate Behavioral Research, 44*, 407–436. [7]

Podsakoff, P. M., MacKenzie, S. B., Lee, J. Y., & Podsakoff, N. P. (2003). Common method biases in behavioral research: A critical review of the literature and recommended remedies. *Journal of Applied Psychology, 88*, 879–903. [5, 6]

Preacher, K. J., & MacCallum, R. C. (2003). Repairing Tom Swift's electric factor analysis machine. *Understanding Statistics, 2*, 13–43. [2]

Price, L. R., Tulsky, D., Millis, S., & Weiss, L. (2002). Redefining the factor structure of the Wechsler Memory Scale–III: Confirmatory factor analysis with cross-validation. *Journal of Clinical and Experimental Neuropsychology, 24*, 574–585. [7]

Rasch, G. (1960). *Probablistic models for some intelligence and attainment tests*. Chicago: University of Chicago Press. [9]

Raudenbush, S. W., & Bryk, A. S. (2002). *Hierarchical linear models: Applications and data analysis methods* (2nd ed.). Newbury Park, CA: Sage. [11]

Raykov, T. (1997). Scale reliability, Cronbach's coefficient alpha, and violations of essential tau-equivalence with fixed congeneric components. *Multivariate Behavioral Research, 32*, 329–353. [8]

Raykov, T. (2001a). Estimation of congeneric scale reliability using covariance structure analysis with nonlinear constraints. *British Journal of Mathematical and Statistical Psychology, 54*, 315–323. [7, 8]

Raykov, T. (2001b). Bias of Cronbach's alpha for fixed congeneric measures with correlated errors. *Applied Psychological Measurement, 25*, 69–76. [8]

Raykov, T. (2002a). Examining group differences in reliability of multiple-component instruments. *British Journal of Mathematical and Statistical Psychology, 55*, 145–158. [8]

Raykov, T. (2002b). Analytic estimation of standard error and confidence interval for scale reliability. *Multivariate Behavioral Research, 37*, 89–103. [8]

Raykov, T. (2004). Behavioral scale reliability and measurement invariance evaluation using latent variable modeling. *Behavior Therapy, 35*, 299–331. [8]

Raykov, T. (2012). Scale construction and development using structural equation modeling. In R. H. Hoyle (Ed.), *Handbook of structural equation modeling* (pp. 472–492). New York: Guilford Press. [7, 8]

Raykov, T., Dimitrov, D. M., & Asparouhov, T. (2010). Evaluation of scale reliability with binary measures. *Structural Equation Modeling, 17*, 122–132. [8]

Raykov, T., & Grayson, D. A. (2003). A test for change of composite reliability in scale development. *Multivariate Behavioral Research, 38*, 143–159. [8]

Raykov, T., & Marcoulides, G. A. (2012). Evaluation of validity and reliability for hierarchical scales using latent variable modeling. *Structural Equation Modeling, 19*, 495–508. [8]

Raykov, T., & Shrout, P. E. (2002). Reliability of scales with general structure: Point and interval estimation using a structural equation modeling approach. *Structural Equation Modeling, 9*, 195–212. [8]

Raykov, T., Tomer, A., & Nesselroade, J. R. (1991). Reporting structural equation modeling results in *Psychology and Aging*: Some proposed guidelines. *Psychology and Aging, 6*, 499–503. [4]

Reise, S. P., Widaman, K. F., & Pugh, R. H. (1993). Confirmatory factor analysis and item response theory: Two approaches for exploring measurement invariance. *Psychological Bulletin, 114*, 552–566. [9]

Rosellini, A. J., & Brown, T. A. (2011). The NEO Five-Factor Inventory: Latent structure and relationships with dimensions of anxiety and depressive disorders in a large clinical sample. *Assessment, 18,* 27–38. [5]

Rosenberg, M. (1965). *Society and the adolescent child.* Princeton, NJ: Princeton University Press. [3, 5]

Rubin, D. B. (1987). *Multiple imputation for nonresponse in surveys.* New York: Wiley. [9]

Russell, J. A. (1979). Affective space is bipolar. *Journal of Personality and Social Psychology, 37,* 1161–1178. [6]

Saris, W. E., & Satorra, A. (1993). Power evaluations in structural equation models. In K. A. Bollen & J. S. Long (Eds.), *Testing structural equation models* (pp. 181–204). Newbury Park, CA: Sage. [10]

Saris, W. E., & Stronkhorst, H. (1984). *Causal modeling in nonexperimental research.* Amsterdam: Sociometric Research Foundation. [10]

Sass, D. A., & Smith, P. L. (2006). The effects of parceling unidimensional scales on structural parameter estimates in structural equation modeling. *Structural Equation Modeling, 13,* 566–586. [9]

Satorra, A. (1989). Alternative test criteria in covariance structure analysis: A unified approach. *Psychometrika, 54,* 131–151. [10]

Satorra, A., & Bentler, P. M. (1994). Corrections to test statistics and standard errors in covariance structure analysis. In A. von Eye & C. C. Clogg (Eds.), *Latent variable analysis: Applications for developmental research* (pp. 399–419). Thousand Oaks, CA: Sage. [9]

Satorra, A., & Bentler, P. M. (2001). A scaled difference chi-square statistic for moment structure analysis. *Psychometrika, 66,* 507–514. [9]

Satorra, A., & Bentler, P. M. (2010). Ensuring positiveness of the scale chi-square test statistic. *Psychometrika, 75,* 243–248. [9]

Satorra, A., & Saris, W. E. (1985). Power of the likelihood ratio test in covariance structure analysis. *Psychometrika, 50,* 83–90. [10]

Schafer, J. L. (1997). *Analysis of incomplete multivariate data.* London: Chapman & Hall. [9]

Schafer, J. L., & Graham, J. W. (2002). Missing data: Our view of the state of the art. *Psychological Methods, 7,* 147–177. [9]

Schmid, J., & Leiman, J. M. (1957). The development of hierarchical factor solutions. *Psychometrika, 22,* 53–61. [8]

Schmitt, N., & Stults, D. M. (1986). Methodology review: Analysis of multitrait–multimethod matrices. *Applied Psychological Measurement, 10,* 1–22. [6]

Schwarz, G. E. (1978). Estimating the dimension of a model. *Annals of Statistics, 6,* 461–464. [5]

Self, S. G., & Liang, K. (1987). Asymptotic properties of maximum likelihood estimators and likelihood ratio tests under nonstandard conditions. *Journal of the American Statistical Association, 82,* 605–610. [4]

Silvia, E. S., & MacCallum, R. C. (1988). Some factors affecting the success of specification searches in covariance structure modeling. *Multivariate Behavioral Research, 23,* 297–326. [4]

Sörbom, D. (1989). Model modification. *Psychometrika, 54,* 371–384. [4]

Spearman, C. (1904). General intelligence, objectively determined and measured. *American Journal of Psychology, 15,* 201–293. [2]

Spearman, C. (1927). *The abilities of man.* New York: Macmillan. [2]

Spiegelhalter, D. J., Best, N. G., Carlin, B. P., & van der Linde, A. (2002). Bayesian measures of model complexity and fit (with discussion). *Journal of the Royal Statistical Society, 64,* 583–639. [11]

Steiger, J. H., & Lind, J. M. (1980, June). *Statistically based tests for the number of common factors.* Paper presented at the meeting of the Psychometric Society, Iowa City, IA. [2, 3]

Stelzl, I. (1986). Changing the causal hypothesis without changing the fit: Some rules for generating equivalent path models. *Multivariate Behavioral Research, 21,* 309–331. [5]

Tabachnick, B. G., & Fidell, L. S. (2013). *Using multivariate statistics* (6th ed.). Boston: Pearson Education. [2, 4, 5, 9]

Takane, Y., & de Leeuw, J. (1987). On the relationship between item response theory and factor analysis of discretized variables. *Psychometrika, 52*, 393–408. [9]

Tanaka, J. S. (1987). "How big is big enough?": Sample size and goodness of fit in structural equation models with latent variables. *Child Development, 58*, 134–146. [10]

Taylor, S. (Ed.). (1999). *Anxiety sensitivity: Theory, research, and treatment of the fear of anxiety.* Mahwah, NJ: Erlbaum. [8]

Thompson, B. (1984). *Canonical correlation analysis: Uses and interpretation.* Newbury Park, CA: Sage. [8]

Thurstone, L. L. (1935). *The vectors of mind.* Chicago: University of Chicago Press. [2]

Thurstone, L. L. (1947). *Multiple-factor analysis.* Chicago: University of Chicago Press. [2]

Tobin, D., Holroyd, K., Reynolds, R., & Wigal, J. (1989). The hierarchical factor structure of the Coping Strategies Inventory. *Cognitive Therapy and Research, 13*, 343–361. [8]

Tomás, J. M., Hontangas, P. M., & Oliver, A. (2000). Linear confirmatory factor models to evaluate multitrait–multimethod matrices: The effects of number of indicators and correlation among methods. *Multivariate Behavioral Research, 35*, 469–499. [6]

Tomás, J. M., & Oliver, A. (1999). Rosenberg's Self-Esteem Scale: Two factors or method effects. *Structural Equation Modeling, 6*, 84–98. [5, 6]

Treiblmaier, H., Bentler, P. M., & Mair, P. (2011). Formative constructs implemented via common factors. *Structural Equation Modeling, 18*, 1–17. [8]

Tucker, L. R., & Lewis, C. (1973). A reliability coefficient for maximum likelihood factor analysis. *Psychometrika, 38*, 1–10. [3]

Tulsky, D. S., & Price, L. R. (2003). The joint WAIS-III and WMS-III factor structure: Development and cross-validation of a six-factor model of cognitive functioning. *Psychological Assessment, 15*, 149–162. [7]

Vandenberg, R. J., & Lance, C. E. (2000). A review and synthesis of the measurement invariance literature: Suggestions, practices, and recommendations for organizational research. *Organizational Research Methods, 3*, 4–69. [7]

Wall, M. M., & Amemiya, Y. (2007). A review of nonlinear factor analysis and nonlinear structural equation modeling. In R. Cudeck & R. C. MacCallum (Eds.), *Factor analysis at 100: Historical developments and future directions* (pp. 337–361). Mahwah, NJ: Erlbaum. [7]

Wang, J., Siegal, H. A., Falck, R. S., & Carlson, R. G. (2001). Factorial structure of Rosenberg's self-esteem scale among crack-cocaine drug users. *Structural Equation Modeling, 8*, 275–286. [5]

Ware, J. E., & Sherbourne, C. D. (1992). The MOS 36-item Short-Form Health Survey (SF-36). I: Conceptual framework and item selection. *Medical Care, 30*, 473–483. [4]

Wechsler, D. (1997). *Wechsler Memory Scale—Third Edition.* San Antonio, TX: Psychological Corporation. [7]

West, S. G., Finch, J. F., & Curran, P. J. (1995). Structural equation models with nonnormal variables: Problems and remedies. In R. H. Hoyle (Ed.), *Structural equation modeling: Concepts, issues, and applications* (pp. 56–75). Thousand Oaks, CA: Sage. [9]

Wheaton, B. (1987). Assessment of fit in overidentified models with latent variables. *Sociological Methods and Research, 16*, 118–154. [3]

Wheaton, B., Muthén, B., Alwin, D., & Summers, G. (1977). Assessing reliability and stability in panel models. In D. R. Heise (Ed.), *Sociological methodology* (pp. 84–136). San Francisco: Jossey-Bass. [3]

Widaman, K. F. (1985). Hierarchically nested covariance structure models for multitrait–multimethod data. *Applied Psychological Measurement, 9*, 1–26. [6]

Widaman, K. F. (1993). Common factor analysis versus principal components analysis: Differential bias in representing model parameters? *Multivariate Behavioral Research, 28*, 263–311. [2]

Wiggins, J. S. (1996). *The five-factor model of personality: Theoretical perspectives.* New York: Guilford Press. [4]

Willett, J. B., Singer, J. D., & Martin, N. C. (1998). The design and analysis of longitudinal studies of development and psychopathology in context: Statistical models and methodological recommendations. *Development and Psychopathology, 10,* 395–426. [3, 4]

Wothke, W. A. (1993). Nonpositive definite matrices in structural modeling. In K. A. Bollen & J. S. Long (Eds.), *Testing structural equation models* (pp. 256–293). Newbury Park, CA: Sage. [3, 4, 5, 9]

Wothke, W. A. (1996). Models for multitrait–multimethod matrix analysis. In G. A. Marcoulides & R. E. Schumacker (Eds.), *Advanced structural equation modeling: Issues and techniques* (pp. 7–56). Mahwah, NJ: Erlbaum. [6]

Wothke, W. A., & Browne, M. W. (1990). The direct product model for the MTMM matrix parameterized as a second order factor analysis model. *Psychometrika, 55,* 255–262. [6]

Wright, S. (1934). The method of path coefficients. *Annals of Mathematical Statistics, 5,* 161–215. [3]

Yu, C. Y. (2002). *Evaluating cutoff criteria of model fit indices for latent variable models with binary and continuous outcomes.* Unpublished doctoral dissertation, University of California, Los Angeles. [3, 9]

Yuan, K. H. (2005). Fit statistics versus test statistics. *Multivariate Behavioral Research, 40,* 115–148. [3]

Yuan, K. H., & Bentler, P. M. (2000). Three likelihood-based methods for mean and covariance structure analysis with nonnormal missing data. In M. E. Sobel & M. P. Becker (Eds.), *Sociological methodology 2000* (pp. 165–200). Washington, DC: American Sociological Association. [9]

Yuan, K. H., Bentler, P. M., & Kano, Y. (1997). On averaging variables in a CFA model. *Behaviormetrika, 24,* 71–83. [9]

Yuan, K. H., & Hayashi, K. (2003). Bootstrap approach to inference and power analysis based on three statistics for covariance structure models. *British Journal of Mathematical and Statistical Psychology, 56,* 93–110. [10]

Yuan, K. H., Marshall, L. L., & Bentler, P. M. (2003). Assessing the effect of model misspecifications on parameter estimates in structural equation models. *Sociological Methodology, 33,* 241–265. [5]

Yuan, Y. C. (2000). *Multiple imputation for missing data: Concepts and new development* (SAS Technical Report No. P267–25). Rockville, MD: SAS Institute. [9]

Yung, Y. F., & Bentler, P. M. (1996). Bootstrapping techniques in analysis of mean and covariance structures. In G. A. Marcoulides & R. E. Schumacker (Eds.), *Advanced structural equation modeling: Issues and techniques* (pp. 195–226). Mahwah, NJ: Erlbaum. [9]

Yung, Y. F., Thissen, D., & McLeod, L. D. (1999). On the relationship between the higher-order factor model and the hierarchical factor model. *Psychometrika, 64,* 113–128. [8]

Zimmerman, D. W. (1972). Test reliability and the Kuder–Richardson formulas: Derivation from probability theory. *Educational and Psychological Measurement, 32,* 939–954. [8]

Zimowski, M., Muraki, E., Mislevy, R., & Bock, D. (2003). *BILOG-MG, Version 3.0* [Computer software]. Chicago: Scientific Software International. [9]

Zinbarg, R. E., Barlow, D. H., & Brown, T. A. (1997). The hierarchical structure and general factor saturation of the Anxiety Sensitivity Index: Evidence and implications. *Psychological Assessment, 9,* 277–284. [8]

Zwick, W. R., & Velicer, W. F. (1986). Factors influencing five rules for determining the number of components to retain. *Psychological Bulletin, 99,* 432–442. [2]

Author Index

445

Subject Index

Page numbers in *italics* indicate pages where a subject is defined or illustrated.

449

About the Author

Timothy A. Brown, PsyD, is Professor in the Department of Psychology and Director of Research at the Center for Anxiety and Related Disorders at Boston University. He has published extensively in the areas of the classification of anxiety and mood disorders, the psychopathology and risk factors of emotional disorders, psychometrics, and applied research methods. In addition to conducting his own grant-supported research, Dr. Brown serves as a statistical investigator or consultant on numerous federally funded research projects. He has been on the editorial boards of several scientific journals, including a longstanding appointment as Associate Editor of the *Journal of Abnormal Psychology*.

Integrated Lasers on Silicon

Advanced Lasers Set

coordinated by
Pierre-Noël Favennec, Frédérique de Fornel, Pascal Besnard

Integrated Lasers on Silicon

Charles Cornet
Yoan Léger
Cédric Robert

First published 2016 in Great Britain and the United States by ISTE Press Ltd and Elsevier Ltd

ISTE Press Ltd
27-37 St George's Road
London SW19 4EU
UK

www.iste.co.uk

Elsevier Ltd
The Boulevard, Langford Lane
Kidlington, Oxford, OX5 1GB
UK

www.elsevier.com

Notices

For information on all our publications visit our website at http://store.elsevier.com/

British Library Cataloguing-in-Publication Data
A CIP record for this book is available from the British Library
Library of Congress Cataloging in Publication Data
A catalog record for this book is available from the Library of Congress
ISBN 978-1-78548-062-1

Printed and bound in the UK and US

Contents

Preface

At first sight, an entire book on integrated lasers on silicon may appear to be a very specialized reading only for experts. But given the tremendous growth of silicon photonics in research and application fields during the last 10 years, we think that a comprehensive presentation of the issues, strategies and realizations of laser sources on silicon with respect to integration constraints will provide useful information for a broad range of readers.

This book may thus be of interest to people both in the industry and academia, from the microelectronics or optoelectronics sector, especially researchers or engineers. It covers a wide range of specialization, from materials research, to laser devices and microprocessor architecture. We try to explain the prerequisites for understanding the issues and goals of laser integration on silicon as simply as possible, so that students may use this book to strengthen their general knowledge on this topic, and benefit from the numerous references provided therein.

Though the material for this book was mainly taken from the large literature in the field, some of it also come from the various interactions that we have regularly with several industrial or academic research groups working on silicon photonics worldwide.

We would like to acknowledge the contribution of colleagues from the 3D-optical-many-cores project supported by CominLabs, especially S. Le Beux, I. O'Connor, X. Letrartre and C. Monat from the Lyon Institute of Nanotechnology (INL), O. Sentieys, D. Chillet and C. Killian from French Institute for Research in Computer Science and Automation (INRIA) with whom integration strategies were widely discussed. We also thank our partners from the OPTOSI ANR Project No. 12-BS03-002-02, especially E. Tournié, J.-B. Rodriguez and L. Cerutti from Institut d'Electronique et des Systèmes (IES), F. Lelarge (Alcatel-III-V Lab) for the fruitful discussions about the integration constraints of lasers on silicon and B. Corbett (Tyndall National Institute) for the discussions about the bonding approaches. Support from "Region Bretagne" is also acknowledged. We finally want to thank all the members of the Optical Functions for Information Technologies (FOTON) laboratory, and especially the researchers from the Optoelectronics, Heteroepitaxy and Materials (OHM) research group of the Institut National des Sciences Appliquées de Rennes (INSA), who made this publication possible. Overall, this book benefited from the constant support of key people – Astrid, Emilie, Clémentine, Lise, Madée and Eilie.

Charles CORNET
Yoan LÉGER
Cédric ROBERT
April 2016

Introduction

With the emergence of the first semiconductor devices in the early 1950s [BAR 48, CHA 54], it was quickly understood that their unique optoelectronic properties would definitely change our everyday lives. Rapidly, silicon became the material of choice for the development of the whole electronic industry, based on the so-called complementary metal oxide semiconductor (CMOS) technology, while III–V semiconductors (such as GaAs or InP) were widely (but independently) used since the 1970s to develop light emitters, e.g. for optical telecommunications, lightning or other photonic devices. At the dawn of the 21st Century, some pioneering research groups realized that the electronic–photonic convergence could bring various advantages in a large field of applications, such as biosensors, light harvesting devices, defense, consumer electronics, medicine, metrology, robotics, environment, imaging and information processing. Ever since a general consensus was reached on the disruptive power of such research field, as evidenced by the hundreds of research groups and thousands of researchers worldwide, either from academic institutions or the industry, presently working on the co-integration of optical functions with the electronic functions of silicon. One of the most exciting promises lies in the integration of photonics, which is at the very heart of today's microprocessors.

Indeed, looking from the information and communication technology (ICT) point of view, the 21st Century is clearly characterized by an explosion of demand for computing, storage and communication capabilities. In this context, the dramatic "bottleneck" is the difficulty in transmitting digital information, from worldwide links to chip-to-chip and intrachip interconnections. While photonics is already found at the heart of today's communication networks, providing enormous performance to the backbone, metro and access systems, at shorter distances, the challenges posed by signal speeds, power consumption, miniaturization and overall costs are still only partially addressed. The ultimate solution to this problem could come from optoelectronic integrated circuits (OEICs), and more specifically from on-chip optical interconnects. In this approach, a photonic layer would be part of the microprocessor for information routing and processing. With lower energy consumption, faster data transmission and processing, the on-chip photonic integration may provide ground-breaking optical computation paradigms. Such a technological revolution is thus considered seriously by IBM, Intel, ST Microelectronics and various companies in the field of semiconductors. But the development of OEICs is, however, not straightforward. Elemental building blocks (waveguides, detectors, modulators, sources) should be first developed independently. One of the most serious challenges that we presently face is the development of an integrated laser source on silicon to feed the OEICs with light. The goal of this book is to comprehensively present the advances that have been achieved toward the development of an ideal integrated laser source on silicon.

1

Laser Integration Challenges

In this chapter, we provide a comprehensive overview of the ultimate properties and performances needed to truly achieve very large-scale integration of laser sources on a silicon chip. Toward this aim, the basic principles of CMOS microprocessors and the different integration schemes of a photonic layer into such architectures are first explained. Then, very simple aspects of semiconductor lasers are presented to provide the minimum prerequisites to the discussion in the next chapters. Finally, an assessment of the required quality that an integrated laser source should ideally possess is given.

1.1. Evolution of microprocessor technologies

The Sparc M7 microprocessor, announced by Oracle in late 2015, shows a record transistor count of 10 billion [AIN 15]. The smallest elements on the chip, the so-called technological node, are 20 nm large and are organized into height clusters of four cores. Thirteen layers of metallization ensure the interconnections between the cores, clock and power drive. We have come a long way from the first integrated circuit developed by Kilby roughly 60 years earlier, which contained two transistors only [KIL 76]. Nowadays, the architecture of microprocessors (MPs) has become so complex that limitations of the complementary metal oxide semiconductor (CMOS) technology progressively reveal themselves. We will see in the following that the major concern of MP designer is presently energy, leading researchers to work on architecture solutions going beyond CMOS. Many leads are explored, from carbon nanotube electronics and spintronics to MP hybridization. In the last case, the introduction of different hardware technologies within a MP gambles on the complementary advantages of information technologies and high yield conversion between different information technology (IT) platforms or computation paradigms. The grand challenge of integrated photonics is to demonstrate the mandatory role that it could play in future generations of microprocessors, for data routing or even specific computation tasks.

This first section is a brief survey of present CMOS MP technologies and their challenges. The interested reader is encouraged to refer to dedicated books for a deeper understanding of microprocessor technologies, such as the Weste and Harris *CMOS VLSI Design* [PEA 16].

1.1.1. *Microprocessor architecture and design*

Very briefly, the architecture of CMOS microprocessors can be described as a silicon substrate or silicon on insulator (SOI) substrate, on which the logic components of the circuitry are processed, i.e. p- and n-type transistors, inverters, etc. Ion implantation and doping diffusion are mainly used to create the good and active layers of the transistors directly into the silicon. On the contrary, the gate oxide and polysilicon parts are deposited onto the substrate. The CMOS components are then connected through multiple metallization layers embedded into silicon dioxide. Transistor processing occurs at the beginning of the production line in the fab. It is called the "front-end of line" process while the final metallization is called the "back-end" process. By extension, front-end and back-end sometimes characterize positions within the chip, as we will see later. A fab line is generally developed for a single technological node as defined by the International Technology Roadmap for Semiconductors (ITRS). For microprocessors, the technological node refers to the lateral size of the transistor polysilicon gates. All the other dimensions of the chip components, from the transistor measurements to the size and spacing of the wires at each metallization level are tabulated following this technological node, as design rules. At the 32 nm node, nine metallization layers are necessary to ensure the interconnections between transistors as shown in Figure 1.1. At the 20 nm node, 13 metallization layers are necessary. This reveals the decisive role that interconnects play in last generations of microprocessors.

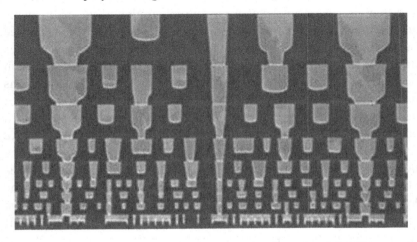

Figure 1.1. *Scanning electron microscopy cross-view of an Intel 32 nm node 9 layer Cu interconnect stack. Only the lowest levels are linked to the transistor plane which is not even spatially resolved on the picture, from Moffat and Josell [MOF 13]. Reprinted with permission from the Electrochemical Society*

On the conceptual side, the design of an MP can be analyzed at different abstraction levels (see Figure 1.2). The higher abstraction level is called the system or architecture level. It describes the prime role of the architecture, its interactions with external devices and also the hierarchy between subsystems of different roles within the chip. An algorithm view of the system with major building blocks is discussed at this level. Next comes the logic and circuit level of abstraction. Its role is to describe in terms of circuitry and logic diagrams the functions of the different subsystems. At this stage, one of the main concerns of developers is to find the most elegant designs to achieve a particular function. The circuit and logic level hides delay and timing considerations, which are scrutinized at a lower abstraction level. By contrast, power and efficiency of the design is already partly determined at this level of abstraction. Closer to the real architecture of the chip is the layout level. It describes the real geometry of the electronic interconnections and position of the CMOS components. The main issues of this level are the co-integration of the components and the compatibility of the design with a fab line. We will indeed see in the following that the lowering of the technological node induces more and more interactions between the chip elements that are often detrimental to the reliability and yields of the chip. Clever designs have to be developed to prevent such problems with the constraint of keeping within a specific process flow. Thermal management is also a concern of this level since MP operation will necessarily induce formation of hot-spots in the architecture, which can affect the behavior of surrounding components [HAM 07]. The layout level is also highly dependent upon the upper and lower abstraction levels: the logic and circuit level and the last so-called physical level. This last level describes the geometry of the CMOS components, for example if the diffusion parts of many transistors are merged together or not. It is mainly compelled by the technological node and the fab capacities. A clever design of the CMOS components will have the crucial and final role on determining the global speed and power of the chip as well as the signal reliability. The commercial introduction of SOI substrates in the late 1990s has deeply modified the design at the physical level. Transistors are then embedded into an insulating oxide, improving the power and delay yields of the components but also inducing novel design complications such as variability of the delays due to the floating body voltage. One especially understands that the physical level will be highly impacted by the shift toward beyond-CMOS technologies where the very nature of transistors will be modified [NIK 13].

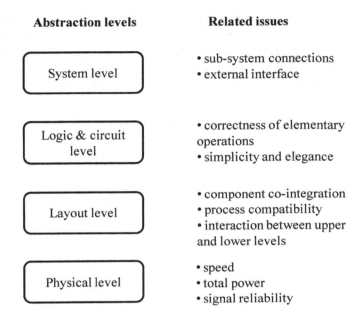

Abstraction levels   **Related issues**

System level

• sub-system connections
• external interface

Logic & circuit level

• correctness of elementary operations
• simplicity and elegance

Layout level

• component co-integration
• process compatibility
• interaction between upper and lower levels

Physical level

• speed
• total power
• signal reliability

Figure 1.2. *Abstraction levels of a microprocessor design and related issues*

1.1.2. *Delay versus power trade-off*

In a microprocessor, the delay basically defines how fast memories are accessed, logic operations are executed and how long signals take to propagate through the chip. In the chip design, this has consequences for each abstraction level. At the physical level, the design of the transistors fixes their RC behavior and thus their dynamics. At the layout level, parasitic capacitances can be avoided by folded geometries where diffusion parts of transistors are shared. Moreover, below the 180 nm technological node, the RC delay related to interconnects exceeds the gate delays. Skin effects and crosstalks between wires also play a detrimental role. At the logic and circuit level, an analysis method called "logical effort" allows designers to choose the best logic function or number of inputs to a specific gate in order to minimize delays. Finally, at the system level, knowing the delay information obtained at lower abstraction level, decisions on the algorithm design of the chip have to be taken. Especially, in the early 2000s, improvements in MP designs brought clock frequencies at a level where designs started to be power-compelled. In order to keep delays decreasing at fixed clock frequencies, parallelism was the only micro-architecture solution and multicore MPs started to be commercialized (see Figure 1.3).

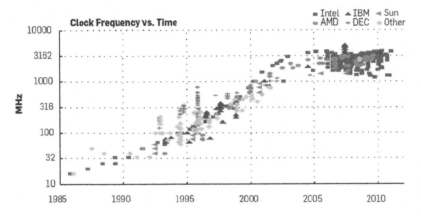

Figure 1.3. *Historical evolution of the microprocessor frequency, From [DAN 12] © 2012 Association for Computing Machinery, Inc. Reprinted by permission. For a color version of the figure, please see www.iste.co.uk/cornet/lasers.zip*

Indeed, by 2004, power consumption of the mono-core architectures had met the 100 W mark. The trade-off between power and delay is highly correlated with the technological node. From the 45 nm node, strong complications arise to develop power-efficient chips. Energy dissipation in MPs has dynamic and static origins coming from the CMOS logic layer and obviously circuitry-related origins. Dynamic dissipation in CMOS comes from the charging and discharging of the load capacities as transistor gates switch. Some short circuit currents are also generated when transistors are partially "on". This dynamic dissipation is directly associated with the computation tasks and can only be decreased by a clever design of the logic components. For example, SOI-based CMOS, through the lowering of parasitic capacitances, induces both shorter delays and lower dynamic dissipation. Still, below the 90 nm node, the dynamic dissipation is not so important as compared with the static dissipation, which represents now about one-third of the MP power budget. Static dissipation is due to subthreshold current leakage in "off" transistors, gate dielectric leakage, junction leakage from source and drain diffusions. To reduce the static dissipation, clever architecture algorithms directly switch off the power supply of functional blocks when they are of no use. But the main actor in the power dissipation of an MP is the interconnect layer. Dynamically, the capacitance of a wire fixes its switching energy. In [PEA 16], a simple estimation of the capacitance-induced dissipation in a single 65-nm node metallization layer is given. For an activity factor of 10%, it represents 48 W. In recent chips, the MP operation algorithms allow for much lower activity factors of the multiple metallization layers but in spite of this, energy dissipation in the interconnects represents half of the power budget of the chip [MIL 09]. The energy issue is definitely the grand challenge of the early 21st-Century microelectronics and photonics could be an elegant way to address it.

1.2. Photonic integration schemes

In the last decade, tremendous advances have been made in the domain of integrated photonics. Boosted by the technological capacity of the microelectronic industry, silicon photonics has become one of the leading areas of optics research. Many building blocks for advanced photonic integrated circuits were developed, which are compatible with CMOS process lines. Merging photonics and microelectronics has been discussed for more than 30 years [GOO 84, FEL 88, MIL 00] but this convergence could become a reality soon. In section 1.1, we underlined the issues and challenges of modern MPs. It is clear that integrated photonics, with technologies like wavelength division multiplexing, advanced modulation schemes, low-loss integrated waveguides, could provide new development possibilities to hybrid microprocessors. All the abstraction levels could be impacted by such a paradigm shift: the very large bandwidth of optical interconnects could enable massive parallelism and much larger activity factors of the MP subsystems. Specific processing tasks could even be executed by photonic functionalities [LAR 12]. But the integration of a photonic layer into MP architectures can also raise issues at both the layout and physical abstraction levels. Everything depends on the photonic integration scheme of the hybrid MP design. This is the purpose of this section.

1.2.1. The photonic layer in 2D MP architectures

The so-called 2D microprocessor architecture is based on a single silicon or SOI wafer. Photonic integration in a 2D architecture thus implies either the monolithic integration (direct growth on silicon) of the photonic layer or the heterogeneous integration (bonding) of photonic dies or full wafers onto the chip. Independently of this integration approach, the photonic layer can be situated at different levels into the chip. Figure 1.4 shows the three integration schemes of a photonic layer into a 2D MP architecture.

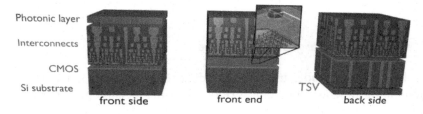

Photonic layer

Interconnects

CMOS

Si substrate

front side   front end   TSV   back side

Figure 1.4. *The three integration schemes of a photonic layer inside a microprocessor. The front-side (or back-end) integration reports the photonic layer far from the CMOS electronics and heat dissipative substrate. The back-side integration requires through silicon vias (TSVs) to couple CMOS and photonic components. The front-end integration causes the issue of combined integration of CMOS electronics and photonic components on the same plane. For a color version of the figure, please see www.iste.co.uk/ cornet/lasers.zip*

The front-side integration scheme occurs at the back-end of line in the MP process flow. This scheme favors off-chip optical communications and also seems to be cost-effective for microelectronics industries since smaller modifications of the CMOS fab-line are required. The front-side scheme is highly compatible with heterogeneous integration of mature III–V semiconductor active photonic devices based on GaAs and InP, making time-to-market probably very short. This is the choice of organizations such as Intel and Luxtera [BOW 89]. However, introducing the less mature functional layer at the end of the process line also generates costly risks of chip degradation, even if photonic components can be tested formerly. Finally, the front-side integration scheme prevents fast communication between the CMOS logic and the photonic layer since the last metallization layers are the slowest ones.

The second integration scheme is the (combined) front-end integration. In this case, the photonic components are processed in the vicinity of the CMOS components. Such a scheme has been chosen by IBM because many building blocks of photonic integrated circuits are compatible with CMOS process and are mature on the silicon platform at telecom wavelengths. This is the case of Si waveguides, Si modulators and Ge photodetectors [ASS 12]. The monolithic integration approach is particularly appealing with this integration scheme, but we will see later that the monolithic front-end integration of a laser source is particularly challenging. This is clearly the present technological bottleneck for this approach. In addition, the combined front-end integration is certainly the integration scheme where most design abstraction levels will undergo modifications, from the physical to the system level. Still, the front-end integration schemes show many advantages and could be seen as the ultimate photonic integration scheme [BAE 12]. It offers favorable thermal management for both the photonic layer and CMOS as long as bulk Si substrates are used, fast interactions between the logic and optical routing components, and it could even limit test costs with an early introduction of the less mature photonic technology in the process line.

Finally, a third option for photonic integration consists in integrating the photonic layer at the back side of the substrate. In this case, so-called through silicon vias (TSVs) should be used to drive the photonic layer, preventing fast communication with the CMOS layer. However, this scheme could show advantages for thermal management and off-chip optical interfacing. Coupling with the heterogeneous integration approach should keep the process-line quite simple since the photonic layer will most likely be implemented at the end of the wafer process. However, using the back-side integration scheme with monolithic devices would require two-sided wafer process [DEB 98].

1.2.2. *Toward wafer-scale 3D MP architectures*

In wafer-scale 3D microprocessor architectures, many wafers are stacked to build a complex architecture where different functionalities and even different computation paradigms operate at each level (see Figure 1.5(a)). In this sense, 3D architectures share similar issues for photonic integration as the front-side and back-side integrations schemes. We can already guess that, depending on its function, the photonic layer in such a 3D architecture could be situated either at one end of the chip or in the inner part, for example in between a logic and a memory layer. Key issues linked to thermal management could then be particularly important. In such 3D architectures, recourse to TSVs is mandatory in order to interconnect the different layers through the bulky silicon substrates. Huge progress has been made in the last decade in order to shorten the 3D vias. The main method consists in face-to-face bonding and substrate removal [FAR 12]. Figure 1.5(b) shows a 3D architecture using such a technology. This method is particularly adapted to the use of SOI substrates where the silicon oxide is efficiently used as an etch stop for the substrate removal [BUR 06]. Such a method allows for a huge reduction of the 3D via length to about 5 μm.

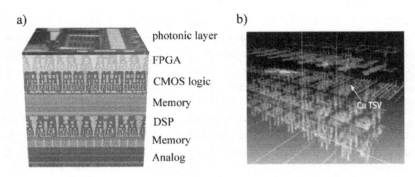

a)

b)

photonic layer

FPGA

CMOS logic

Memory

DSP

Memory

Analog

Cu TSV

Figure 1.5. *a) Sketch of a 3D microprocessor architecture as proposed in Solomon [SOL 08], associating layers of different functionalities and/or computation paradigms. b) From Koyanagi [KOY 15]. X-ray tomographic view of the 3D CMOS architecture of an image sensor chip fabricated by reconfigured wafer-to-wafer technology. Copper TSVs are used to contact the integrated circuits (ICs) of the stacked wafers together. Copyright © 2006 IEICE. For a color version of the figure, please see www.iste.co.uk/cornet/lasers.zip*

Surprisingly, the monolithic integration approach could show a key advantage for 3D architectures. As soon as the 3D stack is realized with full wafers, the use of 200- or 300-mm substrates will be compulsory for any of the functional layers. This will be a strong obstacle for InP- or GaAs-based devices compared to devices directly grown on silicon. This point will be further discussed in Chapter 3.

We can finally imagine that even combined front-end integrated layers could present some interest in 3D architectures, where the CMOS logic co-integrated to the photonic layer would be dedicated to the ultrafast control of the photonic functionalities.

We just discussed the processor architecture and the integration schemes of a photonic layer. The laser source is clearly a cornerstone in the integration scheme. It is also probably the photonic building block causing the most integration problems. Before entering into details in the realization and performance of laser devices on silicon, it is thus mandatory to clarify the main physical concepts (in both optics and materials fields) that are needed to fully understand the achievements and issues presented in the other chapters.

1.3. Semiconductor lasers

In this section, we briefly revisit the general properties of semiconductor lasers. We would like to point out that many outstanding works, publications or books have appeared in the field [BAR 13, AGR 12, GHI 09, SZE 12, CHO 13, ROS 02, COL 12]. Indeed, the first demonstration of semiconductor laser diodes was achieved in 1962 [HAL 62, HOL 62, NAT 62, QUI 62]. Here, we want to give only general considerations about operating a semiconductor laser, so that the reader can easily understand the main concepts developed in the other chapters.

1.3.1. *Operating principle*

Fundamentally, two building blocks are required to operate a semiconductor laser: (1) a gain medium, which can amplify the optical signal created by the initial spontaneous emission, through a stimulated emission process, and (2) a resonant optical cavity, which confines the electromagnetic radiation through the optical modes. The gain medium is generally composed of semiconducting materials, which are expected to provide efficiency, reliability, stability over the years, and an industry-compatible realization scheme.

The physical properties of semiconductors are usually summarized by giving their energy diagram, or "bandstructure" (see Figure 1.6 for an illustration). This gives the allowed energies for an electron, or a missing electron (a "hole") as a function of its wavevector. A photon emission may occur when an electron in the conduction band (at higher energy) recombines with a hole in the valence band (at lower energy), the energy (and therefore the wavelength) of the photon being roughly equal to the energy difference between the conduction and the valence bands, often referred as the "bandgap energy."

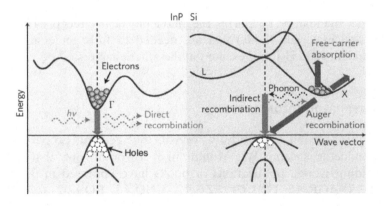

Figure 1.6. *Energy band diagrams and major carrier transition processes in InP and silicon crystals, reprinted by permission from Macmillan Publishers Ltd: Nature Photonics, from ref [LIA 10] copyright 2010. In a direct bandgap structure (such as InP, left), electron–hole recombination almost always results in photon emission, whereas in an indirect band structure (such as Si, right), free-carrier absorption, Auger recombination and indirect recombination exist simultaneously, resulting in weak photon emission*

This recombination must satisfy the momentum conservation rule. This means that the wavevector of the electron must be equal to the wavevector of the hole in absolute value (because the momentum of the photon is generally negligible). In other words, the minimum of the conduction band must coincide with the maximum of the valence band. This is the case for direct bandgap semiconductors (usually at $k = 0$, in the so-called Γ-valley) for which electron–hole radiative recombination is very efficient. Most III–V semiconductors are such direct bandgap semiconductors and are thus considered as ideal systems for semiconductor laser devices. On the other hand, in indirect bandgap semiconductors, the minimum of the conduction band does not coincide with the maximum of the valence band (for example, when the valence band maximum is in the Γ-valley whereas the conduction band minimum is located at a non-zero wavevector state such as the X-valley or L-valley). The electron–hole recombination requires an additional particle (generally a phonon) to satisfy the momentum conservation rule. Such a three-particle process is much less efficient and generally leads to weaker emitted light. It is especially the case in silicon (close to the X-valley), germanium (L-valley) or in GaP (conduction band minimum in the X-valley), which makes these materials not naturally suited for optical emission.

A crucial condition for lasing is known as population inversion. When a gain medium is sufficiently pumped, electrically or optically, the population in the excited state of the lower-energy optical transition becomes larger than the population in the ground state. Emission becomes more likely than absorption and amplification occurs; the dominant emission process is the stimulated emission and the medium shows positive gain. In semiconductor active materials, population inversion is called the Bernard–Duraffourg condition and takes into account both the electron and hole populations in the conduction and valence bands respectively. Optical amplification then occurs when the quasi Fermi level energy difference becomes larger than the transition energy [BER 61]. In a laser, the optical cavity imposes a feedback mechanism to light amplification. The cavity is also an important (and necessary) source of optical losses in the laser device, due to the limited reflectivity of the mirrors that constitute it. It also enhances the light–matter interaction and provides wavelength selectivity through the appearance of confined optical modes. The laser threshold is reached when gain overcomes the optical losses (mirror reflectivity, spontaneous emission, non-radiative recombinations) for a given confined mode. A strong coherent light emission is then observed from this mode, leaking out of the finite reflectivity mirrors. By contrast, the light emitted by the device below threshold, which spreads across all the available modes, is incoherent since spontaneous emission still dominates. The laser threshold is observed when plotting the output optical power of the laser as a function of the electrically or optically injected carriers and consists in a non-linear "kink point" in the optical power dependence. The slope of the curve after threshold is called differential quantum efficiency and is as important as the laser threshold itself to define the best operation point of a given laser source in terms of energy efficiency. As an example, lasers designed for strong output powers often show higher laser thresholds and much larger differential quantum efficiencies than low power lasers.

As shown in Figure 1.7(a), in the case of optically pumped lasers, electrons are photo-excited by an external laser beam, and therefore the threshold behavior is evidenced by plotting the output power as a function of the optical surface power density of the pumping laser. In some specific cases (pulsed operation), the optical excitation is sometimes expressed in energy per pulse, or as a fluence (laser pulse energy per unit area). Optical injection is an important tool to assess the properties of the gain medium and of the optical cavity in terms of structural quality and optical efficiency prior to the development of elements required for electrical injection of carriers (contact, diode design, doping etc.). Different injection conditions can be obtained by tuning the excitation wavelength in order to address specific carrier dynamics issues. Optical injection is thus of great importance for exploratory approaches, which will be discussed intermittently in the next chapters. But it remains a development tool for integrated lasers, which are expected to be driven by co-integrated electronics at the end.

a)   b)

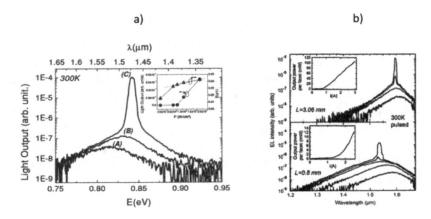

Figure 1.7. *a) Room temperature output spectra of an InP optically pumped quantum dots laser before threshold (A: 3.3 kW.cm^{-2}, B: 8.4 kW.cm^{-2}) and after threshold (C: 18.9 kW.cm^{-2}). The inset shows the energy of the maximum of the light output (triangles) and the light output intensity (circles) versus optical pumping excitation, which unambiguously shows the observation of the lasing threshold, from [PAR 02] © IOP Publishing. Reproduced with permission. All rights reserved. b) Room temperature electroluminescence spectra of an InP electrically pumped quantum dots laser under pulsed operation for several current densities: J = 65, 160, 208, and 220 A.cm^{-2} for different laser cavity lengths. Insets are corresponding light output per facet versus current characteristics, showing the lasing threshold. Reprinted with permission from Caroff et al. [CAR 05], Copyright 2005, AIP Publishing LLC*

As shown in Figure 1.7(b), in the case of electrically pumped lasers, electrons are injected by an external power supply, through metallic contacts. Therefore, the threshold behavior is evidenced by plotting the output power of the laser as a function of the injected current density.

Anyway, in both cases, the observation of a threshold, together with a narrowing of the emission spectrum, is an unambiguous proof of lasing operation. The threshold current density in an electrically pumped laser diode is usually considered as an indicator of materials and carrier injection assets: the lower the better. Typically, efficient

III–V semiconductor laser diodes today have a threshold current density between 50 and 500 A.cm^{-2}, depending on materials, geometries and type of active area. However, as mentioned above, this should be considered prudently since it can be tuned deliberately to obtain targeted specifications of a laser device, related to a given application. In the framework of integrated photonics for information technologies, the maximal wall-plug efficiency (the ratio of the output optical power by the input electrical power) is probably more appropriate to assess the global efficiency of a laser as it also accounts for the differential quantum efficiency of the device. These features will be further discussed in Chapter 5.

1.3.2. *Optical cavity*

The successful operation of a laser requires that the generated electromagnetic field remains confined in the vicinity of the gain region. In semiconductor lasers, this is usually achieved through reflections at dielectric media interfaces. In other words, the electromagnetic field is confined into the material constituting the cavity due to refractive index contrast, a direct consequence of Maxwell's equations. There are different kinds of geometries for semiconductor lasers that will be deeply discussed in Chapter 5, in view of their integration on the silicon chip. The most common (and older) one is the so-called edge-emitting laser. As illustrated in Figure 1.8(a), in this laser, the top-bottom optical confinement is guaranteed by cladding layers, having a lower optical index than the core layer containing the gain medium, so that the active layer acts as a dielectric slab waveguide. The thickness and optical index of the cladding and core layers can be tuned to optimize the optical overlap with the gain area (active region). The device is usually etched to form a ridge waveguide for lateral confinement [HUN 09]. Longitudinal confinement can be simply obtained by cleaving the device perpendicularly to the ridge waveguide. Typically, a 30% internal reflectivity is obtained by using cleaved facets as cavity mirrors.

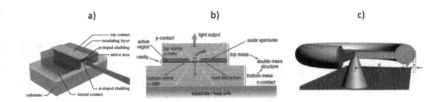

Figure 1.8. *a) Conventional edge-emitting laser geometry, b) Example of a vertical cavity surface emitting laser device [WOL 12]. Reprinted with permission of SPIE. c) Example of a microresonator, allowing the propagation of whispering gallery modes. Reprinted with permission from [XIA 10]. Copyright 2010 by the American Physical Society*

Another common strategy is the development of vertical cavity surface-emitting lasers (VCSELs). In these devices (schematically represented in Figure 1.8(b)), the vertical cavity is designed through distributed Bragg reflectors (DBR), an alternation of semiconductor layers with different optical indices, allowing the selection of a single or a few optical modes [MIC 12]. More recently, microresonators with a large variety of designs were proposed (see Figure 1.8(c)), from whispering gallery modes (WGM) resonators to photonic crystal cavities. In the first case, the electromagnetic waves propagate along the external edges of the device, forming a ring cavity [HEE 08]. On the contrary, photonic crystal cavities offer the strongest optical confinement of all integrated optical cavities, by confining light in all directions at the scale of the wavelength [PAS 13].

For any of these cavity geometries, confined optical modes (Fabry Perot, WGMs, etc.) occur when the cavity full path is equal to an integer multiple of the wavelength. Optical losses induced by the limited mirror reflectivity, scattering losses and parasite absorption allow to define for each confined mode the modal photon cavity lifetime, i.e. the typical time a photon "spends" into the cavity [SVE 12]. A more common approach in semiconductor lasers is the use of the cavity quality factor or Q-factor, which is the number of oscillations of the field before the circulating energy is decreased by a factor $1/e$. Thus a high value of cavity Q implies low losses of the resonant system. In CMOS-compatible WGM micro-resonators,

Q-factors well above 10^8 have been demonstrated [LEE 12b]. The situation is different in typical edge lasers and VCSELs where Q-factors are around 10^2 and 10^3 respectively. The use of high Q-factors is interesting especially when the optical gain of the active area is relatively small, which is the case of some of the materials presented in the next chapters and where low optical power and low-energy budget are needed for an integration architecture design. Finally, the small size of recently developed microcavities greatly enhances the integration potential in a photonic integrated circuit. Unfortunately, reducing the footprint of a laser source also makes electrical injection schemes more challenging.

1.3.3. Carriers injection

The semiconductor laser operation principle and the concept of resonant optical cavity were discussed in the previous parts. It was shown that lasing only occurs when enough electrons (holes) are present in the conduction (valence) band of the active area, the so-called Bernard–Duraffourg (B&D) condition. If optical injection provides some degrees of freedom on the excitation conditions (excitation wavelength, injection power, even polarization) to reach such a regime, electrical injection is considered as a supplementary technological bottleneck in the realization of an integrated laser, raising new issues on the laser design.

First, the laser has to be contacted with metallic connections. Injecting current efficiently through a metal–semiconductor interface is not that easy. The Schottky barrier has to be overcome. With a well-defined succession of metallic layers and annealing steps, an ohmic contact with a low-access electrical resistance can be achieved. It is sometimes proposed to lower the bandgap and increase the doping levels of the surface semiconductor layers to facilitate the current injection. While these processes have been developed and optimized for years with silicon or conventional III–V semiconductors such as InP or GaAs, it is not always the case with the new approaches or materials, which will be presented in this book, and especially the materials with large bandgaps.

Then, at the heart of the semiconductor laser injection is the p–n junction, which is formed by bringing into contact a p-doped and an n-doped semiconductor [AGR 12, SZE 12]. In this configuration, their quasi-Fermi levels equalize through the diffusion of electrons from the n-side to the p-side (and the opposite for holes), forming a depletion region at the p–n interface, where the gain medium is present [KRO 63]. When a direct bias is applied to the diode, electrons and holes massively enter this central region from each side respectively, locally fulfilling the B&D condition in the gain medium. The gain region is often non-intentionally doped to limit the free carrier absorption (see next parts). Here again, while the electronic transport through the p-i-n structures is well known, material research maturity is mandatory to control n- and p- doping levels, as well as non-intentional doping levels, which are critical parameters for efficient carrier injection.

Moreover, once injected into the semiconductor, both electrons and holes should not face any prohibitive potential barriers inside the semiconductor device, related to the complex material sequence imposed by both electrical injection issues (p-i-n junction) and optical confinement (core/cladding sequence, DBRs, etc.). Once carriers have reached the active area, they should even be blocked to form a charge carrier reservoir. This general issue is known as band engineering. This is illustrated in Figure 1.9, with the metamorphic InP-based quantum wells (QWs) laser structure from [GU 15]. Here, the electrons are injected in the conduction band by the n-doped InAlAs region (also serving as cladding layer for the optical confinement). They fall down in the InGaAs-active area, and they are finally trapped by InAs/InGaAs quantum wells where they can recombine with holes. Any electron that could escape these quantum wells would be blocked by the potential barrier induced by the p-doped InAlAs. The situation is similar for holes injected in the valence band.

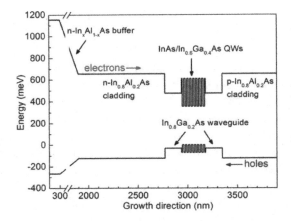

Figure 1.9. *Energy-flat band structure of an InP-based metamorphic type-I quantum wells laser. Injection directions of electrons and holes are indicated. Reprinted with permission from [GU 15]. Copyright 2015, AIP Publishing LLC*

1.3.4. Active area

If some pioneering works report on lasing with bulk semiconductors, it was rapidly understood that using low-dimensional objects with quantum confinement in the active area provide many advantages over bulk materials [VAN 75]. Indeed, if well-designed, the quantum confinement favors the spatial overlap between electrons and holes, and enhances the lasing performances. Most of state-of-the-art laser devices now use quantum wells (QWs) or quantum dots (QDs) as active nanostructures. QWs are an alternation of semiconductor layers, with different bandgap energies. Figure 1.10(a) shows the case of the ideal GaAs–AlAs QW system. Here, the lower bandgap of GaAs allows the electrons and holes to be confined within the emitting GaAs layer. If the quantum well is thin enough, quantum confinement occurs and discretizes the carrier energy levels (reduction of the density of states along the growth direction), offering a supplementary way to tune the bandgap (and therefore the emission wavelength). To further reduce the dimensionality of the nanostructures, and benefit from a true 3D quantum confinement, one can play with the stress of the semiconductor layers and promote the relaxation of strain through the controlled formation of self-assembled QDs [POH 13]. QDs are then small facetted pyramids with typical widths of a few dozen nanometers and height of few nanometers (see Figure 1.10(b) and (c)) for illustration). With adapted growth conditions, the QDs can be grown with a high density at the surface (typically up to 10^{11} cm^{-2}).

a)   b)   c)

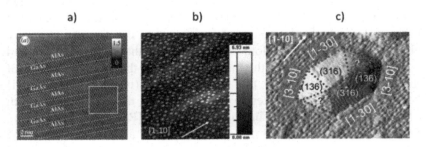

Figure 1.10. a) Cross-sectional transmission electron microscopy imaging of GaAs/AlAs multi-quantum wells, from [LIC 09], reprinted with permission from Royal Society. Scanning tunneling microscopy plan-view imaging of (In,Ga)As/GaP quantum dots, with a 800*800 nm² field b) and at the single quantum dot atomic scale c), Reprinted with permission from [ROB 12a]. Copyright 2012 by the American Physical Society. For a color version of the figure, please see www.iste.co.uk/cornet/lasers.zip

When buried with another semiconductor, with a larger bandgap, QDs provide usually a 3D quantum confinement to electrons and holes that can recombine efficiently. Full carrier confinement provides a lesser sensitivity to the surrounding crystal defects, and therefore to non-radiative recombination channels. Contrary to QWs, the 3D quantum confinement provides true discretized energy levels, and thus "atomic-like" optical transitions within the semiconductor. The lowering of the energy density of states is a crucial advantage for lasing, as it allows us to reach the B&D condition faster and therefore lowers significantly the laser thresholds [BIM 99, UST 03]. The size, shape and composition of the QDs governs the energy levels via quantum confinement directly and therefore the emission wavelength of the operating laser.

If the development of an active area with good structural properties is important, this does not guarantee efficient lasing. Indeed, the optical efficiency of an active area, which directly impacts the optical gain of the structure expressed in cm^{-1} (determined for instance by multisection gain measurements [THO 99] or transmission experiments [ELL 98]), is directly related to the oscillator strength of optical transitions in the nanostructures composing the gain region. Of course, the direct or indirect bandgap character of the optical transitions involved should be carefully considered (a direct bandgap being much more efficient than an indirect one). But the confinement potentials for both electrons and holes inside the nanostructure are also decisive. Indeed, it should first guarantee that both electrons and holes are confined within the nanostructure. In the case where one of them is located outside the nanostructure (type II or type III bandlineup), the spatial overlap is reduced, and the oscillator strength of the optical transition is intrinsically limited [ROB 14]. In addition, the confinement potential for both electrons and holes should be large enough to ensure that thermal excitation of charge carriers will not empty the nanostructures too much by detrapping, from the nanostructure to the barrier, thus hampering population inversion. But in some systems, it should not be too large either, to promote efficient thermal carrier redistribution through the different nanostructures, especially for quantum dots [BIM 99, COR 06]. Finally, we note that the addition of some particular atoms, such as nitrogen (N) [NGU 12] or bismuth (Bi) [PET 08] into semiconductors add specific strong localization properties, which can be of great interest for the development of light-emitting diode (LED) devices on silicon, as they are nearly insensitive to the crystal defects, but these carrier traps are more hardly handled in laser devices, as they nearly freeze the charge carriers into relatively non-optically efficient states.

1.3.5. *Non-radiative recombination and absorption processes*

The ideal picture of the operating laser depicted in the previous sections considered nearly perfect carrier diffusion in the device, and perfect propagation of the light, once generated in the cavity. However, there are many different issues that limit the performance of a semiconductor laser, which will be discussed in the following chapters. First of all, only radiative recombination was considered in the semiconductor. In fact, many non-radiative mechanisms allow transferring electrons from the conduction band to the valence band without any photon emission.

Beyond the different non-radiative processes, the trap-assisted Schokley–Read–Hall (SRH) recombination process is of prime importance for many lasers presented in this book. Indeed, in some particular cases (e.g., the presence of a free surface, or crystal defects in the semiconductor), some energy levels or bands are created in a localized region of the semiconductor. Unfortunately, these energy levels may appear in the forbidden gap of semiconductors. In this situation, carrier transitions between the conduction band and the valence band are made easier through the trap level [AGR 12, GHI 09]. These defects are therefore also called non-radiative recombination centers or deep-level traps. It is one of the most important recombination mechanisms in indirect bandgap semiconductors like Si, Ge or GaP. Indeed, in this situation, the band-to-band radiative transition has a very low probability, so that carrier relaxation through deep level traps becomes more likely. These deep-level traps can have different origins. In many laser devices that will be presented in this book, they come from crystal defects generated during the epitaxial process. Point defects, such as atomic vacancies, or interstitial atomic incorporation are, for instance, generated in all the epitaxial layers, depending on growth conditions and materials. But this non-radiative recombination channel is negligible compared to the impact of defects induced in monolithically integrated lasers. Numerous extended defects are indeed generated at the III–V/Si heterointerface in the III–V monolithic integration approach (Chapter 4) or during the metamorphic growth of group IV semiconductors (Chapter 2) and propagate into the active area. These include especially misfit dislocations that have been shown to severely limit laser diodes performances [EGA 95]. It is usually considered that a dislocation density below $10^6\,\mathrm{cm}^{-2}$ is needed to achieve efficient lasing. Their great impact on laser properties come from the large energy density of deep levels states due to their large extent. The density of deep-level traps, and more particularly of threading misfit dislocations, is therefore a major concern for lasers integrated on silicon. In some cases, additional traps appear when the semiconductor surface is exposed to ambient air (e.g., cleaved facets of an edge-emitting laser) or to a defective or amorphous interface (III–V/Si bonding interface, see Chapter 3). This creates energy levels in the bandgap, and is often referred to as surface or interface recombination. This surface–interface recombination depends as well on the materials, and is usually considered as a limiting process when the defects density in the volume of the device is already low.

The second kind of non-radiative mechanisms that has to be considered is the so-called Auger recombination, schematically explained in Figure 1.6. Typically, an electron located in the conduction band interacts with another. During this interaction, the energy lost by the first electron to reach the valence band is transferred to the other electron to reach higher energy states [AGR 12, GHI 09]. By extension of this simple picture, the Auger processes often refer to all the different many-body interactions that can be imagined in the semiconductor bands that promote the non-radiative recombinations; they are electron- or hole-assisted recombinations. The Auger non-radiative recombination is very sensitive to charge carrier density in the semiconductor. Therefore this process will become prominent at large carrier injection densities in both direct and indirect bandgap semiconductors. In particular, some of the lasers presented in this book have not reached the maturity of low-threshold lasing yet. In these devices, Auger recombination is expected to play an important role. Conversely, in some particular cases (e.g., quantum dots), Auger processes may be helpful in promoting the carrier injection from the barrier to the active nanostructures.

Apart from the competition between radiative and non-radiative recombination dynamics, the photons generated in the active area may be reabsorbed unintentionally in the semiconductor. One of the most important processes for this is the free carrier absorption (FCA). The FCA is schematically depicted in Figure 1.6. Here, an electron (or a hole) in the conduction band (valence band) absorbs one photon, is scattered to a higher energy state within the band and further comes back to the minimum of energy of its band. This process does not provide any advantages for lasing and furthermore induces cavity losses (due to absorbed photons), and additional heating of the lattice. This process strongly depends on the carrier density either due to injection or doping, on the lasing wavelength, together with the bandgap and nature of the semiconductor. Due to the strength of FCA in doped regions, the overlap between the optical mode and the doped layers is usually reduced at its maximum, by designing graded doping profiles, and by using a non-intentionally doped active area. FCA is clearly an issue for many lasers described in this book.

Finally, in some specific operating conditions, two-photon absorption (TPA) can also arise. This non-linear process is the interaction of two incident photons with one electron. By absorbing the energies of the two photons, the electron initially located in the valence band moves at higher energy in the conduction band. This process is usually not considered in most of the conventional laser diodes. But some of the lasers presented in this book operate only under high optical pumping power. In such conditions, the TPA process becomes a non-negligible optical loss contribution in the cavity.

1.3.6. *Influence of the temperature*

The last important parameter that should be discussed before describing the different laser devices on silicon is the temperature. Of course, a laser operating only at cryogenic temperature is not compatible with photonic integration in a CMOS environment. But lowering the temperature has several advantages for device development and materials research. Indeed, at lower temperatures (typically 4 K with helium cooling or 77 K with nitrogen cooling), the charge carrier Fermi distribution gets steeper, favoring occupation of the lower energy levels of the system (thermal excitation to higher energy levels is less probable). Depending on the form of the state density distribution in the gain medium and thus of the dimensionality of the carrier quantum confinement (QDs, QWs or bulk), very different dependences of the threshold current density will be observed. Basically, QW-based (and *a fortiori* bulk-based) laser should show lower thresholds at lower temperatures while QD-based lasers are expected to show no dependence of the threshold current density with temperature. This temperature dependence of the threshold is described by the T_0 parameter. Theoretically, QD-based lasers feature an infinite T_0. Its value is 104°C for bulk-based lasers and 481°C for QW-based lasers [ARA 82]. To this redistribution of carriers, one should add the effects of thermal activation of non-radiative channels. When the carriers are frozen in the optically active nanostructures, they can hardly redistribute to the barrier, which limits non-radiative recombination mechanisms. This further reduces the laser threshold, and improves overall laser performances.

Another effect of temperature often referred to as the "Varshni law" can be of critical importance in lasers undergoing large temperature variations. It describes the decrease of the bandgap energy with increasing temperature, due to temperature-induced variations of the interatomic distance. In some cases, discussed in the next chapters, a deviation from the monotonic Varshni dependence can be observed. This is in particular the case of dilute nitride compounds or QDs with a large confinement potential. Practically, in a laser design, one should carefully assess and compare the energy drift of the gain medium, the one of the confined modes (due to variation of the refractive index) for a given temperature range, as well as the free spectral range of the cavity (the energy spacing between two longitudinal modes) and the spectral width of the gain. As an example, VCSELs generally feature a single longitudinal mode showing much slower energy drift than the gain medium with temperature. Large spectral gain is thus needed to obtain good operation yields in a large temperature range and the design is always optimized for a precise temperature. Still, many semiconductor lasers need a temperature regulation during their operation. This is especially true when drastic constraints are placed on their emission wavelength. In photonic integrated circuits where wavelength-division multiplexing (WDM) is used, advanced thermal management of microlasers should be considered to deal with resonance drifts of microlasers [LEB 14].

During the development of a laser, a common strategy to bypass injection-induced heating of the device is to optically or electrically inject the carriers under pulsed conditions. In this case, carriers are sent to the structure in the form of periodical pulses. This injection regime is interesting, as it allows reaching very high injection levels within a very short time (and thus increases the probability to get the population inversion) but with a reasonable average input power, thus limiting the overall heating of the device. Lasing is thus more easily obtained in these conditions. Here, the duty cycle of the pumping source (either power supply or pumping laser), which is the ratio of the temporal width of one pulse to the total period, is a critical parameter to be chosen.

The temperature also plays an important role in activating the propagation of some defects. This is especially the case of misfit dislocations that appear in metamorphic approaches (Chapter 2 or Chapter 4). Indeed, while subjected to significant thermal heating during the laser operation, some dislocations originally buried in some parts of the device can migrate and extend through the whole device; this is called dislocation gliding. It tends to degrade the performance of the laser over the time, and finally reduces significantly the laser lifetime.

1.3.7. *Ideal performances of the integrated laser*

In the previous sections, we discussed the most basic properties of semiconductor laser devices. But several specific aspects may be considered for the most mature technologies of integrated lasers on silicon: carrier dynamics, modulation rates, linewidth enhancement factors and chirp, gain compression, etc. [GRI 08]. From the previous sections, it is possible to give a picture of the requirements that should follow ideally an integrated laser source on-chip:

– *Electrical injection*: The integrated laser should be electrically driven to benefit from the CMOS electronic environment.

– *Continuous wave operation*: The integrated laser should be able to work under continuous wave conditions, to avoid any capacitive transient process.

– *Room temperature*: The integrated laser should of course being able to run at temperatures between 30 and 150°C, which is the microprocessor temperature range.

– *Emission wavelength:* In a pragmatic approach, any wavelength may be suitable for laser operation as long as the signal can propagate on-chip. However, photonics on silicon have already reached a significant maturity based on Si/SiO_2 waveguides. In this context, the wavelength should be larger than 1.1 µm not to be absorbed by the silicon itself. Particularly, wavelengths of 1.3 or 1.55 µm guarantee very low absorption coefficients in Si waveguides. The wavelength can be extended to the visible range by using a SiN_x waveguide technology. But still, the wavelength should not be too small to avoid absorption. On the contrary, a too large wavelength imposes to use larger optical cavities and waveguides (especially for radius of curvature for in-plane routing of the optical wave), that may drastically enhance the footprint of the photonic layer elemental constituents. Therefore, emitting wavelength between 600 nm and 2.5 µm may be considered as interesting, with a large preference for 1.3 and 1.55 µm.

– *Geometric emission properties*: The integrated laser cavity may preferentially emit in-plane, because in most of integration schemes, the photonic layer is considered as a horizontal sheet. This imposes for vertical emitting devices such as VCSELs to design specific optical reflectors folding the emission direction in the plane. The optical coupling of the laser emission into the photonic circuitry is also an important constraint. It often raises strong alignment issues, in particular in the heterogeneous integration approach, which are generally overcome by processing the III–V components after bonding. In the monolithic approach, localized expitaxy could be of interest. Regardless of the integration approach, when evanescent optical coupling is used, the active area should be located in the vicinity (typically <300 nm) of the photonic routing components.

– *Footprint*: Irrespective of the laser design, photonic footprints are always orders of magnitude above CMOS footprints. This is an important issue of co-integration. On the photonic side, one could imagine that the smaller the laser, the more possibilities architecture designers have. This issue is strongly correlated with global architecture choices and energy considerations. This will be further discussed in Chapter 5.

– *Power and threshold considerations*: One could *a priori* think that on-chip lasers do not need to feature high output powers and that as low as possible threshold currents are worthwhile. The situation is in fact much more complex. Of course, one should target laser solutions showing the best energy yields (wall-plug efficiency). However, output power and threshold considerations are fully dependent on the photonic integrated circuit design, as discussed in Chapter 5. That said, the energy efficiency of a laser will be determined by:

– optically efficient active area (direct bandgap, good quantum confinement, high gain);

– efficient carrier injection through contacts and adapted band engineering;

– low optical losses in the resonant cavity (especially through photons absorption);

– low non-radiative recombination probability (low crystal defects density, and limited Auger recombination).

Group IV Silicon Lasers

In this chapter, we present the different monolithic realizations of light emitters that have been proposed with group IV semiconductors on silicon. These approaches are of great interest as they are intrinsically compatible with photonic integration on-chip from the material point of view. But we are also faced with the inherent difficulty of working with indirect bandgap semiconductors.

2.1. Group IV silicon lasers: issues

In Chapter 1, the interest to integrate light sources and especially lasers on silicon was motivated in the general framework of micro-/nanophotonics integration on-chip. One of the first approaches developed dozens of years ago is the achievement of efficient light emission (and lasing, if possible) directly from group IV semiconductors. This provides number of advantages from the material and technological point of view, which will be explained here, but this approach is also inherently limited by the low optical efficiency of most group IV semiconductors [IYE 93, PAV 12].

Indeed, unlike most III–V semiconductors, silicon is an indirect band-gap semiconductor, therefore light emission is a phonon-mediated process with a low probability (spontaneous recombination lifetimes in the ms range), as illustrated in Figure 1.6. In standard bulk silicon, non-radiative recombination rates are much higher than the radiative rates and most of the generated electron–hole pairs recombine nonradiatively.

This situation leads to a very low internal quantum efficiency (the ratio between the number of photons generated and the charge carriers injected) of 10^{-6} for bulk silicon luminescence [PAV 03]. In addition, carrier density-dependent non-radiative processes such as Auger or free carrier absorption severely impact population inversion for silicon optical transitions at the high pumping rates needed to achieve optical amplification.

2.2. Emission from bulk silicon

Nevertheless, researchers have tried to reduce the influence of non-radiative centers with different strategies. A first approach was proposed by Green *et al.* [GRE 01] using a standard p-n Si architecture. In this work, they proposed lowering the non-radiative recombination rate by reducing the non-radiative centers (high purity Si substrates), surface recombination (passivation by a thermal oxidation of the Si surface and small metallic contacts area) and doping-induced recombinations (by using high-doping layers only nearby the contact region) to the maximum. In this work, electroluminescence at room temperature is obtained, as shown in Figure 2.1(a), with a maximum around 1.1 µm, corresponding to the band-to-band recombination in silicon. The electroluminescence power efficiency reaching 1% is also interesting, but the very long decay time of the indirect band-to-band recombination as compared to the fast free carrier absorption, typical of bulk Si, certainly reduces the hope of reaching lasing conditions (population inversion) following this approach.

A second approach was proposed in [NG 01]. In this original work, instead of acting directly on the non-radiative centers density, it was proposed to limit the carrier diffusion, and thus their ability to reach the non-radiative center, by adding energy barriers, and localizing the carriers in defect-free areas of the bulk silicon in the active area. This was performed through 100 nm-large dislocation loops, as depicted in Figure 2.1(b). Room-temperature electroluminescence around 1.1 µm (being the bandgap of the silicon) was also observed in this device, demonstrating no quantum confinement effect inside the dislocation loops. A 1% external quantum efficiency was claimed in this work. But still, the issues of free carrier absorption and Auger recombination were not addressed in this approach, making the lasing conditions (B&D condition) very unlikely.

Finally, even if interesting light emission properties have been obtained with bulk silicon in both approaches, the 1.1 µm emission wavelength of these devices also raises the problem of losses in silicon-based waveguides. Indeed, the emission energy being resonant with the Si bandgap, the light may be absorbed significantly during the optical signal propagation or processing on the silicon chip. Therefore, considering the emission wavelength of these devices, together with their intrinsic limitations, laser emission from bulk silicon may not being considered as a realistic solution of integrated sources for photonics on silicon.

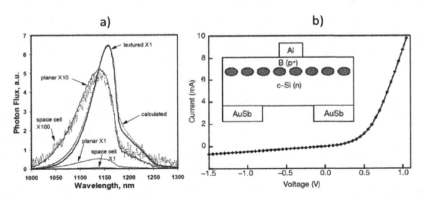

Figure 2.1. *a) Electroluminescence spectra for textured, planar and baseline space cell p-i-n Si diodes under 130 mA bias current at 298 K (diode area is 4 cm²,). Reprinted with permission from Macmillan Publishers Ltd: Nature, from [GRE 01], copyright 2001. b) Current–voltage characteristics of the dislocation loop-based device measured at room temperature. Inset: a schematic of the LED where the grey circles evidenced the defect-free areas delimited by dislocation loops. The 1.1 µm infrared light is emitted through the window left in the bottom contact. Reprinted by permission from Macmillan Publishers Ltd: Nature, from [NG 01], copyright 2001*

2.3. Using quantum confinement

Quantum confinement was then proposed as a method to achieve stimulated emission in Si. The use of Si nanostructures in an air or SiO₂ environment may provide many advantages, as compared to their bulk counterpart: 1) spatial confinement of the charge carriers so that they cannot reach non-radiative recombination centers, 2) a better electron–hole wavefunction overlap together with the appearance of discrete energy levels, leading to larger recombination rates and therefore larger luminescence efficiency.

This approach was first investigated by Canham, which showed that when silicon is partially etched via an electrochemical attack, the surviving structure further called porous silicon consists of small nanocrystals or nanowires emitting a luminescence in the visible range [CAN 90, BIS 00]. This approach has been widely investigated, and electroluminescence of porous silicon LEDs has been demonstrated with external quantum efficiencies around 1% [GEL 00]. But still, despite its interesting optical properties, net optical gain in the porous silicon has been achieved only at low levels under specific conditions [CAZ 04].

This idea was further refined by developing highly homogeneous silicon nanocrystals (Si-nc) or quantum dots, e.g., by negative ion implantation or with the formation of substoichiometric silica films, with a large excess of silicon, in a silica matrix, followed by high-temperature annealing [PAV 03, PAV 00]. The annealing causes a phase separation between the two constituent phases, i.e., silicon and SiO₂ with the formation of small silicon nanocrystals, as shown in Figure 2.2(a). The sizes (typically a few nm) and density (typically around 10^{19} cm^{-3}) of the Si-nc can be controlled by the deposition and the annealing parameters. In this material system, optical gain was first reported by Pavesi *et al.* [PAV 00] in 2000, as shown in Figure 2.2(b), and further confirmed [KHR 01]. The origin of this optical gain together with the fast optical dynamics lies in the presence of different discrete states due to quantum confinement, which drive the transitions dynamics, as described in Figure 2.2(c). Electroluminescence from Si-nc was rapidly obtained, by several groups (see [WAL 05] and [DI 11] and references therein).

a)   b)   c)

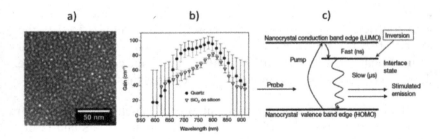

Figure 2.2. *a) Typical Transmission electron microscopy plan-view imaging of Si-nc, from [GUT 15]. Reprinted with permission under CC BY 2.0 license. b) Spectral dependence of the net modal gain in Si-nc synthetized on a quartz wafer (circles) or on a Si wafer (triangles), from [PAV 00] around 800 nm. c) Schematic energy diagram for a nanocrystal showing how population inversion can be reached in this system. Reprinted with permission from Macmillan Publishers Ltd: Nature, from [PAV 00], copyright 2000*

But, even if the observation of the first optically induced laser oscillations in Si-nc was claimed by Nayfeh *et al.* [NAY 02] in 2002, the performances of electrically driven devices were unsatisfactory, and still face reliability issues [FAU 05, PAV 05, PAV 10].

2.4. Raman scattering for lasing

In this approach, scattering of photons on molecules or atoms of the medium is used. Indeed, below the bandgap energy of semiconductors, a non-negligible part of photons still interact with the medium through elastic Rayleigh scattering and in a less probable process through the so-called Raman inelastic scattering [LIA 10]. During Raman scattering, scattered photons can either reduce their energy as compared to those of the incident light (Stokes transition) or increase their energy (anti-Stokes transition). Therefore, a pump beam can be used to excite molecules or atoms in higher energy vibrational states, which then lose energy through a phonon-mediated process, and reemit photons at the Stokes transition energy. This generated beam signal can thus initiate the stimulated resonant excitation of Stokes transitions in the material, with the population inversion guaranteed by the pump beam. This effect is already used in conventional silica fibers, but the control of such a process is difficult to reach in an integrated SOI waveguide. Indeed, while the Raman gain coefficient is significantly larger in Si than in SiO₂ [LIA 10], the losses are also much larger. This, together with the low optical confinement in SOI waveguides confer significant constraints on the design and realization of Raman lasers integrated on silicon.

The idea of taking advantage of stimulated Raman scattering (SRS) in silicon waveguides in order to realize silicon amplifiers and lasers was proposed by Claps *et al.* [CLA 02] in 2002. In this work, the laser pump (1.427 μm, cw, 1.6 W) was well below the Si bandgap to prevent band to band transitions, and free carrier absorption. A Stokes signal was detected at 1.542 μm. But at such a high pumping power, the two-photon absorption was large and prevented the system from reaching the lasing condition. This issue was first solved by using a pulsed laser pump, and Raman laser was demonstrated in 2004 [BOY 04] and 2005 [RON 05b]. The use of a (lateral) p-i-n structure was then adopted to extract free carriers from the active area, as shown in Figure 2.3, and an optically pumped cw-Raman laser was demonstrated in 2005 [RON 05a]. The laser threshold was further reduced by using racetrack ring laser cavities [RON 07]. In this work, the laser had a threshold of 20 mW and a maximum output power of 50 mW under a reverse bias of 25 V.

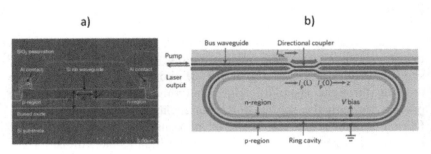

Figure 2.3. *a) Scanning electron microscope view of cross-section image of a silicon rib waveguide with a p-i-n diode structure along the waveguide in which cw-Raman laser was demonstrated. Reprinted by permission from Macmillan Publishers Ltd: Nature, from [RON 05a] copyright 2005 b) Schematic of the low-threshold Si Raman racetrack ring laser with a p-i-n junction design. Ip(0) and Ip(L) are the pump power at the starting point and after a round trip in the cavity, respectively. The light propagation direction is given by z. Reprinted with permission from Macmillan Publishers Ltd: Nature Photonics, from [RON 07], copyright 2007*

The miniaturization of the CW-laser device was recently proposed, by using a photonic-crystal, high-quality-factor nanocavity with an overall footprint of 10 µm on-chip and a 10 µW threshold without the needs for a p-i-n structure [TAK 13].

All these remarkable achievements, in terms of compactness and thresholds, are, however, intrinsically limited for laser integration on-chip. Indeed, the SRS is always induced by the use of external optical pumping, and is therefore difficult to implement in a very large-scale integration scheme.

2.5. Rare-earth doping

The concept of using rare-earth doping for light emission in silicon is not new, as it is the central process of Er-doped fiber all-optical amplifiers [BEC 99]. Here, the silica optical fibers are doped with Er^{3+} ions, whose atomic-like transitions are located within the energy bandgap of the host material (1.54 µm in the specific case of Er). The general strategy adopted when using rare-earth doping is thus the following: carriers are injected from the barrier material (e.g., SiO_2, Si or Si-nc) to the rare-earth atoms, where the relaxation of electrons on atomic-like energy levels allows efficient light emission or amplification. This process is described in Figure 2.4(a), and widely discussed in the literature (see refs. [MIL 10] and [ZHO 15]). Erbium is one of the most popular rare-earth dopants, since its luminescence is around the telecommunication wavelengths of 1.55 µm, but other rare-earth dopants can be used for other wavelengths in the same way [FAN 13b].

But this strategy also faces issues with respect to materials, as the chosen dopant has to be incorporated and activated at high concentrations in Si without the formation of precipitates. The low solid solubility of Er in Si is sometimes counterbalanced by the use of codopants which, through the formation of Er–impurity complexes, avoids Er precipitation [PAV 03]. Then, once excited, Er should decay radiatively. This radiative decay will be in competition with non-radiative relaxation processes, which can be significant due to the long radiative lifetime of Er in Si (2 ms) [PAV 03]. Therefore, combining quantum confinement (via Si-nc) and rare-earth doping may fix this issue. Indeed, multilayer Er-doped silicon nanocrystals (Si-ncs) formed in silicon-rich oxide are now considered as a promising solution due to their higher emission efficiencies [HUD 11, NAC 11]. However, in these structures, non-radiative recombination and free carrier absorption remain high and prevent efficient device operation [ZHO 15]. This limitation was solved, for instance, using lateral injection with the demonstration of electroluminescence [KRZ 12]. But still, laser demonstration with this approach was performed only under optical pumping, with ultrahigh Q-factors cavities, in microdisks, microspheres or toroidal geometries to compensate the low gain of the devices [YAN 03, POL 04, FAN 13a]. Net positive optical gain was not measured with this approach under electrical injection.

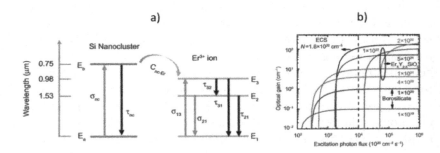

Figure 2.4. a) Energy levels of Er and silicon Si-nc in Er-doped Si-rich materials. Si-ncs and Er3$^+$ are modeled as a two and three level system, respectively. C_{nc-Er} is the transfer coefficient between Si-ncs and Er^{3+}. Reprinted with permission from [MIL 10], Copyright 2010, AIP Publishing LLC. b) Pump-dependent optical gain of different Er-materials. Reprinted with permission from [YIN 12]. Copyright 2012, AIP Publishing LLC. For a color version of the figure, please see www.iste.co.uk/cornet/lasers.zip

To further increase the erbium concentration and therefore the optical gain, nearly stoichiometric erbium compounds were also developed. These alloys include erbium silicates and erbium chloride silicates (ECS), together with the addition of ytterbium and yttrium ions. Photoluminescence properties were strongly improved, benefiting from the activation of most of the Er ions being present in high density, [MIR 07, SAV 08], from the up-conversion limitation, and absence of Er concentration quenching [SUH 10, VAN 10b, WAN 11b]. With these high-density erbium compounds, different groups have succeeded in measuring optical gain, in a variety of alloys, as illustrated in Figure 2.4(b), and already widely commented in [FAN 13b] and [YIN 12], the most promising compounds being the ECS with an Er atoms density around 10^{22} cm^{-3}. Positive material gain has also been measured under electrical injection [WAN 13a] in recent studies. Nevertheless, laser under electrical injection has not been demonstrated yet, one of the main problem being the dielectric properties of these compounds limiting the carrier injection and increasing the threshold voltage of the device [FAN 13b].

Therefore, developing an electrically pumped on-chip laser with rare-earth Si or SiO$_2$ doping remains challenging, because it requires the development of efficient active area (i.e., with a high density of activated and uniformly distributed rare-earth atoms), together with the control of the electrical injection (which is not favorable especially in the case of Er compounds). Recent works show, however, encouraging results and may provide CMOS-compatible solutions [ZHO 15].

2.6. Group IV SiGeSn alloys for lasing

One of the most promising solutions to reach efficient laser emission on silicon is to modify the bandstructure and bandlineups of group IV semiconductors by alloying Si with either Ge or Sn in order to get a direct bandgap alloy. But this is not straightforward, as Ge and Sn have respectively a 4.2 and 19.5% lattice mismatch with Si. Such a lattice mismatch inevitably leads to both roughening of the layers due to strain release and appearance of dislocations if the thickness deposited is typically above dozens of monolayers (see [POH 13] and Chapter 4 for more details). These crystal defects, if propagating in the laser device, act as non-radiative recombination centers, and are of course detrimental for laser operation. Nevertheless, various strategies have been proposed to develop a SiGe(Sn) laser on Si.

2.6.1. *SiGe quantum cascade lasers*

Si and Ge are both semiconductors with an indirect bandgap (X-like fundamental transition for the Si, and L-like fundamental transition for the Ge), but with different bandgap energies (1.12 eV for Si and 0.66 eV for Ge). One way to overcome this limitation is precisely to avoid the use of the interband transition in these materials. To this aim, the quantum confinement imposed by the difference in bandgap energies between Si and SiGe can be used to promote intraband optical transitions, and if the oscillator strength of the involved electronic states transitions are large enough, no limitation to reach lasing conditions exists [SOR 97]. This is the basic principle of the so-called quantum cascade lasers (QCL), which were demonstrated 20 years ago in III–V semiconductors, especially for the mid-IR emission, between 2.5 and 20 µm [BAR 15, GMA 01].

The principle of the device is shown in Figure 2.5(a). The structure is composed of many alternated SiGe(lower bandgap)/Si(larger bandgap) layers, creating delocalized minibands within the conduction or the valence band caused by quantum coupling of the involved states. These minibands allow carrier injection and collection in the structure. A bias is applied to provide a staircase potential profile, so that minibands are not aligned. An intraband optical transition can consequently occur in the active area, and the electromagnetic wave can further be amplified by stimulated emission. Both valence band and conduction band quantum cascade designs can be imagined. But in the specific case of SiGe QCL, the lattice mismatch between Ge and Si does not allow to design a cascade relaxation in the conduction band while staying pseudomorphic. Therefore, quantum cascades are usually considered in the valence band, i.e., with holes [GMA 01]. This is one of the main drawbacks of this approach, as it provides larger effective mass, and a weak confinement potential, which is around 80 meV per 10% Ge concentration [PAV 03]. Nevertheless, electroluminescence around 0.15 eV from SiGe quantum cascade structure grown on Si was demonstrated, and is given in Figure 2.5(b-c) [DEH 00, BOR 02, PAU 10].

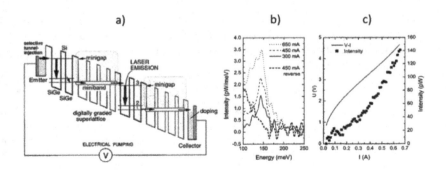

Figure 2.5. *a) Silicon-based intersubband quantum-cascade laser. Reprinted from [SOR 97], copyright 1997, with permission from Elsevier, and from [PAV 03] with permission from SPIE. b) Current-dependent EL spectra in forward bias and spectrum at reverse bias at 80 K. c) I–V curve and integrated EL intensity. Reprinted with permission from [BOR 02], copyright 2002, AIP Publishing LLC*

Therefore, promising results have been obtained toward the realization of SiGe quantum cascade lasers on silicon. But this approach still faces some material issues, imposed by the large lattice mismatch between Ge and Si, and the needs to use the valence band as a cascade path for holes. Efficient lasing on silicon based on this approach still needs large improvements, which are presently under scrutiny in different groups [PAU 10]. We note that this strategy is of high interest for THz emission on silicon [PAU 10], but is unlikely to be adapted to very large-scale optoelectronic integration, as the wavelength produced would require very large device footprints.

2.6.2. *SiGe lasers*

As stated previously, Ge is an indirect bandgap semiconductor, with its minimum of energy in the L-valley, which is responsible for its long radiative lifetime, and thus preventing bulk Ge from efficient lasing. The energy separation between its Γ and L transitions is, however, as small as 136 meV [LIU 14b]. On the other hand, it is well known that strain modifies the bandstructure of semiconductors, while shifting the energy positions of the different valleys (X, L, Γ). An original idea is to apply tensile strain on the Ge, in order to reduce the overall bandgap of Ge and move the Γ-band minimum closer to the L-band one, as depicted in Figure 2.6(b), with the aim to transform the Ge into a possible direct bandgap emitter. The crossover is reached when Ge is tensile-strained at 2% [LIU 07].

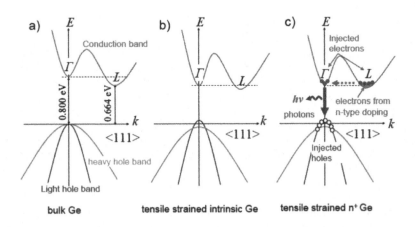

Figure 2.6. *a) Schematic band structure of bulk Ge, showing a 136 meV difference between the direct gap and the indirect gap. b) The difference between the direct and the indirect gaps can be decreased by tensile strain. c) The rest of the difference between direct and indirect gaps in tensile strained Ge can be compensated by filling electrons into the L valleys via n-type doping. Because the energy states below the direct Γ valley in the conduction band are fully occupied by extrinsic electrons from n-type doping, injected electrons are forced into the direct Γ valley and recombine with holes, from [LIU 14b]. Reprinted with permission under CC BY 3.0 license*

Various strategies have been recently tested to increase the level of tensile strain imposed to the Ge layers. The thermal mismatch between Ge and Si was used to produce a 0.3% tensile strain [ISH 03, LIU 04a, HAR 04]. Different buffer layers (e.g., GeSn) or stressor layers (e.g., nitrides) were also proposed [LIU 04b, FAN 07, TAK 08, KUR 11, KER 11, CAP 14]. MEMS have also been proposed to externally control the strain state of the Ge up to 1% of biaxial tensile strain with an increasing of the photoluminescence by a factor of 260 [JAI 12]. More recently, Süess *et al.* achieved a 3.1% uniaxial tensile strain, with a suspended microbridge [SÜE 13]. Finally, a 5.7%

uniaxial tensile strain was reported in recent studies, thus opening the way for true direct bandgap emission [SUK 14].

But still, in most of the works reported here, the direct transition lies just above the indirect L one. Moreover, if the strain is too large, the bandgap energy of Ge becomes smaller than the 0.8 eV target (i.e., 1.55 μm wavelength), which is already well established in silicon photonics technologies. When the tensile-strain is not large enough to reach the indirect–direct crossover, very high n-doping (typically 10^{19} to 10^{20} cm^{-3}) is used to reach lasing or at least efficient optical emission. As described in Figure 2.6(c), n-doping provides enough electrons to fill the lower L-states of Ge. Any subsequently injected electron (pumping of the device) may then reach the next available low-energy available, namely Γ states, giving rise to efficient recombination, and possible stimulated emission (for a detailed explanation, see [LIU 14b]).

a)   b)   c)

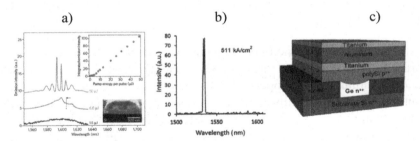

Figure 2.7. a) Optically pumped Ge-on-Si laser demonstrating CW operation at room temperature. Edge-emission spectra of a Fabry–Pérot Ge waveguide under three different levels of optical pumping from a Q-switched laser at 1,064 nm with a pulse duration of 1.5 ns and at a repetition rate of 1 kHz is shown. Top inset: integral emission intensity from the waveguide facet versus optical pump power, showing the lasing threshold. Bottom inset: cross-sectional SEM image of the Ge waveguide. Reprinted with permission from Optical Society of America, from [LIU 10]. b) Ge laser emission spectrum after threshold. c) The cavity length of the waveguide is 333 μm and the waveguide height about 100 nm. Current injection employed pulse widths of 50 μs at 800 Hz and 15°C. Reprinted with permission from Optical Society of America, from [CAM 12]

With this approach, the nature of the direct or indirect bandgap optical transitions were first discussed, and intense photoluminescence and electroluminescence were obtained [SUN 09a, SUN 09b, CHE 09]. A net positive optical gain (>25 cm^{-1}) was claimed, as determined by pump–probe experiments [LIU 09]. Finally, an optically pumped Ge-on-Si laser operating at room temperature was demonstrated in 2010 [LIU 10a], and an electrically pumped device was demonstrated in 2012 near the targeted 1.5 μm wavelength, as shown in Figure 2.7 [CAM 12]. This result was obtained with a thermoelectric cooler regulating at 15°C, not far from room temperature. At this temperature, a lasing threshold of 280 kA.cm^{-2} is determined under electrical injection of 40 μs electrical pulses at 1,000 Hz. Finally, output powers greater than 1 mW were obtained [LIU 12]. We note that beyond this work, an optically pumped laser with amorphized Ge/Si quantum dots was also demonstrated very recently, [GRY 16] which could provide many advantages in future devices [LED 03].

Overall, even if some improvements are needed in this approach, in order to reduce the laser threshold significantly [KOE 15] and operate under continuous wave operation, the demonstration of an electrically pumped laser diode fully integrated on silicon and operating close to room temperature is one of the best laser achievement with group IV semiconductors, and is certainly an interesting path for very large scale integration of photonic sources.

2.6.3. SiGeSn lasers

An alternative technique to achieve a direct bandgap material is to incorporate Sn atoms into a Ge or SiGe matrix. This is expected to reduce the gap at the Γ-point more rapidly than that at the L- or X-point. For a sufficiently high fraction of incorporated Sn, one can expect an indirect-to-direct bandgap transition crossover. This indirect-to-direct transition for relaxed GeSn binaries has been predicted to occur at around 20% Sn, as determined by tight-binding calculations dozens of years ago [JEN 87]. This prediction was recently revised by means of pseudo-potential calculations, and the indirect to direct crossover was computed to arise in the [6–11%] Sn content range

[LOW 12, GUP 13]. Recent results show experimentally that the $Ge_{0.9}Sn_{0.1}$ alloy with a slight compressive strain (1%) present direct bandgap optical behavior [GHE 14]. Despite the promises of this approach, the metastable characteristics of GeSn alloys are by itself a challenge for such integration. Indeed, the Sn equilibrium solubility in Ge is low (<1%) [HE 97].

a)   b)   c)

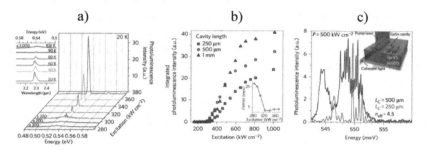

Figure 2.8. *Optically pumped GeSn laser. a) Power-dependent photoluminescence spectra of a 5 µm wide and 1 mm long Fabry–Perot waveguide cavity. Inset: temperature-dependent (20–100 K) photoluminescence spectra at 1,000 kW.cm^{-2} excitation density. b) Integrated photoluminescence intensity as a function of optical excitation for waveguide lengths L_C = 250 µm, 500 µm and 1 mm. Inset: FWHM around the lasing threshold for the 1 mm long GeSn waveguide. c) High-resolution spectra of 250 µm and 500 µm long waveguides taken at 500 kW.cm^{-2}. The mode spacings are 0.50 and 0.27 meV, which correspond to a group refractive index of ~4.5 for the lasing mode. Reprinted by permission from Macmillan Publishers Ltd: Nature Photonics, from [WIR 15], copyright 2015. For a color version of the figure, please see www.iste.co.uk/cornet/lasers.zip*

As stated above, the 15% mismatch between Ge and α-Sn is also challenging. Hence, strategies have been adopted to obtain partially and also fully relaxed GeSn layers on Si [GRZ 12] and on InGaAs [CHE 11a]. Especially, the partial relaxation of GeSn on a Ge/Si(001) virtual substrate was proposed to reach the direct bandgap emission [WIR 13a, WIR 13b]. The strategy adopted in this work is to first relax a large part of the mismatch through the deposition of a thick Ge epilayer on Si, which allows us to bury most of the crystal defects (especially dislocations), and then to control the partial relaxation of GeSn on Ge. With this approach, an optically pumped GeSn laser was demonstrated with a ridge geometry, operating below 90 K, emitting around 2.3 µm (0.55 eV) (see Figure 2.8, from [WIR 15]). Also, the microdisk geometry was considered by the same group, and lasing under optical pumping was also achieved at low temperature [BUC 15]. Electroluminescence was also demonstrated with the same kind of strategy, at room temperature [SCH 15, GAL 15].

Finally, this approach, even if very recent, has shown promising device realizations, but material quality, doping, and bandstructure will have to be carefully optimized to provide efficient electrically injected laser devices with low thresholds [HOM 15]. The authors point out that the emission wavelength (2.3 µm) is compatible with the propagation in SOI waveguide, but may also be of interest for gas detection.

III–V Lasers Bonded on Si

In this chapter, we present the different III–V laser device groups that are heterogeneously integrated on silicon through the bonding techniques. If they benefit from the state-of-the-art performances of group III–V laser devices, they also face technological issues, as well as integration constraints.

3.1. Introduction

The various strategies presented in Chapter 2 to develop a laser using silicon-based group IV materials were mainly motivated by their straightforward CMOS compatibility as proved by the role they played in the development of mature building blocks for silicon photonics (modulators, photodiodes). However, none of these strategies has led to the demonstration of an efficient laser device yet. As discussed previously, the main reason is the inherent limitation of these materials to efficiently emit light (Si and Ge being indirect bandgap semiconductors). By contrast, III–V compounds based on InP and GaAs are direct bandgap semiconductors widely used in commercial laser devices for more than 50 years. The approaches to integrate a III–V laser structure on a silicon substrate have been following two main directions: the monolithic integration approach (direct growth of III–V materials on silicon), which will be presented in Chapter 4, and the bonding approaches now detailed in this chapter.

3.2. Historical flip-chip bonding technology: advantages and drawbacks

The concept of heterogeneous integration of optoelectronic devices on a foreign substrate (silicon) by bonding is not new. Indeed, a large amount of work was performed during the early 1990s to improve the flip-chip solder bonding technique [WAL 90, IM 92, JAC 94] developed by IBM in the late 1960s for the heterogeneous assembly of electronic devices [MIL 69]. The basic fabrication sequence is presented in Figure 3.1. Both host substrate (in our case Si) and the integrated device (laser) are first lithographically patterned with wettable metallic pads (Cr, Ti, Pt, Au) deposited on top of a non-wettable material (dielectric, for example). Then, a solder material (AuSn, PbSn) is deposited on top of the metal pad and reflowed above the melting temperature to form bumps. The integrated device and the substrate are then put in contact with a rough alignment. Because of surface tension forces, both parts tend to self-align (see Figure 3.1(e)). Finally, the structure is cooled down to form a mechanically robust, electrical and thermal connection between the substrate and the device.

The self-alignment is a very important feature of this technique as the coupling between optical devices and waveguides is crucial to limit optical losses. Although a ±1 μm alignment has been claimed [LIN 97], this is generally not ideal as a sub-1 μm tolerance is generally preferred. Second, the spacing between the III–V device and the substrate is generally large due to the thickness of the metal bumps. This also limits the possibilities to couple the light into photonic components situated on the substrate. Moreover, the flip-chip solder bonding technique is complicated, costly and limits the integration density by the size and the pitch of the solders. Finally, because it's a die-per-die process, the technique is too time-consuming to be used in low cost mass production lines.

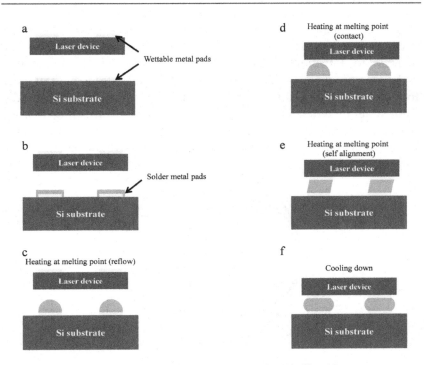

Figure 3.1. *Fabrication sequence for the flip-chip solder bonding technique*

3.3. Die versus wafer bonding

Even though the bonded device tends to self-align on the metal bumps in the flip-chip solder bonding technique, the precision needed to prealign each device individually is still complex. A good way to circumvent alignment tolerance is to bond full wafers or dies of III–V material and process the devices after bonding. In these cases, the alignment procedure is simply transferred to the standard lithographic tools.

Wafer bonding is probably the fastest solution to fabricate multiple III–V lasers on Si in parallel and presents the advantage of requiring standard equipment for handling and cleaning wafers before bonding. Nevertheless, it suffers from some drawbacks:

– The standard size of Si or silicon on insulator (SOI) wafers in CMOS foundries is nowadays 300 mm diameter and may soon increase to 450 mm. Due to their higher brittleness, GaAs and InP wafers are produced with a smaller diameter (generally 6 in. (150 mm) for GaAs and 4 in. (100 mm) for InP). This mismatch (see Figure 3.2) leads to an unavoidable loss of Si material. Moreover, InP and GaAs epilayers are generally much costlier than Si, so that bonding several III–V wafers on a single Si wafer results in a massive overcost.

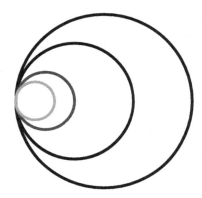

Figure 3.2. *Relative size of 4 and 6 in. III–V wafers (grey lines) and of 300 and 450 mm Si wafers (black lines)*

– Bonding a full costly III–V wafer results in an inefficient use of the III-V material as most of it is lost during the device processing.

– Both wafers must be perfectly clean before bonding to avoid defects such as bubbles. The tolerance is very strict as a particle with a diameter of 1 μm can cause a bubble with a lateral diameter of several millimeters [GÖS 98, PAS 02].

– Efficient laser devices generally require an epitaxial structure with strained multi-quantum wells (InGaAs(P)/InP system, for example). This can result in a large III–V wafer bowing that makes the full wafer bonding more complicated.

A lot of these drawbacks can be avoided by bonding small III–V dies instead of full wafers. Indeed, III–V dies (minimum size 1 mm^2) can be diced and bonded individually at specific positions on the Si wafer (see Figure 3.3). This approach does not suffer from the wafer-size mismatch and thus results in a smarter use of both III–V and Si material. Moreover, it is less sensitive to the III–V wafer bowing as the bonded surface is much smaller. Also, different dies of different epilayers can be bonded on the same Si substrate. Nevertheless, even if a single defect should affect a single die and the yield is larger than with the full-wafer bonding technique, the dicing step can result in small debris and damaged die edges responsible for imperfect bonding. Finally, the time required to handle and bond individually each die (around 30 s/die depending on the transfer technique [LUO 15]) increases the cost and can affect the reproducibility from die to die.

Figure 3.3. *1.2 mm × 1.2 mm, 200 μm thick InP dies bonded on a 200 mm CMOS wafer by CEA Leti (as described in [KOS 05, KOS 06]). Reprinted from Kostrzewa et al. [KOS 06], with permission from Elsevier. For a color version of the figure, please see www.iste.co.uk/cornet/lasers.zip*

Another technique, called massive transfer printing, will be discussed in the following and can avoid the time-consuming individual die bonding. Indeed, by using a patterned elastomeric stamp, thousands of III–V coupons can be transferred in parallel on the Si substrate [JUS 12]. Moreover, the III–V coupons are not diced but defined by etching processes before bonding improving the surface and the edge quality and reducing the size of the III–V material (down to a few micrometers if needed), which results in a strong improvement of the device density.

In the following, we give more technical details on the two main massively parallel techniques able to fabricate III–V lasers bonded on

Si: the full wafer bonding technique and the massive transfer printing technique.

3.4. Basic principles of wafer bonding

The wafer bonding technique consists of contacting two mirror-polished surfaces of two different wafers in order to create covalent bonds between them. Several approaches have been developed but in this part we only focus on the three main methods that are relevant for the development of III–V lasers on Si: the molecular bonding, the polymer-assisted bonding and the metal-assisted bonding.

3.4.1. *Molecular bonding*

The basic sequence is presented in Figure 3.4. To understand the III–V to Si wafer molecular bonding technique, we first explain the simpler example of Si to Si molecular bonding.

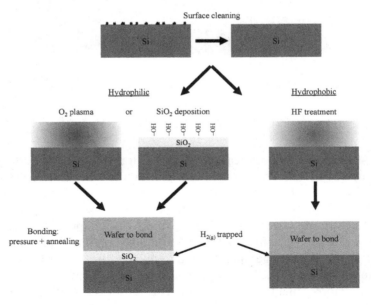

Figure 3.4. *Fabrication sequence for the molecular wafer bonding technique, describing three main approaches (O_2 plasma-assisted bonding, SiO_2-assisted bonding and bonding of hydrophobic surfaces)*

– The first step always requires a rigorous cleaning of the surfaces. This is generally done by a combination of solvents and eventually acids or bases to remove any particles that could be responsible for formation of voids during the bonding step.

– Then the surfaces are treated to be either hydrophilic or hydrophobic. In hydrophilic treatments, surfaces are terminated by hydroxyl bonds –OH. This can be obtained by dipping the wafers in solution such as $NH_4OH:H_2O_2:H_2O$, by exposing the surface to an oxygen plasma or by depositing or growing an oxide layer. In contrast, after hydrophobic treatments (HF dipping, for instance), oxides are completely removed and surfaces are terminated by –H bonds.

– The treated surfaces are then contacted and bonding occurs first through weak Van der Waals forces. In the hydrophilic case, hydroxyl groups start then bridging to form a thin water layer:

$$Si–OH + OH–Si \rightarrow Si–O–Si + H_2O$$

H_2O then oxidized Si through the reaction:

$$Si + 2H_2O \rightarrow SiO_2 + 2H_{2(g)}$$

In the hydrophobic case, the bonding occurs through the hydrogen bridging mechanism:

$$Si–H + H–Si \rightarrow Si–Si +H_{2(g)}$$

In both cases, H_2 gas is created at the interface and is commonly responsible of bubbles. The role of the next step (annealing) is to make the hydrogen diffusing in the Si and to increase the bonding strength to realize true covalent bonds through the chemical reactions described above. In the case of Si to Si wafer bonding, the annealing is generally done at very high temperature (>1,000°C). Air trapping can also be responsible for weak bonding. Thus, bonding and annealing in a controlled atmosphere or in vacuum and with a controlled pressure between the two wafers generally improves the bonding quality.

A strong difference between the hydrophilic and the hydrophobic bonding schemes is the nature of the final bond. In the case of hydrophilic surfaces, there is necessarily an oxide layer at the interface, which can be an issue when one requires driving a current through the junction. In the case of hydrophobic surfaces, the interface is oxide-free and driving electrical current has been demonstrated [TAN 12]. Nevertheless, because the bonding is not assisted by the thin water layer, it is more sensitive to the surfaces roughness and microdefects [PAS 02]. Thus, it is generally more difficult to bond full wafers with hydrophobic surfaces and this is generally employed only for die bonding [TAL 13].

Bonding a III–V wafer to Si follows the same idea with one main difference: the annealing temperature. Indeed, the very high-temperature annealing cannot be used for three main reasons. First, if the wafers are processed before bonding, they are generally incompatible with very high-temperature annealing. Second, As in GaAs or P in InP would significantly desorb at 1,000°C and finally the thermal expansion coefficient of III–V material is around twice that of Si, which would result in wafer cracks. Other strategies have thus been developed to get rid of the bonding gas by-products while keeping low-temperature annealing.

The strategy mastered by the CEA Leti and their collaborators in France and by the A*STAR in Singapore consists in using a thick (several hundreds of nanometers) SiO_2 interface layer [LUO 15, SEA 01, HAT 06]. In this technique, SiO_2 is generally deposited by PECVD on both wafers and eventually planarized by chemical–mechanical polishing to achieve subnanometer RMS roughness. During the bonding, the role of the SiO_2 is to absorb gas by-products and thus to avoid the bubbles at the interface while keeping moderate annealing temperature (200–300°C). However, added to the impossibility to drive electrical current through the insulating interface, such a high dielectric thickness degrades thermal properties and limits light-coupling scheme possibilities between the III–V layers and the Si waveguides. This last drawback can be circumvented by using a thinner oxide layer (below 100 nm) as demonstrated by the works of the UCSB group [LIA 08b, ROE 10].

The second strategy consists of driving the gas by-products out of the interface by etching some channels in the Si layer. Kissinger *et al.* [KIS 91] first demonstrated improved Si wafers bonding by etching shallow horizontal grooves up to the wafer edge in one of the Si wafer prior to bonding (see Figure 3.5(a)). The idea was then adapted by the UCSB group to the bonding of InP wafers on SOI [ROE 10, LIA 08a]. In particular, they demonstrated bonding of a 150 mm InP wafer on SOI by etching holes in the top Si layer to create vertical outgassing channels toward the buried oxide (BOX) layer of the SOI wafer (see Figure 3.5(b)) [LIA 09a]. During bonding, the gas by-products are thus absorbed by the BOX so that only a low temperature 300°C annealing is required.

a)   b)

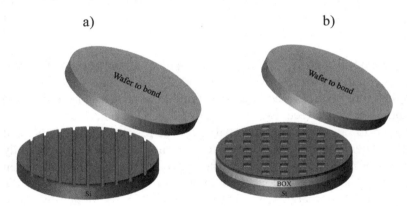

Figure 3.5. *a) Etching of shallow trenches (typical width 2–10 μm, typical depth 100–500 nm, typical pitch 100–500 μm) to evacuate bonding gas by-products as proposed by Kissinger et al. [KIS 91]; b) etching of vertical outgassing channels (typical size 8 × 8 μm², typical pitch 50 μm) in the SOI wafer as proposed by the UCSB group [LIA 09a]*

Finally, important progress has been made by treating the surfaces with O_2 plasma [PAS 02] and this is now a standard step used by every research group to bond III–V and Si wafers. Indeed, the plasma treatment has two effects. First, it brings a cleaning finish to the surface as O_2 plasma is very efficient in removing organic species, which could have been difficult to remove with the first cleaning steps. Second, it leaves a very thin oxide on the surface with a strong hydrophilic property. Thus, spontaneous bonding at room temperature occurs between both surfaces treated by the plasma. If a post-annealing step is still recommended to achieve a strong covalent bonding, this can be done with a low thermal budget (<300°C).

3.4.2. Polymer-assisted bonding

A second well-proven technique to bond III–V wafers to Si to fabricate lasers is the adhesive wafer bonding and, more specifically, the polymer-assisted bonding. In this technique, mastered by a group at the IMEC-Ghent University in Belgium, an intermediate layer of divinysiloxane-bis-benzocyclobutene is used to glue the III–V wafer (or III–V dies) to a Si (or SOI) wafer.

Benzocyclobutene (BCB) is a popular thermosetting polymer commercialized by the Dow Chemical Company and used in many applications (e.g. bonding, passivation). In a standard BCB-assisted bonding process (see Figure 3.6) [STA 13], the surfaces are first cleaned to remove any particles that can be responsible for unbonded areas. Then, the BCB is spin-coated on one of the wafers (generally the Si or SOI wafer). Prior to the BCB spin coating, an adhesion promoter can be spin-coated to improve the adhesion between the BCB and the semiconductor. Then the solvent is evaporated by a soft baking (typical temperature 150°C) in order to avoid bubbles during the bonding. The two wafers are then brought into contact while still at 150°C and a uniform pressure is applied. The next crucial step is the curing (polymerization) of the BCB. Polymerization is a time- and temperature-dependent process in which BCB transforms from liquid to solid phase. After curing at 250°C for 1 h in an oxygen-free controlled atmosphere, the two wafers are cooled down and form a solid stack ready for further processing.

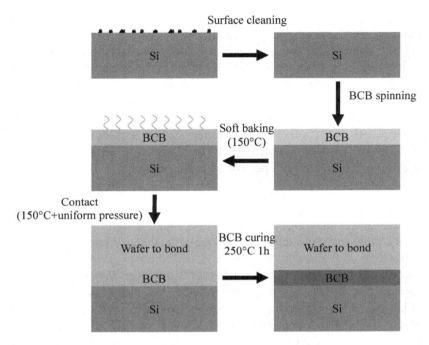

Figure 3.6. *Fabrication sequence for the BCB-assisted wafer bonding technique*

A strong advantage of BCB is that the polymerization reaction does not produce any gas by-products so that reaching a void-free bonded interface does not require the complex processing steps developed in the molecular bonding method (see Figure 3.5). Secondly, BCB is known to easily planarize after spin-coating and baking which is particularly interesting when bonding patterned wafers. A second advantage of this good topography accomodation is that the cleaning procedure can be less stringent than for molecular bonding. Finally, BCB has the advantage of being compatible with most of the standard III–V processing steps (annealing, wet etching).

Nevertheless, the BCB bonding technique suffers from major drawbacks. First, the very low thermal conductivity (0.29 W.K^{-1}.m^{-1}), even lower than the thermal conductivity of SiO_2 (1.4 W.K^{-1}.m^{-1}), is a serious issue for the thermal dissipation in laser devices especially for standard BCB thicknesses (>1 μm). Second, the refractive index is slightly higher than the one of SiO$_2$ (1.54 against 1.45 at 1.55 μm), which results in a higher optical reflection at the semiconductor–BCB interface. These two drawbacks can be limited when reducing the BCB thickness [ROE 10] in the tens of nanometers range at the expense of a much lower tolerance to surface imperfections and consequently to a very strict cleaning of the surfaces. Finally, the lifetime of devices bonded with BCB is still unknown as this technique does not offer a perfect hermetic sealing.

3.4.3. *Metal-assisted bonding*

We have seen in section 3.2 that metallic solder bumps can be used to bond III–V lasers onto Si substrates with the main drawback of high cost due to the tricky alignment between the laser devices and the Si waveguides. In analogy with the adhesive polymer-assisted bonding technique, thin metal layers can also be used as a bonding interface to attach III–V dies or wafers to Si or SOI wafers and release the alignment steps to the standard lithographic tools. In the metal-assisted bonding method, a metal alloy with a specific eutectic point is deposited on one or on both wafers. When brought into contact and heated above the eutectic temperature, the metal melts and fills the voids between the wafers. When cooling back down, the metal solidifies and forms a strong bond.

The choice of the alloy depends on the eutectic point. If this temperature is too high, cracks can appear due to the thermal expansion mismatch between III–V and Si. If it is too low, it limits the thermal budget available for the postbonding processing steps. A possible alloy is AuGeNi (80:10:10), used by Tanabe *et al.* [TAN 10] to bond a GaAs laser structure on a Si substrate. In this case, AuGeNi was deposited on both wafers and bonding was performed at 300°C (for a eutectic point of 280°C) under uniaxial pressure.

The advantages of the metal-assisted bonding are the same as for the solder flip-chip bonding: in addition to a strong bond, it provides an electrical contact and a heat sink. Also, if the surfaces need to be cleaned before bonding, the tolerance is less strict than for molecular bonding. Nevertheless, the technique brings some issues: some classical metals like gold are contaminants for Si devices as they are known to easily diffuse at bonding temperatures. But more importantly, metal layers absorb light in a few nanometers, which complicates light-coupling schemes. However, solutions exist as demonstrated by Creazzo *et al.* [CRE 13] who reported bonding an InP-based laser structure in a recess of an SOI wafer to achieve butt coupling between III–V and the Si/SiO$_2$ waveguide. Another solution has been proposed by Hong *et al.* [HON 10]. Instead of depositing metal on the whole wafers, they can be deposited as AuGeNi/In/Sn metal on a selective area away from the waveguides consequently leaving the possibility of evanescent light coupling between the III–V device and the underlying Si waveguide.

The main advantages and drawbacks of the various bonding techniques are summarized in Table 3.1.

3.5. Basic principles of transfer printing

3.5.1. *Substrate removal versus epitaxial lift-off*

In the bonding techniques presented in the previous section, the III–V wafer is bonded from the frontside. To access to the III–V epilayer and to process the laser structure, the III–V substrate first needs to be removed (see Figure 3.7). This is generally done by mechanical grinding followed by selective wet etching. For this purpose, an etch stop layer is grown on the III–V substrate prior to the growth of the laser structure. One drawback of this approach is that the III–V substrate is lost during the removal. The technique is thus quite costly due to the waste of the pricy III–V material.

Bonding technique	Tolerance to surface defects	Light coupling with possible underlying waveguide	Thermal resistance	Cross-bonding electric drive
Molecular (O₂-plasma assisted)	Low	Yes	Medium to low	No
Molecular (SiO₂-assisted)	Medium for large SiO₂ thickness	Yes for thin SiO₂ thickness	High for large SiO₂ thickness	No
Molecular (hydrophobic surfaces)	Very low (Ok only for die bonding)	Yes	Low	Yes
Polymer-assisted	High for large BCB thickness	Yes for thin BCB thickness	Very high for large BCB thickness	No
Metal-assisted	High	Complicated	Very low	Yes

Table 3.1. *Assessment of the various bonding techniques*

Another approach called epitaxial lift-off is expected to advantageously save the III–V substrate and make it reusable for further epitaxial growth [CHE 13]. In this case, the epitaxial layers are released from the III–V substrate before bonding through the selective wet-etching of a sacrificial layer (see Figure 3.7). To transfer and to bond the few micron-thick released layers on Si, a mechanical support is required. Historically, the III–V stack can be stuck to a temporary substrate (glass, Teflon, Si) via the application of Apiezon wax (known as black wax). The advantage of this wax is its very high resistance to the wet etchants used for the release of the sacrificial layer [YAB 90, PAL 09] so that the III–V epilayers are fully protected. After bonding, the wax is easily removed by chlorinated solvents.

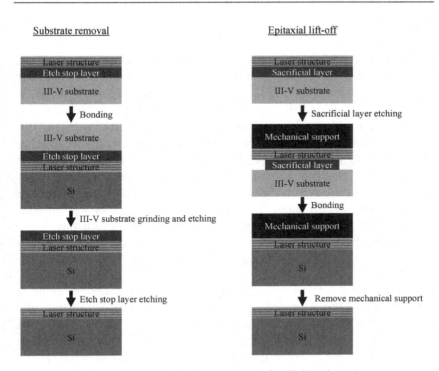

Figure 3.7. *Fabrication sequence for III–V substrate removal and the epitaxial lift-off technique*

The etching time depends on many parameters. Obviously, the etchant solution and the material of the sacrificial layer are important but the geometry of the III–V piece to release also plays a role. This is why the transfer of dies is generally preferred to full wafers. Rectangular dies with one dimension being small are easier to lift off than square dies with the same area if the lateral etching remains fast in the direction of the smallest dimension. In section 3.3, we presented the advantages and the drawbacks of die bonding compared to full-wafer bonding. In particular, we mentioned that die bonding was a time-consuming technique. We will now present a recent and promising approach, the massive transfer printing technique, which benefits from the advantages of epitaxial lift-off of III–V dies while being massively parallel.

3.5.2. *Massive transfer printing*

The basic principles of the massive transfer printing method are presented in Figure 3.8. The III–V structure consists of a laser structure grown on top of a sacrificial layer (a). Coupons of III–V material (typical size 400 µm × 100 µm [JUS 12]) are first defined through standard etching (wet or dry) of the epitaxial layers (a→b). These coupons are then secured by anchoring them to the underlying substrate with a specific resist pattern (c). This step enables us to keep the coupons attached to the substrate during the sacrificial layer undercut (selective wet etching) (c→d). In parallel, an elastomer stamp (PDMS) is prepared. This stamp is used to pick up the coupons from the III–V substrate and transfer them onto the Si substrate. Posts are designed on the stamp by casting the PDMS on a mold with a pattern corresponding to the coupons size and pitch.

The interest of using a PDMS stamp is its kinetic-dependent adhesion [MEI 06]. To release the coupons from the III–V substrate, the stamp is pressed onto the coupons (e) and peel-back quickly (f) so that the adhesion force between the stamp and the coupon is stronger than the adhesion force between the coupon and the III–V substrate. On the contrary, to print the coupon onto the Si substrate, the inked stamp is pressed onto the Si substrate (g) and peeled-back slowly (h) so that the adhesion force between the stamp and the coupon is weaker than the adhesion force between the coupon and the Si substrate. Finally, the resist is removed (i) and the printed coupons are ready for further processing (etching, contact deposition) (j).

Another interest in PDMS is its optical transparency, which facilitates optical alignment. A precise alignment of ±1 µm can be obtained for the transfer of coupons with a size of 10 µm × 10 µm. To couple the laser to a Si waveguide, sub-µm tolerance is required. This can be achieved by processing the laser devices after the transfer printing to align the III–V waveguide with the Si waveguide [JUS 12]. Full fabrication before printing has also been demonstrated [SHE 15] and can be useful when the Si wafer cannot undergo the III–V fabrication steps (etching, annealing). In this case, light coupling schemes must be adapted to release the alignment tolerance.

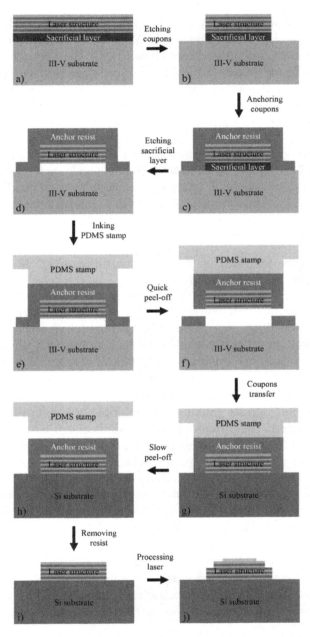

Figure 3.8. *Fabrication sequence for the transfer printing technique*

Similar to wafer bonding, various interfaces between III–V and Si can be used. Direct printing of III–V on Si is advantageous, but it requires strictly smooth surfaces [JUS 12]. On the opposite, polymer-assisted printing releases the tolerance on clean surfaces but at the expense of thermal issues [SHE 15]. Finally, a metal-based interface layer has been demonstrated to provide an efficient heat sinking compatible with good device performance [SHE 15].

Various geometries of lasers on Si can be fabricated with this method. The group in Tyndall, Ireland and their collaborators designed ridge waveguide lasers with etched facets [JUS 12, SHE 15, COR 13], but surface emitting lasers with Si photonic crystals mirrors have also been fabricated by Yang *et al.* [YAN 12].

3.6. Device structures and performances of III–V lasers coupled to SOI waveguides

In this section, we present the various laser structures that have been developed by the world leading research groups. A crucial characteristic of a III–V laser on Si is its coupling to a Si waveguide. Indeed, the laser is just one of the blocks that constitutes a more complex photonic integrated circuits (PIC). In order to integrate the laser source in such a PIC, the emitted light must be guided toward the other photonic blocks (modulators, multiplexers, amplifiers, filters, photodetectors). The different possible architectures will be discussed in detail in Chapter 5. Here we focus on the laser structures for which the coupling to an SOI waveguide has been demonstrated.

SOI waveguides are the most commonly used solution for Si photonics. A SOI substrate consists of three layers (see Figure 3.9). A buried thermal oxide layer (typical thickness between 1 and 2 µm) is sandwiched between the Si wafer and a thin Si top layer (typical thickness between 200 nm and 1 µm). Such a structure has been developed first for electronics (to isolate the transistors from the bulk silicon and thus reduce the parasitic capacitances), but its advantages for photonics are obvious. The refractive index contrast between Si and SiO_2 is very high: around 2 (3.45 for Si, 1.45 for SiO_2 at 1.55 µm). This, combined with a proper etching of the Si top layer, provides a compact optical mode confinement. Moreover, SOI patterning is perfectly compatible with CMOS technology so that fabricating complex waveguide structures on very large wafers (up to 300 mm) is convenient.

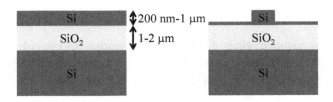

Figure 3.9. *SOI substrate and SOI waveguide*

3.6.1. *Hybrid III–V/Si lasers by molecular bonding*

In the hybrid III–V/Si platform developed by the group of UCSB, III–V layers are bonded by O_2 plasma-assisted molecular bonding onto pre-patterned SOI waveguides. The SOI waveguide and the III–V layers are specially designed so that the optical mode is shared between the III–V gain region and the Si core waveguide (Figure 3.10). Typically, a small fraction of the mode is confined into the III–V active region (a few percent) whereas up to 75% of the mode overlaps with the Si core waveguide.

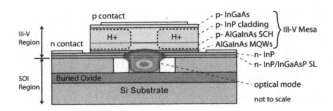

Figure 3.10. *Cross-section of the hybrid III–V/Si laser developed by UCSB. Reprinted with permission from Optical Society of America, from [FAN 06]. For a color version of the figure, please see www.iste.co.uk/cornet/lasers.zip*

A hybrid III–V/Si laser with a simple Faby–Pérot cavity was first demonstrated in 2006 [FAN 06]. The cavity was formed by dicing the processed laser and polishing the facets. A threshold current of 65 mA (at 15°C) in continuous wave (cw) pumping, a maximum output power of 1.8 mW and a maximum lasing temperature of 40°C were obtained. After the demonstration of this proof of concept, more complex cavities have been designed especially to obtain a single-wavelength laser.

Grating-based cavities are a classical option to provide facetless single-wavelength lasers. Distributed Bragg Reflectors (DBR) hybrid III–V/Si lasers have been available since 2008 [FAN 08a]. In this structure, the cavity is defined by fabricating two Bragg mirrors in the Si waveguide prior to bonding (Figure 3.11). To minimize the coupling losses between the hybrid region (where the light is evanescently coupled between the III–V gain region and the Si waveguide) and the Bragg mirror regions, the width of the III–V guide is linearly tapered (Figure 3.11). Lasing at a single-wavelength of 1,596 nm (sidemode suppression ratio of 50 dB) with a threshold current of 65 mA (at 15°C), a maximum output power of 11 mW and a maximum lasing temperature of 45°C has been demonstrated. With the same idea, distributed feedback (DFB) lasers have also been fabricated (Figure 3.12) [FAN 08b] and recently been improved [ZHA 14a] to reach threshold current as low as 8.8 mA (at 20°C), maximum output power of 3.75 mW and sidemode suppression larger than 55 dB.

For ultrashort distance optical interconnects, compact laser sources with low-energy consumption are required. Such sources have also been fabricated with hybrid III–V/Si platform [LIA 09b]. After bonding of III–V layers on SOI, both III–V and Si top layer are patterned to fabricate a III–V microring self-aligned on a Si microdisk (Figure 3.13). The typical size of the disk is 15–50 μm in diameter. Similar to the previously described structures, the microring/microdisk is designed to have a hybrid wave mode shared between the III–V gain region (ring) and the Si disk. A Si bus waveguide is also fabricated in close proximity to extract the light from the disk. With this design, threshold current lower than 4 mA at 10°C, maximum output power of 3.5 mW and maximum lasing temperature of 65°C have been demonstrated [LIA 11].

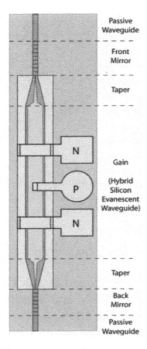

Figure 3.11. *Top view of the DBR hybrid III–V/Si laser developed by UCSB © 2008 IEEE. Reprinted, with permission, from [FAN 08a]*

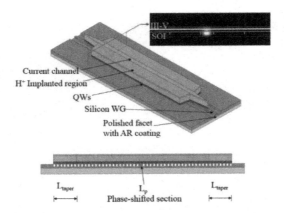

Figure 3.12. *Sketch of the DFB hybrid III–V/Si laser developed by UCSB. Reprinted with permission from Optical Society of America, from [ZHA 14a]*

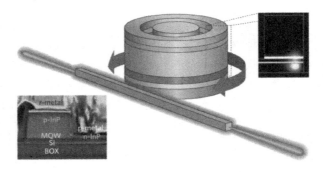

Figure 3.13. *Sketch of the hybrid III–V/Si microring laser developed by UCSB. Reprinted by permission from Macmillan Publishers Ltd: Nature Photonics, from Liang and Bowers [LIA 10]*

3.6.2. *Hybrid III–V/Si lasers by polymer-assisted bonding*

In the hybrid platform developed by the UCSB group, the optical mode is shared between the III–V and Si with only a few percent of the mode overlapping with the III–V gain region. Similar structures have been developed by the group of IMEC-Ghent University in Belgium with a polymer-assisted bonding technique.

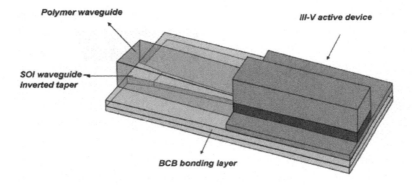

Figure 3.14. *Sketch of the III–V/Si Fabry–Pérot laser coupled to polymer waveguide and SOI waveguide developed by Ghent University. Reprinted with permission from Optical Society of America, from [ROE 06a]*

Before designing complex structures, the group at Ghent University first demonstrated a Fabry–Pérot cavity laser in 2006 [ROE 06a]. At that moment, the BCB layer was too thick (300 nm) to consider direct evanescent coupling between the III–V layer and the SOI waveguide. Thus, butt-coupling between the III–V Fabry–Pérot cavity and a polymer waveguide was used as an intermediate coupler (Figure 3.14). The coupling between the polymer and the SOI waveguides is then provided by a Si taper. Only a high threshold current of 150 mA in pulsed mode (at 20°C) was demonstrated due to the bad quality of the facets and the thick BCB layer responsible for a very high thermal resistance.

Following this first result, the technique has been considerably improved to reduce the thickness of the BCB layer below 50 nm. Stankovic *et al.* fabricated both Fabry–Pérot [STA 11] and DFB [STA 12] lasers comparable with the structures developed by Fang *et al.* with the UCSB platform [FAN 06, FAN 08a], i.e. with a shared optical mode between the III–V gain region and the Si core waveguide (Figure 3.10). They reached a threshold current down to 45 mA under cw operation for the Fabry–Pérot cavity and 25 mA for the DFB laser, output powers of several mW (5.2 mW for Fabry–Pérot cavity, 2.85 mW for DFB) and sidemode suppression of 45 dB for the DFB structure. It should be noted that the laser wavelengths are different for UCSB lasers (1.55 μm) and Ghent lasers (1.3 μm).

3.6.3. *Heterogeneous III–V/Si lasers*

One disadvantage of the aforementioned hybrid platforms is that only a small fraction (a few percent) of the optical mode overlaps with the III–V gain region. This generally leads to long cavities in order to achieve sufficient gains and thus significantly high threshold currents. To reduce threshold currents, the group of CEA Leti in France proposed to engineer the optical mode profile [BEN 11a]. With a combination of inverted tapers fabricated in the III–V materials and in the Si guide, the optical mode is mainly confined in the III–V materials in the gain region whereas it is mainly confined in the Si waveguide outside the gain region (Figure 3.15). This design has been applied with both bonding techniques (BCB-assisted [KEY 13b] and

SiO$_2$ assisted [KEY 13a]). A Fabry–Pérot cavity laser with a low threshold current of 30 mA (at 20°C), a maximum output power of 4 mW and a maximum lasing temperature of 70°C has been demonstrated by Lamponi *et al.* [LAM 12]. Single-wavelength DFB [KEY 13b] and DBR (with ring filters) [KEY 13a] lasers have also been fabricated with this technique with similar performance levels (threshold current of 35–38 mA (at 20°C), a maximum output power of 10–14 mW, a maximum lasing temperature of 60°C and a sidemode suppression ratio of 45–50 dB).

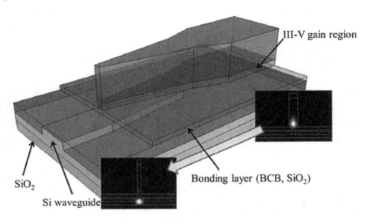

Figure 3.15. *Sketch of the coupling structure transforming the optical mode from the III–V gain region to the Si waveguide © 2012 IEEE. Reprinted, with permission, from Lamponi et al. [LAM 12]. For a color version of the figure, please see www.iste.co.uk/cornet/lasers.zip*

Recently, a complex design based on a resonant mirror fabricated in the Si waveguide was proposed [DEK 15] to avoid the complex fabrication of tapers in the III–V materials. Although only pulsed operation has been demonstrated for now, the very low threshold current of 4 mA of this structure makes it promising for low power consumption PIC.

Finally, compact microdisk lasers have been fabricated on SOI substrate by both BCB-assisted and SiO$_2$-assisted III–V bonding on

Si [ROE 10]. The structure is shown in Figure 3.16. The whispering gallery mode of the III–V disk is evanescently coupled to the underlying Si waveguide. Threshold current as low as 0.45 mA with 145 μW output power has been demonstrated with 7.5 μm diameter microdisks [CAM 08, SPU 13].

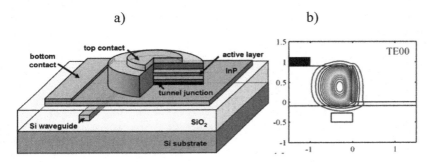

Figure 3.16. *a) Sketch of the III–V microdisk laser fabricated on top of a SOI waveguide. Reprinted with permission from Optical Society of America, from [VAN 07]; b) simulation of the optical mode in the microdisk/waveguide structure. Reprinted with permission from Roelkens et al. [ROE 10], Copyright © 2010 WILEY-VCH Verlag GmbH & Co. KGaA, Weinheim. For a color version of the figure, please see www.iste.co.uk/cornet/lasers.zip*

3.7. Conclusion

Bonding of III–V materials is currently the most developed technique for fabricating lasers on Si. This is now adapted by semiconductor companies to address the medium-volume applications of high-bandwidth data transmissions in servers and data centers [PAN 10]. However, the reliability, performance, uniformity and the costs of these approaches, especially for high-volume very large-scale integration on-chip interconnects, are still unknown [LIA 10, CHE 08].

4

Monolithic III–V Lasers on Silicon

In this chapter, we present the III–V semiconductor devices that have been monolithically integrated on silicon. Although this approach benefits from the excellent optical properties of III–V semiconductors, and a large flexibility for integration, it faces issues with regard to materials.

4.1. The monolithic integration: issues and strategies

In Chapter 3, the interest in using III–V semiconductors for laser integration on silicon was motivated by their excellent optical properties, as compared to group IV lasers (Chapter 2). Instead of integrating III–V lasers with Si by bonding approaches (Chapter 3), a monolithic integration solution can be provided by the direct crystal growth of III–V semiconductor lasers on Si. This less mature approach theoretically enables a low cost and reliable way to integrate the laser source on-chip [LEE 13]. But it also has to deal with complex material issues, which will be presented here.

4.1.1. *Heteroepitaxy: the material issues*

The direct crystal growth (often called "epitaxy") of III–V semiconductors is directly performed on the group IV silicon substrate. For more details about epitaxial processes, and the related physics, we direct the reader to [POH 13]. During this heterogeneous growth process, a number of different crystal defects can be generated at the III–V/Si heterointerface and may propagate in the whole III–V photonic structure, sometimes drastically reducing the overall laser performance (crystal defects being usually considered as non-radiative recombination centers). Therefore, the achievement of a III–V crystal epitaxially grown on silicon substrate having the same defects density than the corresponding epitaxy on the native III–V substrate is considered as the "holy grail" for most of the physicists working in the field.

4.1.1.1. *Lattice mismatch, dislocations and bandgap*

In terms of epitaxy, one of the important parameters is the lattice mismatch between the deposited material (i.e., the "epilayer") and the substrate (in the present work, the silicon wafer). Indeed, as illustrated in Figure 4.1(a), any difference in the lattice parameter between the substrate and the epilayer above a certain thickness ("the critical thickness") will lead to the formation of lines of missing atoms called "dislocations" [POH 13], which should be avoided in the optical active area, as they act as non-radiative recombination centers. Note that under well-defined intermediate crystal growth conditions, the control of the crystal relaxation can be managed through the formation of quantum dots (see Chapter 1 or [BIM 99]), which are interesting for laser devices as a gain medium, but cannot be deposited with thicknesses beyond 10 nm without generating defects.

Moreover, the crystal lattice parameter of the semiconductors varies with the temperature, which is characterized by the thermal expansion coefficient, and is different from one semiconductor to the other [POH 13]. The crystal growth of most III–V semiconductors being typically performed in the 350–1,200°C range, the lattice mismatch at growth and room temperature can therefore be very different. The epilayer can thus experience a large tensile or compressive stress, especially during the postgrowth cooling step, leading to crack formation or delamination. These effects prevent from any further laser device fabrication.

In terms of optical properties, the grown III–V semiconductor should have a bandgap adapted to the emission wavelength targeted within the application. Nitride-based semiconductors are usually considered for UV/visible optical emission, while GaP and GaAs are employed for the visible range. The near infrared emitters are often composed of GaAs- or InP-based semiconductors, and the wavelength can even being pushed to the mid-infrared range by using antimonide compounds. Also, the optical transition efficiency is an important parameter for a laser operation. This efficiency is mainly related to the bandstructure of the semiconductor (see Figure 1.6, Chapter 1). While most of the III–V semiconductors have a direct bandgap transition (i.e., the minimum of the conduction band is located at the Γ-point, and the optical transition is very efficient), few of them are of indirect type such as GaP or AlSb (i.e., the minimum of the conduction band is located at the X- or L-point of the Brillouin zone, and the optical transition is less efficient).

To summarize some of the important parameters to be considered before developing a laser device on silicon, a very common way is to represent the energy bandgap of the different III–V (binary and ternary alloys) and group IV semiconductors as a function of their lattice parameter. This is shown in Figure 4.1(b), extracted from [BAY 13]. With this figure, one can easily identify the lattice mismatch with the silicon substrate (the lattice parameter equals 5.431 Å), which is directly related to the dislocation density, the corresponding expected wavelength and whether the optical efficiency is good or not. Most of the time, there is no ideal solution and a compromise has to be found between these parameters.

4.1.1.2. Defects induced by the different crystal structure between III–V and Si

The crystal structure of both the epilayer and the substrate is also an important parameter to consider. Indeed, most of the III–V semiconductors are stable in a zinc blend (cubic) crystal phase, except for GaN-related alloys, which are also stable in ambient conditions in a wurtzite (hexagonal) crystal phase. On the other hand, group IV elements and especially Si usually present a diamond (cubic) crystal structure [POH 13].

a)   b)

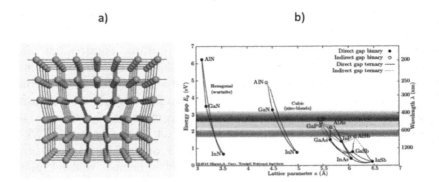

Figure 4.1. *a) Edge dislocation in a cubic crystal. Reprinted by permission from Macmillan Publishers Ltd: Nature, from McNally [MCN 13] copyright 2013. b) Band gap E_g of the technologically important III–V binaries at T = 0K as a function of lattice parameter (in-plane lattice parameter in the case of WZ nitrides). Courtesy of M.A. Caro, reprinted from [BAY 13]. For a color version of the figure, please see www.iste.co.uk/cornet/lasers.zip*

In very few specific cases, an epitaxial relationship between the bulk unstrained III–V material deposited and the substrate can occur. In this situation, some of the atoms of the epilayer crystal are well aligned with some of the atoms of the substrate, allowing a local correspondence of both the crystals. This is, for instance, the case when growing the wurtzite GaN crystal on the diamond Si(111) crystal [NAK 98].

Most of the time, the heterogeneous crystal growth of III–V semiconductors on Si has to deal with polarity inversion of the crystal. Here we will illustrate this idea with two examples. The first one concerns the growth of III–V zinc blend materials on Si (001). The zinc blend structure is polar, as every group III atom is bonded to four group V atoms. On the contrary, the diamond crystal structure of Si is non-polar, as every Si atoms is bonded to four other Si atoms. To simplify the explanation, the formation of polarity-inverted domains during the GaP semiconductor growth on Si(001) is considered in Figure 4.2(a), as the lattice mismatch between GaP and Si is very small (0.37% at room temperature) and allows getting away from dislocation issues. In the ideal case, the Si surface is atomically flat, and the first group III (Ga, here) or group V (P, here) atomic layer is fully completed, allowing a perfect stacking of the atoms, to build the III–V crystal. In real life, the situation is different. The presence of monoatomic steps at the silicon surface is inevitable, and the first deposited III–V layer cannot be controlled perfectly [KRO 87, VOL 11]. As shown in Figure 4.2(a), this induces a breaking of the translational symmetry of the crystal and generates a series of Ga–Ga or P–P false bonds in the crystal, often called antiphase boundaries (APBs). The domain of the crystal with a change of polarity is called an antiphase domain (APD). As observed in Figure 4.2(b), APBs can propagate in the whole structure, and are known to be electrically and optically active, thus degrading the overall device performance [KRO 87, TEA 14]. Moreover, it was shown that APBs induce roughness at the surface during the growth (see Figure 4.2(b)). Such roughness is prohibitive for further laser active area growth (quantum wells or quantum dots). Most of the research teams working in the field use an intentional substrate miscut (typically between 0.1° and 6°) to favor the appearance of double atomic steps at the silicon surface [QUI 13], which is known to reduce antiphase disorder but may make the postgrowth technological processing of the laser device difficult, especially for facets cleavage.

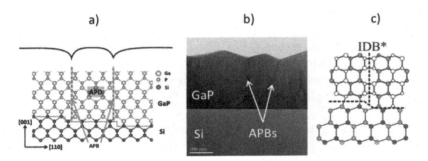

Figure 4.2. *a) Sketch of the antiphase domain (APD) formation during the GaP heterogeneous growth on a Si surface, delimited by two antiphase boundaries (APBs), inducing a roughness at the surface. b) Transmission electron microscopy (TEM) unraveling the propagation of APBs (dark lines or area) to the surface, with roughness induced. Courtesy of M. Bahri, L. Largeau, G. Patriarche, LPN. c) Schematic illustration of inverse polarity domains that coexist through an inversion domain boundary (IDB) emanating from the epitaxial AlN/Si(111) heterointerface. Reprinted with permission from [DIM 05]. Copyright © 2005 WILEY-VCH Verlag GmbH & Co. KGaA, Weinheim. For a color version of the figure, please see www.iste.co.uk/comet/lasers.zip*

The second example that can be given concerns the heteroepitaxy of the wurzite materials (AlN here) on Si(111) substrates and is intrinsically similar to the previous one. The presence of monoatomic steps at the silicon surface once again perturbs the subsequent hexagonal ordering of the III–V atoms, as shown in Figure 4.2(c), from [DIM 05]. Here, the monoatomic step induces an inversion domain boundary (IDB) in the crystal, separating an Al-polar hexagonal phase and a N-polar hexagonal phase [RAD 12]. The precise structure of IDBs is more complex than the one presented in the previous example. Indeed, the configuration of the atoms on either side of the boundary can be located at interchanged positions in the crystalline structure (such as APBs), but IDB can also leave one atomic species invariant while the other switch their tetrahedral site occupation. IDBs and their interaction with silicon steps may favor the crystal relaxation (reduce the strain) in this system. Here again, IDBs are expected to be electrically active and are considered as detrimental for device operation.

Another material issue is related to the twinning of the grown III–V crystal on silicon. Indeed, the ideal stacking of (111) planes in the zinc blend structure can be disturbed in various ways and creates area defects. This is depicted, for instance, in [GOD 04]. Especially, the crystal on one side of a plane can adjoin its twin from the other side. This is called a micro-twin (MT). The MT can also be understood as a rotation of the twin crystal by 60° or 180° around a <111> direction. A model of microtwin in GaP structure is schematically represented in Figure 4.3(a), as given by Skibitzki et al. [SKI 12].

This stacking fault (SF) is bounded by two partial dislocations. Microtwins are generally formed at the III–V/Si interface and propagate in the volume, as illustrated in Figure 4.3(b), at the GaSb–Si interface [MAD 15]. It has to be noticed that the same III–V crystal growth performed on a III–V substrate does not lead to any MT formation. Therefore, its origin can be attributed to the heterogeneous epitaxy (polar on non-polar crystal growth). To be more specific, the three-dimensional growth and the coalescence of the 3D nucleation sites were most of the time considered as the cause of their formation [ERN 88]. Nevertheless, the nucleation of III–V on Si step edges and the quality (absence of contaminants) of the initial Si surface are serious candidates to explain the appearance of these defects [QUI 13, MAD 15, WAN 15]. A clean and flat silicon surface and a careful control of the first III–V monolayer is expected to strongly influence the generation of MTs [VOL 11]. Contrary to antiphase boundaries, MTs do not perturb the subsequent crystal growth and a high-temperature overgrowth tends to naturally limit their density [XIE 90].

a)   b)

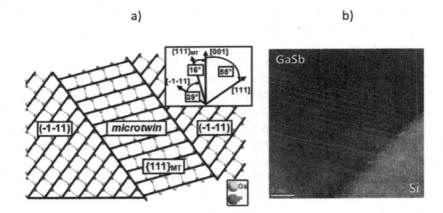

Figure 4.3. a) Schematic representation of microtwin model in zinc-blend GaP structure. Reprinted with permission from [SKI 12], copyright 2012, AIP Publishing LLC. b) High-resolution transmission electron microscopy image of the GaSb–Si interface, with a twinned domain generated at the GaSb/Si interface. Reprinted from Madiomanana et al. [MAD 15], copyright (2015), with permission from Elsevier. For a color version of the figure, please see www.iste.co.uk/cornet/lasers.zip

A final issue we would like to mention here is the appearance of a 3D crystal growth mode even at the beginning of the III–V growth. Indeed, while a conventional III–V semiconductor crystal growth on a III–V substrate leads to a perfectly smooth (2D) surface growth front, the heterogeneous epitaxy of III–V/Si always leads to a 3D island formation. Even if this 3D growth is usually attributed to crystal relaxation due to too large a stress in conventional III–V semiconductors, we note that this effect is also observed for quasi-lattice matched heterogeneous systems such as GaP/Si, [GUO 12] and

GaAs/Ge [PET 86]. Here again, the antiphase boundaries and domains play an important role and explain partially (together with the crystal relaxation, and of course interfacial III–V/Si energy) the observed roughness.

From the previous description of APDs, MTs and 3D-islanding issues, it becomes clear that the III–V/Si interface plays a fundamental role for subsequent high quality III–V overgrowth, and particularly the way silicon atoms and first monolayer III–V atoms organize themselves [SUP 15]. The driving force for this is related to the electrostatic forces induced by the different possible dipoles (Si-III, Si-V, III-V, Si-III-V) on both planar Si or Si atomic step edges [KRO 87]. For illustration, if the first atomic plane on the silicon surface is a continuous Ga plane, the two bonds with the Si lattice would introduce a donor-like charge defect with the charge density equal to 3×10^{14} cm^{-2} [KRO 87]. This huge charge would build up an electric field of 4×10^7 V/cm in the epilayer. Such a large electric field could result in a rearrangement of the atoms at the topmost Si surface and also the rearrangement of III–V atoms at the heterointerface. Therefore it explains why the control of the charge neutrality of the interface through the arrangement of Si atoms at the surface and steps, the nature of deposited III–V atoms and the initial growth conditions used during the III–V/Si nucleation has attracted much attention [BEY 12].

4.1.1.3. *Crystal growth of III–V on silicon: epitaxial setups and procedures*

In the general case, the development of III–V laser devices (on III–V substrates) requires a high-quality crystalline material, which can only be achieved with so-called metal-organic chemical vapor deposition (MOCVD, sometimes referred to as MOVPE or OMVPE), or molecular beam epitaxy (MBE) technique [POH 13, HEN 13, AYE 07].

In the case of MOCVD, the crystal is grown by using a chemical reaction between the molecules and the surface. This technique enables relatively high growth rates, at a high pressure and high temperature on large wafers. With MBE, the crystal is grown by using a physical deposition on the substrate. This technique usually has lower growth rates, at lower temperatures and lower pressures but provides a precise control of interfaces, far-from-thermodynamic equilibrium conditions, *in situ* monitoring through reflection high-energy electron diffraction (RHEED), and low residual doping. Both techniques were industrially developed for III–V laser devices, depending on the targeted applications.

Going deeper in the technological solutions, which can be provided to achieve the successful monolithic integration of III–V lasers on silicon, three main requirements are mandatory: the precise control of the silicon surface on which the III–V is to be grown, the control of the III–V growth parameters during the subsequent hetero-epitaxial growth and the control of electrical properties of the heterogeneous interface through adapted doping (this latter requirement is only needed for electrical carrier injection through the III–V/Si interface, which is not often considered).

This involves being able to grow first doped or undoped homoepitaxial silicon layers, and then to continue the growth with the deposition of doped or undoped III–V semiconductors. There is, however, an incompatibility between Si and III–V codeposition in the same deposition chamber, one being a dopant for the other and vice versa. A two-chamber strategy could then be followed to avoid a cross-contamination. Si surface is, however, very reactive. A direct exposure of the silicon surface to ambient air is catastrophic for any crystalline regrowth, because of the apparition of strong defective bonds such as Si–C or Si–N. One technological solution has been proposed in the last decades: the connection of two deposition chambers (one for the silicon, and the other one for III–V semiconductors) with a transfer area in ultra-high vacuum (UHV) conditions or in controlled atmosphere, which aims to avoid the exposure of the silicon surface to the ambient air. Given the complexity of UHV technologies, the development of such "growth cluster technology" is not straightforward but is now commercially proposed by most of the epitaxial chambers suppliers (e.g., Riber or Veeco).

From all these considerations, most of the research groups seem to consider the epitaxial strategy described below as the ultimate one. At first, good quality silicon wafers should be used (this is not straightforward as there is a large difference in quality from one supplier to the other, and because the quality depends as well on their size; usually the larger the better). These wafers often present a slight misorientation to prevent antiphase domain formation, and promote their annihilation. These wafers may be processed in a clean room, for scientists needing patterns at the silicon surface before III–V overgrowth. The silicon wafer is then usually chemically prepared to remove surface impurities. Here, there are two main strategies: (1) ending the chemical preparation by the formation of a SiO_2 oxide layer at the surface of the silicon, which can be thermally removed at a temperature above 900°C, or ending the chemical preparation with a hydrogenated Si–H surface, which can be thermally removed around 650°C. The sample may then be loaded in the Si growth chamber, and the oxide or hydrogen is removed thermally prior to the crystal growth of the buffer. The thickness of the buffer depends on the growth techniques used (i.e., MOCVD, CVD or MBE, using SiH_4 gas, effusion cells or electron guns). The sample is then transferred under controlled atmosphere or UHV conditions to the III–V chamber (MOCVD or MBE), where the growth begins. Here the control of the first monolayers (typically the first nanometer of deposited III–V material) is a key step for achievement of a high-quality III–V device on the top of it. Note that despite the two separate chamber strategy, a III–V/Si interdiffusion may occur near the interface, depending on the III–V regrowth temperature and the nature of the deposited III–V atoms.

4.1.2. *The different strategies*

The management (and reduction) of defects in III–V/Si laser devices has been widely studied in the literature. Here is an overview of the different strategies proposed to overcome the defects issue in III–V/Si optical devices.

4.1.2.1. *Defects annihilation using the pseudomorphic approach*

In order to avoid the formation of misfit dislocations, one can choose to grow a quasi-lattice matched III–V semiconductor on silicon. From Figure 4.1, this is typically the case of GaP or AlP. The very low lattice mismatch between GaP and Si (0.36% at room temperature) prevents the appearance of misfit dislocations for thicknesses deposited below 100 nm. In this approach, both MT and APB generation can be limited by using carefully adapted growth conditions. In this context, complex two-step growth strategies are usually employed, with a nucleation layer, an alternated growth sequence to keep the 2D character of the crystal growth, and a conventional GaP overgrowth for remaining antiphase domain annihilation. Several groups worldwide have shown impressive structural results. An illustration is given in Figure 4.4, where the good atom alignment between the GaP and Si is demonstrated (Figure 4.4(a)) and where the density of generated MTs is below 5 per micron, and most of the antiphase boundaries are annihilated within 10 nm after the III–V/Si interface (antiphase domains are represented by black areas in the image of Figure 4.4(b)). This approach is very encouraging for reaching a perfect crystal quality very near the III–V/Si interface, which is a great advantage for highly integrated devices on silicon, and photonic/electronic coupling. On the other hand, this approach suffers from the indirect nature of the GaP or AlP bandstructures, which limits the optical efficiency of active area for laser devices. Given the GaP and AlP bandgaps, this approach usually targets visible (around 600–900 nm) or near-infrared applications (900 nm–1.1 μm).

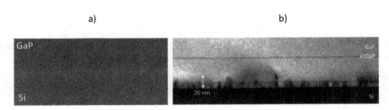

Figure 4.4. *a) High-resolution TEM image of the GaP/Si interface showing the perfect atoms alignment (below critical thickness). b) Cross-sectional dark-field TEM image of a GaP/Si sample, showing the very low MT density and antiphase boundaries annihilated at the early stages of growth (below 10 nm for most of them). Courtesy of J. Stodolna, M. Vallet and A. Ponchet, in relationship with the works presented in [WAN 15, PIN 15]*

4.1.2.2. *Management of dislocations in metamorphic approaches*

On the contrary, for III–V mismatched alloys on Si, the formation of dislocations is inevitable and becomes the main issue. In specific cases, dislocation formation can be self-organized. This is, for instance, true for antimonide compounds (lattice mismatch is around 11% at room temperature between AlSb and Si). Under well-controlled growth conditions, a planar array of 90° misfit dislocations localized at the AlSb/Si interface appears, as illustrated in Figure 4.5(a) and (b) [BAL 05, HUA 08, JAL 07]. In the ideal configuration, dislocations can be buried, and if antiphase boundaries are also annihilated, the subsequent III–V overgrowth may reach a very low density of emerging defects [HUA 08].

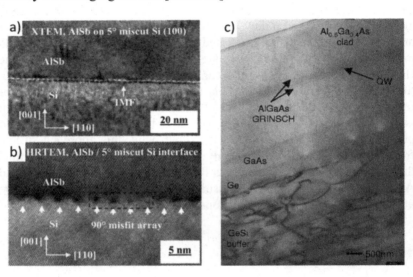

Figure 4.5. *Cross-sectional a) TEM image and b) high-resolution TEM image of the interfacial array of 90° misfit dislocations at the AlSb/Si interface. Reprinted with permission from [HUA 08], Copyright 2008, AIP Publishing LLC. c) TEM cross-section micrograph of a GaAs-based laser structure on Ge/GeSi/Si. Reprinted with permission from [GRO 03], Copyright 2003, AIP Publishing LLC*

These materials provide a number of advantages because of their very low and direct bandgap (for application typically in the 1.55 to 3 μm range). Still, the aforementioned description of the perfectly buried dislocation is far from the real situation, where the density of threading dislocations is still high. Therefore a thick III–V buffer is usually grown before the III–V laser device structure. This is one of the main drawbacks of this approach, in view of the optical coupling between the III–V device and a silicon-based photonic circuitry.

For the management of dislocations, other strategies have been studied, one of them being the progressive adaptation of the lattice parameter through (step)-graded buffers. The idea is to curve the dislocations with strain engineering, the targeted results being the lateral propagation of dislocations and the dislocation-free laser epitaxy on the top of fit. Many different groups have tried, for instance, to integrate GaAs-based laser diodes on silicon. Toward this aim, they have developed $Si_xGe_{(1-x)}/Si$ pseudo-substrates, where the final growth ends with pure Ge, and then the GaAs laser structure being quasi-lattice-matched to Ge is grown on the top of it (see Figure 4.5(c)) [GRO 03]. Another approach was proposed by using $GaAs_xP_{(1-x)}/Si$ graded buffers, where the GaP is first grown pseudomorphically to Si, and the lattice parameter is then progressively increased to reach the GaAs one [GEI 06].

Finally, with the common idea of using strain to curve dislocations and avoid their propagation in the active area, InAs/GaAs quantum dots (QDs) were proposed by Yang *et al.*, as dislocation filter, before the integration of a GaAs-based laser structure [MI 06, YAN 07].

4.1.2.3. *Defect filtering by masking techniques*

Beyond the purely epitaxial strategies described above, the management of dislocations can also be assisted by using selective area growth (SAG) and/or epitaxial lateral overgrowth (ELO) techniques. These techniques usually require additional technological steps, and especially the use of well-defined masking conditions. Here, artificial vertical or horizontal boundaries are added to the crystal with these masks to avoid the propagation of dislocations. This is illustrated in the following with two examples. The basic principle of the ELO strategy is depicted in Figure 4.6(a) [LOU 12]. A III–V seed layer is deposited directly on the silicon. This seed layer usually contains a high density of dislocations (typically above 10^8 cm^{-2}). A mask is then deposited with opened windows, and an epitaxial regrowth is performed on top of it. The growth takes place only at the opening, lattice-matched to the seed layer, and then propagates laterally on top of the masking layer, resulting in a low density of defects, as illustrated in Figure 4.6(b) [LOU 12]. Another example can be given with the SAG approach (Figure 4.6(c)). Here, a SiO₂ masking layer is first generated at the silicon surface. Small apertures (typically at the 100 nm scale) are then etched in the SiO₂ to define small nucleation areas for the III–V growth. We note here that in both ELO or SAG approaches, the size of the apertures in the mask is a critical parameter for successful III–V regrowth, and is also dependent of the technique used to perform the III–V growth. As illustrated in Figure 4.6(c), the defect necking effect [LAN 00] of the narrow holes blocks the propagation of the dislocations coming from the lattice-mismatched interface and a III–V epilayer with a low density of defects is obtained at the mask surface, on which subsequent lattice-matched growth of bulk III–V semiconductors or nanowires can be performed. In both approaches, given the needs for lateral growth in the process that arises above the mask, the III–V growth technique to be employed should not be too far from the thermodynamic equilibrium. This is typically the case for liquid phase epitaxy (LPE). Even if not ideal, MOCVD techniques also proved their efficiency for these approaches. In the specific case of MBE, which is a far-from-equilibrium epitaxial technique, this approach, even if not impossible, is harder to develop.

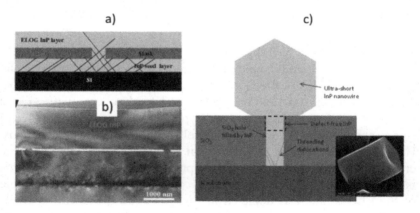

Figure 4.6. *a) Principle of the epitaxial lateral overgrowth process, the dark lines are threading dislocations. b) TEM cross-sectional view of ELOG InP on Si indicating filtering of defect from the seed layer by the mask. Reprinted from [LOU 12], Copyright (2012), with permission from Elsevier. c) Example of an InP nanowire laser monolithically integrated on (001) silicon substrate by using a SAG approach, © 2013 IEEE. Reprinted, with permission, from [WAN 13]*

4.1.2.4. *III–V nanowires/nanopillars on silicon*

The last strategy that has been extensively studied in the past years is the growth of III–V nanowires (NWs) or nanopillars on silicon. In this approach, submicronic III–V core–shell nanowires are vertically grown on silicon substrate, by using for instance droplets of Au catalyst or catalyst-free epitaxial processes [KOB 14, MÅR 04]. An illustration is given in Figure 4.7, [SVE 08] showing the good control of nanowires growth conditions, and the precise control of their positioning at the silicon surface [ROE 06]. With this technique, nanowires are usually considered to be antiphase domain-free [SVE 08]. Moreover, the small cross-section of the NWs gives an increased ability to accommodate lattice strain in semiconductor heterostructures. Especially, it is considered that NWs in axial or core–shell geometries can accommodate 5–10 times more strain than its equivalent planar epilayer, and therefore misfit dislocations are not energetically favorable for small enough radii [KAV 10]. Finally, crystal twinning is often observed in these structures, but it was found that it can be controlled with a precise adjustment of growth conditions [KAV 10, CAR 09].

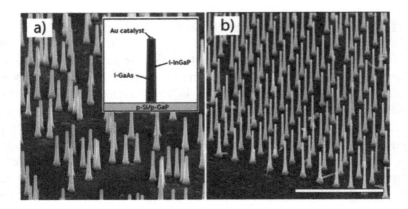

Figure 4.7. *a) Core–shell GaAs/InGaP nanowires grown on Si(111). The epi-structure is shown in the inset. b) The same epi-structure as panel a), but lithographically defined in an array with 1 μm pitch. The scale bar is 5 μm and it applies to both panels, from [SVE 08].* © *IOP Publishing. Reproduced with permission. All rights reserved*

Therefore, this approach has a lot of structural benefits and allows us to locally integrate various different efficient III–V light emitters on silicon. Nevertheless, obtaining vertical nanowires often requires the use of (111) oriented Si substrates, which are usually not convenient for further technological processes in an industrial environment. Moreover, the 3D character of the NW geometry makes technological processes for electrical injection rather complicated, [CHE 11] as compared to their 2D counterpart.

4.2. Monolithic devices

After having described the various material issues that can be encountered, and the possible strategies to overcome them when growing III–V crystals on silicon, the monolithic laser devices and their performances are described below. The classification is proposed by a class of III–V materials integrated on silicon, which roughly gives an idea of the emission wavelength (i.e., GaAs for 700 nm– 1.3 μm, InP for 900 nm–2.5 μm, GaSb for 1.55 μm to mid-infrared, GaN for visible/UV, GaP for visible to 1.1 μm range).

4.2.1. *GaAs/Si-based laser devices*

The GaAs material class was historically the first to be developed for conventional semiconductor lasers. Therefore, it is not surprising that this material was already used in 1985 to demonstrate the first room temperature electrically driven GaAs-based laser monolithically integrated on silicon [WIN 85]. In these pioneering works, lasing was only observed under pulsed operation, and the influence of both strain and crystal quality was already considered as a serious candidate to explain the limited performances of the devices [SAK 87]. Beyond these limitations, the lifetime of the device was also found to be strongly affected by the dislocation gliding, thus increasing the laser threshold and decreasing the efficiency [EGA 95]. In 1999, the first (In,Ga)As/GaAs quantum dots (QDs) laser on silicon was demonstrated by Bhattacharya and coworkers [LIN 99]. QDs, when used in the active area, allow the carriers to be localized more strongly, which, in principle, make the laser less sensitive to dislocations. This approach was further developed using QDs superlattices as dislocation filters [MI 06, MI 05]. Very promising performances were achieved, with a 900 A/cm² laser threshold under pulsed operation at room temperature, with a T_0 up to 278 K, and a – 3dB bandwidth of 5.5 GHz [YAN 07, MI 08]. Finally, these lasers were more recently integrated together with a large variety of photonic functions (for a review, see [MI 09]), and especially with a Si/SiO$_2$ waveguide, thus showing the potential of the monolithic integration definitely.

While in the previous approach, the management of dislocations was performed directly in the III–V epilayer, another approach was successfully developed by Fitzgerald and coworkers, consisting in adapting the lattice parameter with a group IV Ge/Ge$_x$Si$_{(1-x)}$/Si buffer layer of typically 10 μm thickness [CUR 98]. The GaAs is then grown on top of the surrounding Ge layers, ensuring a quasi-perfect in-plane lattice matching. Antiphase boundaries were managed through the use of 6°-off misoriented Si substrates and carefully chosen nucleation conditions. The good structural quality of the GaAs enables the development of an efficient AlGaAs/GaAs laser structure on the top of it (see Figure 4.5(c)). Therefore, lasing from this structure was observed at 858 nm under continuous wave (cw) electrical injection, with a threshold current density of 577 A/cm² [GRO 03]. The T_0 is equal to 61 K in this device. Despite the very encouraging results obtained in this work, many interesting comments were provided in [GRO 03]. At first, the difficulty to process 6°-off misoriented III–V/Si epilayers was underlined, especially because the realization of an edge emitting laser device is strongly dependent on the quality of its mirror facets, usually defined by the cleavage procedure. The 6°-off miscut tends to reduce the mirror reflectivity because of cleaved surface roughening. The performance of the laser device was also found to degrade within 20 min. A complete comparison was performed with the same laser structure grown directly on a GaAs substrate. While the laser on GaAs substrate showed no degradation of its performance, it also presents better characteristics, as shown in Figure 4.8 (threshold current density of 529 A/cm², T_0 of 128 K). The thermal management inside the laser structures was found to be responsible for this, and especially the need to design specific contact geometries for laser devices on silicon is considered in this work as an important issue.

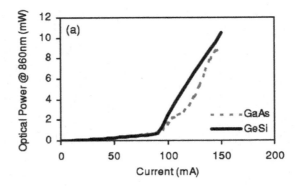

Figure 4.8. *Side-by-side light versus current characteristics for identical GaAs/AlGaAs QW lasers grown on Ge/GeSi/Si and GaAs substrates. Reprinted with permission from [GRO 03], Copyright 2003, AIP Publishing LLC*

The most advanced developments in the field of GaAs-based laser devices grown monolithically on silicon have been proposed since 2011 by the researchers of the University of California at Santa Barbara (USA) and the University College London (UK) [LIU 15]. In these works, they combine different complex growth approaches such as GeSi pseudosubstrates, optimized III–V buffer or dislocation filtering layers, and the use of (In,Ga)As/GaAs QDs as active area. The performance that has been achieved is summarized in Table 4.1.

Reference	I_{th} (mA)/J_{th} (A cm^{-2})	Maximum lasing temperature (°C)	Wavelength (μm)
Wang *et al.* [WAN 11]	1087.5/725 (Pulsed)	42 (Pulsed)	1.3
Lee *et al.* [LEE 12]	45/64.3 (Pulsed), 114/163 (cw)	84 (Pulsed), 30 (cw)	1.26
Chen *et al.* [CHE 14]	150/200 (Pulsed)	111 (Pulsed)	1.25
Liu *et al.* [LIU 14]	16/430 (cw)	119 (cw), >130 (Pulsed)	1.25
Chen *et al.* [CHE 16]	120/62.5 (cw)	75(cw),120 (Pulsed)	1.32

Table 4.1. *Performances of selected (In, Ga)As/GaAs laser devices epitaxially grown on silicon*

In the work of Wang *et al.* [WAN 11], a thick GaAs buffer layer with carefully optimized growth conditions was directly grown on Si, and QDs were further used as dislocation filters. With this structure, lasing was obtained at room temperature under pulsed electrical injection but with relatively high threshold current densities (725 A cm^{-2}). Lee *et al.* [LEE 12] then presented the first continuous-wave InAs/GaAs QDs laser diode on silicon with low-threshold current densities (as low as 64.3 A cm^{-2} under pulsed operation, and 163 A cm^{-2} under cw operation). To achieve such a great performance, they used a 2 μm-thick Ge on Si virtual substrate, and grew a conventional InAs/InGaAs QDs laser architecture on the top of it. The thresholds obtained in this work are comparable to state-of-the-art (In,Ga)As/GaAs QDs lasers on the GaAs substrate (a 10.5 A.cm^{-2} threshold current density was demonstrated with a single QD layer in [DEP 09]). More recently, Chen *et al.* [CHE 14] proposed using InAlAs/GaAs strained-layer superlattices as dislocation filters. While their device performance was not as good as the previous ones, they demonstrated the ability for the device to operate until 111°C. Liu *et al.* [LIU 14] have also used the Ge on Si pseudo-substrate approach to grow InAs/GaAs QDs laser structures. One of the novelties of this work is the use of graded index Al$_x$Ga$_{(1-x)}$As layers, together with the development of modulation p-doped barriers between the QDs. Such a strategy allowed the authors to demonstrate cw operation, with overall reasonable performances (430 A cm^{-2}), but with high-temperature operation range (lasing observed below 119°C in cw operation, and above 130°C in pulsed operation). Finally, very recent results were published by Chen *et al.* [CHE 16], which demonstrate continuous-wave InAs/GaAs quantum dots laser operation with a threshold current density as low as 62.5 A cm^{-2}, a room-temperature output power exceeding 105 mW and operation up to 120°C. In this work, over 3,100 h of continuous-wave operating data have been collected, giving an extrapolated mean time to failure of over 100,158 h. This is explained by the achievement of a low density of threading dislocations on the order of 10^5 cm^{-2} in the III–V epilayers by combining a nucleation layer and dislocation filter layers with *in situ* thermal annealing. This is obviously a huge step toward a realistic photonic integration scheme, as the lifetime of monolithic devices was often considered as one of their main limitation. This is also the best achievement with the monolithic integration approach so far.

In a more original (but less mature) approach, the work of Lu *et al.* [LU 12] can be also considered as a promising contribution in the field. Indeed, the authors use InGaAs/GaAs core–shell nanopillars as active area directly grown on silicon. The nanopillar growth was initiated by spontaneous, catalyst-free nucleation of hexagonal pyramidal nanoneedle seeds, on a non-planar or rough surface followed by layer-by-layer deposition on top of the seeds. The most remarkable idea of this approach is that the laser source was monolithically grown on (001)-silicon-based functional metal-oxide-semiconductor field effect transistors (MOSFETs), as depicted in Figure 4.9(a). The authors carefully compared the performance of MOSFETs before and after the III–V growth. They demonstrated a 3% decrease of the MOSFET threshold voltage after III–V growth, confirming the compatibility of their approach with the CMOS technology, in terms of process temperature.

From the laser device point of view, the work still needs to be developed. Indeed, in this work, the nanopillar was optically pumped under pulsed operation. Lasing at 980 nm was observed at room temperature with a pumping threshold around 5.8 kW cm^{-2} (Figure 4.9(b)). Therefore this approach needs to demonstrate its ability to manage heat (lasing under cw operation), electrical injection and solve the issue of the random orientation of nanopillars, which usually grow along the [1 1 1] directions. A possible tuning of the wavelength was demonstrated by the same authors, and the nature of the optical modes in such complex laser cavity was discussed [CHE 11]. Nevertheless, it provides solutions for very compact laser sources with a very low footprint on-chip.

4.2.2. *InP/Si-based laser devices*

Despite its very suitable bandgap for 1.55 μm wavelength emission, only recent studies have focused on the use of monolithically grown InP for lasing on silicon. In this material class, the most promising results were obtained using masking techniques/localized crystal growth. Huge material developments, including growth by itself, but also chemical mechanical polishing (CMP) techniques, mask designs and etching optimization have been performed, leading to low density of defects in

InP-based materials grown on silicon [KAT 13, MER 14]. Kataria *et al.* [KAT 13] have shown room temperature photoluminescence at 1.53 μm from InP microdisk laser structures by using the ELOG technique (see section 4.1.2.3), and identified the presence of resonant modes inside these microdisks, a promising step for further lasing with InP/Si microdisks.

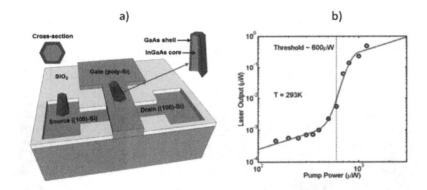

Figure 4.9. *a) Schematic of nanopillars grown on a MOSFET, from [LU 12]. Each nanopillar has a tapered hexagonal shape and consists of an InGaAs core and a GaAs shell. The nanopillars grow on both the gate and source/drain regions with a random orientation. The gate region consists of n-type doped polycrystalline silicon, while the source/drain region is made of n-type doped (001)-silicon. b) L-L curve of a nanopillar laser at room temperature. The circles are experimental data, while the curve is the S-shape fit. The threshold pump power is approximately 600 μW. Reprinted with permission from Optical Society of America, from [LU 12]*

Lasing was also recently demonstrated using InP-based nanoneedles [REN 13]. Here, the approach is similar to that developed with InGaAs/GaAs nanopillars (and performed by the same group from University of California, Berkeley). The growth of high-quality single-phase self-assembled, catalyst-free InP nanoneedles on a Si(111) substrate is achieved. The nanoneedle is then used as a cavity to obtain the laser emission at 828 nm at 4 K under pulsed optical pumping. Still, the laser performance has to be improved to manage the heat (cw operation, room temperature lasing), the electrical injection and to be transposed to the Si(001) substrate used in microelectronics. Here again, this approach would provide a very compact laser source on-chip, which is by itself interesting.

The most advanced works on monolithically integrated InP-based laser sources on silicon is undoubtedly the one proposed by the research groups of IMEC-Ghent University (Belgium). They have first demonstrated room temperature lasing with an InP nanolaser grown on Si (001) at 880 nm [WAN 13].

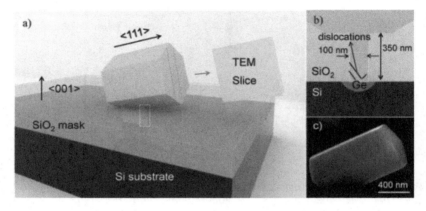

Figure 4.10. *a) A schematic configuration of the InP nanolaser epitaxially grown on (001) silicon; b) A cross-section view of the nanowire bottom (see the dashed rectangle in part a) that is connected to the silicon substrate via a SiO₂ opening and the Ge buffer. c) A tilted scanning electron microcopy (SEM). Reprinted with permission from [WAN 13], copyright 2013 American Chemical Society*

To this aim, they used a prepatterning of the silicon surface, and defined nanotrenches for defect-trapping (see section 4.1.2.3) to grow InP on a thin Ge seed layer. After defects filtering (see Figure 4.10(b)), the emerging InP is used for epitaxial lateral overgrowth of an InP nanowire (Figure 4.10(a) and (c)). With such a structure, they observe lasing at room temperature under pulsed optical pumping, with a threshold around 1.69 pJ at 293 K. Beyond the room temperature lasing, an interesting part of the work is that they observe a broad gain distribution (lasing over a range of 60 nm was achieved) attributed to the polytypic character of the grown nanowires (i.e. presence of both wurtzite and zinc blend crystal phase). The advantages and drawbacks of this approach are the same than the one previously mentioned with InP nanoneedles. But in this work, the growth has been performed directly on Si(001), a huge advantage for the development of silicon photonics. On the other hand, the issue of the (111) growth of nanowires is only partially solved as it does exist an angle between the cavity axis and the substrate surface around 35°. Therefore the light collection from such a device remains tricky.

The same research group has recently demonstrated the most advanced monolithic integration of InP-based lasers on silicon [WAN 15].

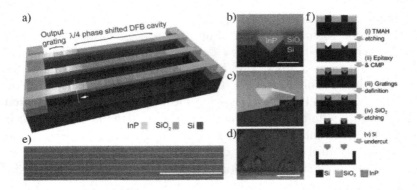

Figure 4.11. *Monolithic integration of InP lasers on silicon as proposed by Wang et al., in [WAN 15]: a) Schematic plot of the monolithically integrated InP DFB lasers on silicon. The laser cavities and the output gratings are labeled. Differently colored output gratings illustrate the tunability of the lasing wavelength. b) False colored SEM cross-section view of an InP-on-silicon waveguide. The scale bar is 500 nm. c) Schematic cross-section plot of the diamond-shaped waveguide, the position where this cross-section is taken is marked by the dashed line in panel a). d) TEM image of a specimen prepared parallel to the InP-on-silicon waveguide. The scale bar is 200 nm long. e) SEM top view of an array of InP-on-silicon waveguides. f) Integration flow used. Cross-sections orthogonal to the laser axis following each separate process step are shown. Reprinted by permission from Macmillan Publishers Ltd: Nature Photonics, from [WAN 15] copyright 2015. For a color version of the figure, please see www.iste.co.uk/cornet/lasers.zip*

As depicted in Figure 4.11(f), from [WAN 15], the InP semiconductor is directly grown on a previously patterned Si wafer.

Here, the approach combines SiO_2 masking layers, for localized InP growth, and V-grooves Si patterns for increasing the quality of the heteroepitaxial InP. The III–V growth thus occurs on Si(111) planes while the substrate remains of (001) type. This is a great advantage of the approach. With a two-step growth process on such (111) Si surfaces, the grown InP is considered as free from antiphase domains and threading dislocations within the first 20 nm (Figure 4.11(d)). A diamond-like cross-sectional shape of the InP waveguide is thus obtained. The SiO_2 is finally removed, together with the silicon substrate beneath the InP-based laser device to ensure a good optical confinement in the 500-nm-thick suspended device (Figure 4.11(a)). Of course, the main drawback of this work is the restriction of the operating conditions to pulsed optical pumping, even if 930 nm lasing is obtained at room temperature. But many aspects of the study are of great interest for further device developments. First, this approach does not require a thick InP buffer layer, a great advantage for photonic integration. Moreover, the demonstration was performed on a 300 mm MOCVD reactor, showing the very large scale integration compatibility with the CMOS industry. The in-plane laser emission is also a great advantage for further light coupling to a waveguide and the precise control of the laser cavity design offers the possibility to reach monomode laser emission. Finally, with such a design, the authors demonstrated the ability to control the emission wavelength by tuning the grating period of the distributed feedback area. This means that this technique offers possible routes for the development of multilaser source arrays integrated on a single chip. This definitely gives a powerful tool to develop wavelength-division multiplexing (WDM) on-chip, thus largely increasing the bandwidth of future optical interconnects. But still, lasers with electrical injection need to be demonstrated with InP-based materials.

4.2.3. GaSb/Si-based laser devices

As compared to GaAs or InP semiconductors, GaSb compounds for laser applications are studied by a limited number of research groups around the world. However, the excellent performances of such laser devices on silicon make this approach very promising for laser integration on-chip. Two groups have mainly contributed to the development of this research field: University of New Mexico (Albuquerque) and IES (Montpellier). The approach is based on the properties of AlSb to generate a misfit dislocation array (see section 4.1.2.2), which is mainly localized at the III–V/Si interface. The use of a thick buffer on the top of it allows reaching a good quality AlSb or GaSb material. Balakrishnan *et al.* [BAL 06] first demonstrated room temperature optically pumped vertical-cavity surface-emitting laser (VCSEL) on a Si(001) substrate. The VCSEL, composed of two vertical AlGaSb/AlSb DBR, is lasing at 1.65 µm at room temperature under pulsed optical pumping. In this work, a 2×10^6 cm^{-2} dislocation density is claimed, which is a very low value for such mismatched systems on silicon. Lasing under pulsed optical pumping was also observed at room temperature within a microresonator geometry [YAN 07].

With the same approach, the first electrically pumped edge-emitting laser was demonstrated in 2007 by Jallipalli *et al.* [JAL 07]. In this work, the device is a conventional electrically driven edge-emitting laser device with a 2 µm-thick GaSb buffer layer prior to the AlGaSb laser structure. The authors observe lasing under pulsed conditions at 1.54 µm below 77 K with a current threshold of 2 kA cm^{-2}.

The first electrically pumped laser at room temperature was finally achieved by Rodriguez *et al.* [ROD 09] with the same approach. Lasing was observed at 2.25 µm under electrically driven pulsed operation. In this work, after the AlSb nucleation layer, the authors use a 1-µm-thick GaSb buffer layer. Here, the authors point out the critical influence of the III–V/Si nucleation on the overall device performance. By developing an active area based on InGaSb QWs, the same authors achieved a room temperature electrically driven lasing under pulsed operation at 1.55 µm [CER 10]. But the threshold current density remains high (5 kA cm^{-2}). Still the paramount importance of the III–V/Si interface was underlined by the authors. The most advanced device demonstration was undoubtedly performed in 2011 by Reboul *et al.* [REB 11]. In this work, the authors demonstrate lasing at room temperature under cw electrical injection, at 2 µm in a conventional ridge laser architecture (see Figure 4.12(b)). To reach such an excellent performance, GaInAsSb QWs were used as active area, and careful band engineering was performed to ensure efficient carrier injection in different configurations.

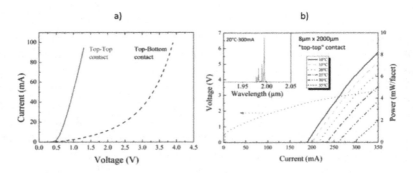

Figure 4.12. *Lasing properties of a GaSb laser diode on Si with a) I–V characteristics of 8 µm*2 mm laser diodes for the top-bottom (dash line) and top-top (solid line) contact designs. b) cw output-power characteristics at 2 µm for different temperatures and the voltage–current curve at 20°C. The inset shows the lasing spectrum taken at 20°C under 300 mA cw injection. Reprinted with permission from [REB 11], copyright 2011, AIP Publishing LLC. For a color version of the figure, please see www.iste.co.uk/cornet/lasers.zip*

A very interesting part of the paper is the comparison of two kinds of geometries for carrier injection. The first one denoted as "top-bottom" corresponds to a conventional top contact on the p-doped cladding layer, together with a bottom contact taken at the back side of the n-doped silicon substrate. In this configuration, the charge carriers have to go through the III–V/Si interface, which can be highly defective. The second one denoted as "top-top" configuration, corresponds to a conventional top contact on the p-doped cladding layer. But in this case the n-contact is directly taken in the GaSb buffer layer, thanks to a thin InAsSb contact layer, which can be reached by using adapted etching procedure. Therefore, the charge carriers do not have to cross the III–V/Si interface. The threshold current densities measured in Figure 4.12(a) (1.5 kA cm^{-2} for the top-bottom configuration, and 900 A cm^{-2} for the top-top) highlights the major role of the III–V/Si interface in the charge carrier injection in the active area. Although improvements are possible with this approach (limiting the density of threading dislocations, management of the III–V/Si interface), the excellent electrical properties of these devices should be considered as a credible route for the development of integrated laser sources on silicon: the only limitation being the needs for a thick buffer layer (typically 1 μm) before subsequent laser device development.

4.2.4. GaN/Si-based laser devices

Historically, the GaN-compound laser devices followed different developments than the other III–V laser devices. The research, initiated in the 1970s, was for a long time considered to be confidential. But the situation changed very suddenly after 2000 with the development of efficient blue lasers, and blue/white LEDs. And therefore, several research groups around the world started working in the field. Of course, its possible integration on the silicon platform was considered. Here, we would like to remind ourselves that contrary to most of the conventional III–V semiconductors, the GaN presents an equilibrium wurtzite crystal phase (see section 4.1.1.2). Therefore, controlling its crystal growth on a diamond Si (001) substrate is very tricky. On the other hand, there does exist an epitaxial relationship with the Si (111) surface, which explains why most of the works presented here have been performed on such substrate orientation. In the general case of bulk integration, the growth is usually initiated with an AlN thin buffer layer, and continued with a GaN laser structure. As in the case of Sb-based laser structures, the interaction between Al and Si is assumed to play an important role for improving the crystal quality.

Pioneering works were performed by Bidnyk et al. [BID 98] where they used selective lateral overgrowth method (see section 4.1.2.3) to form 5 μm-large GaN pyramids on a Si(111) substrate. Lasing of individual pyramids was observed at room temperature under pulsed optical pumping at 370 nm. Despite the very high lasing threshold (25 MW cm^{-2}), these results were already promising. A few attempts were also performed with as-grown (meaning here 2D growth of bulk alloys and QWs) GaN laser structures [YAB 04]. In this work, the authors use a thin AlN buffer layer and grow an InGaN/GaN QWs laser device on top of it. With this approach, lasing was also observed around 450 nm at room temperature under pulsed optical pumping

[LUT 08]. The measured laser thresholds are still high in this study (above 135 kW cm^{-2} in any cases). The relatively low performance with these approaches reveal the inherent difficulty to grow good quality GaN materials on silicon.

Thanks to the improvement of the AlN nucleation layer on silicon [SCH 06] and to the developed technological processes for realizing AlGaN microdisks [CHO 14], the first GaN microdisks laser on silicon was achieved in 2006 [CHO 06]. In this work, the authors used optimized AlGaN/GaN superlattices to improve the crystal quality of the layers, and take advantage of the reduced size of the laser cavity (diameter of 20 μm) as compared to the typical length between thermal cracks in the sample (50–100 μm), due to the thermal mismatch between GaN and Si.

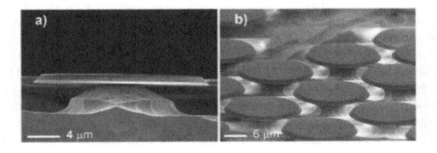

Figure 4.13. *Scanning electron microscopy images showing a) the cross section of a single GaN pivoted microdisk on Si and b) a GaN microdisk array on Si. Reprinted with permission from [CHO 06], copyright 2006, AIP Publishing LLC*

The epitaxial structure is finally etched to define the microdisk (see Figure 4.13) with a thin Si pedestal. With this device, they demonstrated the presence of whispering gallery modes and lasing around 365 nm at 4 K under pulsed optical pumping. The limited performances were partially explained by the low-quality factor (~80) resulting from technological processes. The performances of such devices were recently greatly improved by the same research group with the use of microsphere lithography [ZHA 14]. From the material point of view, the procedure is similar (i.e., AlN buffer layer and subsequent InGaN/GaN QWs laser device), but the etching process was improved to reach 2 μm-diameter microdisks. With this geometry, lasing was achieved around 440 nm at room temperature under pulsed optical pumping, with a lasing threshold of 8.43 mJ cm^{-2}. Finally, with a very similar approach, 1 μm-diameter microdisks were fabricated by using an equivalent silica microsphere approach [ATH 14]. The growth strategy remains the same (optimized AlN buffer layer and subsequent InGaN/GaN QWs laser device on the Si (111) substrate). The single microdisk laser and photoluminescence properties were carefully examined, and the whispering gallery modes were also evidenced. Lasing was observed at 514 nm under continuous-wave optical pumping, and the threshold was estimated around 1 kW cm^{-2}. The concept was recently extended to the deep UV range below 300 nm and lasing was obtained at room temperature with GaN/AlN QWs grown on Si or SOI substrates [SEL 16]. This good performance and compactness of the laser devices also confirm the great potential of the approach. However, the use of the Si(111) substrate, together with the difficulty to electrically drive these devices, remain key issues.

The last approach that was developed for integrating GaN laser devices on silicon is the monolithic growth of nanowires, as already shown in the case of InP and GaAs semiconductors. Here, the growth of catalyst-free GaN nanowires allows reaching a high crystal growth quality. By embedding a single vertical nanowire in a TiO$_2$-based 2D photonic crystal cavity, lasing was first observed at 371 nm in the structure under pulsed optical pumping conditions at room temperature, with a lasing threshold around 120 kW cm^{-2} [HEO 11]. But the most remarkable results were obtained recently with the development of electrically driven devices, and more specifically on Si(001) substrates. Frost et al. [FRO 14], succeeded in demonstrating a laser device with a quantum wire assembly depicted in Figure 4.14.

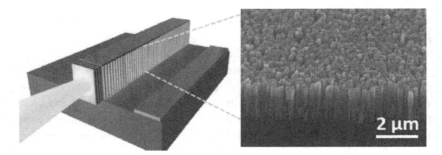

Figure 4.14. *Electrically injected GaN nanowire array edge-emitting laser on Si(001) and corresponding scanning electron microscopy of the GaN-based nanowires assembly. Reprinted with permission from [FRO 14], copyright 2014 American Chemical Society*

In these works (see [FRO 14] and references therein), the authors grow uniform arrays of catalyst-free (Al, Ga, In)N quantum wires with a reduced density of defects along the [001] direction. They manage the growth sequence to insert InGaN/GaN disks as active area, and design adapted cladding layers. The very good structural homogeneity of the quantum wires finally allows processing the whole quantum wire assembly as a conventional edge-emitting laser, with parylene as a filling medium between quantum wires. As a result, lasing is observed at room temperature under electrically driven continuous wave operation at 533 nm. The laser threshold is measured at 1.76 kA cm^{-2}, and T_0 is 232 K. These very promising performances definitely show the compatibility of the nanowires arrays laser sources for its integration on silicon. The dynamic characteristics of such lasers were even evaluated recently with slightly optimized device design [JAH 15]. It was found that such laser device can be modulated with a modulation bandwidth of 3.1 GHz, chirp values of 0.8 Å, α-parameter of 0.2 and a differential gain of 3.1×10^{-17} cm^2. These values are encouraging as they are not far from what can be measured on conventional III–V QDs laser devices. Finally, we would like to mention that there are many aspects that are still not understood. For instance, the light propagation and stimulated emission in a nanowire random cavity is not straightforward and was recently discussed in [LI 15]. In this reference, following a somehow similar approach but on a Si(111) substrate, lasing was achieved at 340 nm with a lasing threshold of 12 A cm^{-2} at low temperature (6 K). In conclusion, these realizations of quantum wires arrays devices become the most advanced demonstration of integrated GaN lasers on silicon.

4.2.5. GaP/Si-based laser devices

As described in section 4.1.2.1, the monolithic integration of GaP on silicon has many structural advantages, because of its low lattice mismatch with the substrate. Different groups have shown that most of the crystal defects can be buried very near (typically 50 nm or even below) the III–V/Si interface. But GaP intrinsically has an indirect bandgap, and this approach suffers from the relative optical inefficiency of GaP-based light emitters. Nevertheless, a laser device was demonstrated by Liebich et al. [LIE 11]. In this work, the authors use a thin GaP buffer layer, followed by BGaP cladding layers, in which a GaAsPN/BGaAsP QWs active area is embedded. Edge-emitting laser operation under optical or electrical pulsed conditions is observed below 150 K between 800 and 900 nm. Interestingly, these performances are not far from the one obtained on the native GaP substrate [HOS 11], which tends to indicate that the quality of the III–V/Si interface is not the main limitation in this system. The main issue is therefore the device design, bandlineups and the development of a true direct bandgap with GaP-based materials. Even GaAs is an indirect band gap semiconductor when strained on GaP. GaAsPN alloy has been considered but remains a pseudo-direct bandgap semiconductor with limited efficiency [ROB 12]. An interesting alternative has been proposed with the use of (In,Ga)As/GaP QDs where the hope for reaching a true direct bandgap seems more realistic [ROB 12, STR 14, HEI 14]. Even if not mature, this approach could provide an interesting solution for highly integrated devices on silicon as it does not require any thick buffer layer.

In conclusion, monolithic III–V lasers on silicon were presented here. Their yields did not reach those of the heterogeneous devices yet. However, a number of encouraging results have been obtained, especially with the GaAs-based materials. If efforts are still pursued on material research to limit crystal defects, reduce the thickness of buffer layers and increase optical efficiencies, the monolithic integration may provide a true solution in the midterm for on-chip integration.

5

Laser Architectures for On-chip Information Technologies

In this chapter, we present different laser integration architectures in view of their use for on-chip information routing and processing. Some of the advantages that may be provided by on-chip photonics are also discussed.

5.1. The role of integrated lasers in hybrid photonic–electronic chips

So far we have described the different strategies for the integration of laser sources on silicon chips. But, beyond the choice of the integration approach, the large range of integrated laser designs developed during the last few decades leads us to wonder about the laser architecture of choice for a given application. In this chapter, we will restrict our work to on-chip information technologies, even if integrated laser sources could, of course, find applications for bio- or motion sensors [SRI 14, EST 12]. We will present the different laser architectures, either integrated on silicon or with potential for integration, and extract a set of relevant parameters that allow us to discuss their advantages and drawbacks in the field of on-chip information routing.

5.1.1. *Information routing*

The convergence of integrated photonics and microelectronics has been debated for more than 30 years [GOO 84, FEL 88, MIL 00]. While microelectronics seemed to perfectly match society's expectations until the end of the 20th Century, it has started showing weaknesses for some years. The world's general purpose computing has now passed the bar of 10^{12} million instructions per second (MIPS) [HIL 11], and some limitations of our current microprocessor architectures are progressively appearing. The Moore law of microelectronics does not hold in the face of the demand for fast and powerful computing. This does not come from the calculation power of the CMOS technology, but rather from the on-chip electrical interconnects: on the one hand, they constitute a bottleneck for chip performances due to RC-induced delays [RAK 12]. On the other hand, they are responsible for the largest energy loss in the power envelope of microprocessors [MIL 09].

Just like optical fibers for telecommunications, optical interconnects have been thought of as the remedy for power losses associated with on-chip information routing and a pathway toward the "More–Moore" scaling of microprocessor performances. Most proposals for hybrid photonic–electronic microprocessors focus on architectures where only one part of the interconnection is performed by photonic components. Future microprocessors should indeed feature several hundreds of Core clusters operating in parallel. At the level of a single cluster, photonics will have difficulties competing with microelectronics because of the large spatial footprint of photonic components. Alternative approaches are being studied in this respect, such as plasmonic interconnects for example [RAK 12]. The situation is different at the scale of intercluster communication. Here, the photonic footprint is no longer a central issue and the technology of wavelength division multiplexing appears as a strong advantage due to its complementarity with multicluster parallel computing.

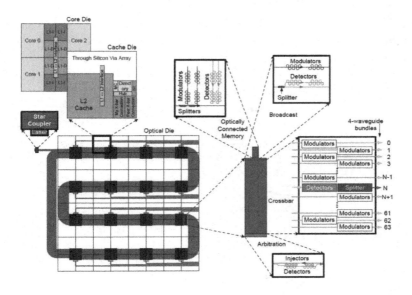

Figure 5.1. *From Vantrease et al. [VAN 08b]; architecture of the 64-cluster Corona microprocessor with an optical crossbar as photonic interconnects. In this hybrid architecture, each photonic node (dark blue squares) allow the connection of four clusters of four cores with the other nodes, either through a destination-cluster-specific channel or through a broadcast channel. Connection to memory and arbitration is also ensured by optical means. One specific feature of this architecture is the use of one or several laser sources situated out of the photonic network, either on- or off-chip. © 2008 IEEE. Reprinted, with permission, from [VAN 08b]. For a color version of the figure, please see www.iste.co.uk/cornet/lasers.zip*

The Corona [VAN 08, AHN 09] and Chameleon [LEB 11, LEB 14] architectures are examples of this paradigm. The Corona architecture, presented in Figure 5.1, is a 8 × 8 cluster architecture. The optical crossbar ensuring intercluster communication is composed of 64 bundles of four waveguides, with each waveguide carrying 64 wavelength channels. Each bundle starts at one node with the injection of optical power and runs through all the other nodes before coming back the departure node, where it ends with a detection device constituted by 64 resonant detectors. At each node, resonant modulators are used to inject information on 256 channels (64 wavelengths on four guides). A given bundle can thus be used to send information to the terminal node by all the other nodes. Considering modulation rates of 10 Gb/s for the modulators, the Corona architecture shows an aggregated bandwidth of 20 TB/s.

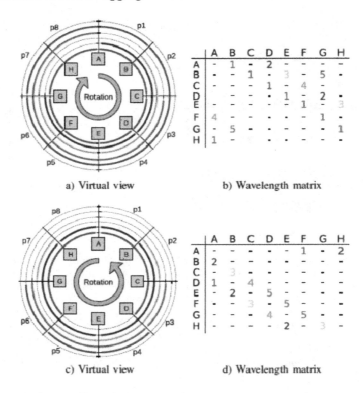

a) Virtual view   b) Wavelength matrix

	A	B	C	D	E	F	G	H
A	-	1	-	2	-	-	-	-
B	-	-	1	-	3	-	5	-
C	-	-	-	1	-	4	-	-
D	-	-	-	-	1	-	2	-
E	-	-	-	-	-	1	-	3
F	4	-	-	-	-	-	1	-
G	-	5	-	-	-	-	-	1
H	1	-	-	-	-	-	-	-

c) Virtual view   d) Wavelength matrix

	A	B	C	D	E	F	G	H
A	-	-	-	-	-	1	-	2
B	2	-	-	-	-	-	-	-
C	-	3	-	-	-	-	-	-
D	1	-	4	-	-	-	-	-
E	-	2	-	5	-	-	-	-
F	-	-	3	-	5	-	-	-
G	-	-	-	4	-	5	-	-
H	-	-	-	-	2	-	3	-

Figure 5.2. *From [LEB 11]; example of an eight-node ring optical network, at the basis of the Chameleon architecture. Plots a), b) and c), d) show the channels used to link the different nodes together as well as the associated connection matrix for (counter-) clockwise propagation of the optical data in the ring network. A given channel can be partitioned into one or several links depending on their length. Moreover, this partition is dynamically reconfigurable. © 2011 IEEE. Reprinted, with permission, from [LEB 11]. For a color version of the figure, please see www.iste.co.uk/cornet/lasers.zip*

The Corona architecture, proposed in 2008, shows distinctive features that highlight the issue of photonic integration. A single WDM 64-channel laser device is needed to power the whole optical crossbar, the optical power of which should be about 0.8 W. For this reason, no solution of on-chip laser source had been proposed at that point in time. We will see that this still remains a major issue for photonic interconnects. Moreover, the use of WDM technology requires the implementation of thousands of resonant filters, modulators and detectors within the chip (e.g., microrings in the Corona architecture). These devices can present fabrication imperfections that shift their operating frequency to the target. Active tuning should thus be added to biasing in the power envelope of the photonic layer. All in all, the Corona architecture shows a power envelope of 48 W, which should be compared with the 100 W Joule losses in current microprocessor electrical interconnects. The reader will notice that such a strategy of optical powering of a photonic layer with an off-chip laser has been successfully demonstrated by IBM in a front-end integrated photonic–electronic WDM transceiver operating at 25 Gb/s [ASS 12].

On the contrary, the Chameleon architecture [LEB 14] is based on a ring bundle of WDM waveguides ensuring communication through all the core clusters. Information can be sent either clockwise or counterclockwise in this ring network. At each communication node, as many laser sources and resonant detectors as there are communication channels in the ring network allow the associated cluster to communicate with the others, as presented in Figure 5.2. The advantages of this architecture are waveguide partitioning (a single wavelength can be used for multiple communication links within the same waveguide), multicast (the same data can be communicated by a node to several), dynamical allocation (depending on the calculation needs, zero, one or several channels can be allocated to communication between two nodes). For an 8 × 8 cluster architecture, a bundle of 16 waveguides operating with 63-wavelength multiplexing is required as well as 64,512 microlasers and the same number of detectors. In this case, the total optical power required to operate the photonic interconnect layer is in the range of 400 mW, and could go down to 10 mW, with optimized photonic components.

However, the power required to actively tune the resonant components in order to balance the fabrication imperfections and temperature fluctuations within the microprocessor has not yet been estimated.

5.1.2. Optical computing

Moving from on-chip optical routing toward optical processing is the long-term goal of the photonics community [WOO 12]. If electronics is the medium of choice for digital computing, photonics shows appealing advantages such as coherence, large nonlinearities, spinor or analogic character etc. for the development of alternative computation paradigms like quantum [LAD 10] or neuromorphic computing [LAR 12]. "More-Moore" microprocessor performances could indeed be improved not only by hybridizing microprocessors at the hardware level (i.e. with photonic and electronic layers) but also at the computational level, by allocating a given task to the best suited computation paradigm [ITR 13]. Many building blocks for on-chip optical computing have been proposed, among them passive [MAS 15] and active devices [LAD 10, BEN 11b], but once again, research for specific integrated light sources is a priority. We can cite in particular quantum-noise-limited sources for continuous variable entanglement [MAS 15], entangled or single photon sources, excitable lasers operating as spiking neurons for neuromorphic computing [COO 11].

As for the design of optical networks on chip, light sources for optical computing should indeed meet precise specifications. As an example, let us sketch the proposal of Coomans *et al.* on optical spiking neurons out [COO 11]. The cornerstone of many optical neuron networks is a network node, which provides a nonlinear response to stimulation by one or several optical inputs. This response has to be strong enough to act itself as a stimulus for the next node (so-called cascadability). To address this challenge, the authors of [COO 11] proposed the development of integrated laser sources operating in a regime close to bistability, the so-called excitability regime.

The basic scheme of their device is presented in Figure 5.3. The optical neuron consists in an electrically pumped on-chip microring semiconductor laser with chiral symmetry. It is coupled to either one or several optical waveguides for the injection of the stimuli and ejection of the response. Due to the particular symmetry of the device, stable lasing operation occurs in one direction only, defined by the sign of the chirality. However, if an optical pulse in the opposite direction enters the cavity through one of the waveguide couplers, the ring laser can switch into the counterpropagating state for a limited time, typically a few tens of nanoseconds before coming back to its stable state. Such an excitability regime only exists in a small injection current span (close to a Takens–Bogdanov bifurcation) and for particular values of parameters such as the backscattering coupling phase and the linewidth enhancement factor [GEL 09]. The two last parameters can be roughly set during the design of the optical neuron by a careful choice of the position of the spiral corner with respect to the waveguide couplers [SPR 90] and the choice of active materials [UKH 04], for example. Specification drift due to fabrication imperfections will thus be more likely to require individual and even dynamical adjustment of the operation parameters of the optical neurons, such as the injection current or the device temperature.

Optical neuron networks operating as aforementioned will most likely require thousands of microlasers, but this is not the case of time-domain neuromorphic networks [LAR 12, PAQ 12]. In the latter case, a laser beam coming from a single source is modulated by the use of a Mach–Zehnder interferometer and sent into a feedback loop composed of a delay line and a mixer, which combines the input data with the output of the delay line and in turn drives the interferometer. The nonlinear nodes of the network are thus consecutive time spots circulating in the feedback loop. The processed optical data is detected at the end of the delay line. Such a seemingly simple architecture has been implemented with standard telecom hardware to demonstrate both voice recognition and time series prediction based on a 400-node time network. If these designs will most likely be extremely application-specific, their performances compared to similar functionalities in the electronic domain could be extremely competitive. An on-chip demonstrator of such a system still needs to be demonstrated.

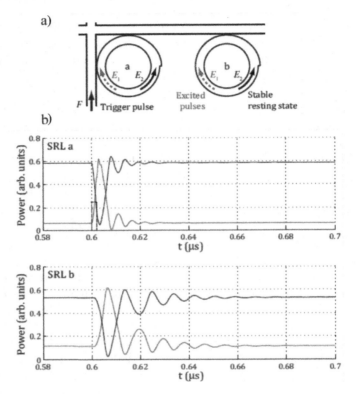

Figure 5.3. *a) Sketch of two optical spiking neurons consisting of chiral microring lasers operating in an excitable regime; b) a stimulation pulse exciting the neuron from the left waveguide. Neuron **a** switches from counterclockwise operation to clockwise operation for about 10 ns. This sends a stimulation along the upper waveguide. Information is then cascaded to neuron **b**. Reprinted with permission from Coomans et al. [COO 11] Copyright 2011 by the American Physical Society*

5.1.3. *A single laser or a multitude*

From this brief overview of the photonic architectures achieving either data routing or processing tasks, we can see that many different light source solutions could probably meet the requirements for the implementation of hybrid photonic–electronic processors, whether for general-purpose computing or application-specific devices.

Several architectures are indeed based on the use of laser sources that power the whole photonic layer in a very similar way to the electronic counterpart. Such sources should provide large output optical powers to stand optical losses in the complex photonic circuitry; for example, the 0.8 W laser required in the Corona architecture. Moreover, these sources should provide WDM signals (up to 64 wavelength channels in the aforementioned works). Nevertheless, such solutions do not require integration of the laser sources on a chip containing powerful CMOS processors, memories and multilevel electronic interconnects. They can rather be integrated on a basic "optical power chip" with limited surrounding electronics and its own thermal management solutions, focused on the optimization of the laser operation. Chip-to-chip optical links should then be implemented to couple light from the optical power chip into the hybrid microprocessor. Of course, this raises new challenges in terms of coupling losses at this scale [HAS 13]. In addition, the reader will notice that photonic routing architectures are always built on the use of resonant components (such as ring filters), whose operation frequencies are extremely dependent on the chip temperature. As electronic microprocessors are subject to local and dynamical temperature fluctuations (sometimes in the range of a few tens of kelvins), tuning between an off-chip laser source and on-chip resonant components can become tricky. We can wonder whether the active tuning solutions proposed by Ahn et al. [AHN 09] (i.e., thermal and carrier injection frequency tuning) would practically be enough: in the proposal, the overall active frequency tuning already represents more than half of the Corona photonic layer power envelope (26 W).

The second option is thus the integration of thousands of microlasers within the microprocessor. In this case, direct modulation of the lasers, as well as the use of resonant detectors made of similar materials as the sources, limiting the need for other resonant devices, could then prevent partly dynamical frequency detuning between the individual components. Still, this does not prevent frequency drift at the microscale, due to hotspots in the architecture [HAM 07, LIU 13].

This option also minimizes optical losses, as optical circuitry between a remote source and modulators is suppressed. However, the footprint of the lasers becomes a crucial parameter. To give an order of magnitude, a 64-cluster architecture has a surface of one square-inch. We can estimate that sources and detectors should not represent more than half of the chip surface. In the case of Chameleon (featuring 64,532 lasers and the same number of detectors), this gives a maximal footprint for the active components of 2.5×10^{-5} cm^2 (50 μm × 50 μm). We will keep this value in mind when comparing the different laser geometries. Finally, the optical power requirement for each laser can be assessed, knowing optical losses in the photonic circuitry, efficiency of the detectors and for a target bit error rate of 10^{-12} [LEB 14]. One can finally estimate the required optical power per microlaser to be in the range of a few microwatts.

5.2. Laser designs for on-chip routing

In this section, we investigate the different laser designs, either integrated on silicon or with a strong integration potential, in the light of the performance requirements of light sources for on-chip data routing and computing mentioned above. Four laser geometries will thus be discussed: ridge lasers, microdisk or microring lasers, photonic crystal lasers and high-contrast-grating (HCG) vertical cavity surface emitting lasers (VCSELs).

5.2.1. *Comparing apples and oranges*

These four laser designs are extremely different from each other: there can be orders of magnitude between their footprints, power consumption and output power. Moreover, most of these designs have been applied on many semiconductor platforms (InP, GaAs, GaSb, Si, etc.), with quantum dots, quantum dashes or quantum wells as active materials, which brings more confusion to the comparison between geometries. Our goal is to provide a set of metrics that one can use as a framework to investigate the potential of lasers for photonic integration. From such a performance chart, we will highlight the specific features of each laser design, independently of material considerations or nanostructure choice.

Table 5.1 reports on several semiconductor lasers investigated after 2011. We only focus here on some lasers with a high integration potential within microprocessors. By this, we mean that:

– they operate under electrical injection;

– they can provide in plane emission (this fact will be further detailed for VCSELs);

– their integration on Si or SOI substrates is either demonstrated or more likely to occur in the next years.

Our investigation of optical pumped lasers indeed revealed that yields can hardly be compared between optically and electrically pumped structures. Indeed, even for a given semiconductor platform and laser geometry, the detailed designs are rarely the same since optical injection relaxes issues such as free carrier absorption in doped layers or carrier distribution management. Moreover, for some devices, like the Si-Raman lasers, for which optically injected demonstrators do not show conceptual difficulties, the transfer to electrical injection seems doubtful.

In this table, some parameters are extremely dependent on material choices. Heterogeneously integrated lasers are generally more mature than monolithically integrated ones, as seen when comparing GaAs ridge lasers, either reported on Si [JUS 12] or grown on GaP/Si pseudo-substrates [HUA 14]. Heterogeneous lasers show lower threshold currents and current densities (see Chapters 3 and 4), higher maximal operation temperatures, but not necessarily higher output powers. The threshold current density is also more dependent on the active-area nanostructures than on the laser geometry itself and is subject to large fluctuations from one research team to another. The usual trend shows that QDs feature smaller threshold current densities than QDashes and QWs. Of course, the temperature dependence of the threshold current density also depends on the dimensionality of the nanostructures as discussed in Chapter 1. The reader will finally notice the large reduction of threshold current density when lasers operate at low temperatures, such as the GaAs photonic crystal laser of Table 5.1, underlining the difficulties sometimes encountered when trying to compare laser devices.

Some other parameters depend much more on the laser design itself: this is the case of the footprint, of course, but also of the total input and output powers and the number N_{ch} of lasing modes supported by the device. To a lesser extent, the modulation rate also depends on the laser design, since the diode capacitance increases with the polarized area. These parameters are also key features when considering lasers from the application point of view: a black-box opto-electronic transceiver. In this case, the energy per bit and the optical power per channel will be preferred to the input power and the total optical power. In particular, the energy per bit can be directly estimated from the ratio of the input power by the number-of-channel modulation-rate products. This calculation is thorough for monomode lasers for which direct modulation can be used. For multimode lasers, independent modulation of each wavelength channel by external modulators into a more complex architecture should in fact be considered, allowing for a higher modulation rate (typically 30 to 50 GHz [SAC 12, THO 12]) but also requiring more power for modulator biasing.

These three parameters allow us to plot the performance chart presented in Figure 5.4. In this chart we also report the associated requirements for photonic integration. We already discussed the requirement on device footprint and minimal optical power per channel in section 5.1.3. The maximal energy per bit can be estimated to 10 fJ/bit for competitiveness to CMOS solutions [MIL 09]. Each laser device is represented as a black dot, on both panels of Figure 5.4, having the same footprint. The left panel gives the corresponding energy per bit while the right panel gives the corresponding output power per channel.

Laser design	Semiconductor platform	References	Active material	Footprint (cm²)	Input power (W)	Threshold current (A)	Threshold current density (A cm⁻²)	Total optical power (W)	N_{ch}	T_{max} (K)	λ (μm)	Modulation rate (Hz)	Energy per bit* (J)
	InP/SOI	[DUA 14]	InGaAsP QWs	4.80E-03	7.50E-02	3.00E-02	833	1.00E-03	1	350	1.55	2.10E+10	3.57E-12
Ridge	InP	[MER 09]	InAs Qdashes	2.00E-03	2.25E-02	1.80E-02	720	3.00E-03	28	300	1.57	–	1.78E-12
	GaSb/Si	[REB 11]	GaInAsSb QWs	1.60E-03	9.00E-01	2.00E-01	900	3.00E-03	10	305	2.00	–	3.00E-04
	GaAs/GaP/Si	[HUA 14]	InGaAs QWs	9.60E-04	1.00E+01	5.00E+00	5600	1.00E-01	30	300	1.04	–	3.33E-03
	GaAs/Si	[JUS 12]	AlGaInAs QDs	4.00E-04	7.20E-02	1.70E-02	1500	5.00E-03	13	370	824	3.00E+09	3.80E-04
Microdisk	GaAs	[MAO 11]	InGaAs QDs	4.00E-06	1.60E-03	4.50E-04	1350	1.10E-05	1	300	1.06	–	1.10E-05
	InP	[VAN 07]	InAsP QWs	9.00E-06	8.00E-03	3.50E-04	909	1.00E-05	5	320	1.58	1.00E+10	2.00E-06
Photonic crystal	GaAs	[ELL 11]	InGaAs QDs	1.60E-05	7e7	2.87E-07	13	2.00E-13	1	150	1.12	1.00E+11	7.00E-18
HCG-VCSEL	InP	[MAT 12]	InGaAsP QWs	1.50E-06	6.80E-05	1.40E-05	1800	6.00E-07	1	370	1.56	1.00E+10	6.80E-15
	InP	[CHA 10]	AlGaInAs QDs	1.00E-04	3.75E-03	1.00E-03	1350	2.50E-04	1	370	1.55	–	3.75E-13
	GaAs	[ANS 13]	InGaAs QWs	1.00E-04	4.00E-03	1.00E-03	1000	2.00E-04	1	340	1.06	–	4.00E-13

Table 5.1. Benchmark of several state-of-the art semiconductor lasers. Only electrically pumped lasers for which full information on the injection specifications and optical output power were available were considered. To compare the optical power of these lasers, a similar operation regime, situated at twice the threshold current, has been arbitrarily chosen

* Modulation rates are not always provided in the literature. In this case, a typical modulation rate of 10 GHz is used to estimate the energy per bit.

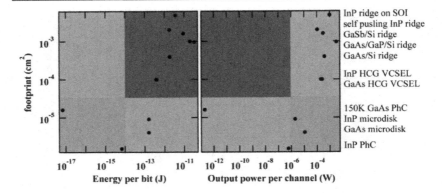

Figure 5.4. *Survey of laser geometries that are or could be integrated on Si. The green/orange/red (pale to dark) areas depict the matching to on-chip optical communication criteria. For a color version of the figure, please see www.iste.co.uk/cornet/lasers.zip*

Figure 5.4 shows that so far, no laser source has been developed to meet the three on-chip communication criteria. Ridges are too cumbersome to be integrated within the microprocessor chip and the energy per bit is too large. However, their use is much more comfortable in terms of optical power per channel than other designs. They would thus constitute the best solution for external optical powering as proposed in the Corona architecture and the IBM optical transceiver [ASS 12, VAN 08]. Microlasers such as microdisks and photonic crystal cavities, while they easily meet the footprint criterion, can hardly fulfill both small bit energy cost and sufficient output powers. We can anticipate that future realizations of these two geometries will be on the borderline of these two criteria. This could be a problem for future implementation of hybrid processors where lasers would limit the yield progression. Finally, VCSELs have intermediate performances. To be competitive with microlasers or ridge lasers, VCSELs should gain at least a decade on two criteria over three. In the last sections of this chapter, we will detail the specific features of state-of-the art examples of each of these four geometries and discuss their evolution potential in the framework of microprocessor hybridization.

5.2.2. Ridge lasers

Ridge laser geometries show a longer gain medium, sometimes of the order of a millimeter. This explains their high output power, in the range of a fraction to a few milliwatts per wavelength channel. Consequently, they also feature the largest power envelope, above 1 pJ/bit. These trends are valid for mature and exploratory semiconductor platforms and different integration approaches, such as heterogeneous and monolithic integration on silicon.

Before recent advances on the heterogeneous integration of III–V active components on silicon, ridge laser specifications were quite restricted because of their simple design: in order to close the optical cavity, the wafer supporting the ridge waveguide of the laser was cleaved at its two ends. The cleaved facets featured small reflectivity, typically between 30 and 90% depending on the deposition of a high reflectivity coating [HUF 98, LIU 05]. The reader will note that this simple cavity fabrication method is still currently used when the laser platform is under development [REB 11, HUA 14]. The low reflectivity of the cavity mirrors, coupled with large gains, favors high output powers but also high current thresholds (see Figure 5.5(b)). Ridge lasers thus show high wall plug efficiencies (up to 75% [LI 08]), which is a strong advantage for optical interconnects. These lasers also generally feature multimode lasing on a few tens of phase-locked longitudinal modes, promoting these sources as good candidates for WDM applications, even if separation of the different wavelength channels is still under development due to partition noise.

The potential of ridge lasers strongly increased during the past few years as a result of the tremendous progress made on silicon photonics, especially in the evanescent coupling of light from a bonded III–V layer to a buried silicon waveguide. Figure 5.5(a) shows such a coupler where tapering optimizes the optical power transfer over a large wavelength range. This strategic building block has allowed researchers to combine state-of-the-art active materials with the diversity of passive silicon photonics components. The III–V layers mostly play the role of the active area while the optical cavity is closed in the silicon photonic layer with the help of counterpropagation Bragg waveguides. Supplementary elements can be added into the structure, such as ring resonator etalons and tunable filters or arrayed-waveguide-gratings for wavelength division multiplexing [DUA 14, BOW 06].

Finally, thermal management is generally good in these structures because of a bulky design and the distribution of gain over a large area. However, heterogeneous integration, involving insulator layers such as silica or BCB, raises an issue on heat dissipation in these structures since it thermally decouples the semiconductor diode from the silicon substrate. This will probably be a major challenge to address in the future years.

The ridge laser design is thus particularly adapted to hybrid microprocessor architectures featuring remote powering light sources [VAN 08]: they show high output powers and multimode emission. If their large footprint is not a central issue in this case, the only remaining challenge is thus the reduction of the optical bit energy cost.

5.2.3. *Microdisk resonators*

Microdisk lasers confine light in the form of whispering gallery modes. The index contrast ratio between the semiconductor microdisk and the capping layer made of BCB or glass is about two, allowing small curvature radii and thus small footprint microdisks. Most microdisk lasers made of mature semiconductor platforms, such as GaAs and InP, are <10 μm in diameter. An advantage of whispering gallery modes is that despite the very small footprint of the device, they feature both small mode volumes and very high quality factors, limited mainly by etching wall roughness, bending losses and evanescent coupling to a bus waveguide.

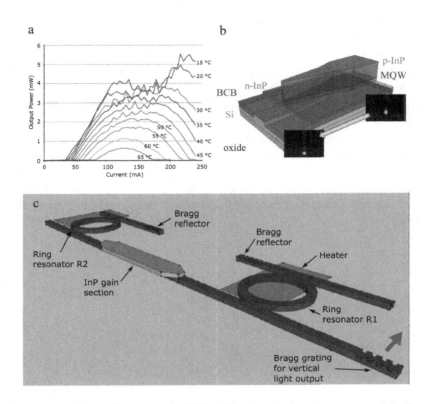

Figure 5.5. *From Duan et al. [DUA 14]; a) output power versus injection current of an hybrid silicon-photonics – III–V ridge laser; b) sketch of a tapered evanescent coupler between the III–V layer and a buried silicon waveguide; c) sketch of the whole hybrid ridge laser design featuring cavity mirrors and intracavity ring resonators etalons in the silicon photonic layer © 2014 IEEE. Reprinted, with permission, from [DUA 14]. For a color version of the figure, please see www.iste.co.uk/cornet/lasers.zip*

At the research level, Q factors in the range of 10^5 could be obtained in suspended GaAs microdisks [PAR 15b]. Due to these small losses, these laser designs reach threshold at small injection currents and thus feature consumption in the range of 0.1pJ/bit, only one order of magnitude above the goal estimated by Miller [MIL 09]. Moreover, huge progress can probably be made on this point. On the architecture of Van Campenhout *et al.* [VAN 07] for example (see Figure 5.6(a)), the choice has been made to optimize optical confinement by suppressing the semiconductor cladding layers from the structure. As a result, a lower diameter top contact is used to prevent absorption of the WGMs in the metal. One can easily understand that this injection architecture does not favor carrier concentration in the WGM volume. Going forward, oxide masks or tunnel junctions, as commonly used in VCSELs to concentrate the carrier flow, might be developed for microdisk geometries in order to improve carrier injection in the vicinity of WGMs. But this solution poses challenges for laser fabrication [VAN 08a], especially for heterogeneously integrated lasers, such as nanostructure regrowth after tunnel junction processing or oxide delamination during the process.

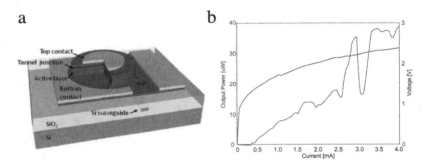

Figure 5.6. *a) Sketch of an InP microdisk laser heterogeneously integrated on a SOI substrate. Light is coupled out of the microdisk by evanescent coupling into a buried Si waveguide; b) typical evolution of output power and polarization voltage as a function of the injection current for such a device. The large power fluctuations observed above 2.5 mA are related to self-heating-induced frequency hopping. Reprinted with permission from Optical Society of America, from [VAN 07]*

The revolution symmetry of microdisks is also responsible for bidirectional operation. This regime has to be distinguished from the two side operation of a ridge laser or a VCSEL. Indeed, in the case of microdisks, such as in all ring cavities, clockwise-propagating and counterclockwise-propagating modes are orthogonal modes that compete for lasing. This sometimes leads to complex stability phase diagrams where unidirectional, bidirectional or even excitability or bistability regions can be found. Practically, this induces large output power fluctuations when modulating the injection current close to stability region boundaries, which can be detrimental for on-chip information routing. This said, such mode competition could also be used to develop more complex optical networks, including optical information processing. Bistability can be used for the realization of wavelength converters or all-optical memories [LIU 10]. Excitability can be applied to the implementation of optical spiking neurons for neuromorphic computing.

Finally, as far as thermal management is concerned, microdisk designs are also more sensitive than ridge lasers to thermal issues. A comparison between Si-integrated ridge and microdisk lasers shows, for example, that the microdisk thermal resistance is one order of magnitude larger than that of ridge lasers [LUC 16]. Microdisks are thus prone to self-heating, which can lead to wavelength and intensity fluctuations if modulation is applied in a nonappropriate current range (see Figure 5.6(b)). If active control of the device temperature by integrated thermal regulation can be implemented to limit intensity fluctuations, frequency drift or frequency hopping [VAN 10a], these solutions unfortunately increase the power budget of these microlasers.

5.2.4. *Photonic crystal cavity lasers*

Photonic crystal cavity lasers usually consist of membrane planar waveguides in which the hole network of the photonic crystal cavity is processed. The goal of this architecture is to maximize the optical confinement through high refractive index contrast. These architectures are also the cavities showing the smallest mode volume, on the order of the cubic wavelength. Both these features contribute to an extreme lowering of the lasing threshold and consequently of the laser power dissipation. However, such geometry also makes electrical injection practically difficult. The first electrically driven PhC laser was demonstrated by Prof. Yong-Hee Lee's group in 2004 by use of InP-based compounds [PAR 04]. Electrical injection was then achieved by use of an annular top metal contact surrounding the PhC and a bottom injection post underneath the cavity core, obtained by partial etching of a sacrificial InP layer. Apart from the challenging practical realization of such devices, the presence of the central injection post was quite strongly detrimental to the quality factor of the PhC cavity ($Q\sim2,500$). The demonstrated threshold current (260 µA in pulsed regime at room temperature) was then only slightly better than typical threshold current of microdisk lasers [VAN 07, MAO 11].

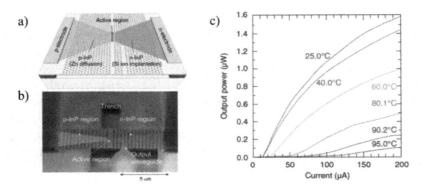

Figure 5.7. *From Matsuo et al. [MAT 13] a) Sketch of a lateral injection PhC laser showing the p- and n-ion-implanted contacts; b) SEM image of such a laser architecture; c) typical power–current curves of the device for several temperatures. Reprinted with permission of SPIE*

In 2011, the demonstration of a novel electrical injection technique was a major milestone of PhC laser on-chip integration. Lateral electrical injection was demonstrated by the use of ion-implanted contacts overlapping the hole network to the vicinity of the cavity core [ELL 11]. The design of these contacts allows optimization of the carrier flow into the cavity core, limiting current losses. Moreover, the membrane geometry is conserved with this method, which permits in principle to keep the quality factors very high (theoretically above 10^5). However, in this first demonstration, losses originating probably from carrier absorption into the doped regions and fabrication roughness drastically degraded the quality factor, down to 1,200.

In Ellis *et al.* [ELL 11], emphasis is placed on the ultralow threshold of these lasers (287 nA at 150 K). Such a low threshold current is indeed impressive as it is four orders of magnitude below all other on-chip lasers (see Figure 5.4). But this result should be put into perspective, as the laser is demonstrated at low temperature. It is well known that carrier thermalization to lower temperatures makes it possible to reach the Bernard–Duraffourg condition at smaller injection currents. This is always observed in lasers featuring quantum well or bulk active materials. This is also most probably the case here, even if quantum dots are used. For quantum dots, the temperature dependence of the threshold current is usually not monotonous: thermal energy first activates relaxation channels from dot to dot, which generally contributes to a reduction of the threshold current with increasing temperature prior to the usual increase (see Chapter 1 for more details) [CAR 05, SMO 07].

Room-temperature lasing of similar PhC laser architectures has then been demonstrated using the InP platform as shown in Figure 5.7. Three orders of magnitude are lost in the bit energy cost of the device as compared to the low-temperature demonstration of Ellis *et al.* [ELL 11], but the specifications of the devices are still the most competitive compared to other laser architectures. The Achille's heel of PhC crystals is a small optical output power. It could hardly be improved, since the choice for high Q-factor cavities to decrease the current threshold imposes small coupling efficiencies to the optical circuitry.

High modulation rates are also an attractive feature of PhC lasers. Above the threshold, lasers' modulation rates are limited by the longer time between the stimulated emission time and the cavity lifetime [PAO 70, BJO 91]. In the case of PhC cavities, which show both very small mode volumes and high Q factor, the stimulated emission rate is boosted by the large Purcell factor, while the cavity lifetime still remains small. Direct modulation rates between 10 and 100 GHz can thus be considered for these laser architectures.

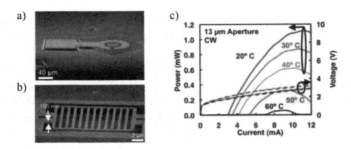

Figure 5.8. *a) SEM image of a HCG VCSEL operating at telecom wavelength under electrical injection; b) close up on the suspended HCG membrane reflector; c) P–I and V–I characteristics of the device. Reprinted with permission from Optical Society of America, from [CHA 10]*

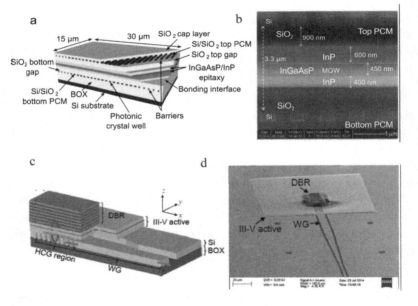

Figure 5.9. *a) and b) are respectively the sketch and transverse SEM image of a Si-integrated double HCG VCSEL operating under optical injection (from Sciancalepore et al. [SCI 12]) © 2012 IEEE. Reprinted, with permission, from [SCI 12]; c) and d) are the sketch and SEM images of a single HCG VCSEL showing efficient coupling into a Si waveguide. Reprinted with permission from [PAR 15a], Copyright © 2015 The Authors. For a color version of the figure, please see www.iste.co.uk/comet/lasers.zip*

5.2.5. *High-contrast-grating VCSEL*

VCSELs were applied as key light sources for many different applications. Compared to edge-emitting lasers, their advantages are many: a better beam quality, a longitudinal monomode character, smaller threshold currents while conserving reasonably high-output powers. The main applications of VCSELs are information routing in DataCom links, video links, sensors and printing/cutting [MIC 12].

Some of the VCSEL features are particularly appealing for on-chip optical interconnects. The reduced volume of the active area provides VCSELs both smaller power consumption and higher modulation rates compared to ridge lasers, as shown in Figure 5.5. The best VCSELs feature maximal wall plug efficiencies in excess of 60% [TAK 10], which appear very advantageous for on-chip optical interconnects. However, their intrinsic vertical emission direction becomes an issue when shifting from board-to-board and chip-to-chip communication to on-chip interconnects. Moreover, the large spatial footprint of DBRs and their fabrication specificity are problems for the device integration on silicon.

An alternative to DBRs was proposed in 2004 [MAT 04]. These so-called high-contrast grating (HCG) reflectors also comprise Bragg reflectors made of high index strips embedded into a low index matrix (or air). However, contrary to DBRs, the HCG periodicity axis is contained in the growth plane, as shown in Figure 5.8. The reflector thickness is thus reduced to a single layer of a few hundred of nanometers. Physically, the high reflectivity of HCGs comes from two elements: first, the periodicity of the grating being smaller than the field wavelength, no mode propagating in the growth plane can be excited – contrary to what happens in waveguide grating couplers [TAI 02]. Second, along the growth axis propagation direction, the HCG can be seen as an array of coupled slab waveguides. The periodicity of the HCG can be adjusted so that the incident field only decomposes into the two first propagating modes of this array. The thickness of the HCG is then calculated to produce constructive interferences of these modes on the inner side of the HCG and destructive interferences on the outer side, leading to reflectivity in excess of 99% [KAR 12].

These HCGs were first developed as a solution to tune the wavelength of VCSELs: the HCGs, being very thin, can indeed be integrated as membranes into MEM architectures at one end of the VCSEL in order to control the cavity length and thus the VCSEL emission wavelength [HUA 08a]. But HCGs could also be a solution to promote VCSELs as good candidates for on-chip optical interconnects. The high reflectivity of HCGs and the coupling properties of waveguide grating couplers [TAI 02] can be mixed so that most optical losses from the reflectors are injected in the growth plane, through a photonic circuitry including the HCG itself [SCI 12, HAT 03, CHU 10]. The first experimental demonstrations of such a coupling were recently reported in optically pumped VCSELs [PAR 15a, VIK 13]. In Park *et al.* [PAR 15a], the in-plane transmittance and reflectivity of such a HCG are calculated to be about 0.3 and 99.5% respectively. Interestingly, [VIK 13] focuses on the ultimate HCG-VCSEL for on-chip optical routing, with the use of two HCGs as cavity mirrors, as shown in Figure 5.9, and highlights the difficulties for electrical injection in these structures.

5.3. Concluding remarks

This survey first shows us that integrating laser sources on silicon, with high standards in terms of spatial and energy footprints, output powers and modulation, is still a long-term goal. So far no conventional laser source, among ridge lasers, microdisk resonators, photonic crystal lasers and VCSELs, can meet these criteria whatever the optical network architecture under scrutiny. The reader will notice that the latest and most promising advances for all types of laser geometries have been precisely obtained by mixing elements of these different laser sources (e.g., inserting ring cavities in ridge lasers or coupling VCSEL geometries to PhCs in the form of planar HCG couplers...). The best light source solution for on-chip optical interconnects might even be at the crossroad between semiconductor photonics and other research fields such as plasmonics [HIL 07, HO 15], or might not even be a laser [KER 15]!

Conclusion

In this book, we have reviewed and analyzed the numerous efforts that have been done in the last 20 years to develop a suitable laser source for integration at the silicon chip level. One of the first conclusions that can be drawn is that at present no single technological approach for integrated laser sources on-chip exists as a panacea, but some approaches offer advantages. If the technological maturity of bonded heterogeneous laser devices makes them a serious candidate in the midterm to provide optical sources to on-chip silicon photonics applications, this could be achieved only if their wall-plug efficiency is significantly improved. In a long-term perspective, the direct monolithic integration of both group IV or group III–V semiconductor lasers may provide ultimate integration opportunities, through full wafer or localized epitaxy. This development still requires significant improvements in material research, to reach efficient lasing properties by limiting the crystal defects, as well as to bring the laser active area closer to the silicon chip for further coupling to the optical layer within the microprocessor.

One of the reasons to explain the uncertainty about the mainstream technological solution that will be chosen is also that it strongly depends on the optical microprocessor architecture considered, which is not fixed yet. For instance, one could imagine integrating thousands of low-power small-footprint lasers on-chip, or could prefer integrating only few lasers with higher output powers, feeding the entire photonic chip. Constraints on laser device performances are, of course, very different depending on the integration scheme. Consequently, it is also hard to exclude *a priori* any laser integration approach and that's why a large overview of the research field was proposed here.

Bibliography

[AGR 12] AGRAWAL G., *Long-Wavelength Semiconductor Lasers*, Springer, Netherlands, 2012.

[AHN 09] AHN J., FIORENTINO M., BEAUSOLEIL R.G. *et al.*, "Devices and architectures for photonic chip-scale integration", *Applied Physics A, Materials Science & Processing*, vol. 95, no. 4, pp. 989–997, February 2009.

[AIN 15] AINGARAN K., JAIRATH S., KONSTADINIDIS G. *et al.*, "M7: Oracle's Next-Generation SPARC Processor", *IEEE Micro*, vol. 35, no. 2, pp. 36–45, March 2015.

[ANS 13] ANSBAEK T., CHUNG I.-S., SEMENOVA E.S. *et al.*, "1060-nm tunable monolithic high index contrast subwavelength grating VCSEL", *IEEE Photonics Technology Letters*, vol. 25, no. 4, pp. 365–367, February 2013.

[ARA 82] ARAKAWA Y., SAKAKI H., "Multidimensional quantum well laser and temperature dependence of its threshold current", *Applied Physics Letters*, vol. 40, no. 11, pp. 939–941, June 1982.

[ASS 12] ASSEFA S., SHANK S., GREEN W. *et al.*, "A 90 nm CMOS integrated nano-photonics technology for 25 Gbps WDM optical communications applications", *IEEE International Electron Devices Meeting (IEDM)*, pp. 33.8.1–33.8.3, 2012.

[ATH 14] ATHANASIOU M., SMITH R., LIU B. *et al.*, "Room temperature continuous – wave green lasing from an InGaN microdisk on silicon", *Scientific Reports*, vol. 4, p. 7250, November 2014.

[AYE 07] AYERS J.E., *Heteroepitaxy of Semiconductors: Theory, Growth, and Characterization*, CRC Press, 2007.

[BAE 12] BAEHR-JONES T., PINGUET T., GUO-QIANG P.L. *et al.*, "Myths and rumours of silicon photonics", *Nature Photonics*, vol. 6, no. 4, pp. 206–208, 2012.

[BAL 05] BALAKRISHNAN G., HUANG S., DAWSON L.R. *et al.*, "Growth mechanisms of highly mismatched AlSb on a Si substrate", *Applied Physics Letters*, vol. 86, no. 3, p. 034105, 2005.

[BAL 06] BALAKRISHNAN G., JALLIPALLI A., ROTELLA P. *et al.*, "Room-temperature optically pumped (Al)GaSb vertical-cavity surface-emitting laser monolithically grown on an Si(1 0 0) substrate," *IEEE Journal of Selected Topics in Quantum Electronics*, vol. 12, no. 6, pp. 1636–1641, November 2006.

[BAR 48] BARDEEN J., BRATTAIN W.H., "The transistor, a semi-conductor triode", *Physical Review*, vol. 74, no. 2, p. 230, July 1948.

[BAR 13] BARANOV A., TOURNIE E., *Semiconductor Lasers: Fundamentals and Applications*, Elsevier, 2013.

[BAR 15] BARANOV A.N., TEISSIER R., "Quantum Cascade Lasers in the InAs/AlSb Material System", *IEEE Journal of Selected Topics in Quantum Electronics*, vol. 21, no. 6, pp. 85–96, November 2015.

[BAY 13] BAYO CARO M.A., Theory of elasticity and electric polarization effects in the group-III nitrides, PhD Thesis, University College Cork, 2013.

[BEC 99] BECKER P.M., OLSSON A.A., SIMPSON J.R., *Erbium-Doped Fiber Amplifiers: Fundamentals and Technology*, Academic Press, 1999.

[BEN 11a] BEN BAKIR B., DESCOS A., OLIVIER N. *et al.*, "Electrically driven hybrid Si/III-V Fabry-Pérot lasers based on adiabatic mode transformers", *Optics Express*, vol. 19, no. 11, p. 10317, May 2011.

[BEN 11b] BENSON O., "Assembly of hybrid photonic architectures from nanophotonic constituents", *Nature*, vol. 480, no. 7376, pp. 193–199, December 2011.

[BER 61] BERNARD M.G.A., DURAFFOURG G., "Laser conditions in semiconductors", *Physica Status Solidi B*, vol. 1, no. 7, pp. 699–703, January 1961.

[BEY 12] BEYER A., OHLMANN J., LIEBICH S. *et al.*, "GaP heteroepitaxy on Si(001): correlation of Si-surface structure, GaP growth conditions, and Si-III/V interface structure", *Journal of Applied Physics*, vol. 111, no. 8, April 2012.

[BID 98] BIDNYK S., LITTLE B.D., CHO Y.H. *et al.*, "Laser action in GaN pyramids grown on (111) silicon by selective lateral overgrowth", *Applied Physics Letters*, vol. 73, no. 16, pp. 2242–2244, October 1998.

[BIM 99] BIMBERG D., GRUNDMANN M., LEDENTSOV N.N., *Quantum Dot Heterostructures*, John Wiley & Sons, 1999.

[BIS 00] BISI O., OSSICINI S., PAVESI L., "Porous silicon: a quantum sponge structure for silicon based optoelectronics", *Surface Science Reports.*, vol. 38, nos. 1–3, pp. 1–126, April 2000.

[BJO 91] BJORK G., YAMAMOTO Y., "Analysis of semiconductor microcavity lasers using rate equations", *IEEE Journal of Quantum Electronics*, vol. 27, no. 11, pp. 2386–2396, November 1991.

[BOR 02] BORMANN I., BRUNNER K., HACKENBUCHNER S. *et al.*, "Midinfrared intersubband electroluminescence of Si/SiGe quantum cascade structures", *Applied Physics Letters*, vol. 80, no. 13, pp. 2260–2262, April 2002.

[BOW 89] BOWERS J., "Trends, possibilities and limitations of silicon photonic integrated circuits and devices", *IEEE Custom Integrated Circuits Conference (CICC)*, pp. 1–89, 2013.

[BOW 06] BOWERS J. E., III-V photonic integration on silicon, Patent no. US8110823 B2, 2006.

[BOY 04] BOYRAZ O., JALALI B., "Demonstration of a silicon Raman laser", *Optics Express*, vol. 12, no. 21, p. 5269, 2004.

[BUC 15] BUCA D., WIRTHS S., STANGE D. *et al.*, "Direct bandgap GeSn alloys for laser application", *Advanced Photonics 2015, OSA Technical Digest*, available at: eprints.whiterose.ac.uk, 2015

[BUR 06] BURNS J.A., AULL B.F., CHEN C.K. *et al.*, "A wafer-scale 3-D circuit integration technology", *IEEE Transactions on Electron Devices*, vol. 53, no. 10, pp. 2507–2516, October 2006.

[CAM 08] CAMPENHOUT J.V., LIU L., ROMEO P.R. *et al.*, "A compact SOI-integrated multiwavelength laser source based on cascaded InP microdisks", *IEEE Photonics Technology Letters*, vol. 20, no. 16, pp. 1345–1347, August 2008.

[CAM 12] CAMACHO-AGUILERA R.E., CAI Y., PATEL N. *et al.*, "An electrically pumped germanium laser", *Optics Express*, vol. 20, no. 10, pp. 11316–11320, May 2012.

[CAN 90] CANHAM L.T., "Silicon quantum wire array fabrication by electrochemical and chemical dissolution of wafers", *Applied Physics Letters.*, vol. 57, no. 10, pp. 1046–1048, Sep. 1990.

[CAP 14] CAPELLINI G., REICH C., GUHA S. *et al.*, "Tensile Ge microstructures for lasing fabricated by means of a silicon complementary metal-oxide-semiconductor process", *Optics Express*, vol. 22, no. 1, p. 399, January 2014.

[CAR 05] CAROFF P., PARANTHOEN C., PLATZ C. *et al.*, "High-gain and low-threshold InAs quantum-dot lasers on InP", *Applied Physics Letters*, vol. 87, no. 24, p. 243107, December 2005.

[CAR 09] CAROFF P., DICK K.A., JOHANSSON J. *et al.*, "Controlled polytypic and twin-plane superlattices in III-V nanowires", *Nature Nanotechnology*, vol. 4, no. 1, pp. 50–55, January 2009.

[CAZ 04] CAZZANELLI M., KOVALEV D., DAL NEGRO L. *et al.*, "Polarized optical gain and polarization-narrowing of heavily oxidized porous silicon", *Physical Review Letters*, vol. 93, no. 20, p. 207402, November 2004.

[CER 10] CERUTTI L., RODRIGUEZ J.B., TOURNIE E., "GaSb-based laser, monolithically grown on silicon substrate, emitting at 1.55 µm at room temperature", *IEEE Photonics Technology Letters*, vol. 22, no. 8, pp. 553–555, April 2010.

[CHA 10] CHASE C., RAO Y., HOFMANN W. *et al.*, "1550 nm high contrast grating VCSEL", *Optics Express*, vol. 18, no. 15, pp. 15461–15466, 2010.

[CHA 54] CHAPIN D.M., FULLER C.S., PEARSON G.L., "A new silicon p-n junction photocell for converting solar radiation into electrical power", *Journal of Applied Physics*, vol. 25, no. 5, p. 676, 1954.

[CHE 08] CHEBEN P., SOREF R., LOCKWOOD D. *et al.*, "Silicon photonics, silicon photonics", *Advances in Optical Technologies Advances in Optical Technologies*, August 2008.

[CHE 09] CHENG S.-L., LU J., SHAMBAT G. *et al.*, "Room temperature 1.6 μm electroluminescence from Ge light emitting diode on Si substrate", *Optics Express*, vol. 17, no. 12, p. 10019, June 2009.

[CHE 11a] CHEN R., LIN H., HUO Y. *et al.*, "Increased photoluminescence of strain-reduced, high-Sn composition Ge1−xSnx alloys grown by molecular beam epitaxy", *Applied Physics Letters*, vol. 99, no. 18, p. 181125, October 2011.

[CHE 11b] CHEN R., TRAN T.-T.D., NG K.W. *et al.*, "Nanolasers grown on silicon," *Nature Photonics*, vol. 5, no. 3, pp. 170–175, 2011.

[CHE 13] CHENG C.-W., SHIU K.-T., LI N. *et al.*, "Epitaxial lift-off process for gallium arsenide substrate reuse and flexible electronics", *Nature Communications*, vol. 4, p. 1577, March 2013.

[CHE 14] CHEN S.M., TANG M.C., WU J. *et al.*, "1.3 μm InAs/GaAs quantum-dot laser monolithically grown on Si substrates operating over 100°C", *Electronics Letters*, vol. 50, no. 20, pp. 1467–1468, September 2014.

[CHE 16] CHEN S., LI W., WU J. *et al.*, "Electrically pumped continuous-wave III–V quantum dot lasers on silicon", *Nature Photonics*, March 2016.

[CHO 06] CHOI H.W., HUI K.N., LAI P.T. *et al.*, "Lasing in GaN microdisks pivoted on Si", *Applied Physics Letters*, vol. 89, no. 21, p. 211101, November 2006.

[CHO 13] CHOW W.W., KOCH S.W., *Semiconductor-Laser Fundamentals: Physics of the Gain Materials*, Springer Science & Business Media, 2013.

[CHO 14] CHOI A.H.W., *Handbook of Optical Microcavities*, CRC Press, 2014.

[CHU 10] CHUNG I.-S., MØRK J., "Silicon-photonics light source realized by III–V/Si-grating-mirror laser", *Applied Physics Letters*, vol. 97, no. 15, p. 151113, October 2010.

[CLA 02] CLAPS R., DIMITROPOULOS D., HAN Y. *et al.*, "Observation of Raman emission in silicon waveguides at 1.54 μm", *Optics Express*, vol. 10, no. 22, p. 1305, November 2002.

[COL 12] COLDREN L.A., CORZINE S.W., MASHANOVITCH M.L., *Diode Lasers and Photonic Integrated Circuits*, John Wiley & Sons, 2012.

[COO 11] COOMANS W., GELENS L., BERI S. *et al.*, "Solitary and coupled semiconductor ring lasers as optical spiking neurons", *Physical Review. E, Statistical, Nonlinear, and Soft Matter Physics*, vol. 84, no. 3, p. 036209, September 2011.

[COR 06] CORNET C., HAYNE M., CAROFF P. *et al.*, "Increase of charge-carrier redistribution efficiency in a laterally organized superlattice of coupled quantum dots", *Physical Review B*, vol. 74, no. 24, p. 245315, December 2006.

[COR 13] CORBETT B., BOWER C., FECIORU A. *et al.*, "Strategies for integration of lasers on silicon", *Semiconductor Science and Technology*, vol. 28, no. 9, p. 094001, 2013.

[CRE 13] CREAZZO T., MARCHENA E., KRASULICK S.B. *et al.*, "Integrated tunable CMOS laser", *Optics Express*, vol. 21, no. 23, p. 28048, November 2013.

[CUR 98] CURRIE M.T., SAMAVEDAM S.B., LANGDO T.A. *et al.*, "Controlling threading dislocation densities in Ge on Si using graded SiGe layers and chemical-mechanical polishing", *Applied Physics Letters*, vol. 72, no. 14, pp. 1718–1720, April 1998.

[DAN 12] DANOWITZ A., KELLEY K., MAO J. *et al.*, "CPU DB: recording microprocessor history", *Communications of the ACM*, vol. 55, no. 4, pp. 55–63, April 2012.

[DEB 98] DEBUSK D.K., PICKELSIMER B.L., Method for forming active devices on and in exposed surfaces of both sides of a silicon wafer, Patent no. US5739067 A, April 1998.

[DEH 00] DEHLINGER G., DIEHL L., GENNSER U. *et al.*, "Intersubband Electroluminescence from Silicon-Based Quantum Cascade Structures", *Science*, vol. 290, no. 5500, pp. 2277–2280, December 2000.

[DEK 15] DE KONINCK Y., ROELKENS G., BAETS R., "Electrically pumped 1550 nm single mode III-V-on-silicon laser with resonant grating cavity mirrors", *Laser & Photonics Reviews.*, vol. 9, no. 2, pp. L6–L10, March 2015.

[DEP 09] DEPPE D.G., SHAVRITRANURUK K., OZGUR G. *et al.*, "Quantum dot laser diode with low threshold and low internal loss", *Electronics Letters*, vol. 45, no. 1, pp. 54–56, January 2009.

[DI 11] DI D., PEREZ-WURFL I., WU L. *et al.*, "Electroluminescence from Si nanocrystal/c-Si heterojunction light-emitting diodes", *Applied Physics Letters*, vol. 99, no. 25, p. 251113, December 2011.

[DIM 05] DIMITRAKOPULOS G.P., SANCHEZ A.M., KOMNINOU P. *et al.*, "Interfacial steps, dislocations, and inversion domain boundaries in the GaN/AlN/Si (0001)/(111) epitaxial system", *Physica Status Solidi B*, vol. 242, no. 8, pp. 1617–1627, July 2005.

[DUA 14] DUAN G.-H., JANY C., LE LIEPVRE A. *et al.*, "Hybrid III–V on silicon lasers for photonic integrated circuits on silicon", *IEEE Journal of Selected Topics in Quantum Electronics*, vol. 20, no. 4, pp. 158–170, July 2014.

[EGA 95] EGAWA T., HASEGAWA Y., JIMBO T. *et al.*, "Degradation mechanism of AlGaAs/GaAs laser diodes grown on Si substrates", *Applied Physics Letters*, vol. 67, no. 20, pp. 2995–2997, November 1995.

[ELL 98] ELLMERS C., HOFMANN M., RÜHLE W.W. *et al.*, "Gain spectra of an (InGa)As single quantum well laser diode", *Physica Status Solidi B*, vol. 206, no. 1, pp. 407–412, March 1998.

[ELL 11] ELLIS B., MAYER M.A., SHAMBAT G. *et al.*, "Ultralow-threshold electrically pumped quantum-dot photonic-crystal nanocavity laser", *Nature Photonics*, vol. 5, no. 5, pp. 297–300, May 2011.

[ERN 88] ERNST F., PIROUZ P., "Formation of planar defects in the epitaxial growth of GaP on Si substrate by metal organic chemical-vapor deposition", *Journal of Applied Physics*, vol. 64, no. 9, pp. 4526–4530, November 1988.

[EST 12] ESTEVEZ M.C., ALVAREZ M., LECHUGA L.M., "Integrated optical devices for lab-on-a-chip biosensing applications", *Laser & Photonics Review*, vol. 6, no. 4, pp. 463–487, July 2012.

[FAN 06] FANG A.W., PARK H., COHEN O., JONES R., PANICCIA M.J., BOWERS J.E., "Electrically pumped hybrid AlGaInAs-silicon evanescent laser", *Optics Express*, vol. 14, no. 20, p. 9203, 2006.

[FAN 07] FANG Y.-Y., TOLLE J., ROUCKA R. *et al.*, "Perfectly tetragonal, tensile-strained Ge on Ge1−ySny buffered Si(100)", *Applied Physics Letters*, vol. 90, no. 6, p. 061915, February 2007.

[FAN 08a] FANG A.W., KOCH B.R., JONES R. *et al.*, "A distributed bragg reflector siliconevanescent laser," *IEEE Photonics Technology Letters*, vol. 20, no. 20, pp. 1667–1669, October 2008.

[FAN 08b] FANG A.W., LIVELY E., KUO Y.-H. *et al.*, "A distributed feedback silicon evanescent laser", *Optics Express*, vol. 16, no. 7, p. 4413, March 2008.

[FAN 13a] FAN H., HUA S., JIANG X. *et al.*, "Demonstration of an erbium-doped microsphere laser on a silicon chip", *Laser Physics Letters*, vol. 10, no. 10, p. 105809, 2013.

[FAN 13b] FANG Z., CHEN Q.Y., ZHAO C.Z., "A review of recent progress in lasers on silicon", *Optics and Laser Technology*, vol. 46, pp. 103–110, March 2013.

[FAR 08] FAROOQ M., HANNON R., IYER S. *et al.*, 3D integrated circuit device fabrication with precisely controllable substrate removal, Patent no. US8129256 B2, August 2008.

[FAU 05] FAUCHET P.M., "Light emission from Si quantum dots", *Materials Today : Proceedings*, vol. 8, no. 1, pp. 26–33, January 2005.

[FEL 88] FELDMAN M.R., ESENER S.C., GUEST C.C. *et al.*, "Comparison between optical and electrical interconnects based on power and speed considerations", *Applied Optics*, vol. 27, no. 9, p. 1742, May 1988.

[FRO 14] FROST T., JAHANGIR S., STARK E. *et al.*, "Monolithic electrically injected nanowire array edge-emitting laser on (001) silicon", *Nano Letters*, vol. 14, no. 8, pp. 4535–4541, August 2014.

[GAL 15] GALLAGHER J.D., SENARATNE C.L., SIMS P. *et al.*, "Electroluminescence from GeSn heterostructure pin diodes at the indirect to direct transition", *Applied Physics Letters*, vol. 106, no. 9, p. 091103, March 2015.

[GEI 06] GEISZ J.F., OLSON J.M., ROMERO M.J. *et al.*, "Lattice-mismatched GaAsP solar cells grown on silicon by OMVPE," *Conference Record of the 2006 IEEE 4th World Conference on Photovoltaic Energy Conversion*, vol. 1, pp. 772–775, 2006.

[GEL 00] GELLOZ B., KOSHIDA N., "Electroluminescence with high and stable quantum efficiency and low threshold voltage from anodically oxidized thin porous silicon diode", *Journal of Applied Physics*, vol. 88, no. 7, pp. 4319–4324, October 2000.

[GEL 09] GELENS L., VAN DER SANDEG., BERI S. *et al.*, "Phase-space approach to directional switching in semiconductor ring lasers", *Physical Review. E, Statistical, Nonlinear, and Soft Matter Physics*, vol. 79, no. 1, p. 016213, January 2009.

[GHE 14] GHETMIRI S.A., DU W., MARGETIS J. *et al.*, "Direct-bandgap GeSn grown on silicon with 2230 nm photoluminescence", *Applied Physics Letters*, vol. 105, no. 15, p. 151109, October 2014.

[GHI 09] GHIONE G., *Semiconductor Devices for High-Speed Optoelectronics*, Cambridge University Press, October 2009.

[GMA 01] GMACHL C., CAPASSO F., SIVCO D.L. *et al.*, "Recent progress in quantum cascade lasers and applications", *Reports on Progress in Physics*, vol. 64, no. 11, p. 1533, 2001.

[GOD 04] GODET J., PIZZAGALLI L., BROCHARD S. *et al.*, "Computer study of microtwins forming from surface steps of silicon", *Computational Materials Science*, vol. 30, no. 1–2, pp. 16–20, May 2004.

[GOO 84] GOODMAN J.W., LEONBERGER F.J., KUNG S.-Y. *et al.*, "Optical interconnections for VLSI systems", *Proceedings of the IEEE*, vol. 72, no. 7, pp. 850–866, July 1984.

[GÖS 98] GÖSELE U., TONG Q.-Y., "Semiconductor wafer bonding", *Annual Review of Materials Research*, vol. 28, no. 1, pp. 215–241, 1998.

[GRE 01] GREEN M.A., ZHAO J., WANG A. *et al.*, "Efficient silicon light-emitting diodes", *Nature*, vol. 412, no. 6849, pp. 805–808, August 2001.

[GRI 08] GRILLOT F., DAGENS B., PROVOST J.G., *et al.*, "Gain compression and above-threshold linewidth enhancement factor in 1.3μm InAs/GaAs quantum-dot lasers", *IEEE Journal of Quantum Electronics*, vol. 44, no. 10, pp. 946–951, October 2008.

[GRO 03] GROENERT M.E., LEITZ C.W., PITERA A.J. *et al.*, "Monolithic integration of room-temperature cw GaAs/AlGaAs lasers on Si substrates via relaxed graded GeSi buffer layers", *Journal of Applied Physics*, vol. 93, no. 1, pp. 362–367, January 2003.

[GRY 16] GRYDLIK M., HACKL F., GROISS H. *et al.*, "Lasing from Glassy Ge Quantum Dots in Crystalline Si", *ACS Photonics*, vol. 3, no. 2, pp. 298–303, February 2016.

[GRZ 12] GRZYBOWSKI G., BEELER R.T., JIANG L. *et al.*, "Next generation of Ge1−ySny (y=0.01–0.09) alloys grown on Si(100) via Ge3H8 and SnD4: reaction kinetics and tunable emission", *Applied Physics Letters*, vol. 101, no. 7, p. 072105, August 2012.

[GU 15] GU Y., ZHANG Y.G., MA Y.J. *et al.*, "InP-based type-I quantum well lasers up to 2.9 μm at 230 K in pulsed mode on a metamorphic buffer", *Applied Physics Letters*, vol. 106, no. 12, p. 121102, March 2015.

[GUO 12] GUO W., BONDI A., CORNET C. *et al.*, "Thermodynamic evolution of antiphase boundaries in GaP/Si epilayers evidenced by advanced X-ray scattering", *Applied Surface Science*, vol. 258, no. 7, pp. 2808–2815, January 2012.

[GUP 13] GUPTA S., MAGYARI-KÖPE B., NISHI Y. *et al.*, "Achieving direct band gap in germanium through integration of Sn alloying and external strain", *Journal of Applied Physics*, vol. 113, no. 7, p. 073707, February 2013.

[GUT 15] GUTSCH S., HILLER D., LAUBE J. *et al.*, "Observing the morphology of single-layered embedded silicon nanocrystals by using temperature-stable TEM membranes", *Beilstein Journal of Nanotechnology*, vol. 6, pp. 964–970, April 2015.

[HAL 62] HALL R.N., FENNER G.E., KINGSLEY J.D. *et al.*, "Coherent light emission from GaAs junctions", *Physical Review Letters*, vol. 9, no. 9, pp. 366–368, November, 1962.

[HAM 07] HAMANN H.F., WEGER A., LACEY J.A. *et al.*, "Hotspot-limited microprocessors: direct temperature and power distribution measurements", *The IEEE Journal of Solid-State Circuits*, vol. 42, no. 1, pp. 56–65, 2007.

[HAR 04] HARTMANN J.M., ABBADIE A., PAPON A.M. *et al.*, "Reduced pressure–chemical vapor deposition of Ge thick layers on Si(001) for 1.3–1.55-μm photodetection", *Journal of Applied Physics*, vol. 95, no. 10, pp. 5905–5913, May 2004.

[HAS 13] HASHARONI K., BENJAMIN S., GERON A. *et al.*, "A high end routing platform for core and edge applications based on chip to chip optical interconnect", *Optical Fiber Communication Conference/National Fiber Optic Engineers Conference*, 2013.

[HAT 03] HATTORI H., LETARTRE X., SEASSAL C. *et al.*, "Analysis of hybrid photonic crystal vertical cavity surface emitting lasers", *Optics Express*, vol. 11, no. 15, p. 1799, July 2003.

[HAT 06] HATTORI H.T., SEASSAL C., TOURAILLE E. *et al.*, "Heterogeneous integration of microdisk lasers on silicon strip waveguides for optical interconnects", *IEEE Photonics Technology Letters*, vol. 18, no. 1, pp. 223–225, January 2006.

[HE 97] HE G., ATWATER H.A., "Interband transitions in SnxGe(1-x) Alloys", *Physical Review Letters*, vol. 79, no. 10, pp. 1937–1940, September 1997.

[HEE 08] HEEBNER J., GROVER R., IBRAHIM T. *et al.*, *Optical Microresonators: Theory, Fabrication, and Applications*, Springer, London, 2008.

[HEI 14] HEIDEMANN M., HÖFLING S., KAMP M., "(In,Ga)As/GaP electrical injection quantum dot laser", *Applied Physics Letters*, vol. 104, no. 1, p. 011113, January 2014.

[HEN 13] HENINI M., *Molecular Beam Epitaxy: From Research to Mass Production*, Elsevier Science, Amsterdam, 2013.

[HEO 11] HEO J., GUO W., BHATTACHARYA P., "Monolithic single GaN nanowire laser with photonic crystal microcavity on silicon", *Applied Physics Letters*, vol. 98, no. 2, p. 021110, January 2011.

[HIL 07] HILL M. T., OEI Y.-S., SMALBRUGGE B. *et al.*, "Lasing in metallic-coated nanocavities", *Nature Photonics*, vol. 1, no. 10, pp. 589–594, October 2007.

[HIL 11] HILBERT M., LÓPEZ P., "The World's technological capacity to store, communicate, and compute information", *Science*, vol. 332, no. 6025, pp. 60–65, April 2011.

[HO 15] HO J., TATEBAYASHI J., SERGENT S. *et al.*, "Low-Threshold near-infrared GaAs–AlGaAs core–shell nanowire plasmon laser", *ACS Photonics*, vol. 2, no. 1, pp. 165–171, January 2015.

[HOL 62] HOLONYAK N., BEVACQUA S.F., "Coherent (visible) light emission from GaAs(1-x)Px junctions", *Applied Physics Letters*, vol. 1, no. 4, pp. 82–83, December 1962.

[HOM 15] HOMEWOOD K.P., LOURENÇO M.A., "Optoelectronics: the rise of the GeSn laser", *Nature Photonics*, vol. 9, no. 2, pp. 78–79, February 2015.

[HON 10] HONG T., RAN G.-Z., CHEN T. *et al.*, G.-G., "A selective-area metal bonding InGaAsP/Si laser", *IEEE Photonics Technology Letters.*, vol. 22, no. 15, pp. 1141–1143, August 2010.

[HOS 11] HOSSAIN N., SWEENEY S.J., ROGOWSKY S. *et al.*, "Reduced threshold current dilute nitride Ga(NAsP)/GaP quantum well lasers grown by MOVPE", *Electronics Letters*, vol. 47, no. 16, pp. 931–933, August 2011.

[HUA 08a] HUANG M.C.Y., ZHOU Y., CHANG-HASNAIN C.J., "A nanoelectromechanical tunable laser", *Nature Photonics*, vol. 2, no. 3, pp. 180–184, March 2008.

[HUA 08b] HUANG S.H., BALAKRISHNAN G., KHOSHAKHLAGH A. *et al.*, "Simultaneous interfacial misfit array formation and antiphase domain suppression on miscut silicon substrate", *Applied Physics Letters*, vol. 93, no. 7, pp. 071102–071102–3, August 2008.

[HUA 14] HUANG X., SONG Y., MASUDA T. *et al.*, "InGaAs/GaAs quantum well lasers grown on exact GaP/Si (001)", *Electronics Letters*, vol. 50, no. 17, pp. 1226–1227, August 2014.

[HUD 11] HUDA M.Q., SUBRINA S., "Multiple excitation of silicon nanoclusters during erbium sensitization process in silicon rich oxide host", *Applied Physics Letters*, vol. 98, no. 11, p. 111905, March 2011.

[HUF 98] HUFFAKER D.L., PARK G., ZOU Z. *et al.*, "1.3 μm room-temperature GaAs-based quantum-dot laser", *Applied Physics Letters*, vol. 73, no. 18, pp. 2564–2566, November 1998.

[HUN 09] HUNSPERGER R.G., *Integrated Optics: Theory and Technology*, 6th ed., Springer, NY, 2009.

[IM 92] IMLER W.R., SCHOLZ K.D., COBARRUVIAZ M. *et al.*, "Precision flip-chip solder bump interconnects for optical packaging", *IEEE Transactions on Components, Hybrids, and Manufacturing Technology*, vol. 15, no. 6, pp. 977–982, December 1992.

[ISH 03] ISHIKAWA Y., WADA K., CANNON D.D. *et al.*, "Strain-induced band gap shrinkage in Ge grown on Si substrate", *Applied Physics Letters*, vol. 82, no. 13, pp. 2044–2046, March 2003.

[ITR 13] ITRS, "2013 Executive Summary", available at: www.itrs2.net/itrs-reports.html, 2013.

[IYE 93] IYER S.S., XIE Y.-H., "Light emission from silicon", *Science*, vol. 260, no. 5104, pp. 40–46, April 1993.

[JAC 94] JACKSON K.P., FLINT E.B., CINA M.F. *et al.*, "A high-density, four-channel, OEIC transceiver module utilizing planar-processed optical waveguides and flip-chip, solder-bump technology", *Journal of Lightwave Technology*, vol. 12, no. 7, pp. 1185–1191, July 1994.

[JAH 15] JAHANGIR S., FROST T., HAZARI A. *et al.*, "Small signal modulation characteristics of red-emitting (λ = 610 nm) III-nitride nanowire array lasers on (001) silicon", *Applied Physics Letters*, vol. 106, no. 7, p. 071108, February 2015.

[JAI 12] JAIN J.R., HRYCIW A., BAER T.M. *et al.*, "A micromachining-based technology for enhancing germanium light emission via tensile strain", *Nature Photonics*, vol. 6, no. 6, pp. 398–405, June 2012.

[JAL 07a] JALLIPALLI A., BALAKRISHNAN G., HUANG S.H. *et al.*, "Atomistic modeling of strain distribution in self-assembled interfacial misfit dislocation (IMF) arrays in highly mismatched III–V semiconductor materials", *Journal of Crystal Growth*, vol. 303, no. 2, pp. 449–455, May 2007.

[JAL 07b] JALLIPALLI A., KUTTY M.N., BALAKRISHNAN G. *et al.*, "1.54 μm GaSb/AlGaSb multi-quantum-well monolithic laser at 77 K grown on miscut Si substrate using interfacial misfit arrays", *Electronics Letters*, vol. 43, no. 22, October 2007.

[JEN 87] JENKINS D.W., DOW J.D., "Electronic properties of metastable GexSn(1-x) alloys", *Physical Review B*, vol. 36, no. 15, pp. 7994–8000, November 1987.

[JUS 12] JUSTICE J., BOWER C., MEITL M. *et al.*, "Wafer-scale integration of group III-V lasers on silicon using transfer printing of epitaxial layers", *Nature Photonics*, vol. 6, no. 9, pp. 610–614, September 2012.

[KAR 12] KARAGODSKY V., CHANG-HASNAIN C.J., "Physics of near-wavelength high contrast gratings", *Optics Express*, vol. 20, no. 10, pp. 10888–10895, 2012.

[KAT 13] KATARIA H., JUNESAND C., WANG Z. *et al.*, "Towards a monolithically integrated III–V laser on silicon: optimization of multi-quantum well growth on InP on Si", *Semiconductor Science and Technology*, vol. 28, no. 9, p. 094008, 2013.

[KAV 10] KAVANAGH K.L., "Misfit dislocations in nanowire heterostructures", *Semiconductor Science and Technology*, vol. 25, no. 2, p. 024006, February 2010.

[KER 11] DE KERSAUSON M., KURDI M.E., DAVID S. *et al.*, "Optical gain in single tensile-strained germanium photonic wire", *Optics Express*, vol. 19, no. 19, p. 17925, September 2011.

[KER 15] KERN J., KULLOCK R., PRANGSMA J. *et al.*, "Electrically driven optical antennas", *Nature Photonics*, vol. 9, no. 9, pp. 582–586, September 2015.

[KEY 13a] KEYVANINIA S., ROELKENS G., VAN THOURHOUT D. *et al.*, "Demonstration of a heterogeneously integrated III-V/SOI single wavelength tunable laser", *Optics Express*, vol. 21, no. 3, p. 3784, February 2013.

[KEY 13b] KEYVANINIA S., VERSTUYFT S., VAN LANDSCHOOT L. *et al.*, "Heterogeneously integrated III-V/silicon distributed feedback lasers", *Optics Express*, vol. 38, no. 24, p. 5434, December 2013.

[KHR 01] KHRIACHTCHEV L., RÄSÄNEN M., NOVIKOV S. *et al.*, "Optical gain in Si/SiO2 lattice: experimental evidence with nanosecond pulses", *Applied Physics Letters*, vol. 79, no. 9, pp. 1249–1251, August 2001.

[KIL 76] KILBY J.S., "Invention of the integrated circuit", *IEEE Transactions on Electron Devices*, vol. 23, no. 7, pp. 648–654, July 1976.

[KIS 91] KISSINGER W., KISSINGER G., KRÜGER J. *et al.*, "High yield in wafer bonding with surface structured wafers", *Materials Letters*, vol. 12, no. 4, pp. 266–269, November 1991.

[KOB 14] KOBLMÜLLER G., ABSTREITER G., "Growth and properties of InGaAs nanowires on silicon", *Physica Status Solidi RRL – Rapid Research Letters*, vol. 8, no. 1, pp. 11–30, January 2014.

[KOE 15] KOERNER R., OEHME M., GOLLHOFER M. *et al.*, "Electrically pumped lasing from Ge Fabry–Perot resonators on Si", *Optics Express*, vol. 23, no. 11, p. 14815, June 2015.

[KOS 05] KOSTRZEWA M., DI CIOCCIO L., FEDELI J.M. *et al.*, "Die-to-Wafer molecular bonding for optical interconnects and packaging", *EMPC*, Brugge, Belgium, 2005.

[KOS 06] KOSTRZEWA M., DI CIOCCIO L., ZUSSY M. *et al.*, "InP dies transferred onto silicon substrate for optical interconnects application", *Sensors and Actuators A: Physical*, vol. 125, no. 2, pp. 411–414, January 2006.

[KOY 15] KOYANAGI M., "Recent progress in 3D integration technology", *IEICE Electronics Express*, vol. 12, no. 7, pp. 20152001–20152001, 2015.

[KRO 63] KROEMER H., "A proposed class of hetero-junction injection lasers", *Proceedings of the IEEE*, vol. 51, no. 12, pp. 1782–1783, December 1963.

[KRO 87] KROEMER H., "Polar-on-nonpolar epitaxy", *Journal of Crystal Growth*, vol. 81, no. 1–4, pp. 193–204, February 1987.

[KRZ 12] KRZYŻANOWSKA H., NI K.S., FU Y. *et al.*, "Electroluminescence from Er-doped SiO2/nc-Si multilayers under lateral carrier injection", *Materials Science and Engineering: B*, vol. 177, no. 17, pp. 1547–1550, October 2012.

[KUR 11] KUROYANAGI R., ISHIKAWA Y., TSUCHIZAWA T. *et al.*, "Controlling strain in Ge on Si for EA modulators", *8th IEEE International Conference on Group IV Photonics (GFP)*, pp. 211–213, 2011.

[LAD 10] LADD T.D., JELEZKO F., LAFLAMME R. *et al.*, "Quantum computers", *Nature*, vol. 464, no. 7285, pp. 45–53, March 2010.

[LAM 12] LAMPONI M., KEYVANINIA S., JANY C. *et al.*, "Low-threshold heterogeneously integrated InP/SOI lasers with a double adiabatic taper coupler", *IEEE Photonics Technology Letters*, vol. 24, no. 1, pp. 76–78, January 2012.

[LAN 00] LANGDO T.A., LEITZ C.W., CURRIE M.T. *et al.*, "High quality Ge on Si by epitaxial necking", *Applied Physics Letters*, vol. 76, no. 25, pp. 3700–3702, June 2000.

[LAR 12] LARGER L., SORIANO M.C., BRUNNER D. *et al.*, "Photonic information processing beyond turing: an optoelectronic implementation of reservoir computing", *Optics Express*, vol. 20, no. 3, p. 3241, January 2012.

[LEB 11] LE BEUX S. TRAJKOVIC J. *et al.*, "Optical Ring Network-on-Chip (ORNoC)", *Design, Automation and Test in Europe Conference and Exhibition (DATE)*, pp. 1–6, 2011

[LEB 14] LE BEUX S., LI H., O'CONNOR I. *et al.*, "Chameleon: channel efficient optical network-on-chip", *Design, Automation and Test in Europe Conference and Exhibition (DATE)*, pp. 1–6, 2014.

[LED 03] LEDENTSOV N.N., "Si-Ge quantum dot laser: what can we learn from III-V experience?", in PAVESI L., GAPONENKO S., NEGRO L.D. (eds), *Towards the First Silicon Laser*, Springer, Netherlands, 2003.

[LEE 12a] LEE A., JIANG Q., TANG M. *et al.*, "Continuous-wave InAs/GaAs quantum-dot laser diodes monolithically grown on Si substrate with low threshold current densities", *Optics Express*, vol. 20, no. 20, pp. 22181–22187, September 2012.

[LEE 12b] LEE H., CHEN T., LI J. *et al.*, "Chemically etched ultrahigh-Q wedge-resonator on a silicon chip", *Nature Photonics*, vol. 6, no. 6, pp. 369–373, May 2012.

[LEE 13] LEE A., LIU H., SEEDS A., "Semiconductor III–V lasers monolithically grown on Si substrates", *Semiconductor Science and Technology.*, vol. 28, no. 1, p. 015027, January 2013.

[LI 08] LI L., LIU G., LI Z. *et al.*, "High-efficiency 808-nm InGaAlAs-AlGaAs double-quantum-well semiconductor lasers with asymmetric waveguide structures", *IEEE Photonics Technology Letters*, vol. 20, no. 8, pp. 566–568, April 2008.

[LI 15] LI K.H., LIU X. *et al.*, "Ultralow-threshold electrically injected AlGaN nanowire ultraviolet lasers on Si operating at low temperature", *Nature Nanotechnology*, vol. 10, no. 2, pp. 140–144, February 2015.

[LIA 08a] LIANG D., BOWERS J. E., "Highly efficient vertical outgassing channels for low-temperature InP-to-silicon direct wafer bonding on the silicon-on-insulator substrate", *Journal of Vacuum Science & Technology B*, vol. 26, no. 4, pp. 1560–1568, July 2008.

[LIA 08b] LIANG D., FANG A.W., PARK H. *et al.*, "Low-temperature, strong SiO_2-SiO_2 covalent wafer bonding for III–V compound semiconductors-to-silicon photonic integrated circuits", *Journal of Electronic Materials*, vol. 37, no. 10, pp. 1552–1559, June 2008.

[LIA 09a] LIANG D., BOWERS J.E., OAKLEY D.C. *et al.*, "High-Quality 150 mm InP-to-silicon epitaxial transfer for silicon photonic integrated circuits", *Electrochemical and Solid-State Letters*, vol. 12, no. 4, pp. H101–H104, January 2009.

[LIA 09b] LIANG D., FIORENTINO M., OKUMURA T. *et al.*, "Electrically-pumped compact hybrid silicon microring lasers for optical interconnects", *Optics Express*, vol. 17, no. 22, p. 20355, October 2009.

[LIA 10] LIANG D., BOWERS J.E., "Recent progress in lasers on silicon", *Nature Photonics*, vol. 4, no. 8, pp. 511–517, 2010.

[LIA 11] LIANG D., FIORENTINO M., SRINIVASAN S. *et al.*, "Low threshold electrically-pumped hybrid silicon microring lasers", *IEEE Journal of Selected Topics in Quantum Electronics*, vol. 17, no. 6, pp. 1528–1533, November 2011.

[LIC 09] LICHTE H., GEIGER D., LINCK M., "Off-axis electron holography in an aberration-corrected transmission electron microscope", *Philosophical Transactions of the Royal Society of London Mathematical, Physical, and Engineering Sciences*, vol. 367, no. 1903, pp. 3773–3793, September 2009.

[LIE 11] LIEBICH S., ZIMPRICH M., BEYER A. *et al.*, "Laser operation of Ga(NAsP) lattice-matched to (001) silicon substrate", *Applied Physics Letters*, vol. 99, no. 7, p. 071109, 2011.

[LIN 97] LINDGREN S., AHLFELDT H., BACKLIN L. *et al.*, "24-GHz modulation bandwidth and passive alignment of flip-chip mounted DFB laser diodes", *IEEE Photonics Technology Letters*, vol. 9, no. 3, pp. 306–308, March 1997.

[LIN 99] LINDER K.K., PHILLIPS J., QASAIMEH O. *et al.*, "Self-organized $In_{0.4}Ga_{0.6}As$ quantum-dot lasers grown on Si substrates", *Applied Physics Letters*, vol. 74, no. 10, pp. 1355–1357, March 1999.

[LIU 04a] LIU J., CANNON D.D., WADA K. *et al.*, "Deformation potential constants of biaxially tensile stressed Ge epitaxial films on Si(100)", *Physical Review B*, vol. 70, no. 15, p. 155309, October 2004.

[LIU 04b] LIU J., CANNON D.D., WADA K. *et al.*, "Silicidation-induced band gap shrinkage in Ge epitaxial films on Si", *Applied Physics Letters*, vol. 84, no. 5, pp. 660–662, February 2004.

[LIU 05] LIU H.Y., CHILDS D.T., BADCOCK T.J. *et al.*, "High-performance three-layer 1.3-micron InAs-GaAs quantum-dot lasers with very low continuous-wave room-temperature threshold currents", *IEEE Photonics Technology Letters*, vol. 17, no. 6, pp. 1139–1141, June 2005.

[LIU 07] LIU J., SUN X., PAN D. *et al.*, "Tensile-strained, n-type Ge as a gain medium for monolithic laser integration on Si", *Optics Express*, vol. 15, no. 18, p. 11272, 2007.

[LIU 09] LIU J., SUN X., KIMERLING L.C. *et al.*, "Direct-gap optical gain of Ge on Si at room temperature", *Optics Letters*, vol. 34, no. 11, p. 1738, June 2009.

[LIU 10a] LIU J., SUN X., CAMACHO-AGUILERA R. *et al.*, "Ge-on-Si laser operating at room temperature", *Optics Letters*, vol. 35, no. 5, p. 679, March 2010.

[LIU 10b] LIU L., KUMAR R., HUYBRECHTS K. *et al.*, "An ultra-small, low-power, all-optical flip-flop memory on a silicon chip", *Nature Photonics*, vol. 4, no. 3, pp. 182–187, March 2010.

[LIU 12] LIU J., KIMERLING L.C., MICHEL J., "Monolithic Ge-on-Si lasers for large-scale electronic–photonic integration", *Semiconductor Science and Technology*, vol. 27, no. 9, p. 094006, September 2012.

[LIU 13] LIU Z., HUANG X., TAN S. X.-D. *et al.*, "Distributed task migration for thermal hot spot reduction in many-core microprocessors", *IEEE 10th International Conference on ASIC (ASICON)*, pp. 1–4, 2013.

[LIU 14a] LIU A.Y., ZHANG C., NORMAN J. *et al.*, "High performance continuous wave 1.3 μm quantum dot lasers on silicon", *Applied Physics Letters*, vol. 104, no. 4, p. 041104, January 2014.

[LIU 14b] LIU J., "Monolithically Integrated Ge-on-Si Active Photonics", *Photonics*, vol. 1, no. 3, pp. 162–197, July 2014.

[LIU 15] LIU A.Y., SRINIVASAN S., NORMAN J. *et al.*, "Quantum dot lasers for silicon photonics", *Photonics Research*, vol. 3, no. 5, pp. B1–B9, October 2015.

[LOU 12] LOURDUDOSS S., "Heteroepitaxy and selective area heteroepitaxy for silicon photonics", *Current Opinion in Solid State and Materials Science*, vol. 16, no. 2, pp. 91–99, April 2012.

[LOW 12] LOW K.L., YANG Y., HAN G. *et al.*, "Electronic band structure and effective mass parameters of Ge1-xSnx alloys", *Journal of Applied Physics*, vol. 112, no. 10, p. 103715, November 2012.

[LU 12] LU F., TRAN T.-T.D., KO W.S. *et al.*, "Nanolasers grown on silicon-based MOSFETs", *Optics Express*, vol. 20, no. 11, pp. 12171–12176, May 2012.

[LUC 16] LUCCI I., CORNET C., BAHRI M. *et al.*, "Thermal management of monolithic vs heterogeneous lasers integrated on silicon", *IEEE Journal of Selected Topics in Quantum Electronics, Special Issue on Silicon Photonics*, in press, 2016.

[LUO 15] LUO X., CAO Y., SONG J. *et al.*, "High-throughput multiple dies-to-wafer bonding technology and III/V-on-Si hybrid lasers for heterogeneous integration of optoelectronic integrated circuits", *Optics and Photonics*, vol. 2, p. 28, 2015.

[LUT 08] LUTSENKO E.V., DANILCHYK A.V., TARASUK N.P. *et al.*, "Laser threshold and optical gain of blue optically pumped InGaN/GaN multiple quantum wells (MQW) grown on Si", *Physica Status Solidi C*, vol. 5, no. 6, pp. 2263–2266, May 2008.

[MAD 15] MADIOMANANA K., BAHRI M., RODRIGUEZ J.B. *et al.*, "Silicon surface preparation for III-V molecular beam epitaxy", *Journal of Crystal Growth*, vol. 413, pp. 17–24, March 2015.

[MAO 11] MAO M.-H., CHIEN H.-C., HONG J.-Z. *et al.*, "Room-temperature low-threshold current-injection InGaAs quantum-dot microdisk lasers with single-mode emission", *Optics Express*, vol. 19, no. 15, pp. 14145–14151, 2011.

[MÅR 04] MÅRTENSSON T., SVENSSON C.P.T., WACASER B.A. *et al.*, "Epitaxial III–V nanowires on silicon," *Nano Letters*, vol. 4, no. 10, pp. 1987–1990, October 2004.

[MAS 15] MASADA G., MIYATA K., POLITI A. *et al.*, "Continuous-variable entanglement on a chip", *Nature Photonics*, vol. 9, no. 5, pp. 316–319, May 2015.

[MAT 04] MATEUS C.F.R., HUANG M.C.Y., DENG Y. *et al.*, "Ultrabroadband mirror using low-index cladded subwavelength grating", *IEEE Photonics Technology Letters*, vol. 16, no. 2, pp. 518–520, February 2004.

[MAT 12] MATSUO S., TAKEDA K., SATO T. *et al.*, "Room-temperature continuous-wave operation of lateral current injection wavelength-scale embedded active-region photonic-crystal laser", *Optics Express*, vol. 20, no. 4, pp. 3773–3780, 2012.

[MAT 13] MATSUO S., SATO T., TAKEDA K., "Electrically-driven photonic crystal laser with ultra-low operating energy", *SPIE Newsroom*, August 2013.

[MCN 13] MCNALLY P.J., "Techniques: 3D imaging of crystal defects", *Nature*, vol. 496, no. 7443, pp. 37–38, April 2013.

[MEI 06] MEITL M.A., ZHU Z.-T., KUMAR V. *et al.*, "Transfer printing by kinetic control of adhesion to an elastomeric stamp", *Nature Materials*, vol. 5, no. 1, pp. 33–38, January 2006.

[MER 09] MERGHEM K., ROSALES R., AZOUIGUI S. *et al.*, "Low noise performance of passively mode locked quantum-dash-based lasers under external optical feedback", *Applied Physics Letters*, vol. 95, no. 13, p. 131111, 2009.

[MER 14] MERCKLING C., WALDRON N., JIANG S. *et al.*, "Heteroepitaxy of InP on Si(001) by selective-area metal organic vapor-phase epitaxy in sub-50 nm width trenches: the role of the nucleation layer and the recess engineering", *Journal of Applied Physics*, vol. 115, no. 2, p. 023710, January 2014.

[MI 05] MI Z., BHATTACHARYA P., YANG J. *et al.*, "Room-temperature self-organised In0.5Ga0.5As quantum dot laser on silicon", *Electronics Letters*, vol. 41, no. 13, pp. 742–744, June 2005.

[MI 06] MI Z., YANG J., BHATTACHARYA P. *et al.*, "Self-organised quantum dots as dislocation filters: the case of GaAs-based lasers on silicon", *Electronics Letters*, vol. 42, no. 2, pp. 121–123, 2006.

[MI 08] MI Z., BHATTACHARYA P., "Pseudomorphic and metamorphic quantum dot heterostructures for long-wavelength lasers on GaAs and Si", *IEEE Journal of Selected Topics in Quantum Electronics*, vol. 14, no. 4, pp. 1171–1179, July 2008.

[MI 09] MI Z., YANG J., BHATTACHARYA P. *et al.*, "High-performance quantum dot lasers and integrated optoelectronics on Si", *Proceedings of the IEEE*, vol. 97, no. 7, pp. 1239–1249, July 2009.

[MIC 12] MICHALZIK R., *VCSELs: Fundamentals, Technology and Applications of Vertical-Cavity Surface-Emitting Lasers*, Springer, 2012.

[MIL 69] MILLER L.F., "Controlled collapse reflow chip joining", *IBM Journal of Research and Development*, vol. 13, no. 3, pp. 239–250, May 1969.

[MIL 00] MILLER D.A.B., "Rationale and challenges for optical interconnects to electronic chips", *Proceedings of the IEEE*, vol. 88, no. 6, pp. 728–749, June 2000.

[MIL 09] MILLER D.A.B., "Device requirements for optical interconnects to silicon chips", *Proceedings of the IEEE*, vol. 97, no. 7, pp. 1166–1185, July 2009.

[MIL 10] MILLER G.M., BRIGGS R.M., ATWATER H.A., "Achieving optical gain in waveguide-confined nanocluster-sensitized erbium by pulsed excitation", *Journal of Applied Physics*, vol. 108, no. 6, p. 063109, September 2010.

[MIR 07] MIRITELLO M., LO SAVIO R., IACONA F. *et al.*, "Efficient Luminescence and Energy Transfer in Erbium Silicate Thin Films", *Advanced Materials*, vol. 19, no. 12, pp. 1582–1588, June 2007.

[MOF 13] MOFFAT T.P., JOSELL D., "Electrochemical processing of interconnects", *Journal of the Electrochemical Society*, vol. 160, no. 12, pp. Y7–Y10, January 2013.

[NAC 11] NACZAS S., AKHTER P., HUANG M., "Enhanced photoluminescence around 1540 nm from erbium doped silicon coimplanted with hydrogen and silver", *Applied Physics Letters*, vol. 98, no. 11, p. 113101, March 2011.

[NAK 98] NAKADA Y., AKSENOV I., OKUMURA H., "GaN heteroepitaxial growth on silicon nitride buffer layers formed on Si (111) surfaces by plasma-assisted molecular beam epitaxy", *Applied Physics Letters*, vol. 73, no. 6, pp. 827–829, August 1998.

[NAT 62] NATHAN M.I., DUMKE W.P., BURNS G. *et al.*, "Stimulated emission of radiation from GaAs p-n junctions", *Applied Physics Letters*, vol. 1, no. 3, pp. 62–64, November 1962.

[NAY 02] NAYFEH M.H., RAO S., BARRY N. *et al.*, "Observation of laser oscillation in aggregates of ultrasmall silicon nanoparticles", *Applied Physics Letters*, vol. 80, no. 1, pp. 121–123, January 2002.

[NG 01] NG W.L., LOURENÇO M.A., GWILLIAM R.M. *et al.*, "An efficient room-temperature silicon-based light-emitting diode", *Nature*, vol. 410, no. 6825, pp. 192–194, March 2001.

[NGU 12] NGUYEN THANH T., ROBERT C. *et al.*, "Structural and optical analyses of GaP/Si and (GaAsPN/GaPN)/GaP/Si nanolayers for integrated photonics on silicon", *Journal of Applied Physics*, vol. 112, no. 5, pp. 053521–053521–8, September 2012.

[NIK 13] NIKONOV D.E., YOUNG I.A., "Overview of beyond-CMOS devices and a uniform methodology for their benchmarking", *Proceedings of the IEEE*, vol. 101, no. 12, pp. 2498–2533, December 2013.

[PAL 09] PALIT S., KIRCH J., TSVID G. *et al.*, "Low-threshold thin-film III-V lasers bonded to silicon with front and back side defined features", *Optics Letters*, vol. 34, no. 18, p. 2802, September 2009.

[PAN 10] PANICCIA M., "Integrating silicon photonics", *Nature Photonics*, vol. 4, no. 8, pp. 498–499, August 2010.

[PAO 70] PAOLI T. L., RIPPER J., "Direct modulation of semiconductor lasers", *Proceedings of IEEE*, vol. 58, no. 10, pp. 1457–1465, October 1970.

[PAQ 12] PAQUOT Y., DUPORT F., SMERIERI A. *et al.*, "Optoelectronic reservoir computing", *Scientific Reports*, vol. 2, p. 287, February 2012.

[PAR 02] PARANTHOEN C., BERTRU N., LAMBERT B. *et al.*, "Room temperature laser emission of 1.5 μm from InAs/InP(311)B quantum dots", *Semiconductor Science and Technology*, vol. 17, no. 2, p. L5, 2002.

[PAR 04] PARK H.-G., KIM S.-H., KWON S.-H. *et al.* "Electrically driven single-cell photonic crystal laser", *Science*, vol. 305, no. 5689, pp. 1444–1447, September 2004.

[PAR 15a] PARK G.C., XUE W., TAGHIZADEH A. *et al.*, "Hybrid vertical-cavity laser with lateral emission into a silicon waveguide", *Laser & Photonics Review*, vol. 9, no. 3, pp. L11–L15, May 2015.

[PAR 15b] PARRAIN D., BAKER C., WANG G. *et al.*, "Origin of optical losses in gallium arsenide disk whispering gallery resonators", *Optics Express*, vol. 23, no. 15, p. 19656, July 2015.

[PAS 02] PASQUARIELLO D., HJORT K., "Plasma-assisted InP-to-Si low temperature wafer bonding", *IEEE Journal of Selected Topics in Quantum Electronics*, vol. 8, no. 1, pp. 118–131, January 2002.

[PAS 13] PASSARO V. (ed.), *Advances in Photonic Crystals*, InTech, 2013.

[PAU 10] PAUL D.J., "The progress towards terahertz quantum cascade lasers on silicon substrates", *Laser Photonics Reviews*, vol. 4, no. 5, pp. 610–632, September 2010.

[PAV 00] PAVESI L., NEGRO L.D., MAZZOLENI C. *et al.*, "Optical gain in silicon nanocrystals", *Nature*, vol. 408, no. 6811, pp. 440–444, November 2000.

[PAV 03] PAVESI L., "A review of the various approaches to a silicon laser", *Proceedings of SPIE*, vol. 4997, pp. 206–220, 2003.

[PAV 05] PAVESI L., "Routes toward silicon-based lasers", *Materials Today: Proceedings*, vol. 8, no. 1, pp. 18–25, January 2005.

[PAV 10] PAVESI L., TURAN R., *Silicon Nanocrystals: Fundamentals, Synthesis and Applications*, John Wiley & Sons, 2010.

[PAV 12] PAVESI L., GAPONENKO S., NEGRO L.D., *Towards the First Silicon Laser*, Springer Science & Business Media, 2012.

[PEA 16] PEARSON HIGHER ED, "CMOS VLSI Design: A Circuits and Systems Perspective, 4/E – Neil Weste & David Harris", available at: http://www.pearsonhighered.com/educator/product/CMOS-VLSI-Design-A-Circuits-and-Systems-Perspective/9780321547743.page, 2016.

[PET 86] PETROFF P.M., "Nucleation and growth of GaAs on Ge and the structure of antiphase boundaries", *Journal of Vacuum Science & Technology B*, vol. 4, no. 4, pp. 874–877, July 1986.

[PET 08] PETTINARI G., POLIMENI A., CAPIZZI M. *et al.*, "Influence of bismuth incorporation on the valence and conduction band edges of GaAs$_{1-x}$Bi$_x$", *Applied Physics Letters*, vol. 92, no. 26, p. 262105, June 2008.

[PIN 15] PING WANG Y., LETOUBLON A., NGUYEN THANH T. *et al.*, "Quantitative evaluation of microtwins and antiphase defects in GaP/Si nanolayers for a III–V photonics platform on silicon using a laboratory X-ray diffraction setup", *Journal of Applied Crystallography*, vol. 48, no. 3, pp. 702–710, June 2015.

[POH 13] POHL U.W., *Epitaxy of Semiconductors*, Springer, Berlin, Heidelberg, 2013.

[POL 04] POLMAN A., MIN B., KALKMAN J. *et al.*, "Ultralow-threshold erbium-implanted toroidal microlaser on silicon", *Applied Physics Letters*, vol. 84, no. 7, pp. 1037–1039, February 2004.

[QUI 13] QUINCI T., KUYYALIL J., THANH T.N. *et al.*, "Defects limitation in epitaxial GaP on bistepped Si surface using UHVCVD–MBE growth cluster", *Journal of Crystal Growth*, vol. 380, pp. 157–162, October 2013.

[QUI 62] QUIST T.M., REDIKER R.H., KEYES R.J. *et al.*, "Semiconductor maser of GaAs", *Applied Physics Letters*, vol. 1, no. 4, pp. 91–92, December 1962.

[RAD 12] RADTKE G., COUILLARD M., BOTTON G.A. *et al.*, "Structure and chemistry of the Si(111)/AlN interface", *Applied Physics Letters*, vol. 100, no. 1, p. 011910, January 2012.

[RAK 12] RAKHEJA S., KUMAR V., "Comparison of electrical, optical and plasmonic on-chip interconnects based on delay and energy considerations", *13th International Symposium on Quality Electronic Design (ISQED)*, pp. 732–739, 2012.

[REB 11] REBOUL J.-R., CERUTTI L., RODRIGUEZ J.-B. *et al.*, "Continuous-wave operation above room temperature of GaSb-based laser diodes grown on Si", *Applied Physics Letters*, vol. 99, no. 12, p. 121113, 2011.

[REN 13] REN F., WEI NG K., LI K. *et al.*, "High-quality InP nanoneedles grown on silicon", *Applied Physics Letters*, vol. 102, no. 1, pp. 012115, January 2013.

[ROB 12a] ROBERT C., CORNET C., TURBAN P. *et al.*, "Electronic, optical, and structural properties of (In,Ga)As/GaP quantum dots", *Physical Review B*, vol. 86, no. 20, p. 205316, November 2012.

[ROB 12b] ROBERT C., PERRIN M., CORNET C. *et al.*, "Atomistic calculations of Ga(NAsP)/GaP(N) quantum wells on silicon substrate: band structure and optical gain", *Applied Physics Letters*, vol. 100, no. 11, pp. 111901–111901–4, March 2012.

[ROB 14] ROBERT C., NESTOKLON M.O., DA SILVA K.P. *et al.*, "Strain-induced fundamental optical transition in (In,Ga)As/GaP quantum dots", *Applied Physics Letters*, vol. 104, no. 1, p. 011908, January 2014.

[ROD 09] RODRIGUEZ J.B., CERUTTI L., GRECH P. *et al.*, "Room-temperature operation of a 2.25 μm electrically pumped laser fabricated on a silicon substrate", *Applied Physics Letters*, vol. 94, no. 6, p. 061124, 2009.

[ROE 06a] ROELKENS G., VAN THOURHOUT D., BAETS R. *et al.*, "Laser emission and photodetection in an InP/InGaAsP layer integrated on and coupled to a silicon-on-Insulator waveguide circuit", *Optics Express*, vol. 14, no. 18, p. 8154, 2006.

[ROE 06b] ROEST A.L., VERHEIJEN M.A., WUNNICKE O. *et al.*, "Position-controlled epitaxial III–V nanowires on silicon", *Nanotechnology*, vol. 17, no. 11, pp. S271–S275, June 2006.

[ROE 10] ROELKENS G., LIU L., LIANG D. *et al.*, "III-V/silicon photonics for on-chip and intra-chip optical interconnects", *Laser & Photonics Reviews.*, vol. 4, no. 6, pp. 751–779, November 2010.

[RON 05a] RONG H., JONES R., LIU A. *et al.*, "A continuous-wave Raman silicon laser", *Nature*, vol. 433, no. 7027, pp. 725–728, February 2005.

[RON 05b] RONG H., LIU A., JONES R. *et al.*, "An all-silicon Raman laser", *Nature*, vol. 433, no. 7023, pp. 292–294, May 2005.

[RON 07] RONG H., XU S., KUO Y.-H. *et al.*, "Low-threshold continuous-wave Raman silicon laser", *Nature Photonics*, vol. 1, no. 4, pp. 232–237, April 2007.

[ROS 02] ROSENCHER E., VINTER B., *Optoelectronics*, Cambridge University Press, 2002.

[SAC 12] SACHER W.D., GREEN W., ASSEFA S. *et al.*, "28 Gb/s silicon microring modulation beyond the linewidth limit by coupling modulation", *OFC*, 2012.

[SAK 87] SAKAI S., SHIRAISHI H., UMENO M., "AlGaAs/GaAs stripe laser diodes fabricated on Si substrates by MOCVD", *IEEE Journal of Quantum Electronics*, vol. 23, no. 6, pp. 1080–1084, June 1987.

[SAV 08] SAVIO R.L., MIRITELLO M., PIRO A.M. *et al.*, "The influence of stoichiometry on the structural stability and on the optical emission of erbium silicate thin films", *Applied Physics Letters*, vol. 93, no. 2, p. 021919, July 2008.

[SCH 06] SCHULZE F., DADGAR A., BLÄSING J. *et al.*, "Metalorganic vapor phase epitaxy grown InGaN/GaN light-emitting diodes on Si(001) substrate," *Applied Physics Letters*, vol. 88, no. 12, p. 121114, March 2006.

[SCH 15] SCHWARTZ B., OEHME M., KOSTECKI K. *et al.*, "Electroluminescence of GeSn/Ge MQW LEDs on Si substrate", *Optics Letters*, vol. 40, no. 13, p. 3209, July 2015.

[SCI 12] SCIANCALEPORE C., BAKIR B.B., LETARTRE X. *et al.*, "CMOS-compatible ultra-compact 1.55-micron emitting VCSELs using double photonic crystal mirrors", *IEEE Photonics Technology Letters*, vol. 24, no. 6, pp. 455–457, March 2012.

[SEA 01] SEASSAL C., ROJO-ROMEO P., LETARTRE X. *et al.*, "InP microdisk lasers on silicon wafer: CW room temperature operation at 1.6 μm", *Electronics Letters*, vol. 37, no. 4, pp. 222–223, February 2001.

[SEL 16] SELLÉS J., BRIMONT C., CASSABOIS G. *et al.*, "Deep-UV nitride-on-silicon microdisk lasers", *Scientific Reports*, vol. 6, p. 21650, February 2016.

[SHE 15] SHENG X., ROBERT C., WANG S. *et al.*, "Transfer printing of fully formed thin-film microscale GaAs lasers on silicon with a thermally conductive interface material", *Laser & Photonics Reviews.*, vol. 9, no. 4, pp. L17–L22, July 2015.

[SKI 12] SKIBITZKI O., HATAMI F., YAMAMOTO Y. *et al.*, "GaP collector development for SiGe heterojunction bipolar transistor performance increase: a heterostructure growth study", *Journal of Applied Physics*, vol. 111, no. 7, pp. 073515–073515–9, April 2012.

[SMO 07] SMOWTON P.M., SANDALL I.C., MOWBRAY D.J. *et al.*, "Temperature-dependent gain and threshold in P-doped quantum dot lasers", *IEEE Journal of Selected Topics in Quantum Electronics*, vol. 13, no. 5, pp. 1261–1266, September 2007.

[SOL 12] SOLOMON N., Three dimensional integrated circuits and methods of fabrication, Patent no. US8136071 B2, March 2012.

[SOR 97] SOREF R.A., "Prospects for novel Si-based optoelectronic devices: unipolar and p–i–p–i lasers", *Thin Solid Films*, vol. 294, nos. 1–2, pp. 325–329, February 1997.

[SPR 90] SPREEUW R.J.C., NEELEN R.C., VAN DRUTEN N.J. *et al.*, "Mode coupling in a He-Ne ring laser with backscattering", *Physical Review. A*, vol. 42, no. 7, pp. 4315–4324, October 1990.

[SPU 13] SPUESENS T., BAUWELINCK J., REGRENY P. *et al.*, "Realization of a compact optical Interconnect on silicon by heterogeneous integration of III-V", *IEEE Photonics Technology Letters*, vol. 25, no. 14, pp. 1332–1335, July 2013.

[SRI 14] SRINIVASAN S., MOREIRA R., BLUMENTHAL D. *et al.*, "Design of integrated hybrid silicon waveguide optical gyroscope", *Optics Express*, vol. 22, no. 21, p. 24988, October 2014.

[STA 11] STANKOVIC S., JONES R., SYSAK M.N. *et al.*, THOURHOUT D.V., "1310-nm Hybrid III-V/Si Fabry–Perot laser based on adhesive bonding", *IEEE Photonics Technology Letters*, vol. 23, no. 23, pp. 1781–1783, December 2011.

[STA 12] STANKOVIC S., JONES R., SYSAK M. N. *et al.*, "Hybrid III-V/Si distributed-feedback laser based on adhesive bonding", *IEEE Photonics Technology Letters*, vol. 24, no. 23, pp. 2155–2158, December 2012.

[STA 13] STANKOVIĆ S., Hybrid III-V/Si DFB lasers based on polymer bonding technology, Ghent University, Belgium, 2013.

[STR 14] STRACKE G., SALA E.M., SELVE S. *et al.*, "Indirect and direct optical transitions in In0.5Ga0.5As/GaP quantum dots", *Applied Physics Letters*, vol. 104, no. 12, p. 123107, March 2014.

[SÜE 13] SÜESS M.J., GEIGER R., MINAMISAWA R.A. *et al.*, "Analysis of enhanced light emission from highly strained germanium microbridges", *Nature Photonics*, vol. 7, no. 6, pp. 466–472, June 2013.

[SUH 10] SUH K., LEE M., CHANG J.S. *et al.*, "Cooperative upconversion and optical gain in ion-beam sputter-deposited $Er_{(x)}Y_{(2-x)}SiO_5$ waveguides", *Optics Express*, vol. 18, no. 8, p. 7724, April 2010.

[SUK 14] SUKHDEO D.S., NAM D., KANG J.-H. *et al.*, "Direct bandgap germanium-on-silicon inferred from 5.7% ⟨100⟩ uniaxial tensile strain [Invited]", *Photonics Research* vol. 2, no. 3, p. A8, June 2014.

[SUN 09a] SUN X., LIU J., KIMERLING L.C. *et al.*, "Direct gap photoluminescence of n-type tensile-strained Ge-on-Si", *Applied Physics Letters*, vol. 95, no. 1, p. 011911, July 2009.

[SUN 09b] SUN X., LIU J., KIMERLING L.C. *et al.*, "Room-temperature direct bandgap electroluminesence from Ge-on-Si light-emitting diodes", *Optics Letters*, vol. 34, no. 8, p. 1198, April 2009.

[SUP 15] SUPPLIE O., MAY M.M., STEINBACH G. *et al.*, "Time-resolved *in situ* spectroscopy during formation of the GaP/Si(100) heterointerface," *The Journal of Physical Chemistry Letters*, vol. 6, no. 3, pp. 464–469, February 2015.

[SVE 08] SVENSSON C.P.T., MÅRTENSSON T., TRÄGÅRDH J. *et al.*, "Monolithic GaAs/InGaP nanowire light emitting diodes on silicon", *Nanotechnology*, vol. 19, no. 30, p. 305201, July 2008.

[SVE 12] SVELTO O., *Principles of Lasers*, Springer Science & Business Media, 2012.

[SZE 12] SZE S.M., *Semiconductor Devices: Physics and Technology*, John Wiley & Sons, Singapore Pte. Limited, 2012.

[TAI 02] TAILLAERT D., BOGAERTS W., BIENSTMAN P. *et al.*, "An out-of-plane grating coupler for efficient butt-coupling between compact planar waveguides and single-mode fibers", *IEEE Journal of Quantum Electronics*, vol. 38, no. 7, pp. 949–955, July 2002.

[TAK 08] TAKEUCHI S., SHIMURA Y., NAKATSUKA O. *et al.*, "Growth of highly strain-relaxed $Ge_{1-x}Sn_x$/virtual Ge by a Sn precipitation controlled compositionally step-graded method", *Applied Physics Letters*, vol. 92, no. 23, p. 231916, June 2008.

[TAK 10] TAKAKI K., IWAI N., KAMIYA S. *et al.*, "Experimental demonstration of low jitter performance and high reliable 1060 nm VCSEL arrays for 10Gbpsx12ch optical interconnection", *SPIE Proceedings*, vol. 7615, p. 761502, 2010.

[TAK 13] TAKAHASHI Y., INUI Y., CHIHARA M. *et al.*, "A micrometre-scale Raman silicon laser with a microwatt threshold", *Nature*, vol. 498, no. 7455, pp. 470–474, June 2013.

[TAL 13] TALNEAU A., ROBLIN C., ITAWI A. *et al.*, "Atomic-plane-thick reconstruction across the interface during heteroepitaxial bonding of InP-clad quantum wells on silicon", *Applied Physics Letters*, vol. 102, no. 21, p. 212101, May 2013.

[TAN 10] TANABE K., GUIMARD D., BORDEL D. *et al.*, "Electrically pumped 1.3 µm room-temperature InAs/GaAs quantum dot lasers on Si substrates by metal-mediated wafer bonding and layer transfer", *Optics Express*, vol. 18, no. 10, p. 10604, May 2010.

[TAN 12] TANABE K., WATANABE K., ARAKAWA Y., "III-V/Si hybrid photonic devices by direct fusion bonding", *Scientific Reports*, vol. 2, p. 349, April 2012.

[TEA 14] TEA E., VIDAL J., PEDESSEAU L. *et al.*, "Theoretical study of optical properties of anti phase domains in GaP", *Journal of Applied Physics*, vol. 115, no. 6, p. 063502, February 2014.

[THO 99] THOMSON J.D., SUMMERS H.D., HULYER P.J. *et al.*, "Determination of single-pass optical gain and internal loss using a multisection device", *Applied Physics Letters*, vol. 75, no. 17, pp. 2527–2529, October 1999.

[THO 12] THOMSON D.J., GARDES F.Y., FEDELI J.-M. *et al.*, "50-Gb/s silicon optical modulator", *IEEE Photonics Technology Letters*, vol. 24, no. 4, pp. 234–236, February 2012.

[UKH 04] UKHANOV A.A., STINTZ A., ELISEEV P.G. *et al.*, "Comparison of the carrier induced refractive index, gain, and linewidth enhancement factor in quantum dot and quantum well lasers", *Applied Physics Letters*, vol. 84, no. 7, pp. 1058–1060, February 2004.

[UST 03] USTINOV V.M., *Quantum Dot Lasers*, Oxford University Press, 2003.

[VAN 07] VAN CAMPENHOUT J., ROJO ROMEO P., REGRENY P. *et al.*, "Electrically pumped InP-based microdisk lasers integrated with a nanophotonic silicon-on-insulator waveguide circuit", *Optics Express*, vol. 15, no. 11, pp. 6744–6749, 2007.

[VAN 08a] VAN CAMPENHOUT J., LIU L., ROMEO P.R. *et al.*, "A compact SOI-integrated multiwavelength laser source based on cascaded InP microdisks", *IEEE Photonics Technology Letters*, vol. 20, no. 16, pp. 1345–1347, August 2008.

[VAN 08b] VANTREASE D., SCHREIBER R., MONCHIERO M. *et al.*, "Corona: system implications of emerging nanophotonic technology", *Proceedings of the 35th Annual International Symposium on Computer Architecture*, pp. 153–164, Washington DC, USA, 2008.

[VAN 10a] VAN THOURHOUT D., SPUESENS T., SELVARAJA S. K. *et al.*, "Nanophotonic devices for optical interconnect", *IEEE Journal of Selected Topics in Quantum Electronics*, vol. 16, no. 5, pp. 1363–1375, September 2010.

[VAN 10b] VANHOUTTE M., WANG B., ZHOU Z. *et al.*, "Direct demonstration of sensitization at 980nm optical excitation in erbium-ytterbium silicates", *7th IEEE International Conference on Group IV Photonics (GFP)*, pp. 308–310, 2010.

[VAN 75] VAN DER ZIEL J.P., DINGLE R., MILLER R.C. *et al.*, "Laser oscillation from quantum states in very thin GaAs–$Al_{0.2}Ga_{0.8}As$ multilayer structures", *Applied Physics Letters*, vol. 26, no. 8, pp. 463–465, April 1975.

[VIK 13] VIKTOROVITCH P., SCIANCALEPORE C., BAKIR B.B. *et al.*, "Double photonic crystal vertical-cavity surface-emitting lasers", *Proceedings of SPIE*, vol. 8633, 2013.

[VOL 11] VOLZ K., BEYER A., WITTE W. *et al.*, "GaP-nucleation on exact Si (0 0 1) substrates for III/V device integration", *Journal of Crystal Growth*, vol. 315, no. 1, pp. 37–47, January 2011.

[WAL 90] WALE M.J., EDGE C., "Self-aligned flip-chip assembly of protonic devices with electrical and optical connections", *IEEE Transactions on Components, Hybrids, and Manufacturing Technology*, vol. 13, no. 4, pp. 780–786, December 1990.

[WAL 05] WALTERS R.J., BOURIANOFF G.I., ATWATER H.A., "Field-effect electroluminescence in silicon nanocrystals", *Nature Materials*, vol. 4, no. 2, pp. 143–146, February 2005.

[WAN 11a] WANG T., LIU H., LEE A. *et al.*, "1.3-μm InAs/GaAs quantum-dot lasers monolithically grown on Si substrates", *Optics Express*, vol. 19, no. 12, p. 11381, June 2011.

[WAN 11b] WANG X.J., WANG B., WANG L. *et al.*, "Extraordinary infrared photoluminescence efficiency of $Er_{0.1}Yb_{1.9}SiO_5$ films on SiO_2/Si substrates", *Applied Physics Letters*, vol. 98, no. 7, p. 071903, February 2011.

[WAN 13a] WANG B., GUO R., WANG X. *et al.*, "Large electroluminescence excitation cross section and strong potential gain of erbium in ErYb silicate", *Journal of Applied Physics*, vol. 113, no. 10, p. 103108, March 2013.

[WAN 13b] WANG Z., TIAN B., PALADUGU M. *et al.*, "Polytypic InP nanolaser monolithically integrated on (001) silicon," *Nano Letters*, vol. 13, no. 11, pp. 5063–5069, November 2013.

[WAN 13c] WANG Z., TIAN B., PALADUGU M. *et al.*, "An ultra-short InP nanowire laser monolithic integrated on (001) silicon substrate", *IEEE Photonics Society Summer Topical Meeting Series*, pp. 23–24, 2013.

[WAN 15a] WANG Y.P., STODOLNA J., BAHRI M. *et al.*, "Abrupt GaP/Si hetero-interface using bistepped Si buffer", *Applied Physics Letters*, vol. 107, no. 19, p. 191603, November 2015.

[WAN 15b] WANG Z., TIAN B., PANTOUVAKI M. *et al.*, "Room-temperature InP distributed feedback laser array directly grown on silicon", *Nature Photonics*, vol. 9, no. 12, pp. 837–842, December 2015.

[WIN 85] WINDHORN T.H., METZE G.M., "Room-temperature operation of GaAs/AlGaAs diode lasers fabricated on a monolithic GaAs/Si substrate", *Applied Physics Letters*, vol. 47, no. 10, pp. 1031–1033, November 1985.

[WIR 13a] WIRTHS S., IKONIC Z., TIEDEMANN A.T. *et al.*, "Tensely strained GeSn alloys as optical gain media", *Applied Physics Letters*, vol. 103, no. 19, p. 192110, November 2013.

[WIR 13b] WIRTHS S., TIEDEMANN A.T., IKONIC Z. *et al.*, "Band engineering and growth of tensile strained Ge/(Si)GeSn heterostructures for tunnel field effect transistors", *Applied Physics Letters*, vol. 102, no. 19, p. 192103, May 2013.

[WIR 15] WIRTHS S., GEIGER R., VON DEN DRIESCH N. *et al.*, "Lasing in direct-bandgap GeSn alloy grown on Si", *Nature Photonics*, vol. 9, no. 2, pp. 88–92, February 2015.

[WOL 12] WOLF P., MOSER P., HOFMANN W. *et al.*, "Abolishing copper interconnects", *SPIE Newsroom*, September 2012.

[WOO 12] WOODS D., NAUGHTON T. J., "Optical computing: photonic neural networks", *Nature Physics*, vol. 8, no. 4, pp. 257–259, April 2012.

[XIA 10] XIAO Y.-F., ZOU C.-L., LI B.-B. *et al.*, "High-Q exterior whispering-gallery modes in a metal-coated microresonator", *Physical Review Letters*, vol. 105, no. 15, p. 153902, October 2010.

[XIE 90] XIE Q.H., FUNG K.K., DING A.J. *et al.*, "Asymmetric distribution of microtwins in a GaAs/Si heterostructure grown by molecular beam epitaxy", *Applied Physics Letters*, vol. 57, no. 26, pp. 2803–2805, December 1990.

[YAB 90] YABLONOVITCH E., HWANG D.M., GMITTER T.J. *et al.*, "Van der Waals bonding of GaAs epitaxial liftoff films onto arbitrary substrates", *Applied Physics Letters*, vol. 56, no. 24, pp. 2419–2421, June 1990.

[YAB 04] YABLONSKII G.P., GURSKII A.L., LUTSENKO E.V. *et al.*, "Optically pumped UV-blue lasers based on InGaN/GaN/Al2O3 and InGaN/GaN/Si heterostructures," in SHUR M.S., ŽUKAUSKAS A. (eds), *UV Solid-State Light Emitters and Detectors*, Springer, Netherlands, 2004.

[YAN 03] YANG L., ARMANI D.K., VAHALA K.J., "Fiber-coupled erbium microlasers on a chip", *Applied Physics Letters*, vol. 83, no. 5, pp. 825–826, August 2003.

[YAN 07a] YANG J., BHATTACHARYA P., MI Z., "High-performance In0.5Ga0.5 As/GaAs quantum-dot lasers on silicon with multiple-layer quantum-dot dislocation filters," *IEEE Transactions on Electron Devices*, vol. 54, no. 11, pp. 2849–2855, November 2007.

[YAN 07b] YANG T., LU L., SHIH M.-H. *et al.*, "Room temperature InGaSb quantum well microcylinder lasers at 2 μm grown monolithically on a silicon substrate", *Journal of Vacuum Science & Technology B*, vol. 25, no. 5, pp. 1622–1625, September 2007.

[YAN 12] YANG H., ZHAO D., CHUWONGIN S. *et al.*, "Transfer-printed stacked nanomembrane lasers on silicon", *Nature Photonics*, vol. 6, no. 9, pp. 615–620, September 2012.

[YIN 12] YIN L., NING H., TURKDOGAN S. *et al.*, "Long lifetime, high density single-crystal erbium compound nanowires as a high optical gain material", *Applied Physics Letters*, vol. 100, no. 24, p. 241905, June 2012.

[ZHA 14a] ZHANG C., SRINIVASAN S., TANG Y. *et al.*, "Low threshold and high speed short cavity distributed feedback hybrid silicon lasers", *Optics Express*, vol. 22, no. 9, p. 10202, May 2014.

[ZHA 14b] ZHANG Y., MA Z., ZHANG X. *et al.*, "Optically pumped whispering-gallery mode lasing from 2-μm GaN micro-disks pivoted on Si", *Applied Physics Letters*, vol. 104, no. 22, p. 221106, June 2014.

[ZHO 15] ZHOU Z., YIN B., MICHEL J., "On-chip light sources for silicon photonics", *Light, Science & Applications*, vol. 4, no. 11, p. e358, November 2015.

Index

Printed in the United States
By Bookmasters